CONSTITUTIVE MODELS FOR RUBBER

PROCEEDINGS OF THE FIRST EUROPEAN CONFERENCE ON CONSTITUTIVE MODELS FOR RUBBER / VIENNA / AUSTRIA / 9 – 10 SEPTEMBER 1999

Constitutive Models for Rubber

Edited by

Al Dorfmann

Institute of Structural Engineering, University of Applied Sciences, Vienna, Austria

Alan Muhr

Tun Abdul Razak Research Centre, MRPRA, Brickendonbury, Hertford, United Kingdom

Taylor & Francis

Taylor & Francis Group

LONDON AND NEW YORK

The texts of the various papers in this volume were set individually by typists under the supervision of each of the authors concerned.

Published by Taylor & Francis,
2 Park Square, Milton Park, Abingdon, Oxon, OX14 4RN
270 Madison Ave, New York NY 10016

Transferred to Digital Printing 2006

ISBN 90 5809 113 9

Publisher's Note
The publisher has gone to great lengths to ensure the quality of this reprint
but points out that some imperfections in the original may be apparent

Printed and bound by CPI Antony Rowe, Eastbourne

Constitutive Models for Rubber, Dorfmann & Muhr (eds)© 1999 Taylor & Francis ISBN 90 5809 113 9

Table of contents

Viscoelasticity

Tyres and friction

Softening phenomena

Applications

Constitutive Models for Rubber, Dorfmann & Muhr (eds) © 1999 Taylor & Francis ISBN 90 5809 113 9

Foreword

The extraordinary stress strain behaviour of rubber has presented an opportunity for inventive engineers and a challenge for scientists since the mid-nineteenth century, and continues to do so today. Major branches of theory, such as the statistical theory of rubber elasticity and finite strain elasticity theory, have been spawned by the properties of rubber. Until recently, however, the theoretical framework for large deformations found little application among rubber engineers because the mathematics rapidly becomes intractable for all but the simplest components. The advent of affordable and powerful computers has changed all this, and brought the challenge of rubber to new sets of people – software engineers and desk top, as opposed to empirical, designers.

The development of the statistical theory of rubber elasticity in the 1940s, of finite strain elasticity theory in the 1950s, and of convenient forms for the strain energy function in the 1970s, all focused on modelling the elastic characteristics of rubber. Although much literature has appeared in recent years following this theme, the Physics of Rubber Elasticity by L.R.G. Treloar (3rd Edition, Clarendon Press, Oxford, 1975) and the proceedings of a Discussion on Rubber Elasticity (Proc. Roy. Soc. London, 1976, A351, No. 1666, 295-406) remain very valuable reviews.

The treatment of rubber as a 'hyperelastic' material – that is, a material modelled by a strain-energy function for finite strains – was implemented into finite strain finite element analysis in the 1980s and is now widely available in commercial software packages.

However, only a few engineering elastomers – such as unfilled natural rubber and some grades of polyurethane – really conform to the 'hyperelastic' ideal. Most other engineering elastomers incorporate 'reinforcing' fillers, needed to confer adequate strength properties and also improving processing characteristics and enabling adjustment of hardness over a wide range. The stress-strain characteristics of such filled elastomers depart significantly from elasticity. While ways of thinking about these departures – such as the 'dynamic static-to-ratio' of rubber springs – may have satisfied a previous generation of design engineers, there is now an opportunity to apply more sophisticated models.

One major current challenge is thus to model these aspects of the inelastic behaviour that are relevant to engineers, and to do this in such a way that the models are implementable in finite element analysis.

Although potentially the involvement of representatives of several disciplines should facilitate progress, this is only the case if they talk to each other. In practice, software engineers might rely on the literature and on desk-top designers as sources of information about rubber, and fail to achieve as good a balance of understanding as they could if they listened also to experimental rubber scientists and empirical designers. Applied mathematicians might develop phenomenological models which address issues of secondary interest to designers, or which misrepresent important aspects of the experimentally observed behaviour. Experimentalists might develop models without reference to the existing framework of continuum mechanics, resulting in internal inconsistencies and difficulty in implementation in software packages. The First European Conference on Constitutive Models for Rubber sprang from the idea of providing a forum for multi-disciplinary discussion, seeking to bring the fragmented strands of recent research together.

Within the UK a start has been made in this direction – through a workshop on Deformation Modelling for Solid Polymers (Oxford University, 1997) and a seminar on Finite Element Analysis of Elastomers (Institution of Mechanical Engineers, London, 1997). The proceedings of the latter are available as a publication of the same name (Professional Engineering Publications, London, 1999). Similarly, in Germany a workshop on Finite Element Analysis – Basics and Future Trends was organised by the Deutsche Institut für Kautschuk Technology (Hanover, 1998). The interest in these essentially national meetings suggested that further cross-fertilisation should be stimulated by providing a European forum for discussion.

The contributions to this Proceedings cover a wide range of subjects. Consistent with the analysis given above, relatively few authors chose to present hyperelastic models for rubber; however, readers interested in this topic will find ample references to earlier work. Several contributions address inelastic effects associated with filled elastomers – such as Mullins' effect and quasi-static hysteresis. For others – most obviously in processing uncured rubber – the interest is in modelling viscoelasticity. In addition to stress-strain behaviour, work is presented on frictional contact and on mechanical failure. Looking at the applications side, computational techniques are addressed and applied to a diverse range of components, including tyres, earthquake isolation bearings and intervertebral discs. Overall, the authors have achieved progress in a wide range of areas – including experimental results, theory, and practical utility. They raise many questions as well, as one might expect from the first forum of this kind.

We would like to thank our colleagues on the Scientific Committee (R.W.Ogden, Chairman; D.Besdo, R.de Borst, K.N.G.Fuller, H.A.Mang, H.Menderez, G.Meschke and H.Rothert) and all the authors who have worked with us to produce this book.

Al Dorfmann
Alan H. Muhr
Vienna/Hertford, June 1999

Constitutive Models for Rubber, Dorfmann & Muhr (eds) © 1999 Taylor & Francis ISBN 90 5809 113 9

Organisation

Al Dorfmann, *Editor*
Institute of Structural Engineering, University of Applied Sciences, Vienna, Austria

Alan Muhr, *Editor*
Tun Abdul Razak Research Centre, MRPRA, Brickendonbury, Hertford, UK

Ulrike Schachinger, *Administration*
Institute of Structural Engineering, University of Applied Sciences, Vienna, Austria

Stefan Burtscher, *Local Organization*
Institute of Structural Concrete, Vienna University of Technology, Austria

SCIENTIFIC COMMITTEE

R.W.Ogden, *Chairman*, University of Glasgow, United Kingdom
D.Besdo, University of Hannover, Germany
R.de Borst, Delft University of Technology, Netherlands
K.N.G.Fuller, Tun Abdul Razak Research Centre, United Kingdom
H.A.Mang, Vienna University of Technology, Austria
H.Menderes, MARC, Analysis Research Corporation, Netherlands
G.Meschke, Ruhr-University Bochum, Germany
H.Rothert, University of Hannover, Germany

Principal Sponsor

MARC
Analysis Research Corporation

Sponsors

ABACOM Software GmbH
HKS. Inc.

COMPAQ

COMPAQ Computer Corporation

RW

REISNER & WOLF Engineering GmbH

vce

Vienna Consulting Engineers

Bank Austria

Vorspann Technik GmbH
Ratingen

fischer

Artur Fischer GmbH & Co. KG

SYHAG CAE-TOOLS GmbH

CITY OF VIENNA

Constitutive and numerical modelling

Constitutive Models for Rubber, Dorfmann & Muhr (eds) © 1999 Taylor & Francis ISBN 90 5809 113 9

Advanced FE analysis of elastomeric automobile components under realistic loading conditions

H. Menderes & A.W.A. Konter
MARC Analysis Research Corporation – Europe, Zoetermeer, Netherlands

ABSTRACT:

In this paper the frequently used constitutive models in the simulation of rubber components will be discussed. Both the simple models, often used in industrial applications, with extensions to visco-elasticity and the more advanced quasi-static models will be reviewed. Attention will also be paid to the techniques for curve fitting of the material parameters for a particular constitutive model. It is shown that errors in parameter determination can easily be made if insufficient experimental data is available. These errors can partially be avoided if good curve fitting tools are available and if they can be used prior to the analysis. The material models will be applied to analysis of simple components under realistic loading conditions. It is demonstrated that both visco-elasticity and inertia effects play a key role in obtaining realistic simulations.

1 INTRODUCTION

Recent advances in FEM technology has resulted in industrial application of simulation tools in the design of elastomeric components. In the first decades of research in mathematical modelling of the material behavior of elastomers and numerical techniques to handle the nearly in-compressible material behavior, the industrial application of simulation tools was still limited. With the availability of simple to use numerical procedures for handling the contact problem, many manufacturers of rubber components such as seals, tires, motor mounts, sport materials have recognized the potential of numerical analysis in the design of rubber components. Currently car manufacturers often demand the results of a numerical simulation when a new design is presented.

For most applications, still a quasi-static deformation analysis is sufficient. The study of the interaction of the rubber with other deforming parts is nowadays a standard application through simple and easy to use contact algorithms, which include self-contact, friction and thermal contact. Recent advances have made it possible to include dynamic effects in the contact algorithm through implicit or explicit transient analysis procedures, which enables studies of the effects in for instance shock absorbers. An other class of problems where dynamic effects have to be included are steady state vibrations subjected to non-linear prestressed structures. Here it usually is sufficient to analyse the behaviour at a particular excitation frequency (or ranges of frequencies), taking into account the appropriate stiffness and damping of the material at that frequency.

Even in a quasi-static analysis the identification of the material parameters to be used for a particular material model based on results of tests on simple test specimen can still be cumbersome. Often curve fitting techniques are required which can show good predictions for particular loading (e.g. a tensile test), but will behave badly when the material is subjected to an equi-biaxial test. Good and easy to use tools to predict and verify the material behavior, prior to the analysis, will avoid failures in the numerical simulation of realistic components.

It is recognized that the material behavior is visco-elastic which displays itself through for instance relaxation of the stresses after closure of the seal, resulting in potential leakage conditions. Also shock absorbers have a stiffness behaviour which can be different depending on the rate of compression.

Another application requiring visco-elasticity models are cyclically loaded structural components in which the material stiffness and the damping are frequency dependent. In addition, the energy dissipation due to the visco-elasticity produces heat. This local heat production will increase the temperature that, in combination with the temperature dependency of the material properties requires a coupled thermo-mechanical analysis.

Recently progress has been made in the study of the acoustic behavior of rubber seals. The medium surrounding the seal is subjected to cyclic pressure variations and the damping characteristics of the seal as well as the potential of exitating the rubber seal in its eigen frequency needs to be analysed. This requires a coupled analysis in which both the pressure in the medium and the deformation of the seal are determined.

2 MATERIAL CHARACTERISATION

Several decades of research in accurate constitutive models for the description of both in-compressible and compressible elastomeric behaviour have resulted in potential accurate models for the description of the material behaviour under arbitrary multiaxial loading conditions. Application of the models to a particular rubber is difficult due various reasons:

- For a specific rubber often insufficient experimental material data is available to determine the parameters for a particular model.
- Frequently only the result of tensile or a compressive tests is available. These results are then used to determine the material parameters in the constitutive model. Depending on the results of the curve fitting process of these parameters the various constitutive models exhibit a behaviour of other homogeneous stress states which can only be judged globally on correctness or verified if results of these tests are available.
- A large class of the available constitutive models are verified or valid for the description of the quasi-static behaviour only.

Realistic simulations of industrial structural components require however constitutive models which include a dependence of:

- temperature dependence on the material properties
- large strain visco-elastic effects
- frequency dependent stiffness and damping

In the first decade of application of numerical simulation techniques to structural components, often these afore mentioned effects have been neglected, mainly since even with these simplification the analysis was already complex enough and a solution could not always be guaranteed.

Progress in simulation techniques, in particular the availability of robust and easy to use techniques for contact analysis, as well as robust finite element technology for incompressible behaviour has proven that numerical simulation of structural components subjected to realistic loading conditions is feasible. This in turn has resulted in the following questions:

- Which model should be used if only limited experimental data is available?
- How accurate is my material model?
- Can one include visco-elastic effects?
- Can one include thermal effects?
- Can one describe damage effects?

2.1 Simple models used in industry

For particular rubbers used in industry often only the Shore hardness A or a linear shear modulus is available. For limited strain ranges the material is often linear in shear and the classical simple models such as the Neo-Hookean or Mooney-Rivlin model behave linear in shear.

A simple logarithmic model to relate the shear modulus G to the Shore A hardness H is described by Batterman/Kohler (1982)

$$G = 0.086 \bullet 1.045^{H} \tag{1}$$

Modifications of this relation or tabular data relating the shore hardness to the modulus of elasticity are described by Lindley (1966), Crawford (1985), Gent (1994) and Gobel (1969).

The material models are often formulated by an elastic energy definition e.g.:

$$W = C_{10}(I_1 - 3) + C_{01}(I_2 - 3) \tag{2}$$

where I_1 and I_2 are the first and second strain invariant, C_{10} and C_{01} are material parameters.

The parameters for the above described most frequently used simple material models can then be obtained from:

Neo Hookean: $G = 2C_{10}$ (3)

Mooney Rivlin: $G = 2(C_{10} + C_{01})$ (4)

where: $C_{01} = 0.2 \ldots 0.25C_{10}$ (5)

In spite of the known limitations to describe particular stress states, several analysts claim to obtain good results using these models for various structural components with local values of the strains up to about 200%.

These models also allow a simple definition of the quasi-static temperature dependency. It suffices to define G or C_{10} and C_{01} as a function of the temperature. Here often tabular data are used. A consequence of temperature change however is often that visco-elastic effects become more dominant and can no longer be neglected.

Non-linear visco-elastic models are often based on modified forms of the general Schapery model. This model allows great flexibility in modelling, but since experimental data to describe all possible effects is limited, often the model lacks application. A much more simple model is given by Simo. In this model the elastic energy is modelled by a N term Prony series expansion similar to a linear visco-elastic model. In the linear visco-elastic model the shear modulus is approximated by:

$$G(t) = G^{\infty} + \sum_{n=1}^{N} G^{n} \exp\left(-\frac{t}{\tau_n}\right) \tag{6}$$

The elastic energy as function of the strain is then given as

$$W(E, t) = W^{\infty} + \sum_{n=1}^{N} W^{n} \exp\left(-\frac{t}{\tau_n}\right) \tag{7}$$

where W^{∞} represents the long term elastic energy (as described by one of the forms mentioned above) and W^{n} an energy contribution which is added to long term elastic energy and corresponds to a time constant τ_n. The model allows different forms of the elastic energy function for each term, but often for simplicity the terms are assumed to have the same shape and differ only by a scalar multiplier.

Temperature effects have a strong influence on the visco-elasticity. This can in most cases accounted for by the so-called thermo-rheologically simple behaviour assumption, in which a shift of the relaxation time τ_n is obtained depending on the local values of the temperature. The most frequently used shift function is the Willams-Landel and Ferry equation.

The visco-elastic model mentioned above allows a transient analysis with arbitrary large deformations. For the study of small amplitude vibrations in non-linear pre-stressed components often a simplification can be made based on the linear elastic storage and loss modulus. The storage modulus $G'(\omega)$ and loss modulus $G''(\omega)$ as function of the frequency ω are obtained from experiments and can either be used directly in tabular data form or through an approximation in a Prony series in the frequency range.

$$G'(\omega) = G^{\infty} + \sum_{n=1}^{N} G^{n} \frac{\omega^2 \tau_n^2}{1 + \omega^2 \tau_n^2}$$
$$G''(\omega) = \sum_{n=1}^{N} G^{n} \frac{\omega \tau_n}{1 + \omega^2 \tau_n^2} \tag{8}$$

The viscoelastic behaviour results in heat production. The energy dissipation per load cycle in a cyclic test with strain amplitude is defined by:

$$\dot{\theta} = G''(\omega)\gamma_0^2/2$$

where γ_0 is the local value of the cyclic strain. It is clear that these simple models do not always provide the correct solution. If sufficient experimental data on simple test-specimens with a homogenous stress state is available, the models can be refined.

In most models the material behavior is assumed to be incompressible or nearly incompressible. This means that the ratio of bulk and shear modulus is approximately $K/G \approx 10000$. Hence for compressible foam materials other models have to be used. Visco-elasticity for the volumetric behaviour is usually neglected.

2.2 More advanced models and the fitting of parameters

Years of research on accurate constitutive modelling has resulted in the availability of a number of models describing the elastic energy as a function of the deformation. An excellent review of these models can be found in e.g. Treloar (1975). The models are either strain invariant or stretch ratio based and have found their way into finite element codes. The general purpose finite element program MARC provides besides the Neo Hookean and Mooney-Rivlin model the following strain invariant models: Signorini, Yeoh, Gent, and other combinations of the general 3rd order models, the stretch ratio based Ogden model and its variant allowing non-linear volumetric behaviour for e.g. foam materials, and the micro-mechanics based (Generalized) Arruda-Boyce model. In addition simple program modifications allow the analyst to define special models, examples of which are e.g. the Kilian (1981) model and the micromechanis based tube model. (Heinrich et al. 1988).

In general the Ogden model has become more popular if large strains have to be considered in the structural component. The tensile stress-strain curve typically has a stress stiffening part near the limiting stretch and the simple modesl fail to capture this stress increase. The fitting of the parameters is purely empirical and often more than one experiment is required. This has led to the development of micromechanics based models such as the Arruda-Boyce which claim to give a good prediction for other stress states purely on the result of a tensile test.

Usually a tensile or compressive test is used as basic test and the parameters in the constitutive model are determined such that the best possible fit is obtained with the results of the tensile test. Common problems with this approach are:

- What is the behaviour of the model for strains larger than used in fitting the experimental result?
- What is the behaviour for other homogeneous stress states, such as simple shear, pure shear, equi-biaxial and volumetric?
- If results of other stress states are availble how should one perform the fit.

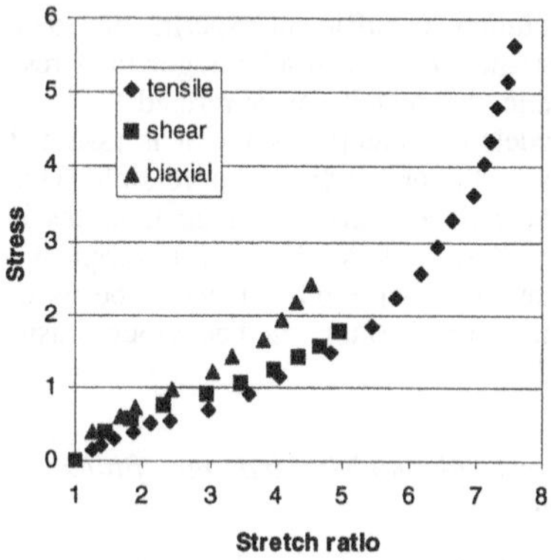

Figure 1 Experimental data from Treloar (1975)

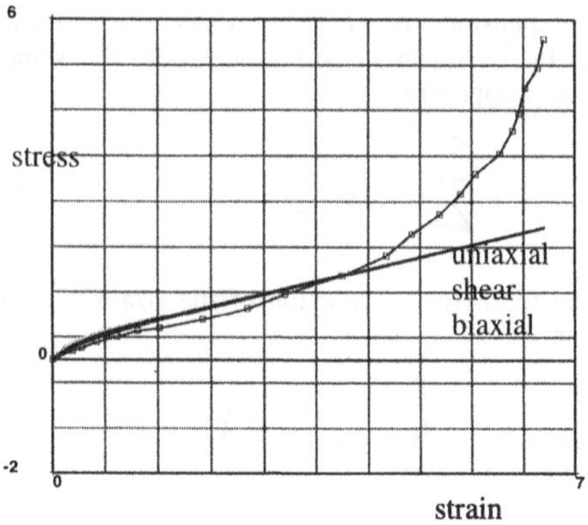

Figure 2 Neo Hookean fit

The most limiting factor in the accuracy of either of these models is still the determination of the material parameters. Undetected errors in the determination of these parameters, result in errors in the simulation of the behaviour of the structural component as well. As an example consider the test data given for a tensile, pure shear and a equi-biaxial test as described by Treloar. The experimental data points are shown in Figure 1.

The results of the uniaxial test have been used in the parameter determination for each model. With these parameters the behaviour for other stress states can simply be predicted and compared with the experimental data of Figure 1.

The best possible fit with a number of models is shown in Figure 2-7.

Close inspection of the figures reveals:

- All models fit the uniaxial curve as close as possible. The Neo-Hookean (and the Mooney Rivlin) model fails to predict correct values of the stesses at higher strain levels.
- The higher order models can easily show a so-called material instability. In particular for the stress states other than the uniaxial one this is often the case. By adding additional constraints to the curve fitting that the constants remain positive this effect can partly be removed.
- Large deviations in predicting the correct stress value for other stress states can be obtained. This can be avoided by performing the curve fit based on all data simultaneously. Often for new materials these experimental data are however not available.
- Micro-mechanics based models claim to give a good prediction for other stress states with curve fitting based on uniaxial data, but it certainly is recommended to verify whether this is true for new materials.
- It is generally recommended to evaluate the behaviour for other stress states than the uniaxial one prior to any finite element analysis. This curve fitting/prediction process has to be part of the pre-processing capabilities.

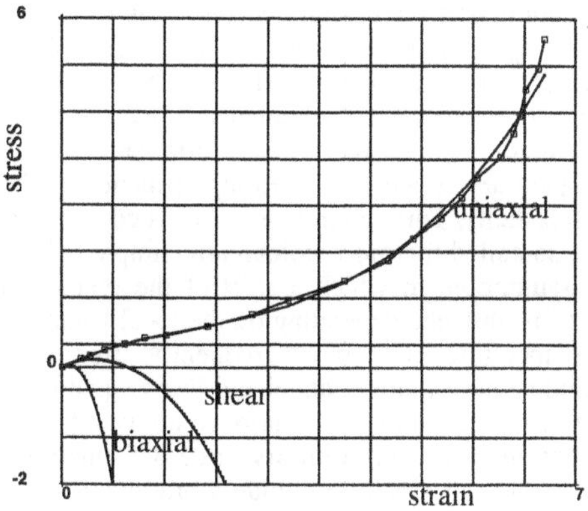

Figure 3 Second Order Model fit

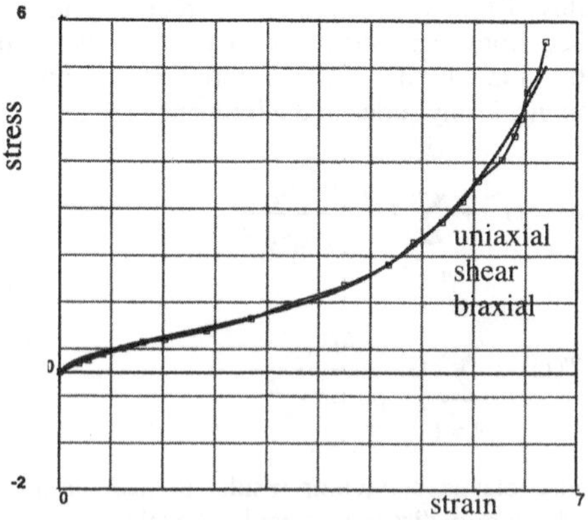

Figure 4 Three Term Ogden model fit

If compressible foam materials have to be described it is hardly impossible to perform the curve fitting based on the results of the uniaxial test only. Specific volumetric data is recommended, for instance by measuring the effective cross-sectional

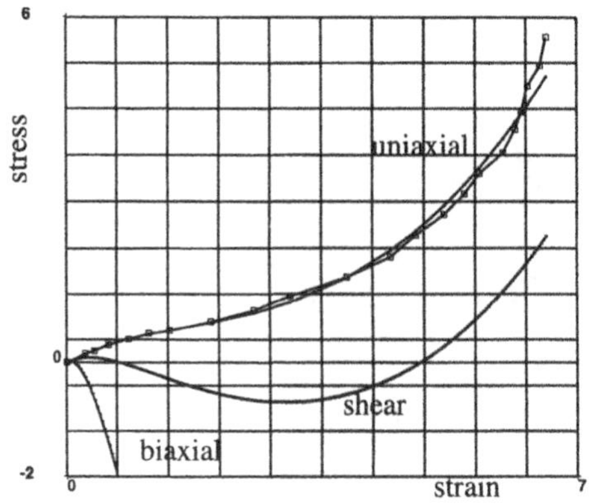

Figure 5 4 Term Ogden model fit

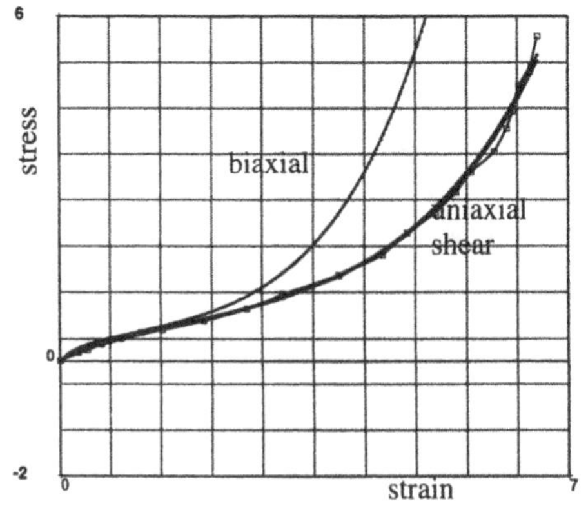

Figure 6 Yeoh model fit

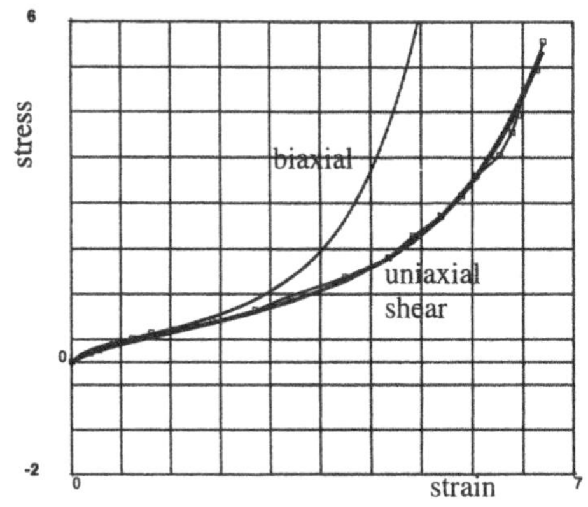

Figure 7 Arruda Boyce model fit

are change in a tensile test, or even better performing a real volumetric test.

The above mentioned advanced material models are very seldom combined with large visco-elasticity, neither in the time domain nor in the frequency domain. Sometimes they are combined with a damage approach in which the elastic energy function is scaled by a scalar which is a function of the strain history. Hence applications of these material models concentrate on quasi-static deformation behaviour of rubber components.

3 SOLUTION PROCEDURES AND ELEMENT TECHNOLOGY

Traditionally rubber elasticity problems have been solved using the Total Lagrange solution procedure. This requires that the constitutive model is formulated in terms of the Green Lagrange strain and the 2nd Piola Kirchoff stress. The material models are usually formulated in terms of the deviatoric behaviour. Hence in the evaluation of the stress-strain behaviour in the finite element method, first a split has to be made between the volumetric and the devi-

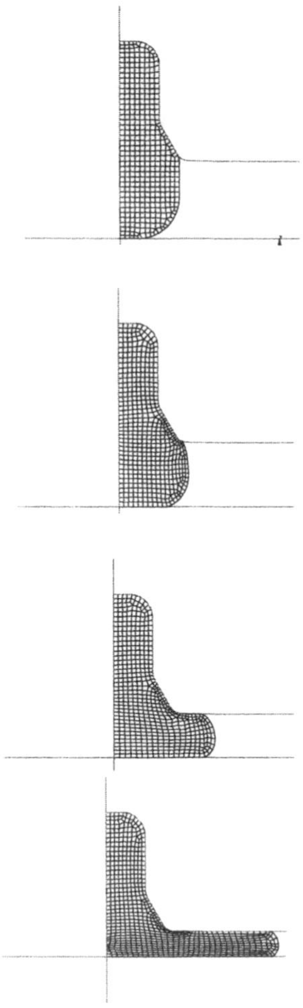

Figure 8 Rubber rezoning

atoric part of the deformation.

A new trend in constitutive modelling is that rubber elasticity problems are formulated in terms of an updated Lagrange formulation. The key benefit of this procedure is that remeshing fits more natural within this approach. Remeshing as shown in Figure 8 is required if the element distortion becomes too large during the analysis. If the element distortion becomes too large often the analysis fails due to large volumetric changes in the integration points of the elements, although the total volumetric strain of the element is still almost zero. In Figure 8 every 5 increments a new mesh has been automatically constructed based on the at that time available outline. Observe that often two different materials are present in one analysis, thus requiring automatic remeshing of the various parts simultaneously.

In the last decades most often 4-noded quadrilateral and 8-noded brick elements were used as the most optimal elements for the combination of incompressible behaviour and contact. Due to the difficulty of automatic meshing with quad and hex elements for arbitrary geometries, a strong need was present to be able to model incompressible behavior with triangular and hexahedral elements. This has resulted in the formulation of enriched triangular and hexahedral elements.

4 APPLICATIONS

4.1 Ring compression

In Figure 9 and Figure 10the undeformed and deformed configuration of a ring compression and subsequent loading by hydrostatic pressure (on the left side of the ring) is shown.
The analysis has also been performed with quadrilateral elements. The resulting reaction force in vertical direction on the top body is compared in Figure 11.
It can be concluded that good comparison beteen both element types can be obtained with these triangular elements. Often however a slightly more refined mesh is needed, in particular if the local deformation is large.

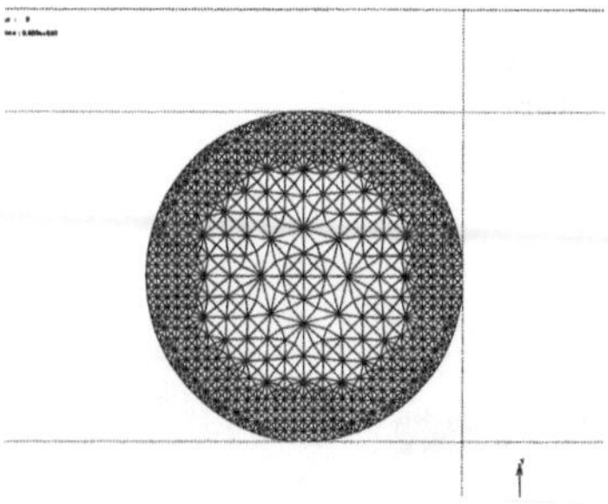

Figure 9 Undeformed ring with triangular elements

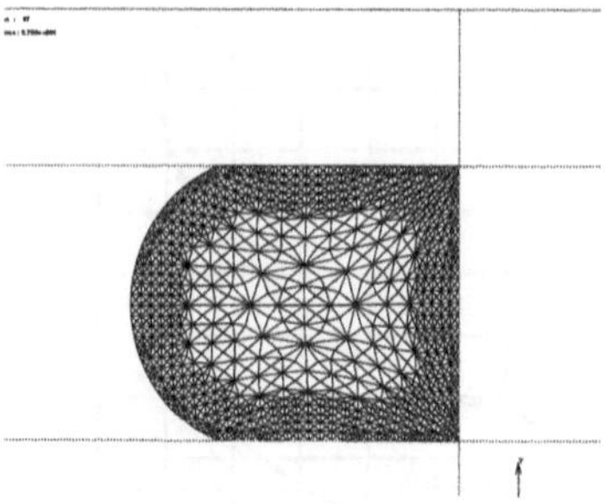

Figure 10 Deformed ring with triangular elements

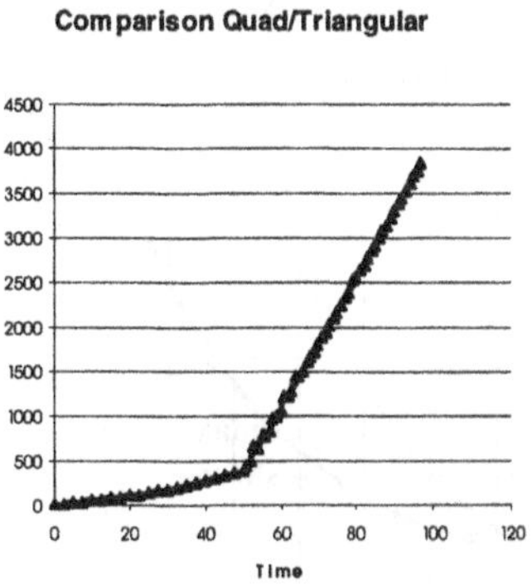

Figure 11 Vertical reaction force on top body

4.2 Seal compression

A flexible rubber seal is mounted to a rigid body and compressed at the top side by another rigid body as shown in Figure 12. The top plate is moving down with a velocity $v=1$ *m/s* during 0.01 *s* and held fixed at its position during 0.01 *s*. The rubber is given the visco-elastic material properties based on a Neo Hookean generalization of the shear moduli. A one term visco-elastic material model has been used.

$$G^{\infty}=4.10^5 \ N/m^2$$

$$G^1=1.10^6 \ N/m^2$$

$$\tau_1 = 0.1 \ s$$

$$\rho=1.10^3 \ kg/m^3$$

In seal design most often a quasi-static approach is used which neglects the visco-elasticity and the inertia effects resulting from the mass density ρ. Here four analyses have been performed.

a. A quasi-static analysis with material properties based on the long term stiffness
b. A quasi-static analysis with visco-elastic material properties
c. A dynamic contact analysis with material proper-

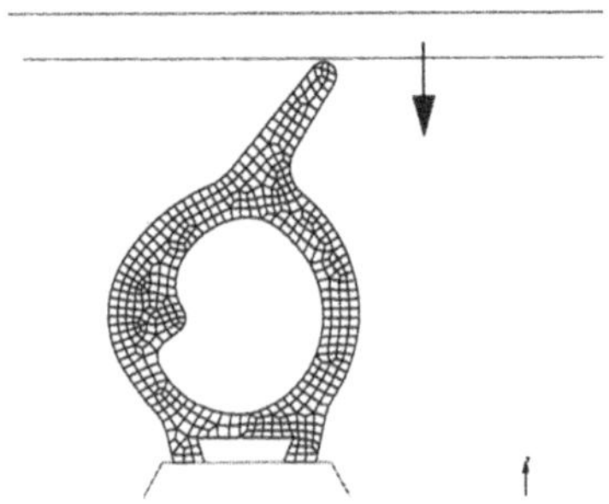

Figure 12 Mounted seal compressed by a rigid plate

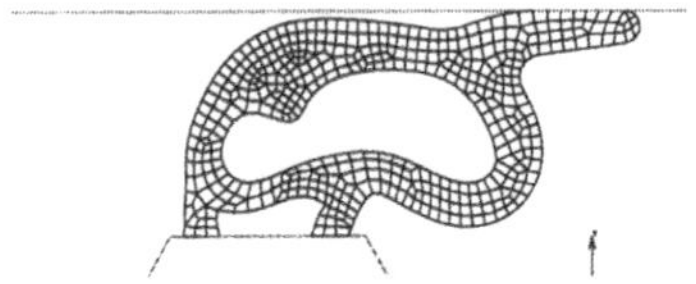

Figure 13 Deformed Seal, quasi static, t=0.01 s

ties based on the long term stiffness

d. A dynamic contact analysis with visco-elastic material properties.

An additional eigen-frequency analysis of the undeformed seal with long term stiffness properties revealed that the lowest eigen frequency was about 80 Hz. Both the visco-elasticity and the deformation have an effect on the eigen-frequency. The load application time however is in the order of magnitude of the lowest eigen period.

The first analysis has been performed using the long term stiffness in a quasi-static approach. The deformed configuration is shown in Figure 13. During the analysis contact is automatically detected using a so-called automatic contact procedure approach. The force in vertical direction on the top body is shown in Figure 14. After 0.01 s the force is constant as can be expected with this material model in a quasi-static approach.

The result of the quasi-static visco-elastic analysis is shown in Figure 14 as well. It is clear that due to the higher stiffness a higher force is obtained and that after 0.01 s the force reduces due to relaxation. (Observe that the application time is relatively short in comparison with the time constant).

The result of the dynamic contact analysis are shown in Figure 15, both for the material modelled with long term stiffness only and for the visco-elastic material. It is clear that in this application both the inertia and the visco-elastic effects have a strong effects on the obtained results and both effects have to be included in the evaluation of the structural behaviour of the seal.

It was shown by Achenbach (1995) that temperature changes can also have a tremendous effect on the visco-elasticity. By using a thermo-rheologically simple material model, with a WLF shift function these effects could easily be included in the analysis.

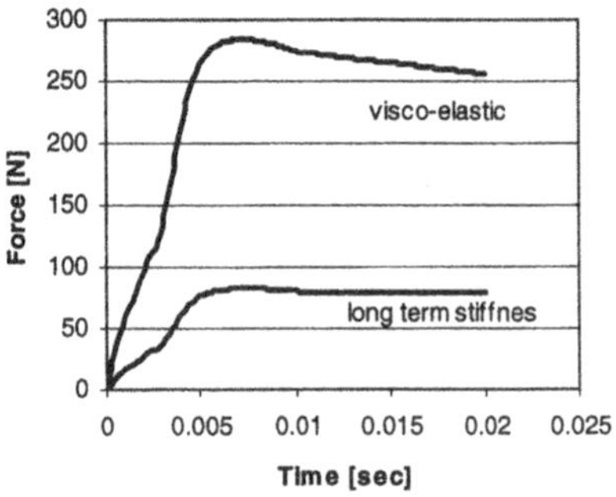

Figure 14 Force on top body versus time (static)

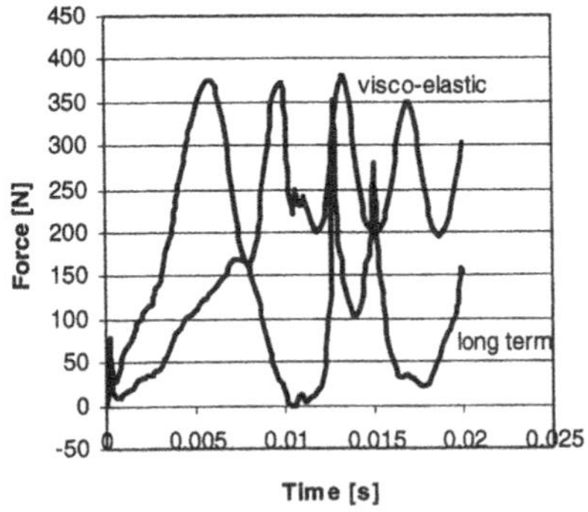

Figure 15 Force on top body versus time (dynamic)

4.3 Coupled structural-acoustic analysis of a seal subjected to harmonic vibrations

The undeformed seal as described in the previous example is surrounded by air. The geometry of the enclosing room cavities is shown in Figure 16. Also indicated is the location of the node at which an harmonic pressure variation is applied. Figure 17 shows a detail near the seal tip. It is clear that in this example leakage is present. The rubber is modelled with 4+1 noded linear displacement elements (including Lagrange multipliers to account for incompressibility).

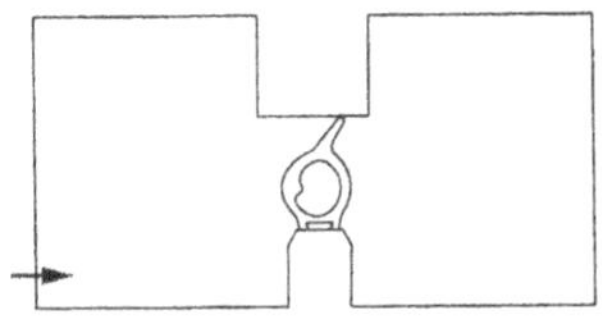

Figure 16 Seal surrounded by air in a cavity

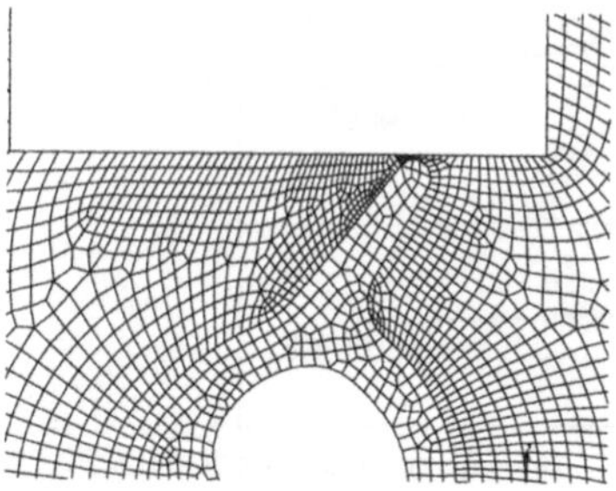

Figure 17 Details air and rubber seal

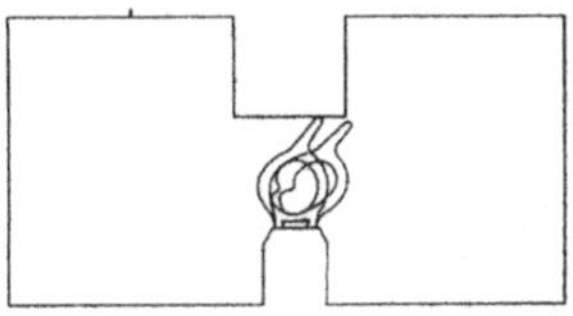

Figure 18 Deformation due to pressure change in air

The air is modelled with 4 noded linear pressure elements. The interfacing between the two is completely modelled through the contact procedure.
The rubber seal is glued to a fixed wall. For all acoustic boundaries fully reflective boundary conditions have been chosen (not a necessary restriction; other boundary conditions will be dependent on the availability of experimental data).

The rubber properties have been described in the previous section. The following material properties have been set for the air:

$$K = 1.10^5 \quad N/m^2$$
$$\rho = 1.2 \quad kg/m^3$$

Two coupled structural-acoustic analysis with a pressure variation in the range 70-150 Hz have been performed. The first coupled analysis is based on a quasi-static material with long term stiffness G^{∞}. In the second coupled analysis a frequency dependent stiffness has been used. An eigen frequency analysis of the seal only indicated a lowest eigen frequency of the seal of 74. Hz. Repetition of the eigen frequency analysis with stiffness based on $G^0 = G^{\infty} + G^1$ indicated a shift of this frequency to 137 Hz.

Figure 18 shows the (enlarged) deformation of the rubber seal due to the pressure variation. The result of the seal tip displacement for both materials is shown in Figure 19.
It is clear that the material for which only the long term stiffness is used passes the eigen-frequency at 74 Hz while for the visco-elastic material the eigen frequency is passed around 134 Hz.

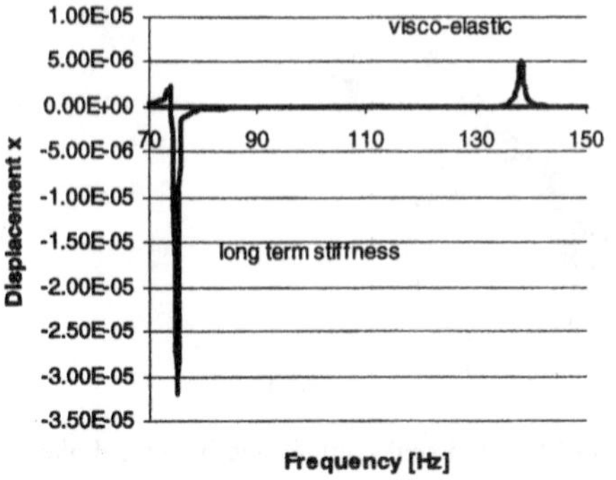

Figure 19 Seal tip displacement versus frequency

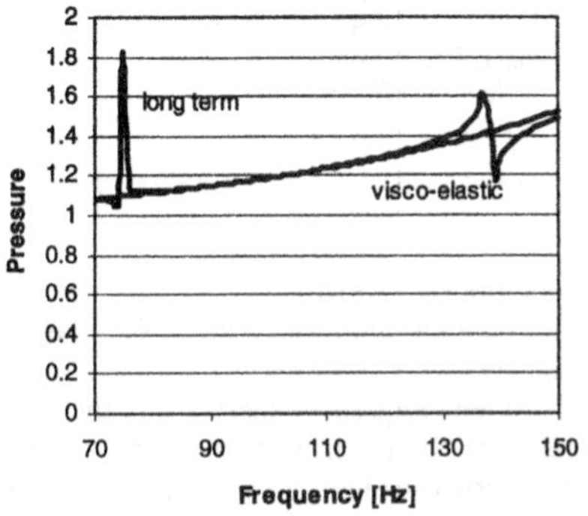

Figure 20 Response pressure versus frequency

The same behaviour is observed in the pressure variation at some point in the right room as displayed in Figure 20.
For a correct evaluation of the noise transfer function of a seal, visco-elastic effects have to be included and an analysis based on the long term stiffness only provides incorrect results. The effects of seal compression, in combination with acoustics requires automatic remeshing of the air.

5 CONCLUSIONS

Techniques for finite element analysis of rubber components have become robust and easy to use. With the applied models realistic results in a simulation of the behaviour of structural components, taking into account realistic loading conditions, can be obtained.

A key factor in the obtained accuracy is still the material model used in the simulation. This is not only due to the limiting capabilities of a model in describing physical phenomena. Often insufficient experimental data is available to determine the parameters required for a particular model, or the fit has to be based on e.g. a tensile test only, potentially

producing unreliable results for other stress states. An easy to use graphical prediction of the behaviour for other stress states, or for extrapolated data outside the region of available experimental stress-strain data, prior to the finite element analysis can avoid many unsuccessful analysis.
It is also shown that neglecting visco-elastic and inertia effects can have a large effect on the obtained results. This is also true if temperature effects are neglected, since they can have a strong effect on the visco-elasticity.

REFERENCES

Achenbach, M. & Stehmans H., 1995, Simulation of time and temperature effects of elastomeric materials or How fast can a seal close an suddenly occurring gap?, Proc. European MARC Users conference, Dusseldorf, MARC Analysis Europe, Zoetermeer.

Arruda, E.M. and Boyce, M.C., 1993, A three-dimensional constitutive model for the large stretch behaviour of rubber elastic materials. *J. Mech. Phys. Solids*, Vol. 41, N0. 2, pp 389-412.

Batterman, W. and Kohler, R., 1982, *Elastomer Federung, Elastische Lagerungen*. Wilhelm Ernst & Sohn, Berlin.

MARC: 1997, Theory and user information, version K7, 1997. MARC Analysis Research Corporation, Palo Alto, CA, USA.

Crawford, R, 1985. *Plastics and Rubbers - engineering design and applications.*

Gent, A.N., 1994, Rubber elasticity: Basic Concepts and behaviour. In Eirich Mark, Erman, editor, *Sience and Technology of Rubber*. Academic Press Inc.

Gobel, E.F., 1969, *Gummifedern - Berechnung und Gestaltung*, 3. Auflage, Springer Verlag, Berlin.

Heinrich, G., Straube, E., Helmis, G., 1988, *Adv. Polym. Sci.*, 85, pp. 243.

Kilian, H.G., 1981, *Polymer*, 22, 209

Lindley, P.B., 1966. The stiffness of rubber springs. In Payne Allen, Lindley editor, *Use of Rubber in Engineering*, McLaren and Sons LTD.

Ogden, R.W., 1972, Large Deformation isotropic elasticity. On the correlation of the theory and experiment for compressible rubberlike materials, *Proc. R. Soc. London*, A. 326, 565.

Ogden, R.W., 1984, *Non-linear elastic deformations*. EllisHorwood: Chisester

Simo, J.C. 1987, On a fully Three-Dimensional Finite Strain Viscoelastic Damage Model: Formulation and Computational Aspects, *Comp. Meth. Appl. Mechs. Eng.*, 60., 153.

Treloar, L.R.G., 1975. *The Physics of Rubber Elasticity*, Oxford University Press, Oxford

Constitutive Models for Rubber, Dorfmann & Muhr (eds) © 1999 Taylor & Francis ISBN 90 5809 113 9

Modelling of the thermo-mechanical material behaviour of rubber-like polymers – Micromechanical motivation and numerical simulation

S. Reese & P. Wriggers
Institute for Structural and Computational Mechanics, University of Hannover, Germany

ABSTRACT: The material behaviour of rubber on micro level is usually described by means of statistical mechanics. In particular, the Neo-Hooke model has been derived in this fashion. We show that a similar concept can be applied in order to include also viscoelastic effects. This results in a continuum mechanical model of finite viscoelasticity which is based on the multiplicative decomposition of the deformation gradient. Due to the latter aspect, the implementation of the model into a finite element code is suitably carried out in a way analogous to finite elastoplasticity. For the integration of the material equations for instance, we use the exponential mapping algorithm which has been proven to be very efficient from the computational point of view. Test calculations show that the formulation is appropiate to model the physical behaviour in practical applications realistically. Also thermo-mechanical coupling effects are incorporated.

1 INTRODUCTION

It is well-known that methods of statistical mechanics are appropiate to describe the thermoelastic material behaviour of rubber-like polymers. See for example the derivation of the Neo-Hooke model (Kuhn (1934), Treloar (1943)) More complicated is such a procedure, if also inelastic effects, e. g. viscoelasticity, need to be included. An appropiate approach for this purpose is the transient network concept (Green & Tobolsky (1946)) which is based on the assumption that chains are steadily breaking and reforming. The above authors have utilized this idea in order to formulate a model of *finite linear* viscoelasticity. The latter accounts for large deformations but only small deformation rates. Thus, only processes close to thermodynamic equilibrium can be considered.

In this contribution, we show how the transient network theory can be extended in order to obtain a more general concept of viscoelasticity. One important issue is that the new model is realistic also for states far away from thermodynamic equilibrium. In contrast to earlier approaches (see e. g. Simo (1987), Holzapfel (1996)), it is based on deformation-like internal variables. Crucial to the method is the fact that the transient network concept includes the idea of a stress-free intermediate configuration. This leads in the continuum mechanical context directly to the multiplicative decomposition of the deformation gradient. Note that in previous works, the multiplicative split had the status of a purely continuum mechanical assumption (see Sidoroff (1974), Lubliner (1985), Lion (1997 a,b), Reese & Govindjee (1998 a,b) and Keck & Miehe (1997)). Using the transient network theory, we are able to motivate this approach on micromechanical level. This reveals an important analogy to common models in finite elastoplasticity. The latter are micromechanically motivated by the observation that the deformation in single crystals can be decomposed into the (plastic) slip on the crystallographic slip planes and (elastic) lattice distortions and rigid rotations.

The present model has the additional advantage that it can be easily extended to include also fully-coupled thermomechanical effects. It should be emphasized that the usual thermo-mechanical split as proposed by Lu & Pister (1975) is not appropiate. This is due to the fact that the energetic contribution to the stresses is over-estimated, whereas the entropic part turns out to be negligible. It is, however, known from very early literature (see e. g. Kuhn & Grün (1942), Treloar (1943)) that the stresses in rubber are mainly entropic. We propose in this work a procedure which is consistent with the previously discussed thermomechanical behaviour and is above that straightforward from the continuum mechanical point of view.

Concerning the implementation of the model into a finite element code, one has to take special care of the time integration of the evolution equation in every Gauss point and the time integration of the energy balance carried out on global level. The local integration is carried out using the exponential mapping algorithm derived originally in the context of elastoplastic problems (see Weber & Anand (1990)). This algorithm has two main advantages. First of all, we can work with the spectral representation of the evolution equation which leads to high computational efficiency in particular for isotropic problems. Secondly, the algorithm preserves the symmetry of the material tangent. Note that this would not be the case for the standard backward Euler algorithm.

Concerning the element formulation, we prefer isoparametric low-order elements due to their robustness and simplicity. In order to avoid locking, a special element technology based on reduced integration plus hourglass stabilization has been developed (see Reese et al. (1999 a,b)).

The paper is structured as follows. In Section 2, we review the transient network theory. The transition to the continuum mechanical level follows in Section 3. Section 4 contains the general incorporation of thermomechanical effects. In Section 5, we state the weak form of the balance equation. The discussion of the numerical aspects follows in Section 6. In the final section, one example is presented in order to validate the presented approach.

2 MICROMECHANICAL CONSIDERATIONS

2.1 *Chain statistics*

In contrast to the standard static network theory, the so-called transient network theory is based on the assumption that chains are steadily breaking and reforming. Thus, we have chains which are elastically active and inactive. In order to make the differences between the two theories more clear, let us consider a rheological model with several parallel springs (see Figure 1). In the static theory, these springs remain always intact. If the length is held constant, consequently, also the

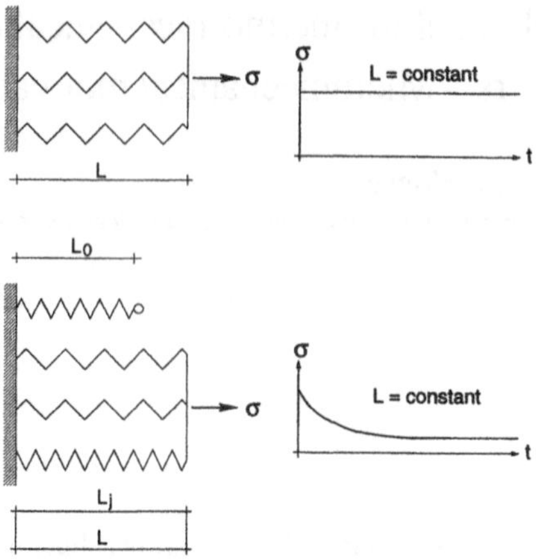

Figure 1. Relaxation test: (a) static theory, (b) transient theory.

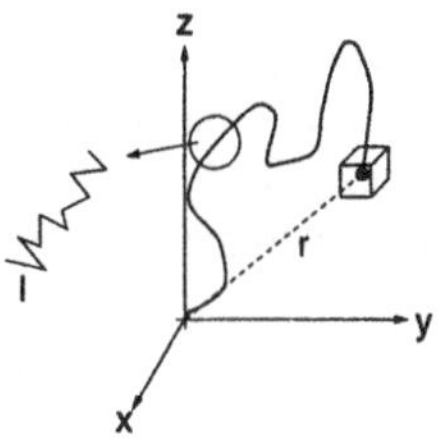

Figure 2. Average chain.

stress is constant. The situation is different in the transient case. It is a physically reasonable assumption that a chain reforms in the *stress-free* state. If we further assume that the total number of chains remains constant, we observe a steady decrease of the stress over time. This is the typical effect of stress relaxation.

To build up a statistical theory, one derives the probability for the end-to-end distance of an average chain lying between r and $r + dr$ (see Figure 2). The quantity l represents the mean length of a chain segment. The propability *density* based on the Gaussian distribution is given by

$$\hat{p}\,(r) = \Big(\frac{\sqrt{3}}{\sqrt{2\,n}\,\sqrt{\pi}\,l}\Big)^3 \exp\,[-\frac{3\,r^2}{2\,n\,l^2}]. \tag{1}$$

According to Boltzmann's law (k Boltzmann's constant), the entropy of a single chain reads

$$\eta_i = k \ln\left(\hat{p}\,(r)\,dV\right) = \widetilde{C} - \frac{3}{2\,n\,l^2}\,r^2. \tag{2}$$

Neglecting the internal energy e_i in the relation $\Psi_i = e_i - \Theta\,\eta_i$ for the Helmholtz free energy (Θ

absolute temperature), we come to

$$\Psi_i = \frac{3\,\Theta}{2\,n\,l^2}\,r^2 + C, \tag{3}$$

where C as well as $\widetilde{C}$ denote constants and n is the number of chain segments.

2.2 *Static network theory*

The next step is to derive the Helmholtz free energy for the whole network (for the whole ensemble of chains). We describe the undeformed state of a rubber test piece with the coordinates x_0, y_0 and z_0. If any chain in the network deforms like the bulk rubber (affinity assumption, see Figure 3), we obtain the coordinates for the deformed configuration as

$$x = \lambda_1\,x_0, \quad y = \lambda_2\,y_0, \quad z = \lambda_3\,z_0. \tag{4}$$

λ_A ($A = 1,2,3$) represent the stretches in the three principal directions. Thus, in the deformed configuration, we may state $r^2 = \lambda_1^2 x_0^2 + \lambda_2^2 y_0^2 + \lambda_3^2 z_0^2$, where λ_A ($A = 1,2,3$) represent principal stretches.

The free energy of the whole network is given by the relation $\Psi = \int \hat{\Psi}_i(r)\,dN$, where N is the number of chains per reference volume. Using (3) and carrying out the latter integration leads finally to the well-known Neo-Hooke model

$$\Psi = \frac{1}{2}\,N\,k\,\Theta\,(\lambda_1^2 + \lambda_2^2 + \lambda_3^2). \tag{5}$$

If we exploit further the assumption of incompressibility $\lambda_1\,\lambda_2\,\lambda_3 = 1$ and investigate the special deformation state $\lambda_2 = \lambda_3 = \frac{1}{\sqrt{\lambda_1}}$ (uniaxial tension), we arrive at $\Psi = \frac{1}{2}\,N\,k\,\Theta\,(\lambda_1^2 + \frac{2}{\lambda_1} - 3)$. The Cauchy stress is then derived by

$$\begin{aligned} \sigma &= \lambda_1\,\frac{\partial\Psi}{\partial\lambda_1} = N\,k\,\Theta\,(\lambda_1^2 - \frac{1}{\lambda_1}) \\ &= N\,k\,\Theta\,(\frac{L^2}{L_0^2} - \frac{L_0}{L}), \end{aligned} \tag{6}$$

where L is computed by $L = \lambda_1\,L_0$.

2.3 *Transient network theory (classical concept)*

Consider now the *transient* network concept according to Green & Tobolsky (1946). The latter authors start from the formula

$$\begin{aligned} \sigma &= k\,\Theta\sum_{(j)} N_j\,(\frac{L^2}{L_j^2} - \frac{L_j}{L}) \\ &:= N\,k\,\Theta\left(\frac{L^2}{q^2} - \frac{m}{L}\right) \end{aligned} \tag{7}$$

In the rheological model (Figure 1b), the quantity L_j represents the length, for which the chain type j has reformed. N_j is the number of chains (per reference volume) belonging to chain type j. The comparison with (6) shows, that in the transient network theory, q^2 and m take the place of L_0^2 and L_0, respectively. Note that the "internal lengths" L_j represent additional unknowns which have to be determined by additional equations. For this purpose, Green & Tobolsky (1946) state an evolution law for the breaking of chains:

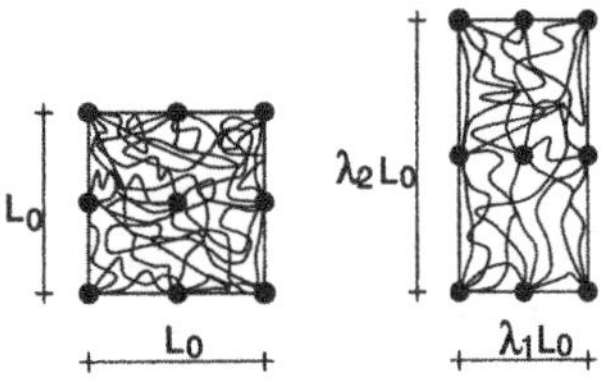

Figure 3. Undeformed and deformed configuration.

$$\dot{M} = -\alpha\,M \quad\Rightarrow\quad M = M_0 \exp\left[-\alpha\,(t - t_0)\right] \tag{8}$$

In the latter formula, α denotes the probability *per time increment* for the chain breakage, and M a certain number of chains. Interestingly, α can be interpreted as the inverse of the relaxation time τ. If the system relaxes immediately ($\tau \to 0$), the probability α goes to infinity ($\delta t \to 0 \Rightarrow \alpha \to \infty$). On the other hand, for very large relaxation times ($\tau \to \infty$), α tends to zero. Green & Tobolsky (1946) finally end up with a stress relation which is similar to the common one-dimensional stress relation in linear viscoelasticity. In order to recognize this analogy, the term $\frac{L^2}{q^2} - \frac{m}{L}$ has to be identified with the elastic logarithmic strain $\varepsilon_e = \varepsilon - \varepsilon_i$ $\varepsilon := \ln\lambda$). In linear viscoelasticity, the additive split of the *linearized* strain measure ε_L into elastic and inelastic parts is well accepted. Thus, interestingly, Green & Tobolsky's approach tacitly implies the additive split of $\varepsilon = \ln\lambda$. The latter results automaticly into the multiplicative split of the stretch λ. But due to the fact that the evolution law for the chain breakage is *linear* their concept is only appropiate for applications in finite linear viscoelasticity, i. e. small perturbations away from thermodynamic equilibrium. Moreover, the approach has been achieved in the context of a one-dimensional consideration. For the purpose of developing a finite fully three-dimensional theory, several additional steps are necessary.

2.4 *Transient network theory (new approach)*

We consider now *three* deformation states, firstly again the undeformed configuration as in Section 2.2. Secondly, we define the coordinates of a so-called *intermediate* configuration by

$$x_j = \lambda_{1j}\, x_0, \quad y_j = \lambda_{2j}\, y_0, \quad z_j = \lambda_{3j}\, z_0. \tag{9}$$

The intermediate configuration for the chain type j represents the configuration, in which this chain type has reformed (stress-free state). The general deformed configuration is then given by the coordinates

$$x = (\lambda_1 \lambda_{1j}^{-1})\, x_j, \quad y = (\lambda_2 \lambda_{2j}^{-1})\, y_j, \quad z = (\lambda_3 \lambda_{3j}^{-1})\, z_j.$$

If we carry out the same procedure as in Section 2.2, we obtain the free energy of the transient network with

$$\begin{aligned} \Psi &= \sum_{(j)} \frac{N_j\, k\, \Theta}{2} \Big((\lambda_1 \lambda_{1j}^{-1})^2 + (\lambda_2 \lambda_{2j}^{-1})^2 \\ &\quad + (\lambda_3 \lambda_{3j}^{-1})^2 - 3 \Big) \end{aligned}$$

Unknown (for each chain type) are here still the number of chains N_j and the so-called "internal" stretches λ_{Aj} $(A = 1, 2, 3)$. Green & Tobolsky (1946) could have started also from this relation in order to derive (7). But at this point, we go beyond the work of the latter authors. The new idea is to replace the dependence on N_j and λ_{Aj} $(A = 1, 2, 3)$ by means of a distribution function $f(\lambda_{Aj}, t)$ Then, we write for the free energy

$$\begin{aligned} \Psi &= \int_1^{\lambda_1} \frac{k\,\Theta}{2} \Big((\lambda_1 \lambda_{1j}^{-1})^2 - 1 \Big) \hat{f}(\lambda_{1j}, t)\, d\lambda_{1j} \\ &+ \int_1^{\lambda_2} \frac{k\,\Theta}{2} \Big((\lambda_2 \lambda_{2j}^{-1})^2 - 1 \Big) \hat{f}(\lambda_{2j}, t)\, d\lambda_{2j} \\ &+ \int_1^{\lambda_3} \frac{k\,\Theta}{2} \Big((\lambda_3 \lambda_{3j}^{-1})^2 - 1 \Big) \hat{f}(\lambda_{3j}, t)\, d\lambda_{3j}, \end{aligned}$$

where the distribution function fulfills

$$\int_1^{\lambda_A} f(\lambda_{Aj}, t)\, d\lambda_{Aj} = \hat{N}(t). \tag{10}$$

The integral gives the current number of chains per reference volume. Due to the fact that the stretches λ_{Aj} describe a certain real but past configuration, the internal stretches have to lie in the interval $[1, \lambda_A]$. The fact that we deal with *isotropic* material behaviour, is included by taking the same distribution function for each direction. The distribution, however, is time- as well as deformation-dependent.

In order to overcome the difficulty of finding an appropiate distribution function, we again con-

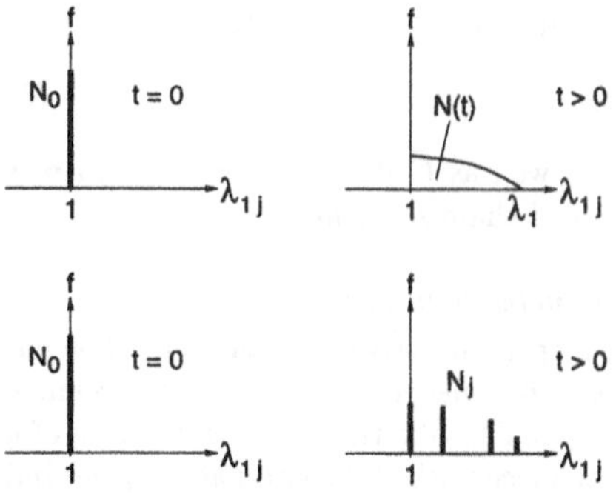

Figure 4. Continuous and discrete distribution.

sider first the one-dimensional case. In general, the distribution function in the undeformed configuration will be described by (see also Figure 4)

$$\hat{f}(\lambda_{1j}, t_0) = \begin{cases} N_0 & \text{for } \lambda_{1j} = 1 \\ 0 & \text{else} \end{cases} \tag{11}$$

Since it is not possible to make any statement about the continuous distribution in the deformed configuration, we study rather the discrete case. Physically, this means that we do not consider every chain type seperately but include several chain types into a few "average" chain types. Consequently, the corresponding average intermediate configuration does not represent necessarily a *real* deformed state. The internal stretches of such a fictive intermediate configuration are denoted by $\lambda_{A\star} = \hat{\lambda}_{A\star}(t)$:

$$\hat{f}(\lambda_{1j}, t) = \begin{cases} N_\infty & \text{for } \lambda_{1j} = 1 \\ N_2 & \text{for } \lambda_{1j} = \lambda_{1\ldots} \\ N_\star & \text{for } \lambda_{1j} = \hat{\lambda}_{1\star}(t) \\ N_4 & \text{for } \lambda_{1j} = \lambda_{1\ldots} \\ 0 & \text{else} \end{cases} \tag{12}$$

Working with a two-type model yields the free energy

$$\begin{aligned} \Psi &= \frac{N_\infty\, k\, \Theta}{2} \left(\lambda_1^2 + \lambda_2^2 + \lambda_3^2 - 3 \right) \qquad (13) \\ &+ \frac{N_\star\, k\, \Theta}{2} \left((\lambda_1 \lambda_{1\star}^{-1})^2 + (\lambda_2 \lambda_{2\star}^{-1})^2 + (\lambda_3 \lambda_{3\star}^{-1})^2 - 3 \right), \end{aligned}$$

where N_∞ as well as $N_\star$ represent material parameters.

For the purpose of an easier understanding of (13), we look at the rheological model plotted in Figure 5. The first part of (13) represents the free energy (strain energy) in the upper spring. This term models the rate-independent part (equilibrium part) of the material behaviour. The second part denotes the strain energy of the second spring.

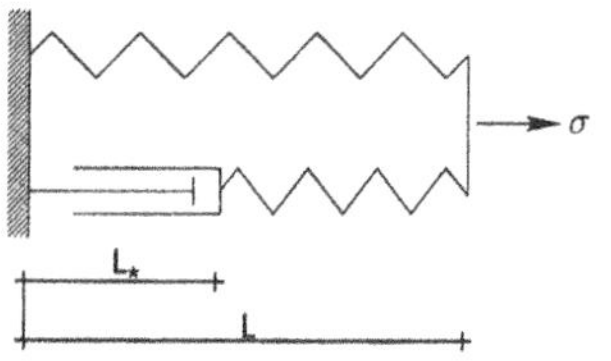

Figure 5. Rheological model for viscoelasticity

The stress resulting from the latter contribution is usually termed over-stress. Note that the rheological model certainly allows only a one-dimensional investigation. The present theory, however, is fully three-dimensional.

3 CONTINUUM MECHANICAL MODELING

The notion of a fictive *average* intermediate configuration brings us already to the continuum mechanical level. It is, however, still unclear for what kind of *continuum mechanical* model (13) stands for.

In order to gain a better understanding of this point, let us start again from the purely continuum mechanical point of view. Using the concept of elastic isomorphism (see in the context of elastoplasticity Bertram (1993), Svendsen (1998)), we arrive at the Helmholtz free energy

$$\Psi = \hat{\Psi}_\infty(\mathbf{C}, \Theta) + \hat{\Psi}_\star(\mathbf{F}_\star^{-T} \cdot \mathbf{C} \cdot \mathbf{F}_\star^{-1}, \Theta). \tag{14}$$

In the latter relation, $\mathbf{F}_\star$ takes the role of an internal variable. Physically, the tensor $\mathbf{F}_\star$ represents the *inelastic* part of the deformation. The *elastic* part can be easily defined by $\mathbf{F}_e := \mathbf{F} \cdot \mathbf{F}_\star^{-1}$.

In the case of isotropic material behaviour, Ψ_∞ and $\Psi_\star$ represent isotropic functions of Θ, $\mathbf{C}$ and $\mathbf{C}_e := \mathbf{F}_e^T \cdot \mathbf{F}_e$, respectively. $\mathbf{F}_\star$ enters the formulation only via $\mathbf{C}_\star := \mathbf{F}_\star^T \cdot \mathbf{F}_\star$. Thus, in the isotropic case, $\mathbf{C}_\star$ is considered as internal variable. We end up with the functions

$$\Psi_\infty = \hat{\Psi}_\infty(\lambda_1^2, \lambda_2^2, \lambda_3^2, \Theta) \tag{15}$$

$$\Psi_\infty = \hat{\Psi}_\star((\lambda_1 \lambda_{1\star}^{-1})^2, (\lambda_2 \lambda_{2\star}^{-1})^2, (\lambda_3 \lambda_{3\star})^2, \Theta). \tag{16}$$

It is not difficult to see that (13) represents a special case of the continuum mechanical form in the isotropic case. To be more precise, (13) is based on the Gaussian distribution (1) and additionally on the assumption that the stresses are of purely entropic origin.

The last open point concerns the derivation of a physically reasonable evolution equation for the internal variable $\mathbf{C}_\star$. As usual, we have to choose a form which is thermodynamically consistent. This means, it has to fulfill the second law of thermodynamics. An appropiate evolution equation is for instance

$$\frac{1}{2} \overset{\triangle}{\mathbf{b}}_e \cdot \mathbf{b}_e^{-1} = \frac{1}{3\,V_{\text{vol}}} \operatorname{vol} \boldsymbol{\tau}_\star + \frac{1}{2\,V_{\text{dev}}} \operatorname{dev} \boldsymbol{\tau}_\star, \tag{17}$$

where the Kirchhoff stress tensor $\boldsymbol{\tau}_\star$ is derived from ($\mathbf{b}_e := \mathbf{F}_e \cdot \mathbf{F}_e^T$)

$$\boldsymbol{\tau}_\star = 2\,\mathbf{b}_e \cdot \frac{\partial \Psi_\star}{\partial \mathbf{b}_e}. \tag{18}$$

and $\overset{\triangle}{\mathbf{b}}_e$ is given by

$$\overset{\triangle}{\mathbf{b}}_e := \mathbf{b}_e \cdot \mathbf{F}^{-T} \cdot \dot{\mathbf{C}}_i \cdot \mathbf{F}^{-1} \cdot \mathbf{b}_e. \tag{19}$$

V_{vol} and V_{dev} represent the deviatoric and the volumetric viscosity, respectively. In the special case of small deviations away from thermodynamic equilibrium, the evolution equation (17) reduces to

$$\dot{\mathbf{C}}_\star = \frac{1}{\tau}(\mathbf{C} - \mathbf{C}_\star). \tag{20}$$

But the main point is here that (13) implies on continuum mechanical level the multiplicative decomposition of the deformation gradient. The application of this split in the context of viscoelasticity leads to a new class of models which have been shown to be well suited for experimental validation as well as for numerical calculations (see Lion (1997 a,b), Reese & Govindjee (1998 a,b), Keck & Miehe (1997)).

4 THERMOMECHANICAL EFFECTS

4.1 *Generalized approach*

It has been shown in Section 2 that neglecting the internal energy leads to a linear dependence of (13) on the absolute temperature. This implies the statement that also the mechanical material parameters like the shear modulus or the bulk modulus depend only linearly on the temperature. Experimental observations, however, show that this assumption is not always realistic (Treloar (1975), Nowinski (1978)). Chadwick (1974) have derived the the relation

$$\Psi = \Psi_0 \frac{\Theta}{\Theta_0} + e_0\,(1 - \frac{\Theta}{\Theta_0}) + \int_{\Theta_0}^{\Theta} c\,(1 - \frac{\Theta}{\tilde{\Theta}})\,d\tilde{\Theta},$$

where the index 0 characterizes quantities evaluated at an arbitrary reference temperature Θ_0. $c = -\Theta \frac{\partial^2 \Psi}{\partial \Theta^2}$ denotes the heat capacity which is usually assumed to be constant. Then, only a *linear* dependence of the stresses on the temperature can be considered. Thus, in order to include more general cases, the heat capacity must be deformation-dependent. Taking this into account, we arrive at

$$\Psi = (\frac{\Theta}{\Theta_0} + g_+)\,\Psi_0 + (1 - \frac{\Theta}{\Theta_0} + h_+)\,e_0 + t\,c_0, \quad (21)$$

where the short hand notations

$$\hat{g}_+(\Theta) \;:=\; \hat{g}(\Theta) - \frac{\partial g}{\partial \Theta}\bigg|_{\Theta_0}(\Theta - \Theta_0)$$

$$\hat{h}_+(\Theta) \;:=\; \hat{h}(\Theta) - \frac{\partial g}{\partial \Theta}\bigg|_{\Theta_0}(\Theta - \Theta_0)$$

$$\hat{t}(\Theta) \;:=\; \Theta - \Theta_0 - \Theta \ln \frac{\Theta}{\Theta_0} \quad (22)$$

have been used. Appropiate choices for the functions g_+ and h_+ are for instance

$$g_+ = a_1 \left((\frac{\Theta}{\Theta_0})^{a_2} - 1 \right) \quad (23)$$

and

$$h_+ = b_1 \left((\frac{\Theta}{\Theta_0})^{b_2} - 1 \right). \quad (24)$$

See for more details Reese & Govindjee (1998 b).

4.2 *Thermomechanical split*

In many papers about thermomechanical models, the so-called thermomechanical split according to Lu & Pister (1975) is utilized. In the latter essay, the multiplicative split of

$$\mathbf{F} = \mathbf{F}_{\mathrm{M}} \cdot \mathbf{F}_{\Theta} \quad (25)$$

into mechanical and thermal parts is carried out. The thermal part is defined by $\mathbf{F}_\Theta = (\det \mathbf{F}_\Theta)^{\frac{1}{3}}\,\mathbf{1}$. For simplicity, we restrict ourselves in the following to thermo-elastic problems. Many models are based on the assumption that Ψ depends only on the mechanical part of $\mathbf{F}$ through $\mathbf{b}_{\mathrm{M}} := \mathbf{F}_{\mathrm{M}} \cdot \mathbf{F}_{\mathrm{M}}^T$. The entropy is then given by

$$\eta = \frac{2}{3}\,\alpha_T\,\Theta\,\mathrm{tr}\,\boldsymbol{\tau}. \quad (26)$$

The value of this expression is small in comparison with the contribution from the internal energy. Thus, we conclude that the thermomechanical split should be applied only in the context of metals, where the energetic contribution dominates the entropic part. For rubber, however, the opposite is the case, such that one should rather take the generalized approach.

5 WEAK FORM OF BALANCE EQUATIONS

In order to close the system of equations, we still need to state the balance equations. Since the balance of mass and the balance of angular momentum are locally fulfilled, we have to formulate only the balance of linear momentum and the balance of energy in weak form. The weak form of the balance of linear momentum reads

$$g_{\mathrm{M}} \;=\; \int_{B_t} \frac{1}{2}\boldsymbol{\sigma} : (\mathbf{F}^{-T} \cdot \delta\mathbf{C} \cdot \mathbf{F}^{-1})\,dv - \int_{B_t} \rho\,\ddot{\mathbf{u}} \cdot \delta\mathbf{u}\,dv + g_{\mathrm{M}}^{\mathrm{ext}} = 0$$

The quantity $\rho\,\ddot{\mathbf{u}}$ represents the inertia force per reference volume. Moreoever, the short hand notation $g_{\mathrm{M}}^{\mathrm{ext}}$ has been used to indicate the contribution of the external loading. The thermomechanical coupling, merely the influence of the temperatur on the deformation, is visible in the temperature dependence of the Cauchy stress tensor $\boldsymbol{\sigma}$.

The weak form of the balance of energy is written as

$$g_{\mathrm{T}} \;=\; \int_{B_t} \mathbf{q} \cdot \mathrm{grad}\,\delta\Theta\,dv + \int_{B_t} (-w_{\mathrm{int}} + w_{\mathrm{ext}} - c\,\dot{\Theta})\,\delta\Theta\,dv - \int_{\partial B_q} \mathbf{q} \cdot \mathbf{n}\,\delta\Theta\,da = 0, \quad (27)$$

where $\mathbf{q}$ represents the spatial heat flux and w_{int} and w_{ext} denote short hand notations for two energy dissipation terms which are not specified here further. The thermomechanical coupling, i. e. the influence of the temperature on the deformation, is here included in the deformation dependence of the dissipation terms and the heat capacity as well as the fact that the spatial heat flux depends on the *spatial* temperature gradient. Due to the deformation dependence of the dissipation terms we observe especially in the case of cyclic loading the typical thermomechanical heating.

6 NUMERICAL ASPECTS

The spatial discretization of a structure is shown in Figure 6. The material modeling takes place exclusively on local level (Gauss point level). Note that for the integration of the evolution equation, we apply the exponential mapping algorithm according to Weber & Anand (1990). Concerning the finite element formulation, we have at every node displacement degrees-of-freedom as well as one temperature degree-of-freedom. Accordingly, on the boundary not only forces and displacements

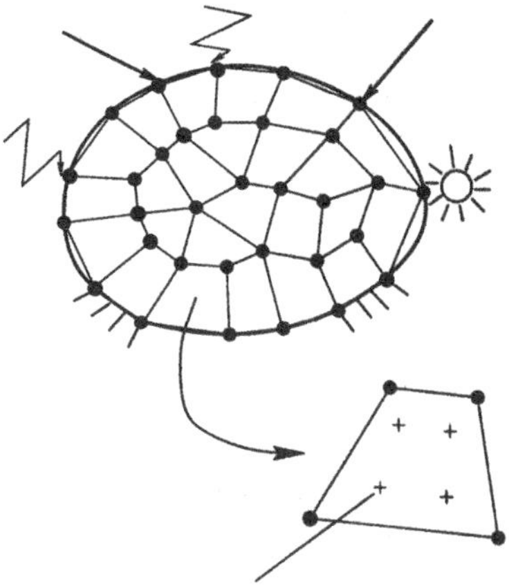

Figure 6: Spatial discreziation

but also heat fluxes and temperatures are prescribed.

On global level we arrive at the semi-discrete coupled initial boundary-value-problem

$$\begin{aligned} \hat{\boldsymbol{R}}_M(\boldsymbol{V}_M, \boldsymbol{V}_T) + \boldsymbol{M}\,\ddot{\boldsymbol{V}}_M &= \hat{\boldsymbol{P}}_M(t) \\ \hat{\boldsymbol{R}}_T(\boldsymbol{V}_M, \boldsymbol{V}_T, \dot{\boldsymbol{V}}_M, \dot{\boldsymbol{V}}_T) &= \hat{\boldsymbol{P}}_T(t) \end{aligned} \quad (28)$$

where the vectors $\boldsymbol{R}_M$ and $\boldsymbol{R}_T$ contain the mechanical and thermal contributation to the vector of inner "forces", respectively. Analogously, the mechanical and thermal degress-of-freedom are written into the vectors $\boldsymbol{V}_M$ and $\boldsymbol{V}_T$, respectively. $\boldsymbol{P}_M$ and $\boldsymbol{P}_T$ are load vectors and $\boldsymbol{M}$ represents the mass matrix. The inertia terms can be neglected for the present applications such that (28) reduces to a coupled differential equation system of first order. After having carryied out the time discretization by means of an appropiate method (here the backward Euler algorithm is used), we obtain a non-linear equation system for $\boldsymbol{V}_M$ and $\boldsymbol{V}_T$. We use the Newton's method to solve this system. Note, however, that due to the thermomechanical contributions, the tangential stiffness matrix is not symmetric.

7 EXAMPLE

Rubber-like materials are often used for bearings. The stiffness of such constructions is commonly increased by steel layers. For both parts of the free energy, Ψ_∞ and $\Psi_\star$, we take the Ogden form (see e. g. Ogden (1984)). The material parameters are given below:

Rubber:

$\mu_1 = 2.1377\ \mathrm{N/mm^2}$	$\alpha_1 = 1.5$
$\mu_2 = -0.00162\ \mathrm{N/mm^2}$	$\alpha_2 = -7.5$
$\mu_3 = 0.000355\ \mathrm{N/mm^2}$	$\alpha_3 = 12.0$
$\Lambda \approx 500\ \mathrm{N/mm^2}$	
$\mu = 2.4\ \mathrm{N/mm^2}$	$\alpha = 2$
$\tau = 20.8\ \mathrm{s}$	
$\alpha_T = 6.36 \cdot 10^{-4}\ \mathrm{1/K}$	$K = 0.15\ \mathrm{N/(K\ s)}$
$c_0 = 1.507\ \mathrm{N/(mm^2\,K)}$	

Steel:

$\mu = 80769\ \mathrm{N/mm^2}$	$\alpha = 2$
$\Lambda = 121154\ \mathrm{N/mm^2}$	
$\sigma_Y = 250\ \mathrm{N/mm^2}$	$H = 30\ \mathrm{N/mm^2}$
$\alpha_T = 1.2 \cdot 10^{-5}\ \mathrm{1/K}$	$K = 45\ \mathrm{N/(K\ s)}$
$c_0 = 3.768\ \mathrm{N/(mm^2\,K)}$	

The first example is the 2D bearing plotted in Figure 7. The displacements are fully constrained at the bottom of the structure. At the top, we control the horizontal displacement (sinusoidal loading). The top plate is allowed to move in vertical direction.

We study the evolution of the temperature for two different loading frequencies. In the first calculation, we choose f such that $1\,s$ for one cycle is needed. Thus, after approximately 20 cycles, we are in the range of the relaxation time ($\tau = 20.8\,s$). This means, that in the 1., 3. and 7. cycle, the inelastic deformation in the rubber parts has not been fully developed. The tempera-

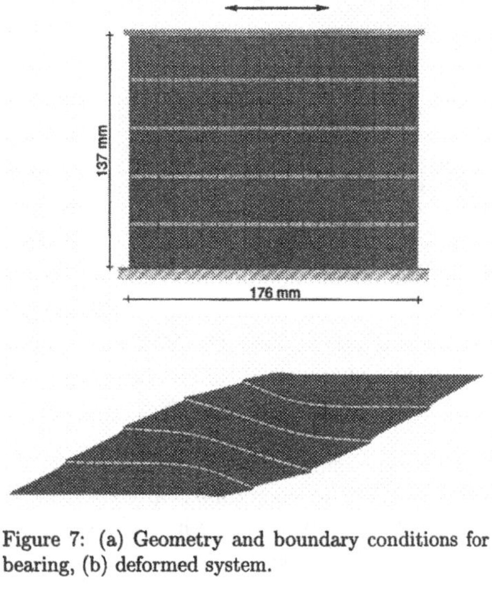

Figure 7: (a) Geometry and boundary conditions for 2D bearing, (b) deformed system.

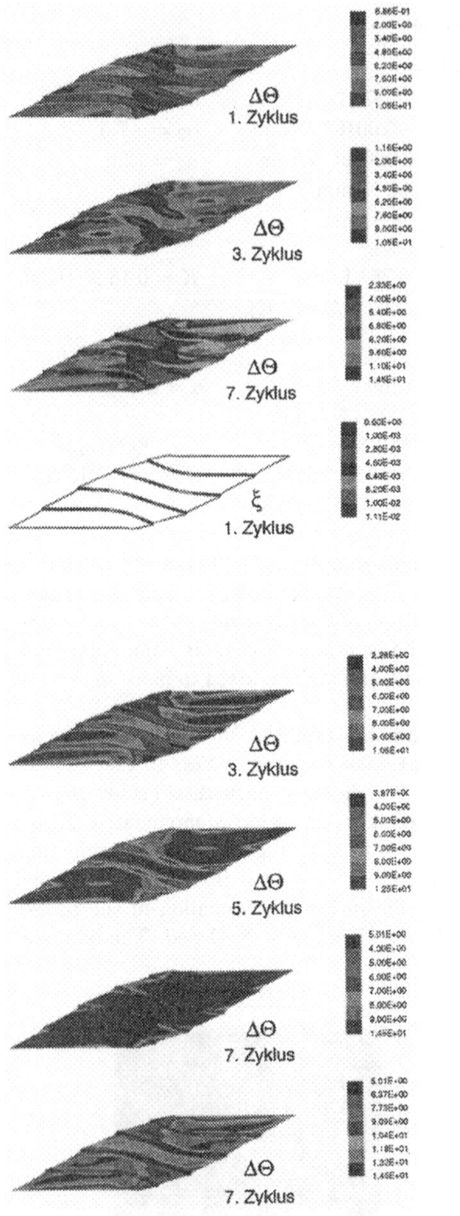

Figure 9: (a-d) Temperature evolution for $f = 0.2\,\mathrm{Hz}$

ture remains relatively low. The steel layers, however, show already noticeable plastic deformation. Therefore, the temperature increase concentrates on these parts of the construction.

If we choose a smaller frequency ($f = 0.2\,\mathrm{Hz}$), the range of the relaxation time is already reached after about four cycles. Then, the inelastic deformation in the rubber becomes relevant, and, consequently, the temperature in the rubber parts increases noticeably (Figure 9). In contrast to the first calculation, the heating in the rubber becomes dominant. The steel shows the same behaviour as before, which is certainly due to the fact that the latter material behaviour is rate-indenpent and thus independent of the loading rate.

8 CONCLUSIONS

In the present paper, we have shown that the transient network theory develops into a continuum mechanical theory of viscoelasticity based on the multiplicative decomposition of the deformation gradient. Thus, this kind of split, which had to be considered before as a purely consitutive assumption, can be motivated on micromechanical level. The resulting model accounts for large deformation as well as large deformation rates. It therefore represents a true *finite* model for viscoelastic material behaviour. Important is also, that thermomechanical coupling phenomeno can be incorporated in a thermodynamically consistent way. Experimental validation of such a model has been carried out by Lion (1997 a,b). Important issues concerning the numerical implementation are the choice of the time step and the finite element formulation. For the latter we use a recently developed stabilization technique which avoids locking completely.

REFERENCES

Bertram, A. 1993. Description of finite inelastic deformations. In Eds. A. Benallal, R. Billardon & D. Marquis *MECAMAT '92, Multiaxial Plasticity* 821-835.

Chadwick, P. 1974. Thermomechanics of rubber-like materials. *Philosophical Transactions of the Royal Society of London, Series A* 276:371-403.

Green, M. S. & A. V. Tobolsky 1946. A new approach to the theory of relaxing polymeric media. *Journal of Chemical Physics* 14:80-92.

Holzapfel, G. 1996. On large strain viscoelasticity: continuum formulation and finite element applications to elastomeric structures. *International Journal for Numerical Methods in Engineering* 39: 3903-3926.

Keck, J. & C. Miehe 1997. An Eulerian over-stress type viscoplastic constitutive model in spectral form. Formulation and numerical implementation. In Eds. D. R. J. Owen, E. Onate, E. Hinton. *Computational Plasticity, Fundamentals and Applications*, Barcelona, Spain.

Kuhn, W. 1934. Über die Gestalt fadenförmiger Moleküle in Lösungen. *Kolloidzeitschrift* 68:2-15.

Kuhn, W. & F. Grün 1942. Beziehungen zwischen elastischen Konstanten und Dehnungsdoppelbrechnung hochelastischer Stoffe. *Kolloidzeitschrift* 101:248-271.

Lion, A. 1997a. A physically based method to represent the thermo-mechanical behaviour of elastomers. *Acta Mechanica* 123:1-25.

Lion, A. 1997b. On the large deformation behaviour of reinforced rubber at different temperatures. *Journal of the Mechanics and Physics of Solids* 45:1805-1834.

Lu, S. C. H. & K. S. Pister 1975. Decomposition of deformation and representation of the free energy function for isotropic thermoelastic solids. *International Journal of Solids and Structures* 11:927-934.

Lubliner, J. 1985. A model of rubber viscoelasticity. *Mechanics Research Communications* 12:93-99.

Nowinski, J. L. 1978. *Theory of Thermoelasticity with Applications.* Sijthoff and Noordhoff, Alphen aan den Rijn.

Ogden, R. W. 1984. *Nonlinear elastic deformations.* Ellis Horwood, Chichester.

Reese, S. & S. Govindjee 1998. A theory of finite viscoelasticity and numerical aspects. *International Journal of Solids and Structures* 35:3455-3482.

Reese, S. & S. Govindjee 1998. Theoretical and numerical aspects in the thermo-viscoelastic material behaviour of rubber-like polymers. *Mechanics of Time-dependent Materials* 1:357-396.

Reese, S., M. Küssner & B. D. Reddy 1999. A new stabilization technique for finite elements in nonlinear elasticity. *International Journal for Numerical Methods in Engineering* 44:1617-1652.

Reese, S., P. Wriggers & B. D. Reddy 1999. A new locking-free brick element technique for large deformations in elasticity. *Computers & Structures* to be published.

Sidoroff, F. 1974. Un modele viscoelastique nonlineaire avec configuration intermediaire. *Journal de Mecanique* 13:679-713.

Simo, J. C. 1987. On a fully three-dimensional finite-strain viscoelastic damage model: formulation, numerical analysis and implementation. *Computer Methods in Applied Mechanics and Engineering* 60:153-173.

Svendsen, B. 1998. A thermodynamic formulation of finite-deformation elastoplasticity with hardening based on the concept of material isomorphism. *International Journal of Plasticity* 14:473-488.

Treloar, L. R. G. 1943. II: The elasticity of a network of long-chain-molecules. *Transactions of the Faraday Society* 39:241-246.

Treloar, L. R. G. 1975. *The Physics of Rubber Elasticity.* Clarendon Press, Oxford.

Weber, G. & L. Anand 1990. Finite deformation constitutive equations and a time integration procedure for isotropic hyperelastic-viscoplastic solids. *Computer Methods in Applied Mechanics and Engineering* 79:173-202.

Constitutive Models for Rubber, Dorfmann & Muhr (eds) © 1999 Taylor & Francis ISBN 90 5809 113 9

An energy-based model of the Mullins effect

R.W.Ogden
Department of Mathematics, University of Glasgow, UK
D.G.Roxburgh
Department of Engineering Mathematics, University of Newcastle upon Tyne, UK

ABSTRACT: In a recent paper (Ogden & Roxburgh 1999) the authors developed a simple energy-based phenomenological approach to the (quasi-static) modelling of the stress-softening feature of the Mullins effect observed in filled rubbers. The residual strain (permanent set) associated with the Mullins effect was not included in the model, however. The model is based on the theory of incompressible isotropic elasticity amended by inclusion of a damage parameter, the role of which is to enable the form of the strain-energy function to change when unloading is initiated from a point on a primary loading path from the virgin state. The resulting theory, in which a primary loading path is characterized by a strain-energy function and each unloading (and subsequent re-loading) path is characterized by a different strain-energy function, is referred to as *pseudo-elasticity*. In the present paper we summarize this theory and then show how, by a simple modification, residual strain may be incorporated.

1 INTRODUCTION

In two recent papers (Lazopoulos & Ogden 1998, 1999) a quasi-static theory of *pseudo-elasticity* was developed through the introduction of additional variables into the energy function of an elastic material. These additional variables provide a means for changing the form of the energy function at certain critical values of the deformation or energy. This theory was then adapted (Ogden & Roxburgh 1999) with a single additional variable (referred to as a *damage variable*) in order to provide a phenomenological description of an idealized form of the (damage induced) Mullins effect (Mullins 1947) occurring in certain rubberlike solids and used as the basis for modelling by previous authors (Govindjee & Simo 1992, for example).

The idealized effect is depicted in Figure 1 for the case of simple tension, with the nominal stress t plotted as a function of the stretch λ. Consider the primary loading path abb' from the virgin state, where b' an arbitrary point at which unloading is initiated and follows the path $b'Ba$. On re-loading the latter path is retraced, and on further loading beyond b' the path $b'c$ is traced in continuation of the primary loading path $abb'cc'd$ (which is the path that would be followed in the absence of unloading). If loading is next terminated at c' then the path $c'Ca$ is followed on unloading and retraced back to c' on re-loading. If no further loading beyond c' is applied then the curve aCc' represents the subsequent material response, which is then elastic, as described in (Ogden & Roxburgh 1999). For loading beyond c' the primary path is again followed and the pattern of unloading/reloading is repeated. A key feature is the *stress softening* on unloading relative to the primary loading path, that is the value of t on aBb' or aCc' is less than that on $abb'cc'$ for the same value of λ.

In this application the criterion for switching the form of the strain-energy function is the onset of unloading from a point (in deformation space) on a primary loading path from the virgin state of the material. In a simple tension test, for example, this amounts to initiation of reduction of the stretch achieved on the primary (tensile) loading path. The damage variable is regarded as constant on the primary loading path and reduces monoton-

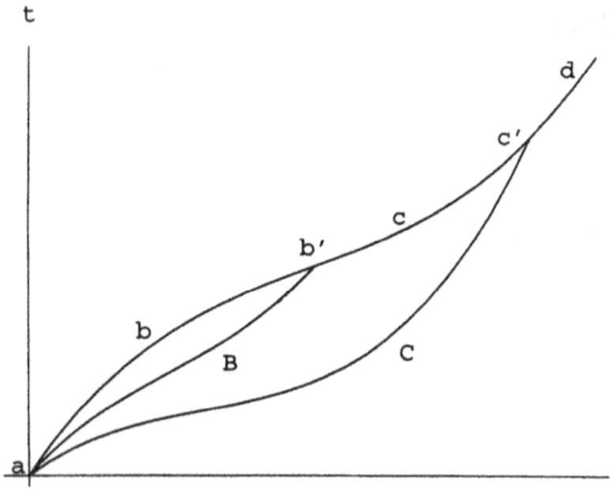

Figure 1. Idealized Mullins effect: schematic loading/unloading curves in simple tension with nominal stress t plotted against the stretch λ.

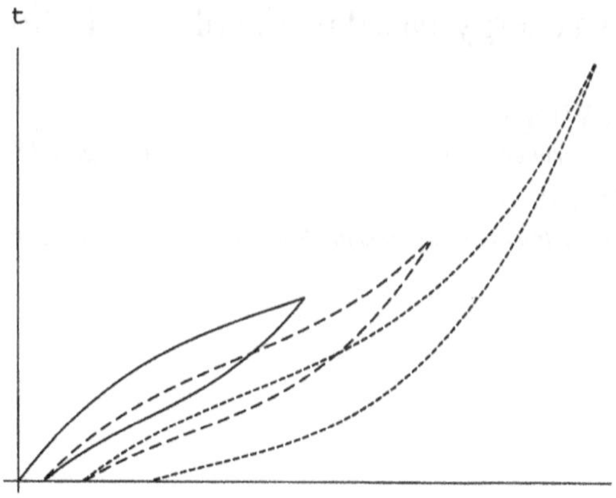

Figure 2. Schematic representation of the nominal stress t as a function of the stretch λ based on simple tension experiments: continuous curve—first loading/unloading cycle; dashed curve—second loading/unloading cycle; short-dashed curve—third loading/unloading cycle.

ically on unloading, thus reflecting the stress softening observed in experiments. In (Ogden & Roxburgh 1999) the theory was shown to give excellent agreement with data from simple tension tests (Mullins & Tobin 1957), and we refer to (Ogden & Roxburgh 1999) for detailed discussion. We emphasize, however, that the theory did not include prediction of residual strains (permanent set), details of which were not available from the tests. In practice the Mullins effect is more complicated than the idealized version shown in Figure 1. This is illustrated in Figure 2, which shows schematically the residual strain and also the departure of the re-loading path from the unloading path for three loading/unloading cycles. The curves in Figure 2 reflect the results of typical simple tension experiments such as those carried out recently at the Tun Abdul Razak Research Centre (Roxburgh 1999), although the magnitude of the residual strain has been exaggerated in Figure 2 in order to avoid congestion. The main features shown in Figure 2 are essentially quasi-static. We refer to a recent paper by (Miehe & Keck 1999) for further discussion of experimental results and the relevant literature. In the present paper we first summarize the pseudo-elasticity theory used by (Ogden & Roxburgh 1999) and then, by a slight modification, apply this theory in such a way that it accounts for residual strain, although we retain the idealization that re-loading coincides with unloading. Thus, the situation we model is intermediate between those shown in Figures 1 and 2 and is depicted in Figure 3. Figure 3 differs from Figure 1 only in that residual stretches associated with the two unloading paths are shown at the points B′ and C′. Subsequent work will include modelling of the full effect shown in Figure 2.

2 PSEUDO-ELASTICITY

In (Ogden & Roxburgh 1999) a *pseudo-elastic* material was defined by incorporating an additional variable into the energy function of an elastic material, thereby yielding what is referred to as a *pseudo-elastic energy function.* Here, we consider the isotropic specialization of this theory and write the pseudo-energy function as $\mathcal{W}(\lambda_1, \lambda_2, \lambda_3, \eta)$, where $\lambda_1, \lambda_2, \lambda_3$ are the principal stretches of the deformation and η is the additional (continuous) variable, here referred to as a *damage variable.* The role of η is to provide a means of changing the form of the energy function according to some well-defined criterion. It may be either active or inactive. When it is inactive it may be set at a constant value. When it is active it is taken to depend on the deformation. This has the effect of changing the form of strain-energy function when η is switched on or off. The manner in which η may be chosen to depend on the deformation is essentially arbitrary and, as in (Ogden & Roxburgh 1999), it is convenient to adopt the (implicit) form

$$\frac{\partial \bar{\mathcal{W}}}{\partial \eta}(\lambda_1, \lambda_2, \lambda_3, \eta) = 0, \tag{1}$$

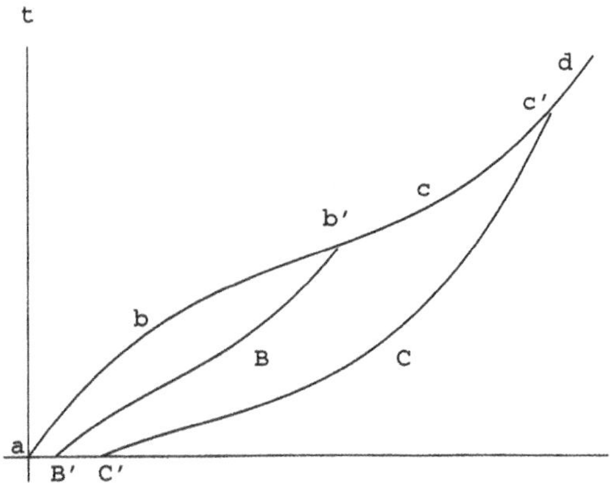

Figure 3. Depiction of the idealized Mullins effect for simple tension with nominal stress t plotted against stretch λ and residual strain incorporated.

with the dependence of $\bar{W}$ on η remaining at our disposal. In this paper we restrict attention to incompressible materials, and the incompressiblity constraint takes the form

$$\lambda_1\lambda_2\lambda_3 = 1 \qquad (2)$$

in terms of the principal stretches.

The principal values σ_1, σ_2, σ_3 of the Cauchy stress are then given by

$$\sigma_i = \lambda_i \frac{\partial \bar{W}}{\partial \lambda_i}(\lambda_1,\, \lambda_2,\, \lambda_3,\, \eta) - p \quad (i = 1,\, 2,\, 3). \quad (3)$$

Equation (2) is used to eliminate λ_3 in terms of λ_1 and λ_2, and the pseudo-energy function is then expressed as a function of λ_1 and λ_2 through

$$W(\lambda_1,\, \lambda_2,\, \eta) \equiv \bar{W}(\lambda_1,\, \lambda_2,\, \lambda_1^{-1}\lambda_2^{-1},\, \eta), \quad (4)$$

which defines the notation W. Substitution of (4) into (3) and elimination of p yields

$$\sigma_1 - \sigma_3 = \lambda_1 W_1\, , \qquad \sigma_2 - \sigma_3 = \lambda_2 W_2, \quad (5)$$

where $W_\alpha = \partial W/\partial \lambda_\alpha$ $(\alpha = 1,\, 2)$, and (1) simplifies to

$$\frac{\partial W}{\partial \eta}(\lambda_1,\, \lambda_2,\, \eta) = 0, \quad (6)$$

which determines η implicitly in terms of λ_1 and λ_2. In this paper we consider only homogeneous deformations, so that the equilibrium equations are not needed.

For the specialization to simple tension we take $\sigma_2 = \sigma_3 = 0$ and write $\sigma_1 = \sigma$ and $\lambda_1 = \lambda$, so that $\lambda_2 = \lambda_3 = \lambda^{-1/2}$, and define $\hat{W}$ by

$$\hat{W}(\lambda, \eta) \equiv W(\lambda, \lambda^{-1/2}, \eta). \quad (7)$$

Equations (5) and (6) then simplify to

$$\sigma = \lambda \hat{W}_\lambda(\lambda, \eta) \equiv \lambda t, \quad \hat{W}_\eta(\lambda, \eta) = 0 \quad (8)$$

respectively, wherein the (nominal) stress $t\,(= t_1)$ is defined and the subscripts signify partial derivatives. Corresponding equations for pure shear were given in (Ogden & Roxburgh 1999).

3 RESIDUAL STRAIN MODEL

As in (Ogden & Roxburgh 1999) we consider the pseudo-energy function $W(\lambda_1, \lambda_2, \eta)$ appropriate for biaxial deformations, so that equations (5) and (6) apply. When η is inactive, as indicated in Section 2, it is set to a constant value. Without loss of generality we set this constant to be unity and define the function $\tilde{W}(\lambda_1,\, \lambda_2)$ via

$$\tilde{W}(\lambda_1,\, \lambda_2) \equiv W(\lambda_1,\, \lambda_2,\, 1) \quad (9)$$

in this case. This is the energy function of the perfectly elastic material for which the primary loading path is also the unloading path. We therefore take $\tilde{W}(\lambda_1,\, \lambda_2)$ to characterize any *primary* loading path in $(\lambda_1,\, \lambda_2)$-space, exemplified by the path *abb′cc′d* in Figure 1 in the simple tension specialization. Standard forms of strain-energy function, such as the neo-Hookean or Ogden forms (see, for example, Ogden 1972, 1982, 1984, 1986), may then be used as representative of $\tilde{W}(\lambda_1,\, \lambda_2)$.

From (5) the specialization (9) yields the stresses

$$\tilde{\sigma}_\alpha - \tilde{\sigma}_3 = \lambda_\alpha \tilde{W}_\alpha \qquad (\alpha = 1, 2), \quad (10)$$

where a superposed tilde refers to a primary loading path and (6) is not operative.

The strain-energy function $\tilde{W}$ should satisfy the usual requirements

$$\tilde{W}(1,1) = 0, \quad \tilde{W}_\alpha(1,1) = 0 \quad (\alpha = 1,2), \quad (11)$$

$$\tilde{W}_{11}(1,1) = \tilde{W}_{22}(1,1) = 2\tilde{W}_{12}(1,1) = 4\mu, \quad (12)$$

where μ (positive) is the shear modulus of the material in the ground state and $\tilde{W}_{\alpha\beta}$ denotes $\partial^2\tilde{W}/\partial\lambda_\alpha\partial\lambda_\beta$ $(\alpha, \beta \in \{1,2\})$.

In addition to (11) and (12) we require that $\tilde{W}$ has a *global minimum* of zero at (1, 1) and that it has no other stationary points in $(\lambda_1,\, \lambda_2)$-space, as discussed in (Ogden & Roxburgh 1999). A primary loading path in $(\lambda_1,\, \lambda_2)$-space is defined as a path starting from (1, 1) on which $\tilde{W}$ is *increasing*. This point was discussed in detail in (Ogden

& Roxburgh 1999) and we remark here that for many standard forms of strain-energy function it can be confirmed that, in fact, $\tilde{W}$ increases along *any straight-line path* from (1, 1).

Unloading may take place from any point on a primary loading path, and when unloading is initiated η is activated. For unloading and subsequent sub-maximal loading and unloading paths, the value of η then varies in accordance with equation (6). It is assumed here that equation (6) can be solved explicitly for η and we write

$$\eta = \chi(\lambda_1, \lambda_2) = \chi(\lambda_2, \lambda_1). \quad (13)$$

Then, an energy function for unloading, symmetrical in (λ_1, λ_2) and denoted $w(\lambda_1, \lambda_2)$, is defined by

$$w(\lambda_1, \lambda_2) \equiv W(\lambda_1, \lambda_2, \chi(\lambda_1, \lambda_2)). \quad (14)$$

From equations (5) and (6) it follows that

$$\sigma_\alpha - \sigma_3 = \lambda_\alpha \frac{\partial w}{\partial \lambda_\alpha} = \lambda_\alpha \frac{\partial W}{\partial \lambda_\alpha} \quad (\alpha = 1, 2). \quad (15)$$

Let $(\lambda_{1m}, \lambda_{2m})$ be the values of (λ_1, λ_2) at a point at which unloading begins. Then, $\chi(\lambda_{1m}, \lambda_{2m}) = 1$. We emphasize that this implies that the function χ, and hence w, depends on the point from which unloading starts. This dependence is made more explicit in what follows.

At this point there is no restriction placed on the form of the energy function $W(\lambda_1, \lambda_2, \eta)$. Now, however, following (Ogden & Roxburgh 1999), we specialize the form of constitutive law in the form

$$W(\lambda_1, \lambda_2, \eta) = \eta \tilde{W}(\lambda_1, \lambda_2) + \phi(\eta), \quad (16)$$

where $\phi(\eta)$, referred to as the *damage function*, is taken to be a smooth function of its argument and, for consistency with (9), is subject to

$$\phi(1) = 0. \quad (17)$$

From (15) and (10) the stresses are calculated as

$$\sigma_\alpha - \sigma_3 = \eta \lambda_\alpha \frac{\partial \tilde{W}}{\partial \lambda_\alpha} = \eta(\tilde{\sigma}_\alpha - \tilde{\sigma}_3) \quad (\alpha = 1, 2). \quad (18)$$

We note that the model (16) is still very general since both $\tilde{W}(\lambda_1, \lambda_2)$ and $\phi(\eta)$ remain general, so that considerable flexibility is retained.

The simple tension specialization of (18), obtained on the basis of equations (7) and (8), is, in terms of the nominal stress,

$$t = \eta \hat{W}_\lambda(\lambda, 1) = \eta \tilde{t}, \quad (19)$$

where $\tilde{t}$ is the nominal stress on the primary loading path at the same value of λ. For (19) to predict stress softening we must have $\eta \le 1$ on the unloading path, with equality only at the point where unloading begins. In (Ogden & Roxburgh 1999) we took $\eta > 0$, which ensures that t remains positive on unloading until $\lambda = 1$ is reached. This restriction, however, does not permit the prediction of residual strain. Thus, in the present work we admit the possibility of vanishing η, as explained below.

On substitution of (16) into equation (6), we obtain

$$-\phi'(\eta) = \tilde{W}(\lambda_1, \lambda_2), \quad (20)$$

which, implicitly, defines the damage parameter η in terms of the deformation. The simple tension specialization of (20) is

$$-\phi'(\eta) = \hat{W}(\lambda, 1). \quad (21)$$

We emphasize that, in general, the value of η derived from (20) will depend on the values of the principal stretches λ_{1m} and λ_{2m} attained on a primary loading path as well as on the specific forms of $\tilde{W}(\lambda_1, \lambda_2)$ and $\phi(\eta)$ used. Since $\eta = 1$ at any point on the primary loading path from which unloading is initiated, it follows from equation (20) that

$$-\phi'(1) = \tilde{W}(\lambda_{1m}, \lambda_{2m}) \equiv W_m, \quad (22)$$

which defines the notation W_m. In accordance with the properties of $\tilde{W}$, W_m increases along a primary loading path. Clearly, the function ϕ depends on the point where unloading started.

As in (Ogden & Roxburgh 1999), we associate unloading with decreasing η and hence the inequality

$$\phi''(\eta) < 0, \quad (23)$$

follows. Thus, $\phi'(\eta)$ is a monotonic decreasing function of η and η is therefore uniquely determined from (20) as a function of $\tilde{W}(\lambda_1, \lambda_2)$. Indeed, the function χ, as described in (13), is the inverse of the function ϕ'.

During unloading η decreases until it reaches the value zero, at which point, as follows from (18), the stresses vanish, although the deformation will not in general have returned to the undeformed state corresponding to $\lambda_1 = \lambda_2 = 1$. Thus, there will be residual deformation, and we denote the values of λ_1 and λ_2 where $\eta = 0$ as λ_{1r} and λ_{2r} respectively.

From equation (20) we then obtain

$$\phi'(0) = -\tilde{W}(\lambda_{1r}, \lambda_{2r}) = -W_r, \qquad (24)$$

which defines the associated value W_r of $\tilde{W}$. Since the function ϕ depends on W_m it follows from (24) that W_r depends on W_m and hence that the residual deformation depends on the deformation at the end of the primary loading process.

In view of the assumed continuity of η, when the material is re-loaded from any point on an unloading path its properties are again governed by the strain-energy function (14) until η reaches the value unity. At this point, either unloading is repeated or further primary loading is initiated, governed by the strain-energy function (9).

When the damaged material is in a fully unstressed state, so that $\eta = 0$, the pseudo-energy function (16) has the residual value

$$w(\lambda_{1r}, \lambda_{2r}) = W(\lambda_{1r}, \lambda_{2r}, 0) = \phi(0). \qquad (25)$$

Thus, the residual (non-recoverable) energy $\phi(0)$ may be interpreted as a measure of the energy required to cause the damage in the material. In a uniaxial test such as simple tension $\phi(0)$ is the area between the primary loading curve and the relevant unloading curve and above the axis $t = 0$. This is analogous to the interpretation given in (Ogden & Roxburgh 1999) in the case in which there is no residual strain.

We recall that in view of the condition (22) the function ϕ depends, through W_m, on the point at which unloading starts. In (Ogden & Roxburgh 1999) this was accommodated by introducing a function f, *independent* of W_m, such that

$$\phi'(\eta) + W_m = \phi'(\eta) - \phi'(1) = f(\eta). \qquad (26)$$

Specification of the form of the function f is then equivalent to specifying the constitutive function ϕ subject to (17), (22) and (23). A specific form of f was adopted in (Ogden & Roxburgh 1999) and gave an excellent fit to the data in (Mullins & Tobin 1957) but is not discussed here since, when residual stresses are accounted for, f must depend on W_m and satisfy

$$f(1) = 0, \quad f(0) = W_m - W_r. \qquad (27)$$

Thus, the form of f used by (Ogden & Roxburgh 1999) is inappropriate for the analysis of residual strains and requires modification. Details of this will be reported elsewhere, and here we merely outline the theory.

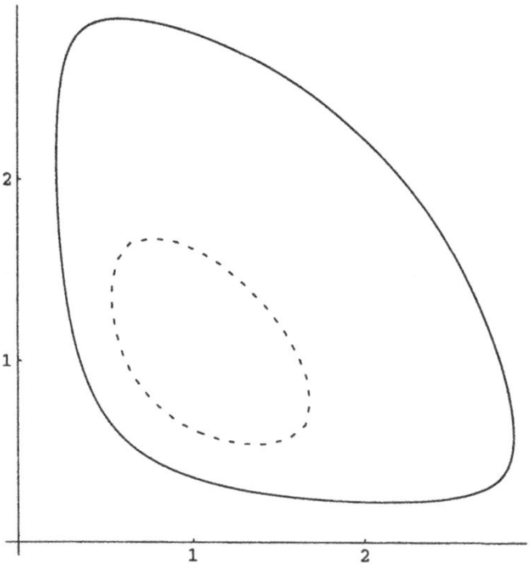

Figure 4. Typical contour plots of $\tilde{W}(\lambda_1, \lambda_2) = W_m$ (continuous curve) and $\tilde{W}(\lambda_1, \lambda_2) = W_r$ (dashed curve) in (λ_1, λ_2)-space.

As mentioned previously, it is the energy on the primary path rather than the particular deformation corresponding to that energy that is important. In the case of general biaxial deformations it is the value of the energy maximum $W_m = \tilde{W}(\lambda_{1m}, \lambda_{2m})$ on the primary loading path rather than the specific $(\lambda_{1m}, \lambda_{2m})$ pair that governs the unloading response. Thus, any other pair of (λ_1, λ_2) values corresponding to the same value of W_m could equally be taken as the starting point for unloading. Such pairs satisfy the equation

$$\tilde{W}(\lambda_1, \lambda_2) = W_m, \qquad (28)$$

which, for given (constant) W_m, describes a closed contour in (λ_1, λ_2)-space. For illustration, this is depicted in Figure 4 in respect of the neo-Hookean strain-energy function

$$\tilde{W}(\lambda_1, \lambda_2) = \frac{\mu}{2}(\lambda_1^2 + \lambda_2^2 + \lambda_1^{-2}\lambda_2^{-2} - 3), \qquad (29)$$

where μ is the shear modulus appearing in (12).

The contour defined by the current maximum value W_m represents the current damage threshold, and $\eta = 1$ at any point on this contour. For any deformation path within this contour $\eta < 1$ and no further damage occurs. An example of such a contour is shown in Figure 4 (continuous curve) together with the contour

$$\tilde{W}(\lambda_1, \lambda_2) = W_r, \qquad (30)$$

corresponding to $\eta = 0$ (dashed curve) defining the residual deformations induced by W_m. The energy required on the primary loading path to cause the damage is not required on subsequent loading up to the contour boundary. However, if the deformation path crosses the contour primary loading is again activated, $\eta = 1$ and further damage will occur. The value of $\tilde{W}(\lambda_1, \lambda_2)$ will increase until the next maximum value W_m is reached, at which point loading terminates and a new threshold contour is established (which encloses the previous one). Similarly, subsequent residual deformation contours enclose each previous one.

4 DISCUSSION

We emphasize that although a primary loading path is *represented* here by a strain-energy function this path is nevertheless associated with dissipation. The dissipation is calculated retrospectively using information about unloading provided by specification of the function ϕ and is measured by the non-recoverable energy $\phi(0)$. On the other hand, there is no dissipation on unloading and submaximal re-loading associated with the present model. An explicit calculation of the dissipation was given by (Ogden & Roxburgh 1999) for the case of no residual strain. For the model considered in the present paper this calculation requires modification and also specification of a different form of the function ϕ. Details of this will be discussed in a separate paper dealing with a model which takes account of differences between the unloading and re-loading paths in accordance with Figure 2.

ACKNOWLEDGEMENTS

This work was supported by a research grant from the UK Engineering and Physical Sciences Research Council and by the Tun Abdul Razak Research Centre. The authors are grateful to Professor Miehe for providing a pre-print of his paper cited below.

REFERENCES

Govindjee, S. & J.C. Simo 1992. Transition from micro-mechanics to computationally efficient phenomenology: carbon black filled rubbers incorporating Mullins' effect. *J. Mech. Phys. Solids* 40:213-233.

Lazopoulos, K. A. & R.W. Ogden 1998. Nonlinear elasticity theory with discontinuous internal variables. *Math. Mech. Solids* 3:29-51.

Lazopoulos, K. A. & R.W. Ogden 1999. Spherically-symmetric solutions for a spherical shell in finite pseudo-elasticity. *Int. J. Nonlinear Mech.*, to appear.

Miehe, C. & J. Keck 1999. Superimposed finite elastic-viscoelastic-plastoelastic stress response with damage in filled rubbery polymers. Experiments, modelling and algorithmic implementation. *J. Mech. Phys. Solids*, to appear.

Mullins, L. 1947. Effect of stretching on the properties of rubber. *J. Rubber Research* 16:275-289.

Mullins, L. & N.R. Tobin 1957. Theoretical model for the elastic behaviour of filler-reinforced vulcanized rubbers. *Rubber Chem. Technol.* 30:551-571.

Ogden, R. W. 1972. Large deformation isotropic elasticity: on the correlation of theory and experiment for incompressible rubberlike solids. *Proc. R. Soc. Lond.* A326:565-584.

Ogden, R. W. 1982. Elastic deformations of rubberlike solids. In H. G. Hopkins & M. J. Sewell (eds), *Mechanics of Solids*:499-537. Oxford: Pergamon Press.

Ogden, R. W. 1984. *Non-linear elastic deformations.* Chichester: Ellis Horwood.

Ogden, R. W. 1986. Recent advances in the phenomenological theory of rubber elasticity. *Rubber Chem. Technol.* 59:361-383.

Ogden, R.W. & D.G. Roxburgh 1999. A pseudo-elastic model for the Mullins effect in filled rubber. *Proc. R. Soc. Lond.* A, to appear.

Roxburgh, D.G. 1999, personal communication.

Constitutive Models for Rubber, Dorfmann & Muhr (eds) © 1999 Taylor & Francis ISBN 90 5809 113 9

Material law selection in the Finite Element simulation of rubber-like materials and its practical application in the industrial design process

F.J.H. Peeters
Hibbitt, Karlsson and Sorensen Incorporated, Aachen, Germany

M. Küssner
ABACOM Software GmbH, Aachen, Germany

ABSTRACT: The improving performance of modern computers enables engineers to solve nonlinear problems of increasing size. Non-linear material laws for rubber-like materials are nowadays well-established and form a natural part of the material library of commercial nonlinear Finite Element codes. But these well-established material models may cause serious problems in practical applications, if insufficient testing is done when calibrating the material constants. New constitutive laws for rubber-like materials have been developed that are specially designed to tackle this difficulty.These more phenomenological material laws are implemented in the Finite Element program ABAQUS and are particular compared with the more traditional approach of using hyperelastic strain energy functions in view of their practical application in the industrial design process.

1 INTRODUCTION

A large variety of fine constitutive models for rubber-like materials are available already since quite some time. Material laws like e.g. the Ogden or the Polynomial model are well acknowledged and substantially tested and discussed in literature. Commercial nonlinear Finite Element programs have incorporated these hyperelastic material laws and made this technology easy accessable for engineers. Good agreement of numerical simulations with comparative tests increases the demand for commercial nonlinear Finite Element software in the simulation of rubber-like material.

But the use of hyperelastic material laws in numerical simulations is not always without it's problems.

The nonlinearity from the material part adds up to the geometric nonlinearity and possible contact constraints that requires a robust iteration procedure in an implicit integration scheme.

Additional to the aspect of nonlinearity and it's inherent problem of structural stability, considerations of material stability are potential sources of difficulties (Reese et al., 1995). Obviously, this is a stability aspect unknown when using linear material laws.

The above mentioned problems of nonlinearity and stability are not limited to the Finite Element method and need to be considered in other solution procedures eitherl. An unfortunate speciality of the Finite Element method is the occurance of so-called hourglass modes that have been observed with underintegrated quadrilateral and hexagonal elements, since a long time. But even with fully integrated elements, spurious modes can possibly occur when using a nonlinear material law. This problem is subject to current research (Reese et. al., 1999) and will not further be discussed in this paper. Nevertheless, it requires additional care when using the Finite Element method with hyperelastic material laws.

So far potential difficulties with the use of hyperelastic material laws have been mentioned which affect both the scientific as well as the industrial Finite Element community. But the focus of this paper will be a problem of very practical nature: the reliable calibration of material constants. The above mentioned traditional hyperelastic material laws have been proven to be mechanically sound - if, and only if, the material constants are determined correctly. Unfortunately, the experimental requirements for a sufficient accurrate calibration are somewhat in contrast to the demands of an industrial design process: the experiments are expensive and time consuming.

In this paper it will be pointed out that reducing costs and time by reducing the number of experiments is an option that might lead to inacurate material constants and therefore to poor results. Knowing the needs of an industrial design process, new material laws have been included in the Finite Element program ABAQUS. These perform sufficiently accurate even when they are based on limited experimental test data.

In section 2 the method of material data calibration is briefly discussed. The problem of single test data usage is demonstrated in section 3. In section 4 some more recently developed powerful material laws are introduced and compared to the more traditional approach. In section 5 the extension to viscoelastic large-strain elasticity is presented. Two examples for practical application are given in section 6. Finally some conclusions are drawn.

2 CALIBRATION OF MATERIAL CONSTANTS FROM TEST DATA

The idea of the calibration of material test data is straightforward: perform several experiments, each with a different significant deformation mode, measure forces and displacement and reduce an appropriate norm of error between stresses calculated from the material model that has to be calibrated and the experimental stresses. Tests with a significant deformation mode are for example the uniaxial test, the equibiaxial test, the planar test or the volumetric test. The nature of tests with one significant deformation mode allows for proper identification of all other deformation modes.

Figure 1 shows the sketch of an experiment for uniaxial test data.

Clearly, the stretch ratios in all 3 directions can easily be determined by the siginificant deformation mode in 1-direction

$$\lambda_2 = \lambda_3 = \lambda_1^{-1/2} = \lambda_U^{-1/2} \qquad (1)$$

where λ_U is the nominal stretch ratio. The variation of a strain energy function U with respect to the nominal principal stretch λ_U

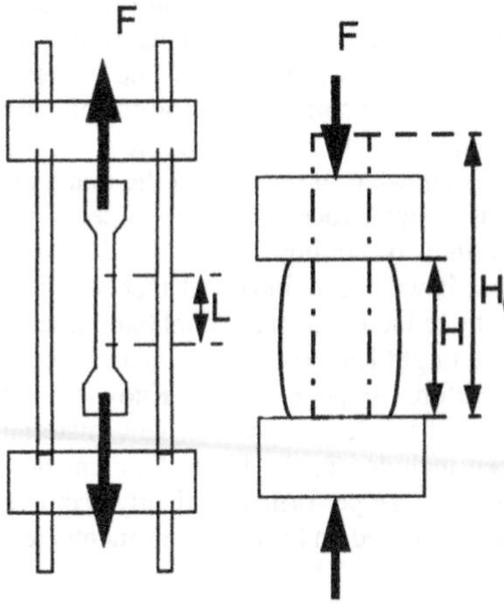

Figure 1. Experiment for uniaxial tension and compression

$$\delta U = T_U \delta\lambda_U \qquad (2)$$

results in the uniaxial nominal stress T_U :

$$T_U = \frac{\partial U}{\partial \lambda_U} = 2(1-\lambda_U^{-3})\left(\lambda_U \frac{\partial U}{\partial \bar{I}_1} + \frac{\partial U}{\partial \bar{I}_2}\right) \qquad (3)$$

Knowing the nominal principal stretch from the experiment and knowing the experimental principal nominal stresses, the material parameter can be calibrated in such a way, that the principal nominal stress T_U from the strain energy function matches best with the experimental principal nominal stress T_{EX}.

Significant importance in the calibration process has the question: what defines the norm for the 'best' agreement between experimental and theoretical stresses? Minimizing the relative error in stress for example can lead to significant different material coefficients than an error norm formulated in the absolute error in stress (Yeoh, 1995), due to a different weighting of the experimental data.

In ABAQUS the relative error in the stresses is minimized in a linear least square fit for strain energy potentials that are linear in it's coefficients (ABAQUS Theory Manual, 1998).

$$E_{rel} = \sum_{i=1}^{n} \left(1 - \frac{T_U^i}{T_{EX}^i}\right)^2 \qquad (4)$$

In case of nonlinear material coefficients a nonlinear curve fitting procedure is applied (Twizell et al., 1986).

Obviously, the accuracy of the model data calibration improves with increasing number n of test data. Equally spaced test data contribute as well to an enhanced accuracy of the calibration procedure. More aspects of accuracy will be discussed in the following section.

3 PROBLEMS OF SINGLE TEST USAGE

The use of hyperelastic constitutive models has been proven to be robust and precise if the material coefficients have been calibrated with sufficient accuracy. In recent time, lot of attention was paid to the analysis of the accuracy of test data calibration based on a single test.

The industrial design process has very different demands compared to the academic world. A 'just-in-time' - strategy does not just apply to the delivering and assembling of products. In the design process all participating disciplines need to deliver their contributions in time. An extensive testing of a new material does not always fit into the industrial design process because of time and/or financial constraints.

In the following a few examples are given for the potentially disasterous results that may be obtained

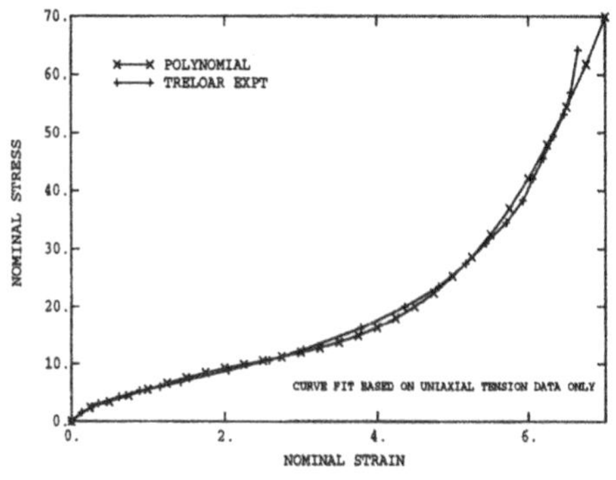

UNIAXIAL TENSION - POLYNOMIAL N=2 (5 TERMS)

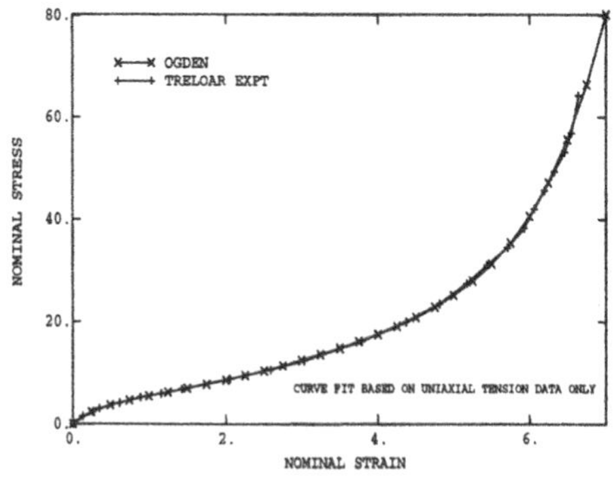

UNIAXIAL TENSION - OGDEN N=3 (3 TERMS)

Figure 2. Uniaxial test based on uniaxial test data

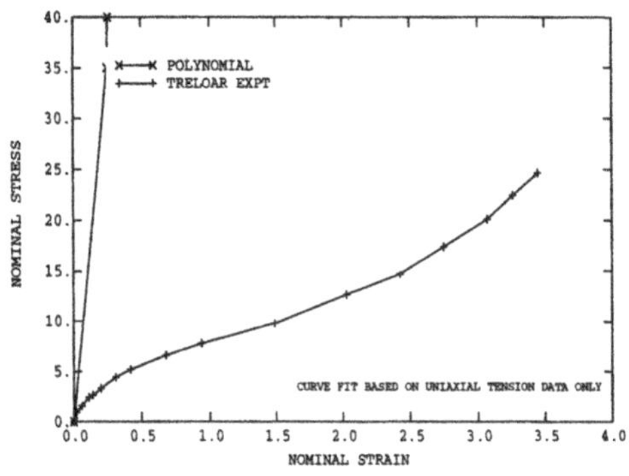

BIAXIAL TENSION - POLYNOMIAL N=2 (5 TERMS)

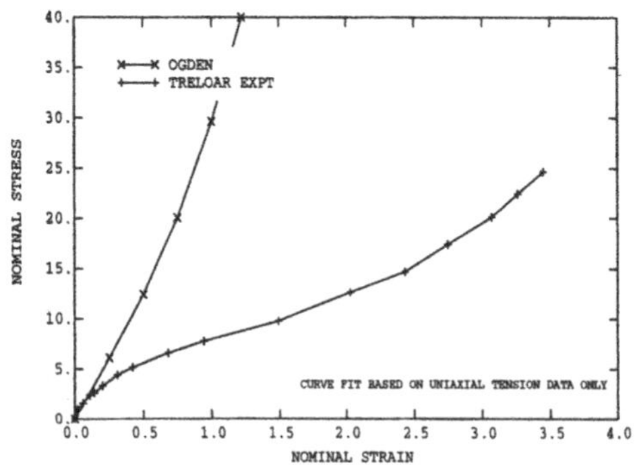

BIAXIAL TENSION - OGDEN N=3 (3 TERMS)

Figure 3. Biaxial test based on uniaxial test data

when performing calculations with an insufficient material data calibration. The inaccuracy is demonstrated with the Polynomial and the Ogden model with material coefficients calibrated on uniaxial test data only.

Figure 2 shows a Finite Element calculation for a uniaxial tension test. A Polynomial model with N=2 and an Ogden model with N=3 are compared to an experiment by Treloar. Both hyperelastic models have been calibrated with uniaxial test data only. Both material laws conform excellently with the experimental results.

It follows that material coefficient calibration is sufficient with a single experiment if the material model is used with the same deformation mode as exists in the experiment.

In contrast to the accurate results shown in Figure 2 the same material models are tested in a biaxial test, as shown in Figure 3.

Again, the results are compared to a biaxial experiment made by Treloar. The material constants have been calibrated with the same uniaxial tension test data that performed very well in Figure 2. It can easily be observed that the material model calibration is useless for any other than the uniaxial deformation mode. This unacceptable behavior has it's only cause in an insufficient material constant calibration. It is not a necessary weakness of the mentioned material models, but solely a calibration problem. This effect can be observed with many traditional hyperelastic material models.

It is an unfortunate situation for the user of Finite Element programs that, once the material has been calibrated, no direct control about the accuracy of the material constant calibration is available any more.

4 NEW HYPERELASTIC MATERIAL MODELS

In the previous sections it has been illustrated that well accepted material laws are not useful in the industrial design process if time or financial restrictions do not allow a proper identification of material parameters. Triggered by the challenge to avoid this problem, a number of fairly new material laws has been introduced.

Before the problem in the material data calibration is investigated, the following important observation must be made. The Polynomial model, written here in the compressible form, depends on the first and the second invariant $\bar{I}_1$ and $\bar{I}_2$ of the left deviatoric Cauchy-Green tensor

$$U = \sum_{i+j=1}^{N} C_{ij}(\bar{I}_1-3)^i(\bar{I}_2-3)^j + \sum_{i=1}^{N} \frac{1}{D_i}(J_{el}-1)^{2i} \tag{5}$$

while the Neo-Hooke model is written only in terms of the first deviatoric invariant:

$$U = C_{10}(\bar{I}_1-3) + \frac{1}{D_1}(J_{el}-1). \tag{6}$$

The strain energy functions are written with the deviatoric invariants in order to have a separate notation for the volumetric part.

Please note that this direct comparison is done between the Polynomial model and the Neo-Hooke model since both depend on invariants of the left Cauchy-Green tensor. The Ogden model depends on the eigenvalues of the finite strain tensor (principle stretches). Looking at the same biaxial problem that had been examined in Figure 2, the Neo-Hooke model performs much better than the Polynomial model or the Ogden model when the material data calibration has been carried out with an uniaxial experiment only.

This example demonstrates what was recently widely discussed and acknowledged in literature (Kaliske, et al.1997, Yeoh 1993) about the second invariant of the left Cauchy-Green tensor:

- the weak dependency on the second invariant if calibrated correctly
- its difficulty for proper calibration
- its potential to spoil the solution.

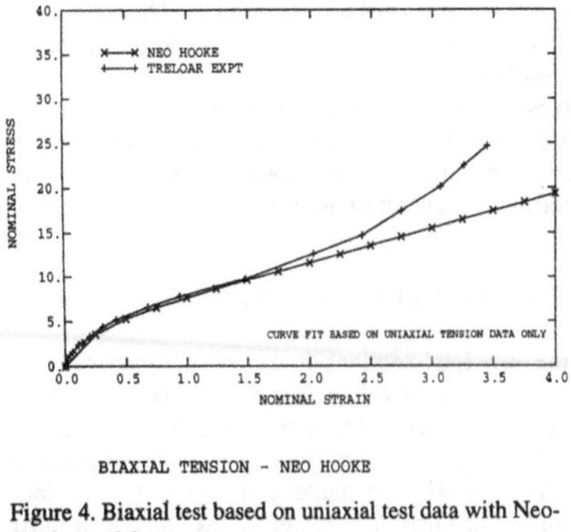

Figure 4. Biaxial test based on uniaxial test data with Neo-Hooke model

Knowing the difficulties that are caused by the second invariant, many new material models have recently been introduced, all more or less focusing on the first invariant. In the following, the new hyperelastic material models that are implemented in the Finite Element program ABAQUS are explained and tested.

4.1 *The Reduced Polynomial Model*

The Reduced Polynomial model follows straightforward from the Polynomial Model by just ignoring the effects of the second invariant of the left Cauchy-Green tensor:

$$U = \sum_{i=1}^{N} C_{i0}(\bar{I}_1-3)^i + \sum_{i=1}^{N} \frac{1}{D_i}(J_{el}-1)^{2i} \tag{7}$$

It can be easily observed that the Reduced Polynomial model is identical to the Neo-Hooke model for N=1.

4.2 *The Yeoh model*

The Yeoh strain energy function is a special case of the Reduced Polynomial model with $N = 3$:

$$U = \sum_{i=1}^{3} C_{i0}(\bar{I}_1-3)^i + \sum_{i=1}^{3} \frac{1}{D_i}(J_{el}-1)^{2i} \tag{8}$$

For the Yeoh form the initial shear modulus and bulk modulus are given by

$$\mu_0 = 2C_{10}, \quad K_0 = \frac{2}{D_1} \tag{9}$$

The Yeoh model typically provides a good fit over a large strain range, see Figure 5.

4.3 *The Arruda-Boyce model*

The Arruda-Boyce model does also depend on the first invariant of the left deviatoric Cauchy-Green tensor, neglecting any influence from the second invariant on the solution (Arruda et al., 1993)

$$U = \mu \sum_{i=1}^{5} \frac{C_i}{\lambda_m^{2i-2}}(\bar{I}_1^i - 3^i) + \frac{1}{D}\left(\frac{J_{el}^2-1}{2} - \ln(J_{el})\right) \tag{10}$$

with

$$C_1 = \frac{1}{2}, C_2 = \frac{1}{20}, C_3 = \frac{11}{1050}$$

$$C_4 = \frac{19}{7000}, C_5 = \frac{519}{673750} \tag{11}$$

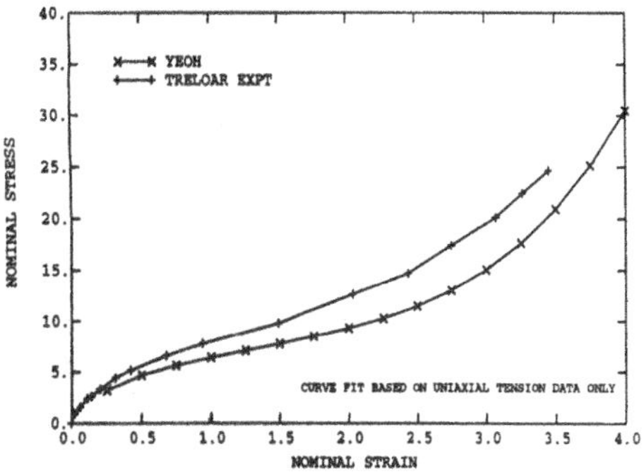

Figure 5. Biaxial test based on uniaxial test data with the Yeoh model

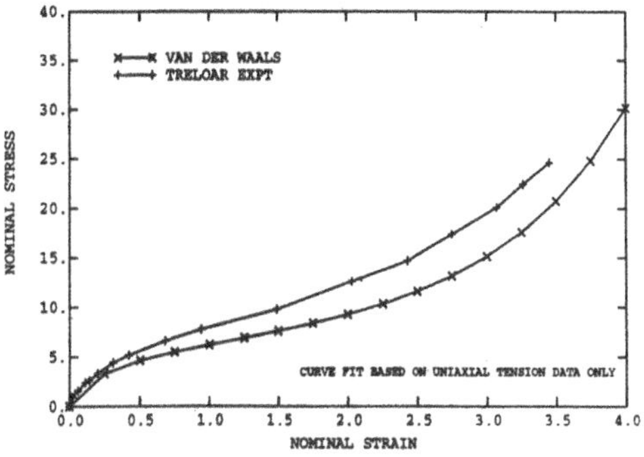

BIAXIAL TENSION - VAN DER WAALS

Figure 6. Biaxial test based on uniaxial test data with the Arruda-Boyce model

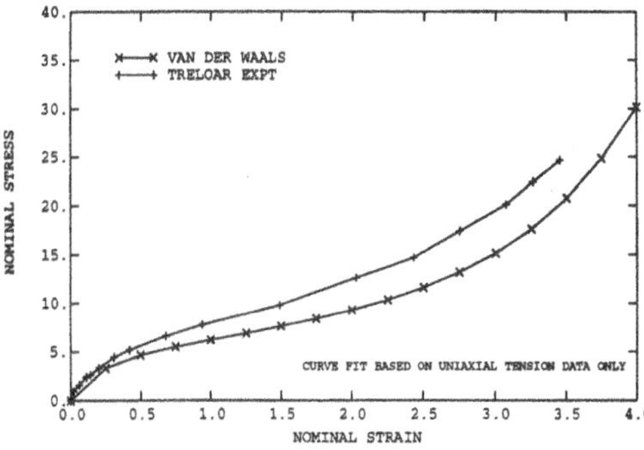

BIAXIAL TENSION - VAN DER WAALS

Figure 7. Biaxial test based on uniaxial test data with Van der Waals model.

The initial shear modulus is μ. A locking stretch λ_m is introduced, marking the stretch in the model's stress-strain curve at which the curve will rise (stiffen) significantly.

This approach is a much more phenomenological one than the traditional hyperelastic models. Neglecting the second invariant of the left Cauchy-Green tensor helps to produce more accurate solutions in the range of smaller strains, but the typical S-shape of the curve cannot be achieved, as could be observed with the Neo-Hooke model. With the additional locking stretch parameter, a sufficient accuracy in both small and large strain is achieved, as is illustrated in figure 6.

4.4 *The Van der Waals - model*

The Van der Waals model (Kilian, 1984)

$$U = \mu\left\{-(\lambda_m^2-3)[\ln(1-\eta)+\eta]-\frac{2}{3}a\left(\frac{\tilde{I}-3}{2}\right)^{\frac{3}{2}}\right\}+$$

$$\frac{1}{D}\left(\frac{J_{el}^2-1}{2}-\ln(J_{el})\right) \qquad (12)$$

with β representing the linear mixture parameter in

$$\tilde{I} = (1-\beta)\bar{I}_1+\beta\bar{I}_2 \qquad (13)$$

uses the combination of the two invariants of the left deviatoric Cauchy Green tensor in one parameter $\tilde{I}$. If β equals zero, only the effects of the first invariant are taken into account. Therefore, if only one type of test data is available, this parameter is recommended to be set to zero.

The interaction parameter a is difficult to estimate. It typically lies between 0.1 and 0.3. The structure of the Van der Waals model is such that it cannot be used when the deformation of the material creates stretches larger than the locking stretch λ_m. Figure 7 shows an example of the Van der Waals - model, again a biaxial tension test with material calibration from a uniaxial tension only.

5 FINITE-STRAIN VISCOELASTICITY

The new hyperelastic material laws can also be included in a finite-strain viscoelastic formulation. The shear modulus and the measure for compressibility are defined by the following Prony series:

$$\mu(\tau) = \mu^0\left(1-\sum_{i=1}^{n}\bar{g}_i^P(1-e^{(-\tau)/\tau_i})\right) \qquad (14)$$

$$\frac{1}{D(\tau)} = \frac{1}{D^0}\left(1-\sum_{i=1}^{n}\bar{g}_i^P(1-e^{(-\tau)/\tau_i})\right) \qquad (15)$$

6 EXAMPLES

6.1 Example 1

In the first example a coupled temperature-displacement Finite Element analysis is discussed. A rigid part is pressed into a ring made of hyperelastic material. The contour plot of the initial temperature and shape is shown in Figure 8, the temperature distribution at the stage of self-contact is shown in Figure 9 in the deformed configuration.

The material model that has been used for this analysis was calibrated from test data. The Polynomial model with N=2 was calibrated with uniaxial as well as multiple test data. The Arruda-Boyce and the Yeoh model were calibrated with uniaxial test data only.

Figure 10 shows the vertical reaction force of the indenting body. At 94 percent of the displacement, the self-contact can be observed with a strong stiffening of the structure.

The Polynomial model with multiple test data calibration has identical results as the Arruda-Boyce model with uniaxial test data only. This result also corresponds very well to the Ogden model with multiple test data calibration. The Yeoh model with uniaxial test data is still acceptable while the Polynomial model with N=2 performs very poor when calibrated with uniaxial test data only.

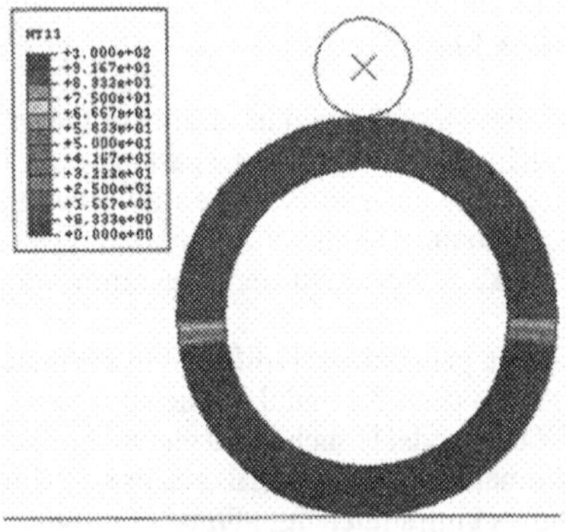

Figure 8. Temperature distribution in the undeformed state

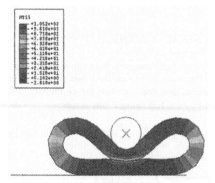

Figure 9. Temperature distribution in the deformed state

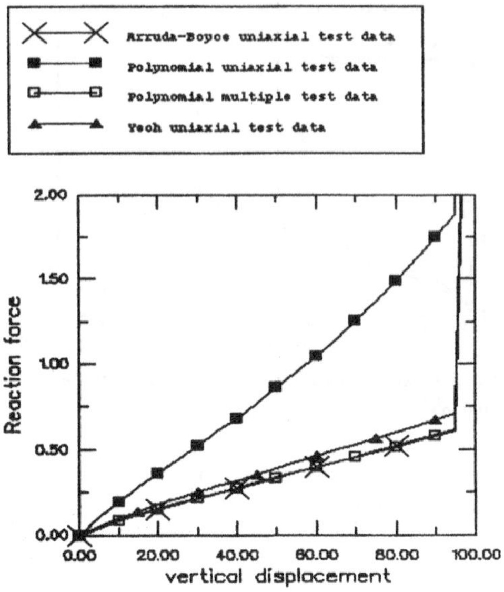

Figure 10. The vertical reaction forces of the indenting body with different material laws

6.2 Example 2

The second example explains typical steps of an industrial application involving hyperelastic materials. In contrast to academic codes, a commercial code must be able to target different objectives of a model in one subsequent chain of analysis. A tire model is tested that demonstrates the flexibility of a commercial code.

Step by step, the Finite Element simulation of a tire inflation, footprint, steady-state analysis and impact is demonstrated.

In Figure 11 an axisymmetric model for the rim mounting and the inflation is sketched. To obtain the footprint of the tire, half a 3D model is generated from the axisymmetric model and the results are transferred

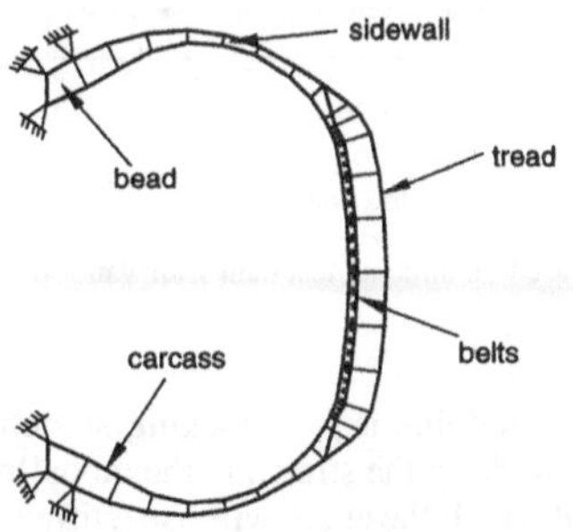

Figure 11. Axisymmetric rim mounting and inflation

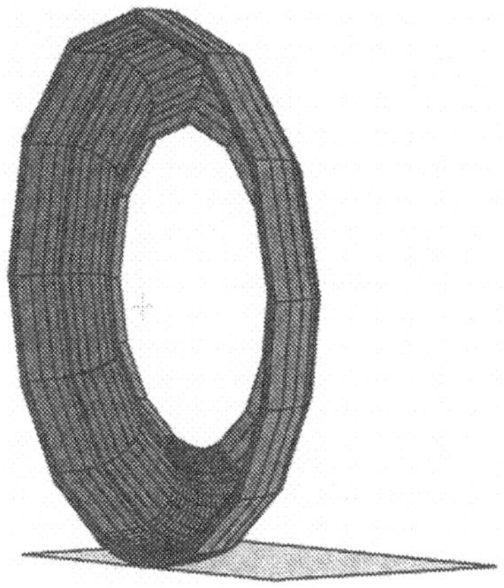

Figure 12. 3D result transfer

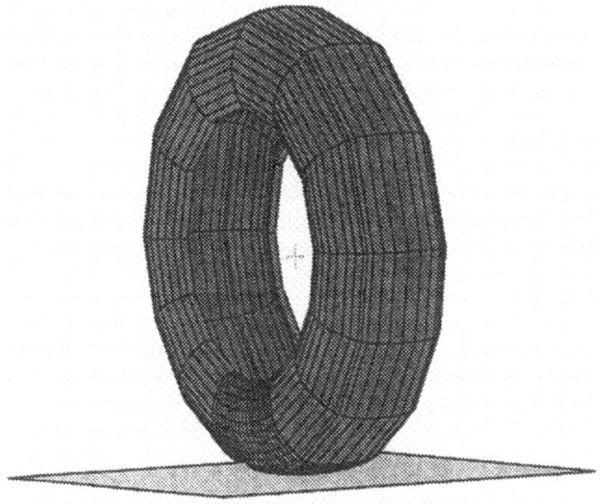

Figure 13. Full 3D model for steady-state analysis

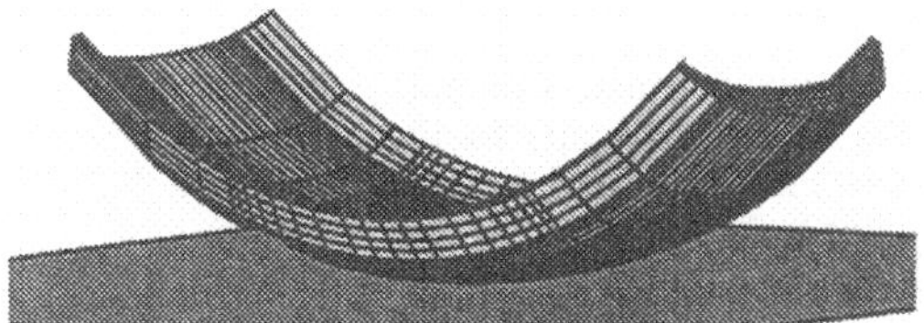

Figure 14. Submodelling of the footprint area

to the 3D model, as shown in Figure 12.

For a steady-state rolling analysis, another automatic results transfer has to be done to obtain the full 3D model, as shown in Figure 13.

A detailed investigation of stresses in the footprint area can be obtained via submodelling.

Now a detailed analysis of the footprint and the stresses in the rebars and carcass can be done, as shown in Figure 15 to 17.

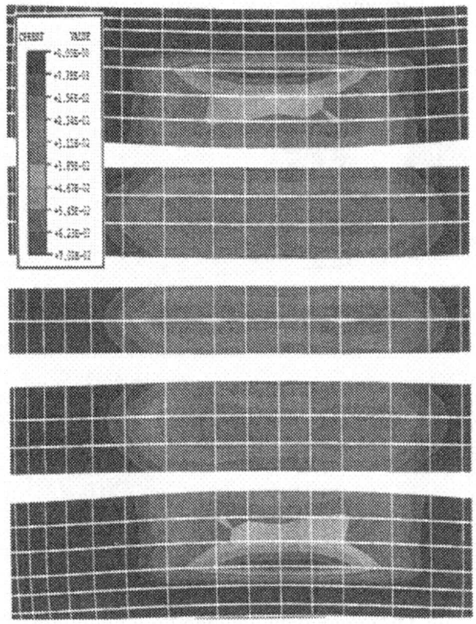

Figure 15. Contact pressure at tire footprint

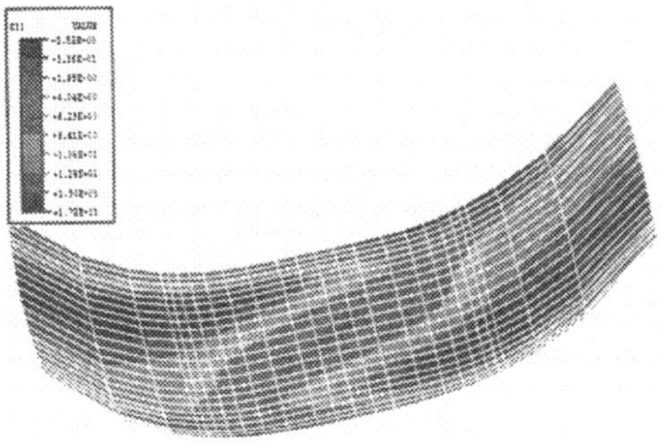

Figure 16. Stresses in rebars

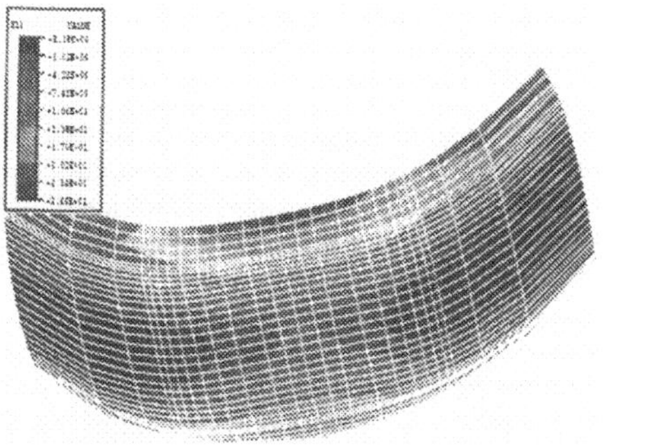

Figure 17. Stresses in the carcass

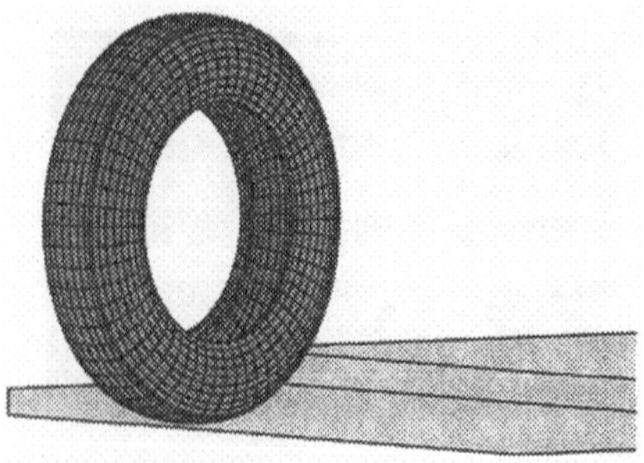

Figure 18. Steady-state rolling of a tire before impact

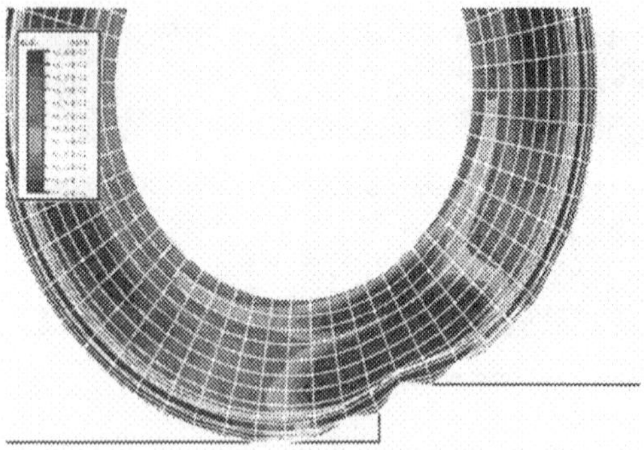

Figure 19. Mises stresses during impact

The 3D steady state rolling analysis can be combined with an explicit integration procedure when the tire is subject to an impact, as demonstrated in Figure 18.

The results from steady-state rolling are transferred to ABAQUS /Explicit for the impact analysis, as shown in figure 19.

The given tire problem shows the flexibility of commercial software to switch between 2D and 3D models, between static and steady-state analysis and between implicit and explicit time integration schemes.

7 CONCLUSION

Since the industrial desgin process is subject to time and financial constraints, a proper material data calibration cannot always be ensured. Traditional hyperelastic material models like the Polynomial or Ogden model perform very well for many application as long as the material data calibration is done with sufficient precision. Recent developments in hyperelastic material model formulations tackle this problem and give pleasing results for most general problems, even though only uniaxial test data has been used for calibration. The hyperelastic material models of Arruda-Boyce, Van der Waals and the Reduced Polynomial model, that are implemented in the program ABAQUS, have been discussed. The results underline the superior behavior of these new material models in case of limited testings, as shown in different examples. Typical steps of a hyperelastic analysis demonstrate which type of analysis functionality needs to be available in commercial Finite Element software for effective use in an industrial environment.

REFERENCES

ABAQUS Standard Theory Manual, Version 5.8, 1998, *Hibbitt, Karlsson & Sorensen Inc.*

Arruda, E. & Boyce, M., 1993. A three-dimensional constitutive model for the large stretch behavior of rubber elastic materials, *J. Mech. Phys. Solids,* Vol 41, No 2., pp 389 - 412

Kaliske, M. & Rothert, H. 1997, On the Fnite Element implementation of rubber-like materials at finite strains, *Engineering Computations,* Vol. 14, No, 2, pp. 216 - 232

Kawabata, S., Yamashita Y., & Ooyama, S. 1995, Mechanism of carbon-black reinforcement of rubber vulcanizate, *Rubber Chemistry anbd Technology, Vol. 68,* pp.311 - 329.

Kilian, H.-G. & Vilgis, T., 1984, Fundamental aspects of rubber-elasticity in real networks, *Colloid & Polymer Science,* Vol.. 262, No. 1, pp.15-21

Reese, S & Wriggers, P. 1997. Material instabilities of an incompressible elastic cube under triaxial tension,
Int. J. Solids Structures, Vol. 34, No. 26, pp. 3433 - 3454

Reese, S., Küssner M. & Reddy, D., 1999. A new Stabilization technique for Finite Elements in non-linear Elasticity, *Int. J. Numer. Meth.* Engng. 44, 1617 - 1652

Twizell, E. H. & Ogden, R. W. 1986. Non-linear optimization of the material constants in Ogden's Stress-Deformation Function for Incompressible Isotropic Elastomers, *J. Austral. Math. Soc. Ser., B*24, pp. 424 - 434

Yeoh, O. H. 1993. Some forms of the strain energy function for Rubber, *Rubber Chemistry anbd Technology, Vol. 66,* pp.754 - 771.

Yeoh, O. H. 1995. On the Ogden strain energy function, *Rubber Chemistry anbd Technology, Vol. 70,* pp.176 - 182.

Constitutive Models for Rubber, Dorfmann & Muhr (eds) © 1999 Taylor & Francis ISBN 90 5809 113 9

The limited static load in finite elasticity

I.A. Brigadnov
North-Western Polytechnical Institute, St. Petersburg, Russia

ABSTRACT: The variational formulation of the elastostatic bounary-value problem for hyperelastic materials is considered. For elastic potentials having the linear growth in modulus of the distortion tensor, the limited analysis problem is formulated. In the framework of this problem the strength of nonlinear elastic solid is estimated. From the mathematical point of view this problem is non-correct and, therefore, needs a relaxation. The partial relaxation is descibed for the limited analysis problem. It is based on the special discontinuous finite-element approximation with functions having breaks of the sliding type. The numerical results show that this technique has qualitative advantages over standard continuous finite-element approximations.

1 INTRODUCTION

The solution of elasticity boundary-value problem (BVP) is of particular interest in both theory and practice. At present there are many models of elasticity in the framework of the finite deformations theory (Green & Zerna 1975, Bartenev & Zelenev 1976, Ciarlet 1988, Lurie 1990). Adequacy and the field of application of every model must be found only by the correlation between experimental data and solutions of appropriate BVPs. Therefore, the analysis of mathematical correctness and the treatment of numerical methods for these problems is very important (Brigadnov 1993–1998).

In this paper the finite elasticity BVP is formulated as the variational problem for the displacement (Ciarlet 1988). For materials with ideal saturation the elastic potential $\Phi(x, \nabla u)$ of which has the linear growth in $|\nabla u|$, where $x \mapsto u \in R^3$ is the map, the existence of the limited static load (such external static forces with no solution of BVP) and discontinuous maps with breaks of the sliding type was proved by the author (Brigadnov 1993–1998). From the physical point of view these effects are treated as the destruction of a solid.

The limited analysis problem for nonlinear elastic solid is formulated. In the framework of this problem the estimation from below for the limited static load is calculated. As a result, the shape optimization problem is also formulated for the nonlinear elastic solid of the maximum strength. It is demonstrated, using the simplest example, that the limited analysis problem has discontinuous solutions with breaks of the sliding type.

From the mathematical point of view the limited analysis problem is non-correct (Ekeland & Temam 1976, Fučik & Kufner 1980, Temam 1983, Giusti 1984) and, therefore, needs a relaxation. We use the partial relaxation which is based on the special discontinuous finite-element approximation (FEA) (Repin 1989, Brigadnov 1998). After this discontinuous FEA the limited analysis problem is transformed into the non-linear system of algebraic equations which is badly determined, because the global stiffness matrix has lines with significantly different factors. Therefore, for the numerical solution the decomposition method of adaptive block relaxation is used (Brigadnov 1996c–1998). Its main idea consists of iterative improvement of zones with "proportional" fields by special decomposition of variables, and separate calculation on these variables.

The numerical results show that for the estimation of the limited static load, the proposed technique has qualitative advantages over standard continuous finite-element approximations.

The results of this paper are new and, therefore, need more theoretical and experimental research.

2 THE LIMITED ANALYSIS PROBLEM IN NONLINEAR ELASTICITY

Let a rigid body in the undeformed reference configuration occupy a domain $\Omega \subset \mathbb{R}^3$. In the deformed configuration each point $x \in \overline{\Omega}$ moves into a position $u(x) = v(x) + x \in \mathbb{R}^3$, where u and v are the map and displacement, respectively. Here and in what follows we use the Lagrangian coordinates. We consider locally invertible and orientation-preserving maps $u : \overline{\Omega} \to \mathbb{R}^3$ with gradient (the distortion tensor) $Q(u) = \nabla u : \Omega \to \mathbb{M}^3$ such that $det(Q) > 0$ in Ω (Green & Zerna 1975, Ciarlet 1988, Lurie 1990), where the symbol $\mathbb{M}^3$ denotes the space of real 3×3 matrices.

The finite deformation of materials is described by the energy pair (Q, Σ), where $\Sigma = \{\Sigma_i^\alpha\}$ is the first non-symmetrical Piola-Kirchhoff stress tensor. It is known that the Cauchy stress tensor $\sigma = \{\sigma^{\alpha\beta}\}$ has the components $\sigma^{\alpha\beta} = (det(Q))^{-1}\Sigma_i^\alpha Q_i^\beta$. Here and in what follows the Roman lower and Greek upper indeces correspond to the reference and deformed configurations, respectively, and the addition over repeating indeces is used.

Elastic materials are characterized by the response function (*the stress-strain relation*) $\Sigma = \Sigma(x, Q)$ such that $\Sigma(x, I) = O$ where $I = Diag(1,1,1)$ and O is the zero matrix. For hyperelastic materials the scalar function (*elastic potential*) $W : \Omega \times \mathbb{M}^3 \to \mathbb{R}_+$ exists such that $\Sigma_i^\alpha(x, Q) = \partial W(x, Q)/\partial Q_i^\alpha$ for every $Q \in \mathbb{M}^3$ and almost every $x \in \Omega$. If a material is incompressible, then $det(Q) = 1$, but for a compressible material $|\Sigma(x, Q)| \to \infty$, $W(x, Q) \to +\infty$ as $det(Q) \to +0$.

We consider the following boundary-value problem. The quasi-static influences acting on the body are: a mass force with density f in Ω, a surface force with density F on a portion Γ^2 of the boundary, and a displacement v^0 of a portion Γ^2 of the boundary is also given. Here $\Gamma^1 \cup \Gamma^2 = \partial\Omega$, $\Gamma^1 \cap \Gamma^2 = \emptyset$ and $area(\Gamma^1) > 0$.

For hyperelastic materials the finite elasticity BVP is formulated as the following variational problem (Ciarlet 1988, Brigadnov 1993–1998)

$$v^* = \arg\left(\inf\{I(v) : v \in V\}\right), \qquad (1)$$

where

$$I(v) = \int_\Omega \Phi(x, \nabla v(x) + I)\,dx - A(v),$$

$$A(v) = \int_\Omega \langle f, v\rangle\,dx + \int_{\Gamma^2} \langle F, v\rangle\,d\gamma,$$

$$\langle g, v\rangle(x) = \int_0^{v(x)} g^\alpha(x, v)\,dv^\alpha.$$

Here $V = \{v : \overline{\Omega} \to \mathbb{R}^3;\ v(x) = v^0(x),\ x \in \Gamma^1\}$ is the set of admissible displacements, $\langle *, v\rangle$ is the specific and $A(v)$ is the full work of the outside forces under the displacement v. It must be marked that even for "dead" forces, i.e. $f, F = const(v, \nabla v)$, the specific work has the form $\langle g, v\rangle(x) = g^\alpha(x)\,v^\alpha(x)$ only in the Descartes coordinates (Ciarlet 1988, Brigadnov 1997).

According to the general theory (Ekeland & Temam 1976, Fučik & Kufner 1980, Temam 1983, Giusti 1984), for potentials of linear growth the set of kinematically admissible displacements is a subset of non-reflexive Sobolev's space $W^{1,1}(\Omega, \mathbb{R}^3)$

$$V = \{v \in W^{1,1}(\Omega, \mathbb{R}^3) : \\ v(x) = v^0(x),\ x \in \Gamma^1\}. \qquad (2)$$

We remind the definition of the limited static load (Brigadnov 1993). For this reason we introduce the set of addmissible "dead" outside forces for which the functional $I(v)$ is bounded from below on V and, therefore, a solution of the problem (1) exists

$$B = \{(f, F) \in L^\infty(\Omega, \mathbb{R}^3) \times L^\infty(\Gamma^2, \mathbb{R}^3) : \\ \inf\{I(v) : v \in V\} > -\infty\}.$$

This set is non-empty because for small outside influences the problem (1) is transformed into the classical variational problem of linear elasticity (Lurie 1990) which always has a solution (Ciarlet 1988; Brigadnov 1996a, 1997).

Definition 1. For outside forces $(f, F) \in B$ we examine the sequence of forces which are proportional to the real parameter $t \geq 0$. The number $t_* \geq 0$ is named *the limited parameter of loading* and $t_*(f, F)$ is named *the limited static load*, if $t(f, F) \in B$ for $0 \leq t \leq t_*$ and $t(f, F) \notin B$ for $t > t_*$.

The limited analysis problem is the investigation of the set of positive parameters t, for which the functional

$$I_t(v) = \int_\Omega \Phi(x, \nabla v(x) + I)\,dx - t\,A(v)$$

is bounded from below on the set of kinematically admissible displacements (2).

In practice the estimation from below for the limited static load is interesting because this information is sufficient for estimation of the strength of nonlinear elastic solid.

Statement 2. For the limited parameter of loading the following estimation from below is true

$$t_* \geq t_- = \inf \{K(v) : v \in V, A(v) = 1\}, \quad (3)$$

where

$$K(v) = \int_\Omega |\nabla v(x)| \varphi(x) \, dx,$$

$$\varphi(x) = \inf \left\{ \frac{\Phi(x, Q)}{|Q|} : Q \in \mathbf{M}^3 \right\}.$$

Proof. From the definition of function $\varphi(x)$ the simplest estimation follows

$$\int_\Omega \Phi(x, \nabla v(x) + I) \, dx \geq$$

$$\geq \int_\Omega |\nabla v(x)| \varphi(x) \, dx + const \geq$$

$$\geq t_- + const$$

for every $v \in V$ such that $A(v) = 1$. If $0 \leq t \leq t_-$ then the functional $I_t(v)$ is bounded from below on V, i.e. $t(f, F) \in B$.

According to the sense, $\varphi(x)$ is *the function of saturation*, it being known that $\varphi(x) > 0$ and for a homogeneous solid $\varphi = const(x)$.

From the definition 1 it follows that for $t_- < 1$ the elastostatic variational problem (1) can have no solution. From the physical point of view this effect is treated as the destruction of solid. According to this sense we can formulate *the shape optimization problem* for the nonlinear elastic solid of the maximum strength

$$\Omega_* = \arg \left(\sup \{ t_-(\Omega) > 1 : \Omega \in C^{0,1}, \right.$$

$$\left. vol(\Omega) = \omega \} \right), \quad (4)$$

where $t_-(\Omega) > 1$ is the solution of the problem (3) on a domain Ω, ω is the prescribed solid's bulk and $C^{0,1}$ is the set of domains with the Lipschitz boundary (Fučik & Kufner 1980, Ciarlet 1988).

3 EXISTENCE OF DISCONTINUOUS SOLUTIONS

We consider the following problem. Suppose that a circular hole of length l and radius b ($l \gg b$) is drilled through an absolutely rigid body and in the hole an absolutely rigid rod of smaller radius a is placed co-axially. The free space of the hole is filled with a deformable material that is attached rigidly to the entire length of the rod and to the inside surface of the hole. The rod is pushed statically along the axis by a given force P.

We assume that the filler is subject to the Bartenev-Hazanovich elastic potential (Bartenev & Zelenev 1976)

$$\Phi(Q) = 2\mu \, (\lambda_1(Q) + \lambda_2(Q) + \lambda_3(Q) - 3),$$

$$det(Q) = \lambda_1 \lambda_2 \lambda_3 = 1, \quad (5)$$

where $\mu > 0$ is the shear modulus under small deformations and $\lambda_k(Q) > 0$ are the eigenvalues of the matrix $(Q^i_\alpha Q^j_\alpha)^{1/2}$ (the principal elasticity coefficients). It is easily seen that this potential has a linear growth in modulus of the distortion tensor because $\lim_{|Q| \to \infty} \Phi(Q)/|Q| \leq 2\sqrt{3}\mu$.

If the filler is homogeneous and isotropic, then, in view of the axial symmetry of the problem, its incompressible deformed configuration can be described in the framework of the model of an antiplanar shear (Lurie 1990) in the reference cylindrical coordinates by the following relation

$$u(\rho, \varphi, z) = x(a\rho, \varphi, az + aw(\rho)),$$

$$\nabla u = \nabla v + I = \begin{pmatrix} 0 & 0 & w' \\ 0 & 0 & 0 \\ 0 & 0 & 0 \end{pmatrix} + I,$$

where $\rho \in [1, \eta]$ and $\eta = b/a$ is the geometrical parameter.

For the elastic potential (5) the variational problem (1) assumes the form

$$w_* = \arg \left(\inf \{ I(w) : w \in V \} \right), \quad (6)$$

where

$$I(w) = \int_1^\eta \left(4 + w'^2 \right)^{1/2} \rho \, d\rho - t w(1),$$

$$V = \{ w \in W^{1,1}(1, \eta) : w(\eta) = 0 \}$$

and $t = P/(4\pi\mu a l)$.

In view of the convexity of the problem in both the functional and the set of admissible functions, a local extremal (if it exists) coincides with a global minimizer (Ekeland & Temam 1976, Fučik & Kufner 1980).

From the condition of stationarity of $I(w)$ we find the local extremal

$$w_*(\rho) = t\left[\cosh^{-1}(\eta/t) - \cosh^{-1}(\rho/t)\right],$$

which exists only for $|t| \leq 1$.

By taking $t > 1$ and the sequence of admissble displacements having the simplest form $w_m \equiv m(\eta-\rho)^m/(\eta-1)^m$ $(m \in \mathbb{N})$, we can see easily that the energy functional is unbounded from below on V, that is, $I(w_m) \to -\infty$ as $m \to +\infty$. This phenomenon shows that for $t > 1$ the variational problem (1) has no solution. Moreover, from the definition 1 it follows that $t_* \leq 1$. The general theorem about existence of the limited static load for hyperelastic materials was proved by the author (Brigadnov 1993).

For elastic potential (5) the limited analysis problem (3) assumes the form

$$t_- = \inf\left\{\int_1^\eta |w'|\rho\, d\rho : \; w \in V,\; w(1) = 1\right\}. \quad (7)$$

Using the equality $|a| = \sup\{ab : \; |b| \leq 1\}$ we can rewrite the problem (7) in the following way

$$t_- = \inf\{\sup(\, L(v,w) : \; v \in V^*\,) :$$
$$w \in V,\; w(1) = 1\},$$

where

$$L(v,w) = \int_1^\eta v w' \rho\, d\rho,$$

$$V^* = \left\{v \in W^{1,\infty}(1,\eta) : \; |v| \leq 1\right\}.$$

From the classical inequality $inf\, sup\,[L(v,w)] \geq sup\, inf\,[L(v,w)]$ (Ekeland & Temam 1976) the estimation follows

$$t_- \geq \sup\{J(v) : \; v \in V^*\},$$

where the duality functional

$$J(v) = -v(1) -$$
$$-\sup\left\{\int_1^\eta (\rho v)' w\, d\rho : \; w \in V,\; w(1) = 1\right\}$$

is proper (i.e. $J(v) \not\equiv -\infty$) on V^* if the duality function has the form $v(\rho) = C/\rho$ for $\rho \in [1,\eta]$ with an indefinite constant C. As a result, we receive the estimation from below for the limited parameter of loading

$$t_* \geq t_- \geq \sup\{-C : \; |C| \leq 1\} = 1.$$

Combining this estimation with estimation from above $t_* \leq 1$, we obtain $t_* = 1$.

It is easily seen that for minimizing sequence of the simplest form $w_m(\rho) = (\eta-\rho)^m/(\eta-1)^m$ $(m \in \mathbb{N})$ the infimum of functional in the problem (7) is reached on the function with break of the sliding type because $w_m(\rho) \to 1 - H(\rho - 1)$ for almost every $\rho \in (1,\eta)$, where H is the Heaviside function of bounded variation (Giusti 1984).

4 DISCONTINUOUS FEA AND THE PARTIAL RELAXATION

From the previous section it follows that for elastic potentials of linear growth the appropriate limited analysis problems need a relaxation. Its main idea consists of the following (Temam 1983).

Let V be the Banach space with norm $\|\cdot\|$ and $I : V \to R$ be the coercive on V functional, i.e. $I(u) \to +\infty$ as $\|u\| \to \infty$. The standard minimization problem $inf\{I(u) : u \in V\}$ is considered. The solution of this problem can be absent. But the sequence $\{u_k\} \subset V$ can exist such that $u_k \to u_0$ almost everywhere and $I(u_k) \to I_0 < +\infty$ as $k \to \infty$, where the limited element $u_0 \notin V$. In this case we can construct a continuation of functional I into the class of functions $\overline{V} \supset V$ such that $\overline{I}(u_0) = I_0$ and $\overline{I}(u) = I(u)$ for every $u \in V$.

For variational problems with the multiple integral functional of linear growth the appropriate space $\overline{V}$ equals the BD space of vector-functions with bounded variations and generalized derivatives as the bounded Radon's measures (Temam 1983, Giusti 1984). In the numerical analysis only finite dimension subspace of BD is used. Therefore, for the limited analysis problem (3) we will use *the partial relaxation* which is based on the special FEA with functions having breaks of the sliding type along ribs of simpleces (Repin 1989, Brigadnov 1998).

Here we examine the plane limited analysis problem. Let $\Omega \subset \mathbb{R}^2$ and $\Omega_h = \cup T_h$ such that $area(\Omega\backslash\Omega_h) \to 0$ and $length(\partial\Omega\backslash\partial\Omega_h) \to 0$ as $h \to +0$, where T_h is the triangle and h is the characteristic step of the regular approximation (Ciarlet 1980). Every FEA is described by the set of nodes $\{x^\alpha\}_{\alpha=1}^m$ and the set of ribs $G_h = \{r_{\alpha\beta} \equiv [x^\alpha, x^\beta]\}$ including inside ribs and ribs on a portion Γ_h^1 of the boundary $\partial\Omega_h$.

For the displacement the following spacial piecewise continuous approximation is used

$$v_h(x) = U_{\alpha\beta}\Psi_{\alpha\beta}(x)$$

where $\alpha, \beta = 1, 2, ..., m$ such that $r_{\alpha\beta} \in G_h$, $U_{\alpha\beta}$ is the component of the displacement in the node x^α which is perpendicular to the rib $r_{\alpha\beta}$, $\Psi_{\alpha\beta} : \Omega_h \to \mathbb{R}$ is the piecewise linear discontinuous function such that $\Psi_{\alpha\beta}(x^\gamma) = \delta_{\alpha\gamma}$ $(\alpha, \beta, \gamma = 1, 2, ..., m)$ and $\Psi_{\alpha\beta} \neq \Psi_{\beta\alpha}$. The $supp\,(\Psi_{\alpha\beta}) = supp\,(\Psi_{\beta\alpha})$ consists of two triangles having the rib $r_{\alpha\beta}$ as common. If a rib $r_{\alpha\beta} \in \Gamma_h^1$ then the $supp\,(\Psi_{\alpha\beta})$ consists of the only triangle.

In this case the subspace $V \subset W^{1,1}(\Omega, \mathbb{R}^2)$ is approximated by the subspace $V_h \subset BD(\Omega, \mathbb{R}^2)$ which is isomorphous to $\mathbb{R}^{2M}$, where M is the number of ribs in the set G_h.

The described FEA possesses the following properties. The component of the displacement, which is perpendicular to an appropriate rib, is continuous; but the tangent projection on this rib has a finite break. As a result, we have the special FEA with functions having breaks of the sliding type along ribs of triangles.

The relaxated problem for the limited analysis problem (3) has the following form

$$t_h = \min\{\, K_h(v) + R_h(v) : \\ v \in V_h,\ A_h(v) = 1\,\}, \qquad (8)$$

where

$$K_h(v) = \int_{\Omega_h} |\nabla v(x)| \varphi(x)\, dx,$$

$$R_h(v) = \sum_{r_{\alpha\beta} \in G_h} \int_{x^\alpha}^{x^\beta} \left| v_\tau^+ \varphi^+ - v_\tau^- \varphi^- \right| d\gamma,$$

$$A_h(v) = \int_{\Omega_h} \langle f_h, v \rangle\, dx + \int_{\Gamma_h^2} \langle F_h, v \rangle\, d\gamma,$$

Here indeces "+" and "-" correspond to the displacement and the function of saturation on the triangles T_h^+ and T_h^- having the common rib $r_{\alpha\beta}$, index τ corresponds to the tangent projection of displacement on this rib, and for ribs on Γ_h^1 the outside displacement is fixed, for example, $v_\tau^- = v_h^0$. Functions (v_h^0, f_h, F_h) are the standard spacial piecewise linear continuous FEAs of outside influences. According to the properties of FEA (Ciarlet 1980) and the results of paper (Repin 1989) we have $t_h \searrow t_-$ as $h \to +0$ regularly.

From the computational point of view the functional in the problem (8) is singular because it has no the classical derivative. Therefore, using the results from (Repin 1989) and the simplest approximation of the modulus $|z| \approx (z^2 + \varepsilon^2)^{1/2}$ with *the regularization parameter* $\varepsilon \ll 1$ (Brigadnov 1998), the following problem is proposed for a numerical solution

$$t_h^\varepsilon = \min\{\, K_h^\varepsilon(v) + R_h^\varepsilon(v) : \\ v \in V_h,\ A_h(v) = 1\,\}, \qquad (9)$$

where

$$K_h^\varepsilon(v) = \int_{\Omega_h} \left(|\nabla v(x)|^2 + \varepsilon^2\right)^{1/2} \varphi(x)\, dx,$$

$$R_h^\varepsilon(v) = \sum_{r_{\alpha\beta} \in G_h} l_{\alpha\beta} \left\{ \left[\left(w_\alpha^+ - w_\alpha^-\right)^2 + \varepsilon^2 \right]^{1/2} + \right. \\ \left. + \left[\left(w_\beta^+ - w_\beta^-\right)^2 + \varepsilon^2 \right]^{1/2} \right\},$$

Here $l_{\alpha\beta} = |x_\alpha - x_\beta|$ and $w_\gamma^\pm = v_\tau^\pm(x^\gamma)\varphi^\pm(x^\gamma)$ $(\gamma = \alpha, \beta)$.

By the necessary condition of stationarity, the problem (9) transforms into a non-linear system of algebraic equations of the form $S(U) \cdot U = B$, where $U \in \mathbb{R}^{2M}$ is the global vector of free variables, $B \in \mathbb{R}^{2M}$ is the global vector of outside influences and $S \in \mathbb{M}^{2M}$ is the global stiffness matrix.

It was proved that for potentials of linear growth the FE approximating algebraic systems can be *badly determined* (Brigadnov 1996c–1998). The main cause of this phenomenon consists of the following: the global stiffness matrix has lines with significantly different factors if the displacement has a large gradient or breaks of the sliding type. For the parameter $\varepsilon \ll 1$ in (9) this situation is more difficult. As a result, for the numerical solution the decomposition method of adaptive block relaxation is used, because it practically disregards the condition number of the global stiffness matrix. The main idea of this method consists of iterative improvement of zones with "proportional" deformation by special decomposition of variables and separate calculation on these variables.

5 NUMERICAL RESULTS

In the numerical experiments, the following boundary value problem was considered: a finite round rod is axial symmetrically stretched in the test machine by a given axial force P (Figure 1). In this case, the map is described by the following relation in the reference cylindrical coordinates

$$u(\rho,\varphi,z) = x\,(a\rho + a\,r(\rho,z),\varphi, lz + lw(\rho,z)),$$

$$C(r,w) = \nabla u - I = \begin{pmatrix} r'_\rho & 0 & \eta\, w'_\rho \\ 0 & r/\rho & 0 \\ \eta^{-1} r'_z & 0 & w'_z \end{pmatrix},$$

where $\rho \in [0,1]$, $\varphi \in [0,2\pi)$, $z \in [0,1]$, a and l are the radius of section and semilength of the rod, respectively; $\eta = l/a$ is the geometrical parameter.

The material of rod is homogeneous and compressible. It is subject to the elastic potential (Brigadnov 1996–1998)

$$\Phi(Q) = \sqrt{3}\mu\,(|Q| - |I|) + k_0\, g(det(Q)),$$

where $\mu > 0$ and $k_0 > 0$ are the shear and bulk moduli under small deformations, respectively; $g : \mathbb{R}_+ \to \mathbb{R}_+$ is the smooth function of compressibility such that $g(1) = g'(1) = 0$ and $g(t) \to +\infty$ as $t \to +0$ or $t \to +\infty$.

For the limited stretching force P_* the estimation from below $P_* \geq \sqrt{3}\mu\pi a^2 t_-$ is true, where the parameter of loading t_- is the solution of the following limited analysis problem

$$t_- = \inf\{\, K(r,w) : \ (r,w) \in V_1 \,\}, \qquad (10)$$

where

$$K(r,w) = 2\int_0^1\!\!\int_0^1 |C(r,w)|\rho\, d\rho dz,$$

$$V_1 = \Big\{ (r,w) \in \left(W^{1,1}((0,1)\times(0,1))\right)^2 :$$

$$r(0,z) = 0,\ r'_z(\rho,0) = 0,\ w(\rho,0) = 0,$$

$$r(\rho,1) = 0,\ w(\rho,1) = 1 \Big\}.$$

According to the convexity of domain, axial symmetry of the problem (10) and continuity of the axial component of dispalcement, the minimizer may have a break of the sliding type along the only line $z = 1$. This break is defined by a finite break of the function $r(\rho,1)$. Therefore, in the set V_1 the condition $r(\rho,1) = 0$ is ignored, but in the functional the appropriate penal item is used

$$R(r) = 2\eta^{-1}\int_0^1 |r(\rho,1)|\,\rho\, d\rho.$$

In the computational experiments the regular $N \times N$ triangulation of the domain $(0,1)\times(0,1)$

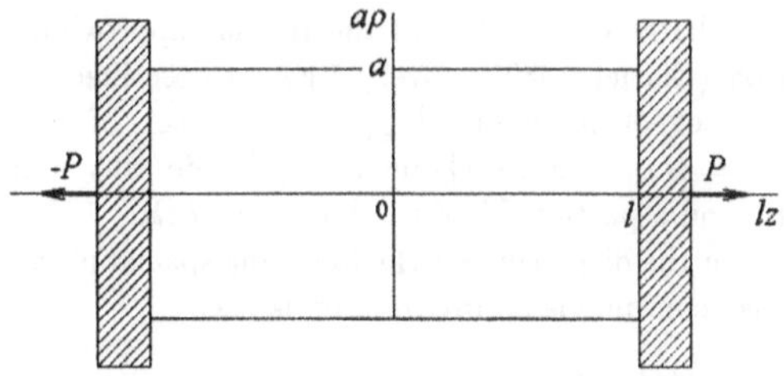

Figure 1. Stretch of the finite round rod by a given axial force in the test machine of the rigid type.

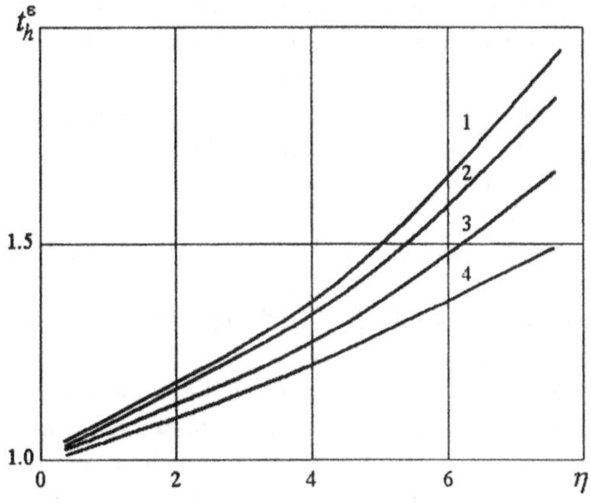

Figure 2. The experimental relations between the geometrical parameter $\eta = l/a$ and the limited parameter of loading t_h^ε for different FEAs.

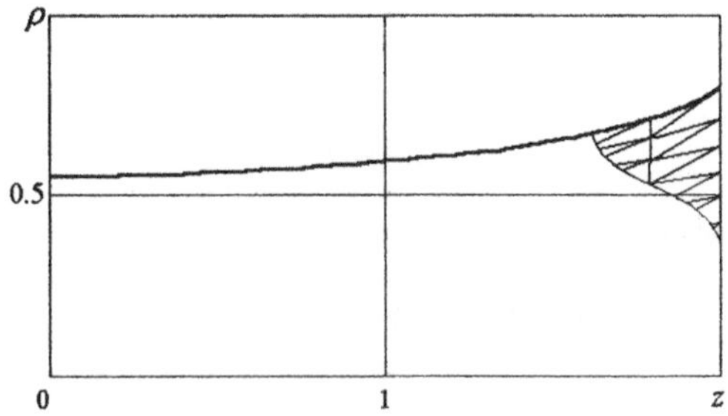

Figure 3. The discontinuous minimizing map of the limited analysis problem for parameter $\eta = 4$ and FEA with $N = 10$ corresponding the curve 4 in Figure 2.

and the regularization parameter $\varepsilon = 10^{-1}$ in (9) were used.

In Figure 2 the experimental relations $\eta \mapsto t_h^\varepsilon$ are shown. Lines 1, 2 and 3 correspond to the continuous FEA with $N = 10$, $N = 20$ and $N = 40$, respectively. Line 4 corresponds to the discontinuous FEA with $N = 10$. The appropriate discontinuous minimizing map for parameter $\eta = 4$ is shown in Figure 3.

It is easily seen that continuous solutions converge to the discontinuous solution with increase of domain's discretization. The decrease of the regularization parameter ε until 10^{-3} practically does not improve either continuous or discontinuous solutions.

6 CONCLUSIONS

1. The limited analysis problem in nonlinear elasticity is formulated. In the framework of this problem the estimation from below for the limited static load is calculated, which is very important for the estimation of the practical strength of nonlinear elastic solids. The shape optimization problem is also formulated for the nonlinear elastic solid of the maximum strength.

2. It is proved that the limited analysis problem has discontinuous solutions with breaks of the sliding type and, therefore, needs a relaxation.

3. The partial relaxation of the limited analysis problem is described. This relaxation is based on the special discontinuous finite-element approximation.

4. The numerical results show that for the estimation of the limited static load, the proposed partial relaxation has qualitative advantages over standard continuous finite-element methods.

ACKNOWLEDGEMENT

I would like to thank the Organising Committee and sponsors of the First European Conference on Constitutive Models for Rubber (ECCMR'99) for their support of my visit to Vienna, Austria.

REFERENCES

Bartenev, G.M. & Yu.V. Zelenev 1976. *A course in the physics of polimers*. Leningrad: Chemistry (Russian).

Brigadnov, I.A. 1993. On the existence of a limiting load in some problems of hyperelasticity. *Mech. of Solids* 5: 46-51.

Brigadnov, I.A. 1996a. Existence theorems for boundary value problems of hyperelasticity. *Sbornik: Mathematics* 187(1): 1-14.

Brigadnov, I.A. 1996b. On mathematical correctness of static boundary value problems for hyperelastic materials. *Mech. of Solids* 6: 37-46.

Brigadnov, I.A. 1996c. Numerical methods in nonlinear elasticity. In: J.-A. Desideri, P. Le Tallec et al. (eds), *Numerical Methods in Engineering'96. Proc. 2nd ECCOMAS Conf., Paris, 9-13 September 1996*: 158-163. Chichester: Wiley.

Brigadnov, I.A. 1997. *Mathematical Methods for Boundary Value Problems of Plasticity and Non-Linear Elasticity*. D.Sci. Thesis, St. Petersburg: St. Petersburg State University (Russian).

Brigadnov, I.A. 1998. Discontinuous solutions and their finite element approximation in non-linear elasticity. In: R. Van Keer, B. Verhegghe et al. (eds), *ACOMEN'98 — Advanced Computational Methods in Engineering, Proc. 1st Int. Conf. ACOMEN'98, Gent, 2-4 September 1998*: 141-148. Maastricht: Shaker Publishing B.V.

Ciarlet, Ph.G. 1980. *The Finite Element Method for Elliptic Problems*. Amsterdam: North-Holland Publ. Co.

Ciarlet, Ph.G. 1988. *Mathematical Elasticity. Vol.1: Three-Dimensional Elasticity*. Amsterdam: North-Holland Publ. Co.

Ekeland, I. & R. Temam 1976. *Convex Analysis and Variational Problems*. Amsterdam: North-Holland Publ. Co.

Fučik, S. & A. Kufner 1980. *Nonlinear Differential Equations*. Amsterdam: Elsevier Sci. Publ. Co.

Giusti, E. 1984. *Minimal Surfaces and Functions of Bounded Variations*. Boston: Birkhäuser.

Green, A.E. & W. Zerna 1975. *Theoretical Elasticity*. Oxford: Oxford University Press.

Lurie, A.I. 1990. *Nonlinear Theory of Elasticity*. Amsterdam: North-Holland Publ. Co.

Repin, S.I. 1989. A variational-difference method for solving problems with functionals of linear growth. *U.S.S.R. Comput. Math. and Math. Phys.* 29(5): 693-708.

Temam, R. 1983. *Problèmes Mathématiques en Plasticité*. Paris: Gauthier-Villars.

Constitutive Models for Rubber, Dorfmann & Muhr (eds) © 1999 Taylor & Francis ISBN 90 5809 113 9

A strain energy function for filled and unfilled elastomers

M.H.B.M.Shariff
School of Computing and Mathematics, University of Teesside, Middlesbrough, UK

ABSTRACT: A general form of strain energy function with linear parameters is developed so that if a standard curve fitting method is used only a handful of linear equations need to be solved; the proposed function is formulated in such a way that only a few parameters are required to characterised the elastomers. One of the parameters is the Young's modulus and this could faciliate a curve fitting algorithm if the value of the Young's modulus is known. A single constant strain energy function is proposed. The predicted curves agree well with numerous experimental data.

1 INTRODUCTION

We assume that a rubberlike material is isotropic relative to a ground (stress free) state and that its isothermal mechanical properties can be represented in terms of an elastic strain energy function S. For simplicity, and because λ_1, λ_2 and λ_3 have clear physical meaning (Ogden 1984,1986), we write S as a function of λ_1, λ_2 and λ_3, i.e. $S(\lambda_1, \lambda_2, \lambda_3)$, where λ_1, λ_2 and λ_3 are the principal stretches of the deformation. Due to isotropy, $S(\lambda_1, \lambda_2, \lambda_3)$, measured per unit volume in the ground state, must be symmetric with respect to interchange of any two of λ_1, λ_2 and λ_3, i.e.,

$$S(\lambda_1, \lambda_2, \lambda_3) = S(\lambda_1, \lambda_3, \lambda_2) = S(\lambda_3, \lambda_1, \lambda_2) \quad (1)$$

We consider only incompressible materials and adopt the incompressibility constraint,

$$\lambda_1\lambda_2\lambda_3 = 1 \quad (2)$$

In view of Equation (2) it is convenient to eliminate λ_3 in favour of λ_1 and λ_2 and the strain energy function takes the form,

$$W(\lambda_1, \lambda_2) = S(\lambda_1, \lambda_2, 1/(\lambda_1\lambda_2)) \quad (3)$$

noting the symmetry,

$$W(\lambda_1, \lambda_2) = W(\lambda_2, \lambda_1) \quad (4)$$

in view of the symmetry given in Equation (1). We shall also assume that the strain energy function is of the form

$$S(\lambda_1, \lambda_2, \lambda_3) = r(\lambda_1) + r(\lambda_2) + r(\lambda_3), \quad (5)$$

according to the hypothesis of Valanis & Landel (1967). It was shown in the literature that for a range of strains the form of S given in equation (5) describes many types of rubberlike materials.

Many good strain energy functions (in the sense that they can represent a wide range of incompressible rubberlike materials and a wide range of strains) of the form given by equation (5) have been developed in the past. The parameters of these functions are commonly determined via a curve fitting method such as the least squares method. However, a curve fitting technique normally requires, for good separable form of strain energy functions such as Ogden's (1927) and Tobisch's (1981), amongst others (i.e. strain energy functions which are nonlinear in their parameters), a solution of a nonlinear system of equations. This could lead to increase computer time and storage when a nonlinear optimization method is used, and the convergence to a global minimum of such a method could be difficult to obtain. The intention of the present work is different from previous workers in the sense that a general strain energy density function for filled and unfilled elastomers is developed which is linear in its parameters ; when a curve fitting method is (sensibly) applied on an experimental data the value of our parameters are obtained uniquely via a linear positive definite system of equations. In addition we also intend to describe rubberlike materials with as few parameters as possible; one of the parameters is a physical parameter , i.e., the Young's modulus E. We note that the parameters of previous forms of strain energy functions are implicitly related (often nonlinearly related) to the Young's modulus; if the Young's modulus of an elastomer is known or accurately obtained, a curve fitting method to obtain the values of the paramaters could be complicated by an additional constrained equation relating the parameters to the Young's modulus. Whereas the strain energy function presented in this paper is facilitated by a known Young's modulus in the sense that only the remaining values of the parameters have to be obtained via a curve fitting method.

2 STRAIN ENERGY FUNCTION

We start by quoting a few fundamental results in elasticity which can be found in Ogden's (1984) book. Let $\sigma_i\,(i=1,2,3)$ be the principal components of the Cauchy (true) stress. In an incompressible material they are related to the strain energy, S by the relation

$$\sigma_i = \lambda_i \frac{\partial S}{\partial \lambda_i} - p \quad (i = 1,2,3), \tag{6}$$

where p is the arbitrary hydrostatic stress required because of the incompressibility constraint (2). It follows from (3) and (5) that

$$\sigma_1 - \sigma_3 = \lambda_1 \frac{\partial W}{\partial \lambda_1}, \quad \sigma_2 - \sigma_3 = \lambda_2 \frac{\partial W}{\partial \lambda_2} \tag{7}$$

We note that equations (7) are unaffected by any superposed hydrostatic stress. Since W depends on only λ_1 and λ_2, equations (7) are sufficient to determine W from experiments in which the principal stretches, λ_1 and λ_2 are varied independently. Equations (7) can be rearranged to give,

$$\sigma_1 - \sigma_2 = \lambda_1 \frac{\partial W}{\partial \lambda_1} - \lambda_2 \frac{\partial W}{\partial \lambda_2}. \tag{8}$$

To be consistent with the classical linear theory of incompressible isotropic elasticity, appropriate for infinitesimal deformations, we must have the relations (Ogden 1984),

$$W(1,1) = 0, \quad \frac{\partial W}{\partial \lambda_1}(1,1) = \frac{\partial W}{\partial \lambda_2}(1,1) = 0$$

$$\frac{\partial^2 W}{\partial \lambda_1 \partial \lambda_2}(1,1) = \frac{2E}{3}, \quad \frac{\partial^2 W}{\partial \lambda_k^2}(1,1) = \frac{4E}{3},$$

$$k = 1,2. \tag{9}$$

For a separable form of strain energy function given in equation (5) the difference of stress expressed in equations (7) and (8) can be simplified, i.e.,

$$\sigma_1 - \sigma_3 = \lambda_1 r'(\lambda_1) - \frac{1}{\lambda_1\lambda_2} r'(\frac{1}{\lambda_1\lambda_2}), \tag{10}$$

$$\sigma_1 - \sigma_2 = \lambda_1 r'(\lambda_1) - \lambda_2 r'(\lambda_2), \tag{11}$$

where the prime signifies differentiation with respect to the argument of the function in question. In view of equations (3) and (5) we only need to establish the functional form of r to establish W (or S). Without loss of generality and for simplicity we restrict the function r so that $r'(1) = 0$. This simplifies equation (11) for $\lambda_2 = 1$, i.e.,

$$\sigma_1 - \sigma_2 = \lambda_1 r'(\lambda_1). \tag{12}$$

The relations in equation (9) are satisfied if $r(1) = 0$ and $r''(1) = 2E/3$.

Recently Shariff (1997) developed a single constant (Young's modulus) strain energy function which characterised a variety of different rubber-like materials for moderate strains. We extend and improve this work for moderate and larger strains by approximating the expression $\lambda r'(\lambda)(= f(\lambda), say)$ by a finite series shown below

$$f(\lambda) = \lambda r'(\lambda) = E \sum_{i=0}^{n} \alpha_i \phi_i(\lambda), \tag{13}$$

where α_i are the parameters of the strain energy function, $\alpha_0 = 1$ and ϕ_i are sufficiently smooth function such that $\phi_0(1) = 0, \phi_0'(1) = \frac{2}{3}, \phi_i(1) = \phi_i'(1) = 0, i = 1,2\ldots,n$. Thus $f(1) = 0$ and $f'(1) = \frac{2E}{3}$. We seek the functional forms of ϕ_i by considering uniaxial, pure shear, equibiaxial and biaxial deformations. In uniaxial, pure shear and equibiaxial extensions the first principle component of the Cauchy stress can be simply written as

$$\sigma_1 = f(\lambda) - f(\frac{1}{\lambda^m}), \quad m = \frac{1}{2}, 1, 2, \tag{14}$$

where $\lambda = \lambda_1 > 1$, and $m = \frac{1}{2}, 1$ and 2 correspond to a uniaxial extension, a pure shear deformation and an equibiaxial extension respectively. For a biaxial deformation we consider the stress

$$\sigma_b = \sigma_1 - \sigma_2 = f(\lambda) - f(\lambda_2). \tag{15}$$

In view of equations (14) and (15) it is interesting to note that if a material can be characterised by a separable form (as expressed in equation (5)) the

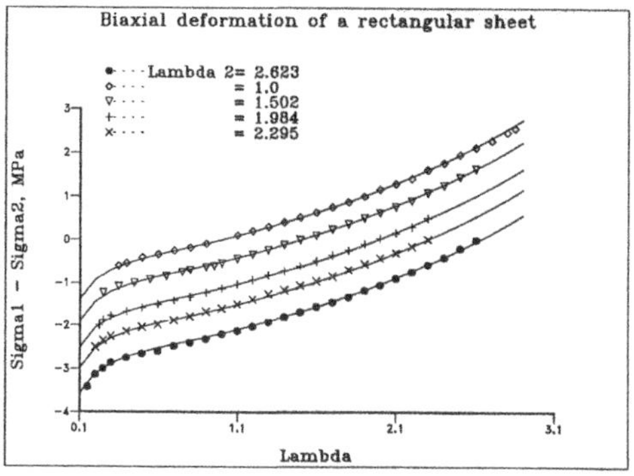

Figure 1. Comparison of theoretical curves with the data of Jones & Treloar (1975):Biaxial deformation of a rectangular sheet. Lambda = λ_1, Sigmal = σ_1, Sigma2 = σ_2. $\sigma_1 - \sigma_2$ v. λ_1 at fixed values of λ_2: E = 1.34356 MPa, $\alpha_1 = 0.8223, \alpha_2 = 0.1257$. No weighting.

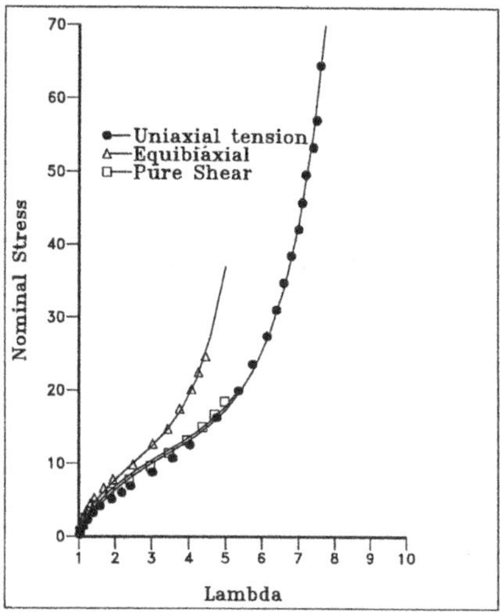

Figure 2. Comparison of theoretical curves with the data of Treloar (1944). Lambda = λ_1, Nominal stress = $\frac{\sigma_1}{\lambda_1}$:$E = 11.0009kg/cm^2, \alpha_1 = 1.37334, \alpha_2 = 0.471163 \times 10^{-1}, \alpha_3 = 0.841383 \times 10^{-4}$ No weighting.

stress σ_1 in a uniaxial or a pure shear or an equibiaxial deformation is related to the biaxial stress σ_b in the following manner:

For any fixed λ_2 at $\lambda = \lambda_o > 1$

$$\sigma_1(\lambda_o) = \sigma_b(\lambda_o) - \sigma_b(\frac{1}{\lambda_o^m}). \qquad (16)$$

Hence if we assume a material to have a separable form of the strain energy function the σ_1 stress for a uniaxial or a pure shear or an equibiaxial deformation can be predicted using the stress σ_b in a biaxial data. This is one way to validate if the material can be characterised by a separable form.

In view of the above and by observing the experimental data of Jones & Treloar (1975) and Treloar (1944) (see Figs. 1 and 2) we choose ϕ_i to take form:

$$\phi_0(\lambda) = \frac{2ln(\lambda)}{3}$$

$$\phi_1(\lambda) = e^{(1-\lambda)} + \lambda - 2$$

$$\phi_2(\lambda) = e^{(\lambda-1)} - \lambda$$

$$\phi_3(\lambda) = \frac{(\lambda-1)^3}{\lambda^{3.6}}$$

$$\phi_j = (\lambda-1)^{j-1}, \quad j = 4, 5, \ldots, n. \qquad (17)$$

We note in passing that any sufficiently smooth function $g(\lambda)$ with $g'(1) = g(1) = 0$ can be approximated as closely as possible by the series

$$\sum_{i=2}^{n} \gamma_{i-1}(\lambda-1)^i \qquad (18)$$

Hence $f(\lambda)$ can be approximated by the series $\gamma_0\psi_0 + \sum_{i=2}^{n} \gamma_{i-1}(\lambda-1)^i$, where ψ_0 is sufficiently smooth with property $\psi_0(1) = 0, \psi_0'(1) = \frac{2}{3}$. We have experimented with the series $f(\lambda) = \gamma_0\psi_0 + \sum_{i=2}^{n} \gamma_{i-1}(\lambda-1)^i$ (including when $\psi_0 = \phi_0$) and found that in general a lot of parameters are needed to fit published experimental data. Our intention, however, as mentioned earlier, is to (linearly) characterise rubberlike materials with as few parameters as possible and it is found that (and shown in Section IV) the expression given by equation (13) (together with equation (17)) seems to serve our intention. It is also noted that the $(\lambda-1)^2$ term in series (13) is absorbed in the term $\phi_2 = e^{(\lambda-1)} - \lambda$ and hence does not appear explicitly in the expression; $\phi_i, i = 0, 1, \ldots, n$ are are infinitely differentiable for $\lambda > 0$ and f can be approximated as closely as possible by the series expressed in equation (13).The function $r(\lambda)$ for the strain energy function is then obtained from the integral given below

$$r(\lambda) = \int_1^{\lambda} \frac{f(s)}{s} ds \qquad (19)$$

3 CONSTITUTIVE INEQUALITIES

To ensure physically reasonable responses the strain energy density function must satisfy certain restrictions (Truesdell 1956). For example, a class of inequalities was proposed by Hill (1970) for examination of isotropic elasticity and he has shown that one and only one of the inequalities (referred to as Hill's inequality) admits incompressibility. For incompressible solids Hill's inequality asserts that the scalar product $tr(\hat{\mathbf{T}}\mathbf{E})$ (where tr denotes the trace of second order tensor) is positive at all strains and for arbitrary non-zero strain-rates. The notations $\mathbf{T}$, $\mathbf{E}$ and $\hat{\mathbf{T}}$ represent the Cauchy stress, the Eulerian strain-rate and the rigid-body derivative (the rate of change on axes rotating rigidly with the local body spin) of the Cauchy stress respectively. It is shown in Ogden (1972) that, relative to the principal axes of the Eulerian strain ellipsoid, the component form of the scalar product $tr(\hat{\mathbf{T}}\mathbf{E})$ is expressible as $\hat{\sigma}_{ij}\epsilon_{ij}$(from now onwards summation is implied by repeated suffixes unless otherwise stated), where $\hat{\sigma}_{ij}$ denote the components of the rigid-body derivative of the Cauchy stress on the axes of the Eulerian strain ellipsoid and ϵ_{ij} are components of the Eulerian strain-rate on the same axes. Then by the use of equations (5) and (6) and taking note $\frac{\dot{\lambda}_i}{\lambda_i} = \epsilon_{ii}$ [see Ogden (1972)] we obtain

$$\hat{\sigma}_{ii} \equiv \dot{\sigma}_i = \lambda_i f'(\lambda_i)\epsilon_{ii} - \dot{p} \;\; (i = 1, 2, 3)$$

$$(i \;\; not\ summed), \tag{20}$$

and

$$\hat{\sigma}_{ij} = \frac{{\lambda_i}^2 + {\lambda_j}^2}{\lambda_i^2 - \lambda_j^2}(f(\lambda_i) - f(\lambda_j))\epsilon_{ij} \;\; (i \neq j),$$

$$(i, j \;\; not\ summed), \tag{21}$$

where the superposed dot denotes the material time derivative. It follows that the scalar product $tr(\hat{\mathbf{T}}\mathbf{E})$ is expressible as (taking note that $\epsilon_{ii} = 0$)

$$\hat{\sigma}_{ij}\epsilon_{ij} \equiv \lambda_i f'(\lambda_i)\epsilon_{ii}^2 + \frac{{\lambda_i}^2 + {\lambda_j}^2}{\lambda_i^2 - \lambda_j^2}(f(\lambda_i) - f(\lambda_j))\epsilon_{ij}^2 \tag{22}$$

In the second group of the right-hand side of (22) the summation over i and j is restricted to $i \neq j$ and $\lambda_i \neq \lambda_j$.

In the present paper we shall only attempt to derive sufficient conditions on the parameters α_i ($i = 1, 2 \ldots, n$) that ensures the expression (22) is positive. We start with the second group of terms in equation (22) and this is positive if $f(\lambda_i) > f(\lambda_j)$ for $\lambda_i > \lambda_j$. Hence we need to obtain the range of α_i ($i = 1, 2, \ldots, n$) for the function f to be strictly monotone. First consider the function f for $0 < \lambda \leq 1$; f is stictly increasing if $f'(\lambda) > 0$ (end point derivative is assumed at $\lambda = 1$). We note that

$$\phi_0'(\lambda) = \frac{2}{3\lambda}$$

$$\phi_1'(\lambda) = 1 - e^{1-\lambda}$$

$$\phi_2'(\lambda) = e^{\lambda-1} - 1$$

$$\phi_3'(\lambda) = 3\frac{(\lambda-1)^2}{\lambda^{3.6}} - 3.6\frac{(\lambda-1)^3}{\lambda^{4.6}}$$

$$\phi_k'(\lambda) = (k-1)(\lambda-1)^{k-2}, \;\; k = 4, 5, \ldots, n. \tag{23}$$

In the range $0 < \lambda \leq 1$, $\phi_0'(\lambda), \phi_3'(\lambda)$ and $\phi_k'(\lambda)$ ($k = 2j+2$, $j = 1, 2, \ldots$) are non-negative, $\phi_1'(\lambda) \leq 0, \phi_2'(\lambda) \leq 0$ and $\phi_k'(\lambda) \leq 0$ ($k = 2j+3$, $j = 1, 2, \ldots$). We impose $\alpha_3 \geq 0$ and $\alpha_k \geq 0 (k = 2j+2,\ j = 1, 2, \ldots)$. Hence the values of α_1, α_2 and α_k ($k = 2j+3$, $j = 1, 2, \ldots$) have to be restricted to ensure that $f'(\lambda) > 0$ To obtain these restricted values for $\lambda \in (0, 1]$ we first consider the following inequalites (which can be easily verified)

$$\frac{1}{4\lambda} \geq 1 - \lambda \geq 1 - e^{\lambda-1}, \tag{24}$$

$$\frac{c}{4\lambda} \geq c(1-\lambda) \geq e^{1-\lambda} - 1, \tag{25}$$

where $c = e^1 - 1$. For $d > 0$ and for $0 < \hat{\alpha} < \frac{8d}{3(cd+1)}$ we have, from combining the above inequalities, the relation

$$\frac{2}{3\lambda} > \hat{\alpha}\frac{cd+1}{4d\lambda} \geq -\hat{\alpha}\phi_1'(\lambda) - \frac{\hat{\alpha}}{d}\phi_2'(\lambda). \tag{26}$$

It follows that for $\alpha_1 \in [0, \hat{\alpha}]$ and $\alpha_2 \in [0, \frac{\hat{\alpha}}{d}]$

$$\frac{2}{3\lambda} + \alpha_1\phi_1'(\lambda) + \alpha_2\phi_2'(\lambda) \geq \frac{2}{3\lambda} + \hat{\alpha}\phi_1'(\lambda) +$$

$$\frac{\hat{\alpha}}{d}\phi_2'(\lambda) > 0. \tag{27}$$

It can be easily verified that for $\lambda \in (0, 1]$

$$(\lambda-1)^k \geq -(\lambda-1)^{k+1}, \;\; k = 2j, \;\; j = 1, 2, \ldots \tag{28}$$

and

$$\alpha_{k+2}(\lambda-1)^k + \alpha_{k+3}(\lambda-1)^{k+1} \geq 0, \; k = 2j,$$

$$j = 1, 2, \ldots \tag{29}$$

if $\alpha_{k+2} \geq 0$ and $\alpha_{k+3} \in [0, \frac{k+1}{k+2}\alpha_{k+2}]$. Hence for $\lambda \in (0,1]$, $\alpha_1 \in [0, \hat{\alpha}]$, $\alpha_2 \in [0, \frac{\hat{\alpha}}{d}]$, $\alpha_3 \geq 0, \alpha_{k+2} \geq 0$ and $\alpha_{k+3} \in [0, \frac{k+1}{k+2}\alpha_{k+2}]$ $(k = 2j,\ j = 1,2,\ldots)$ we have $f'(\lambda) > 0$.

Next, we consider $f'(\lambda)$ for $\lambda > 1$. We note that the only expression in equation (23) which is not positive is $-\dfrac{(\lambda-1)^3}{\lambda^{4.6}}$. In order to find the range of $\alpha_3 \geq 0$ so that $f'(\lambda) > 0$ for $\lambda > 1$ we first consider the equation

$$y = \frac{1}{8.28}\lambda^{4.6} - (\lambda - 1)^2. \tag{30}$$

Since $y'(1) = 1/1.8$ and $y''(\lambda) > 0$, $y'(\lambda)$ is monotonically increasing and hence $y'(\lambda) > 0$ for $\lambda > 1$. Thus for this range of λ, $y(\lambda)$ is monotonically increasing and in view of $y(1) = \frac{1}{8.28}$ we conclude that $y(\lambda) > 0$, i.e. $\frac{\lambda^{4.6}}{8.28} > (\lambda-1)^2$. It follows that for $\lambda - 1 > 0$, $e^{\lambda-1} - 1 > (\lambda - 1) > \frac{8.28(\lambda-1)^3}{\lambda^{4.6}}$. Hence for $\alpha_2 \geq 0$ and $\alpha_3 \in [0, 2.3\alpha_2]$ we obtain the relation $\alpha_2(e^{\lambda-1} - 1) \geq 3.6\alpha_3\frac{(\lambda-1)^3}{\lambda^{4.6}}$. Combining the above relevant relations for both $\lambda \in (0,1]$ and $\lambda > 1$ we deduce that $f'(\lambda) > 0$ for $\lambda > 0$ (and hence $\lambda f'(\lambda) > 0$ for the first group of terms in equation (22)) if

$$\alpha_1 \in [0, \hat{\alpha}], \quad \alpha_2 \in [0, \frac{\hat{\alpha}}{d}], \quad \alpha_3 \in [0, 2.3\alpha_2],$$

$$\alpha_{k+2} \geq 0, \quad \alpha_{k+3} \in [0, \frac{k+1}{k+2}\alpha_{k+2}]$$

$$k = 2j, \quad j = 1,2,\ldots \tag{31}$$

It is clear that the range of values of α_i's, $(i = 1,2,\ldots,n)$ given in equation (31) are sufficient for the expression (22) to be positive.

Less rigorously, an alternative sufficient condition for the expression (22) to be positive can be derived via Figure 3 which indicates the relation

$$\frac{2}{3\lambda} + 1.9\phi_1'(\lambda) + 0.15\phi_2'(\lambda) > 0. \tag{32}$$

Hence, following our previous argument, $f'(\lambda) > 0$ for $\lambda > 0$ if

$$\alpha_1 \in [0, 1.9], \quad \alpha_2 \in [0, 0.15], \quad \alpha_3 \in [0, 2.3\alpha_2],$$

$$\alpha_{k+2} \geq 0, \quad \alpha_{k+3} \in [0, \frac{k+1}{k+2}\alpha_{k+2}]$$

$$k = 2j, \quad j = 1,2,\ldots \tag{33}$$

and the above range of values of α_i's $(i = 1,2,\ldots,n)$ are sufficient for the expression (22) to be positive. In a similar way to the above we can derive various sufficient conditions for the expression (22) to be positive. However, we shall not not discuss the matter further here. The first sufficient condition allow the upper bound of α_2 to be greater than that of α_1.

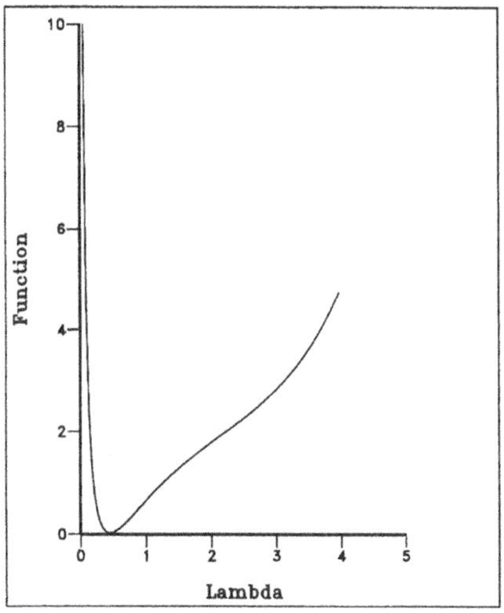

Figure 3. The plot of function $\frac{2}{3\lambda} + 1.9\phi_1'(\lambda) + 0.15\phi_2'(\lambda)$. Lambda = λ.

4 COMPARISION WITH PUBLISHED EXPERIMENTAL DATA

There are many ways to fit a theoretical curve to an experimental data. In the present paper we only consider the standard least squares fit, with and without weighting, and occasionally we use "visual" fitting. The weighting is used to give a higher weight at low stress. We used the weighting function (Johannknecht et al. 1999)

$$g_j = \frac{L}{c + t_j^2} / \sum_j \frac{1}{c + t_j^2}, \tag{34}$$

where L is the number of data points, t_j are the values of the experimental stress (Cauchy stress or Nominal stress or stress difference) at various strains and $c \geq 0$ is a constant. We note the values of the parameters are obtained from a positive definite symmetric system of linear equations. To obtain the values of $\alpha_i (i = 1,2,\ldots,n)$ we initially calculate the values of $\beta_i (= E\alpha_i)$ $(i = 0,1,\ldots,n)$ where $E = \beta_0$ and $\alpha_i = \frac{\beta_i}{E}, (i = 1,2,\ldots,n)$. If the value of E is given we only need to obtain the values of $\beta_i, (i = 1,2,\ldots,n)$.

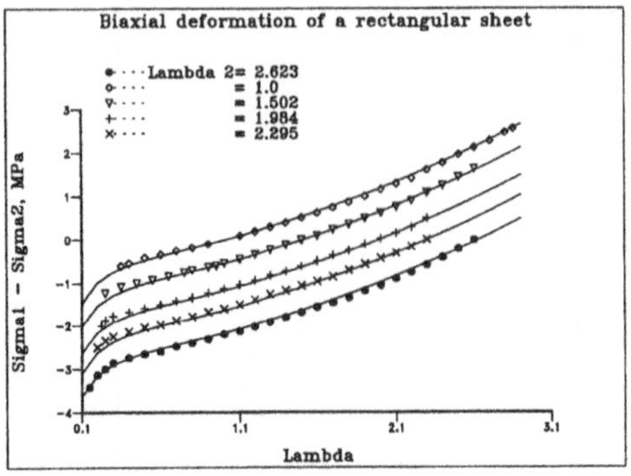

Figure 4. $E = 1.4421$ MPa (given), $\alpha_1 = 0.880941, \alpha_2 = 0.0544614$

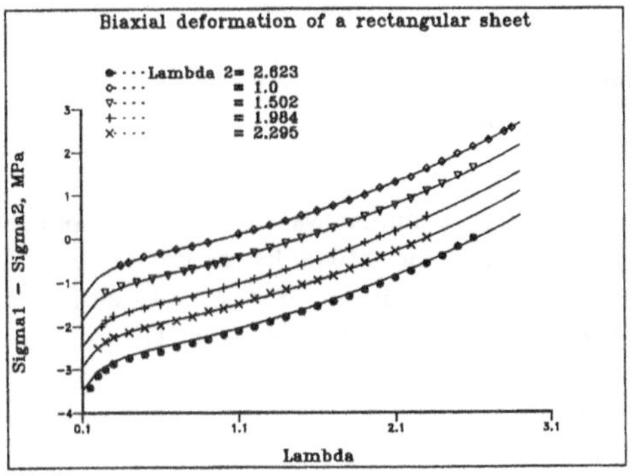

Figure 5. $E = 1.34395$ MPa, $\alpha_1 = 0.9070, \alpha_2 = 0.08578$. 1 set of data

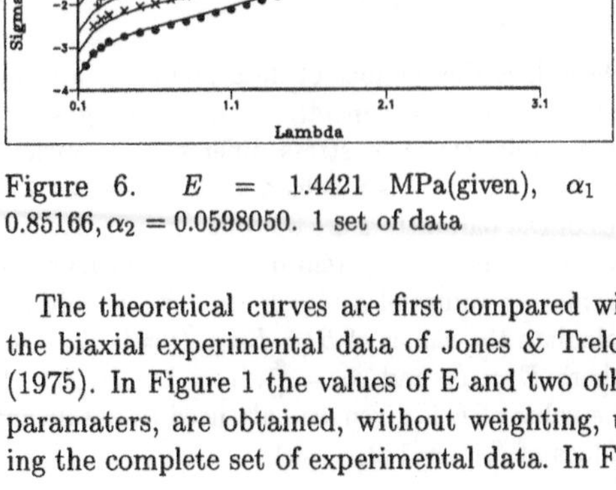

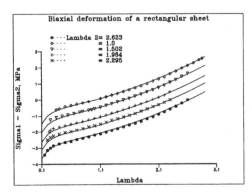

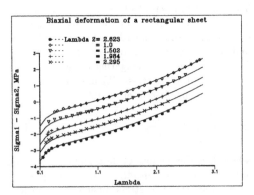

Figure 6. $E = 1.4421$ MPa(given), $\alpha_1 = 0.85166, \alpha_2 = 0.0598050$. 1 set of data

The theoretical curves are first compared with the biaxial experimental data of Jones & Treloar (1975). In Figure 1 the values of E and two other paramaters, are obtained, without weighting, using the complete set of experimental data. In Figure 4 , for given $E = 1.4421$ MPa, two values of the parameters are obtained, without weighting, using the complete set of data. This value of Young's modulus is the same as that used by Ogden (1986) to fit this data. In view of equation (12), $f(\lambda)$ could, in principle, be obtained from only the set of data when $\lambda_2 = 1$. In Figures 5 and 6 the values of three and two parameter, respectively, are obtained (without weighting) using this set of data. In all cases the theoretical curves compare well with the experimental data. The data from Treloar's (1944) experiments have been used subsequently by a number of authors as the basis for comparison of the theory and experiments. In Figures 2, 7 and 8 the theoretical curves are compared with this data. The least square fit is applied simultaneously to all three (simple tension, pure shear and equibiaxial) deformations. Fitting without weighting is shown in Figures 2 and 7 , where four and six terms, respectively, are used to fit the data. In Figure 8 we use six terms and the weighting function given by equation (34) with $c = 0$ to fit the data.

For this experimental data we also calculate the sum of square of errors S, i.e.,

$$S = \sum_j (t_j - f(x_j) + f(\frac{1}{x_j^m}))^2, \tag{35}$$

where x_j are the experimental data values of the principle stretch λ_1. For the four term strain energy function shown in Figure 2, S=24.8222. For the six term function shown in Figures 7 and 8, S=12.6242 and S= 31.6773, respectively. We note that Twizell & Ogden (1983) obtained the error S=15.9 when they used six parameters of Ogden's (1972) material to fit this data via a nonlinear optimization method. Using an *ad hoc* method Ogden's obtained six specific values of his parameters to fit Treloar's data and it is shown (Twizell & Ogden 1983) that for these values S=308.41. Chadwick et al. (1977) obtained two set of values for the six parameter Ogden's material to fit Treloar's data. The errors for the two sets are S=33.55 and S=21.96. In terms of the error S we conclude that our six parameter function fits this data better than the six parameter Ogden's function. We also note that our four parameter strain energy function fits better than Ogden's six parameter strain energy function for some particular values of his six parameters. The Young's modulus is expected to be more accurately calculated when weighting is used in the least square fit.

The proposed form of strain energy function is

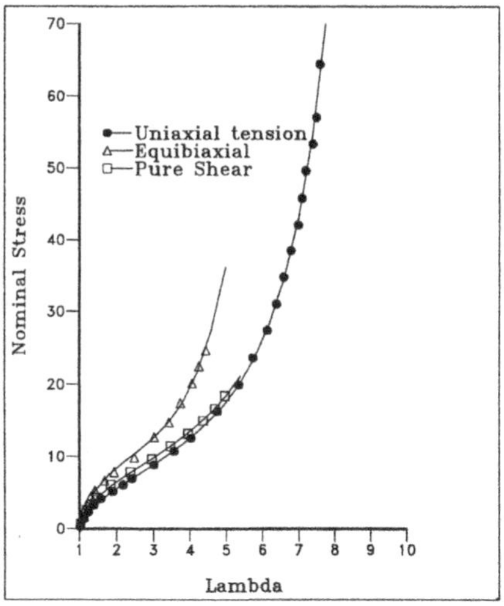

Figure 7. $E = 11.1104 kg/cm^2, \alpha_1 = -0.888029 \times 10^{-2}, \alpha_2 = 0.677477 \times 10^{-1}, \alpha_3 = 0.476368 \times 10^{-1}, \alpha_4 = 0.645226, \alpha_5 = 0.740910 \times 10^{-4}$ No weighting.

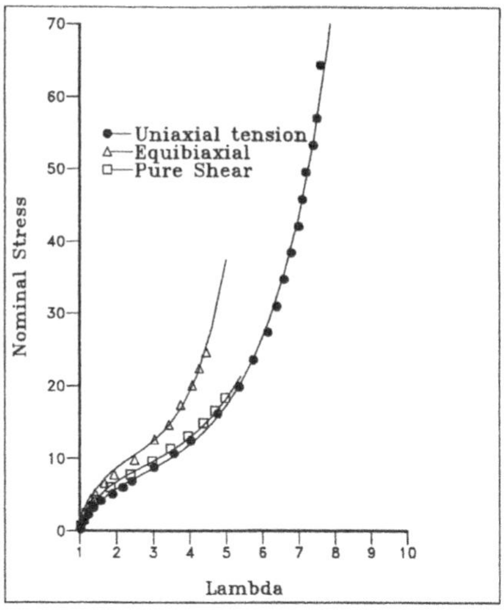

Figure 8. $E = 12.3137 kg/cm^2, \alpha_1 = 0.703299 \times 10^{-2}, \alpha_2 = 0.909616 \times 10^{-2}, \alpha_3 = 0.209726 \times 10^{-1}, \alpha_4 = 0.707786, \alpha_5 = 0.725643 \times 10^{-4}$ Weighting with $c = 0$.

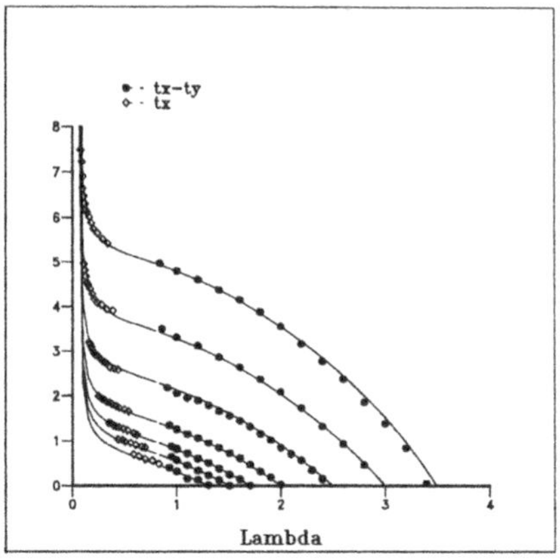

Figure 9. Comparison of the theoretical curves with the experimental data of James et al. (1975). No weighting: $E = 1.37429$ MPa, $\alpha_1 = 0.928269, \alpha_2 = 0.135773, \alpha_3 = 0.156614 \times 10^{-3}$. Biaxial deformation of a rectangular sheet for fixed values of $\lambda_1 = 1.3, 1.5, 1.7, 2.0, 2.5, 3.0, 3.5$. tx= σ_1, ty= σ_2; • Lambda = λ_2, λ_1 = constant; ◇ Lambda = $\lambda_3 = \frac{1}{\lambda_2\lambda_1}, \lambda_1$ = constant.

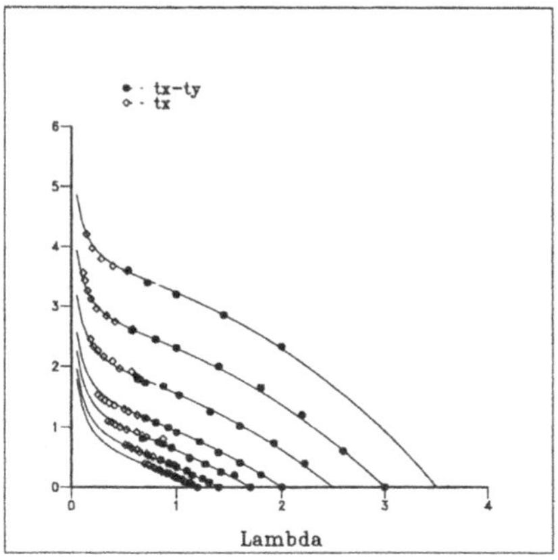

Figure 10. Comparison of the theoretical curves with the data of Kawabata & Kawai (1977). No weighting: $E = 1.12758$ MPa, $\alpha_1 = 0.833904, \alpha_2 = 0.0816449$. Biaxial deformation of a rectangular sheet for fixed values of $\lambda_1 = 1.2, 1.4, 1.7, 2.0, 2.5, 3.0, 3.5$. tx= σ_1, ty= σ_2; • Lambda = λ_2, λ_1 = constant; ◇ Lambda = $\lambda_3 = \frac{1}{\lambda_2\lambda_1}, \lambda_1$ = constant.

tested further by comparing the predicted curves with the data of James et al. (1975) and of Kawabata & Kawai (1977) both for unfilled natural rubber vulcanizates. For prediction we use 4 terms for the data of James et al. and 3 terms for the data of Kawabata and Kawai. Figures 9 and 10 show that the proposed strain energy function can represent these two rubberlike materials satisfactorily. The data of James et al. can also be predicted satisfactorily using three terms but we shall not depict the results here. To further justify our proposed form of strain energy function the engineering stress $\frac{\sigma_1}{\lambda_1}$ for uniaxial elongation, pure

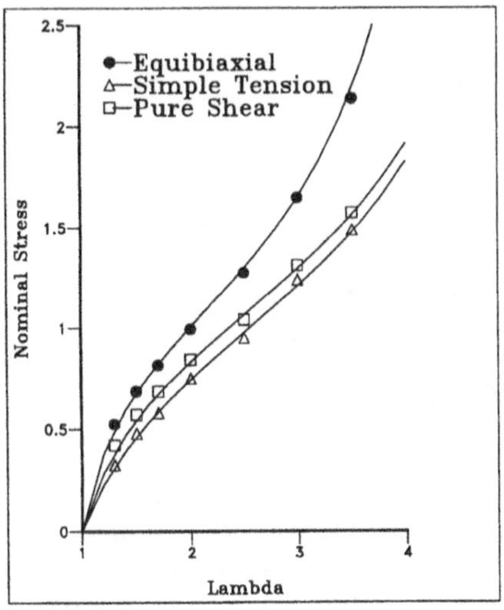

Figure 11. Experimental data of James et al. (1975). No weighting. $E = 1.33486$ MPa, $\alpha_1 = 0.997974, \alpha_2 = 0.133918, \alpha_3 = 0.0203973$: Lambda $= \lambda_1$, Nominal Stress $= \frac{\sigma_1}{\lambda}$.

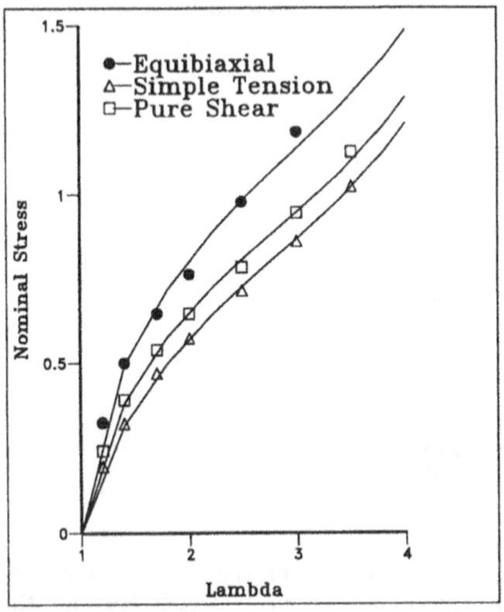

Figure 12. Comparison with the data of Kawabata & Kawai (1977). No weighting. $E = 1.12758$ MPa, $\alpha_1 = 0.833904, \alpha_2 = 0.0816449$ Lambda $= \lambda_1$, Nominal Stress $= \frac{\sigma_1}{\lambda}$.

shear and equibiaxial extension is depicted in Figures 11 and 12, using the values of parameters obtained when fitting the biaxial data of James et al. (1975) and Kawabata & Kawai (1977) All of the above mentioned experimental data are for unfilled elastomers. In Figures 13, 14 and 15 the experimental data of James & Green (1975) and Becker & Rademacher (1962) for filled rubber are compared with the predicted curves. The data of James & Green is fitted using 4 parameters. However, not all of the values of the 4 parameters are obtained via the least squares method. This data is fitted by assuming $E = 2.6$ MPa, the values of α_1 and α_2 are obtained using the least squares method and based on visual fitting we gave α_3 the value of 0.002. We note that Tobisch (1981) only fitted this data for $\lambda_1 = 1.2, 1.4$ and 1.6 (The data for $\lambda_1 = 1.8$ and 2.0 are excluded in his figure). He claimed that relaxation phenomena disturbed his agreement at higher elongation. Our biaxial curves are shown in Figure 13 and the engineering stress $\frac{\sigma_1}{\lambda_1}$ is depicted in Figure 14. The natural rubber in James & Green experiment contained 40 parts by weight of HAF black per 100 parts of natural rubber. The stress-strain behaviour of natural rubber vulcanizates containing 0-43 percent titanium dioxide filler is depicted in Figure 15. Three parameters are used to fit the data and their values are given in the table below: V_1 and V_2 are volumes of the rubber and of the filler, respectively. The parameter values for $V_1/V_2 = 0.1$ are obtained via visual fitting. For $V_2/V_1 = 0$, E is assumed the value of 1 MPa and the values of α_1 and α_2 are obtained using the least squares method. A strain energy function with only a single physical parameter to describe a rubberlike material has attractive features in the sense that the parameter is eas-

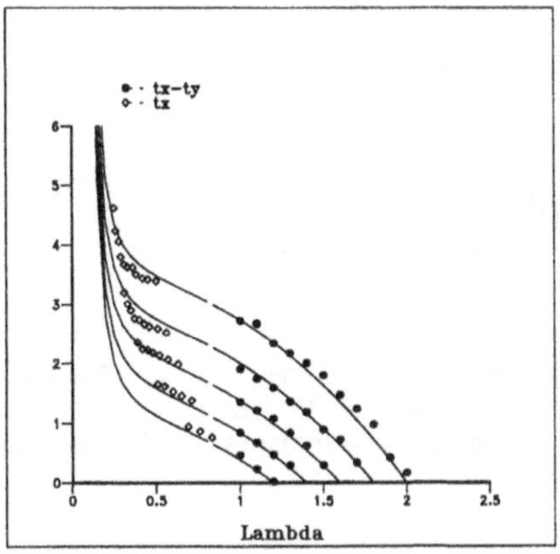

Figure 13. Comparison with data of James & Green (1975). No weighting. $E = 2.6$ MPa, $\alpha_1 = 0.942719, \alpha_2 = 0.335778, \alpha_3 = 0.002$. Biaxial deformation of a rectangular sheet for fixed values of $\lambda_1 = 1.2, 1.4, 1.6, 1.8$ and 2.0. tx$= \sigma_1$, ty$= \sigma_2$; • Lambda $= \lambda_2, \lambda_1 =$ constant; ◇ Lambda $= \lambda_3 = \frac{1}{\lambda_2\lambda_1}, \lambda_1 =$ constant.

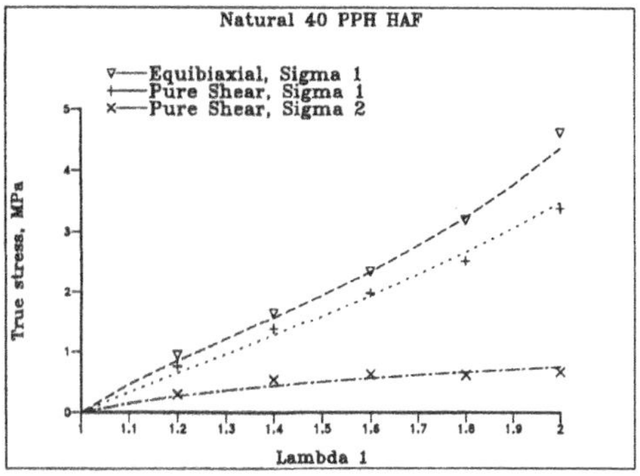

Figure 14. Comparison with the data of James & Green. $E = 2.6$ MPa, $\alpha_1 = 0.942719, \alpha_2 = 0.335778, \alpha_3 = 0.02$: Lambda 1 $= \lambda_1$, Nominal stress $= \frac{\sigma_1}{\lambda_1}$.

Table 1. 4 parameters or three

V_2/V_1	E (MPa)	α_1	α_2
0.0	1.00	0.91647	0.0156354
0.1	1.2	1.7	0.04
0.2	1.48931	1.86513	0.104212
0.3	2.14580	1.50930	0.127694

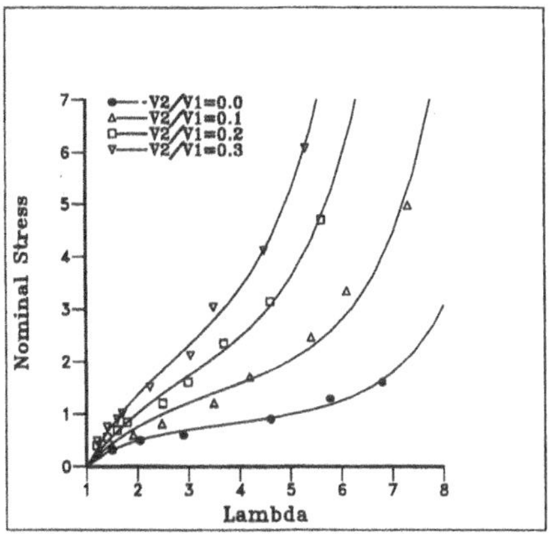

Figure 15. Stress-strain behaviour of elastomers with varying filler contents. Points: Experimental data of Becker & Rademacher (1962). Nominal stress $= \frac{\sigma_1}{\lambda_1}$. Lambda $= \lambda_1$.

ier to obtain experimentally and it is physically related. For the moderate strains the proposed three term strain energy function can be considered as a single constant (the Young's modulus) strain energy function. We do this by fixing the value of $\alpha_1 = 0.9396, \alpha_2 = 0.0435$ for all the different rubbers concerned and use only the parameter E to fit the data. The proposed single constant strain energy function compares well with various different types of experimental data of different filled and unfilled rubbers; an improvement over an earlier work of Shariff(1997) and Shariff & Stalker (1999) on a single constant strain energy function. This is indicated in Figures 16-21. We note that all the fitted parameter values (except for those obtained for the data in Figure 7) given in this paper satisfy the sufficient conditions given in Section III for the expression (22) to be positive. However, it can be easily verified that, for the values obtained to fit the data in Figure 7, $f'(\lambda) > 0$ for $\lambda > 0$. Hence these values are also sufficient to make the expression (22) positive. The predicted curves compare well with all the experiment data shown here.

5 CONCLUSIONS

Linear parameters and single constant strain energy functions are developed. The proposed functions are easily used in curve fitting methods and agree well with published experimental data of filled and unfilled rubberlike materials. The extent of the proposed strain energy function applicability to other rubbers needs to be assessed by comparing it with experimental data of a much wider class of rubberlike materials.

6 ACKNOWLEDGEMENT

The author is grateful to Professor R.W. Ogden for supplying details of the experimental data from Treloar (1944).

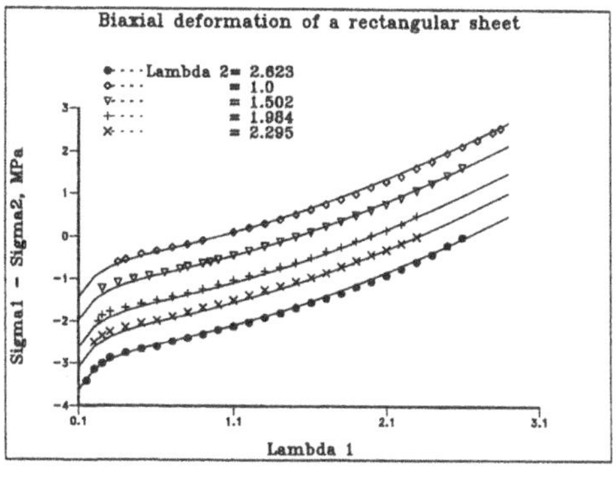

Figure 16. Comparison of theoretical curves with the data of Jones & Treloar (1975):Biaxial deformation of a rectangular sheet. Lambda $= \lambda_1$, Sigma1 $= \sigma_1$, Sigma2 $= \sigma_2$. $\sigma_1 - \sigma_2$ v. λ_1 at fixed values of λ_2:$E =$ 1.4421 MPa

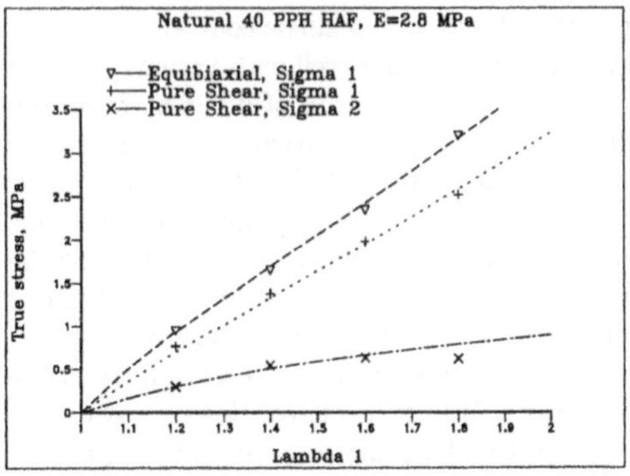

Figure 17. Comparison of predicted curves with the experimental data of James & Green (1975) for natural rubber filled with 40 parts HAF (N330) phr. $E = 2.8$ MPa Sigma 1= σ_1, Sigma 2 = σ_2, Lambda 1 = λ_1.

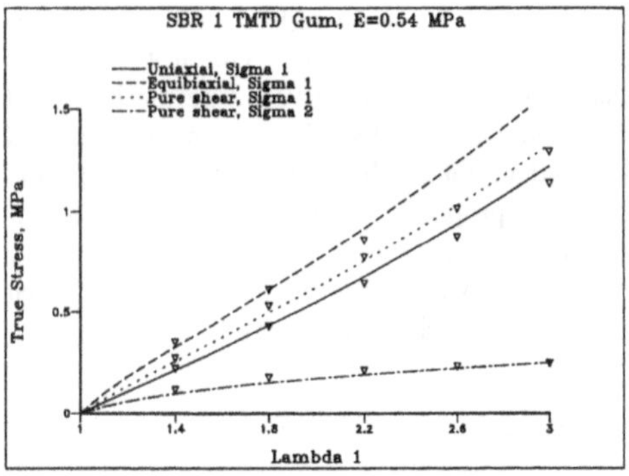

Figure 18. Comparison of predicted values with the experimental data of Tsuge et al. (1978) for SBR1 (gum SBR, TMTD cure). $E = 0.54$ MPa.

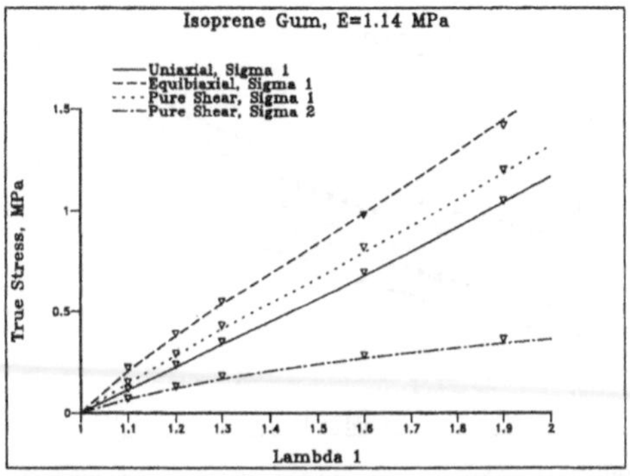

Figure 19. Comparison of predicted values with the experimental data of Kawabata et al. (1981) for isoprene gum. $E = 1.14$ MPa

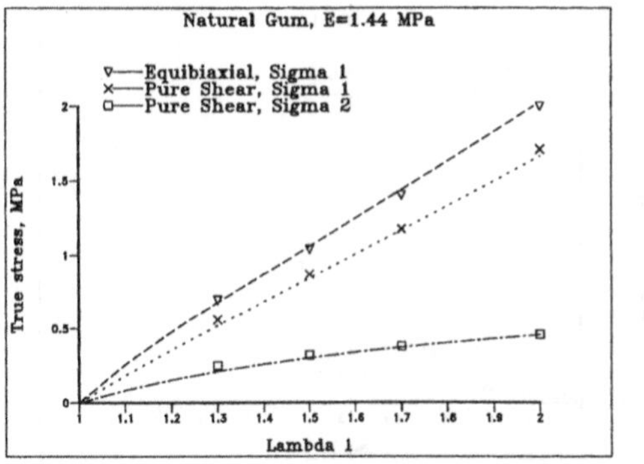

Figure 20. Comparison of predicted values with the experimental data of James et al. (1975). $E = 1.44$ MPa

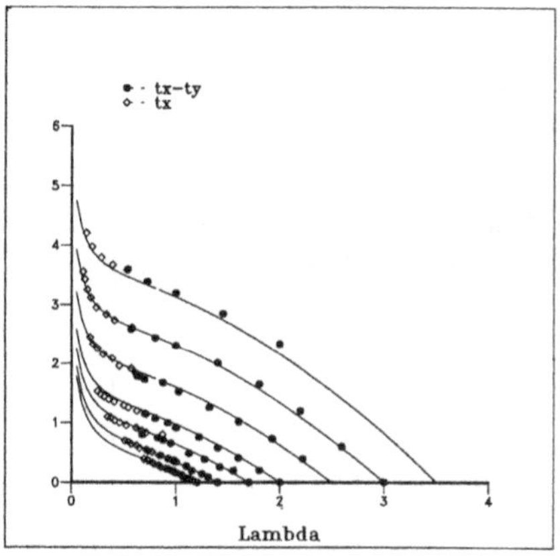

Figure 21. Comparison of the theoretical curves with the data of Kawabata & Kawai (1977). No weighting:E = 1.16 MPa. Biaxial deformation of a rectangular sheet for fixed values of $\lambda_1 = 1.2, 1.4, 1.7, 2.0, 2.5, 3.0, 3.5$. tx= σ_1, ty= σ_2; • Lambda = λ_2, λ_1 = constant; ⋄ Lambda = $\lambda_3 = \frac{1}{\lambda_2\lambda_1}, \lambda_1$ = constant.

REFERENCES

Becker, G.W. & H.J. Rademacher 1962. Mechanical behaviour of high polymers under deformations of different time function, type and magnitude. *J. Appl. Polym. Sci.*. 58:621-631.

Chadwick, P., Creasy, C.F.M. & V.G. Hart 1977. The deformation of rubber cylinders and tubes by rotation. *J. Aust. Math. Soc.*. Ser B 20:62-96.

Hill, R. 1970. Constitutive inequalities for isotropic elastic solids under finite strain. *Proc. R. Soc. Lond.*. A 314:457-472.

James, A.G., Green, A. & G.M. Simpson 1975. Strain

energy functions of rubber. I. Characterisation of gum vulcanizates. *J. Appl. Polym. Sci.*. 19:2033-2058.

James, A.G. & A. Green. 1975. Strain energy functions of rubber. II. Characterisation of filled vulanizates *J. Appl. Polym. Sci.*. 19:2319-2330.

Johannknecht, R., Jerrams, S. & G. Clauss 1999. The uncertainty of implemented curve-fitting procedures in finite element software. In: D. Boast & V. Coveney (eds).*Finite Element Analysis of Elastomers*:141-151. London: Professional Engineering Publishing.

Jones, D.F. & L.R.G. Treloar 1975. The properties of rubber in pure homogeneous strain. *J. Phys. D.*. 8:1285-1304.

Kawabata, S. & H. Kawai 1977. Strain energy density function of rubber vulcanizates from biaxial extensions. *Adv. Polym. Sci.*. 24:89-124.

Kawabata, S., Matsuda, M., Tei, K. & H. Kawai 1981. Experimental survey of the strain energy density function of isoprene rubber vulcanisate.*Macromol.*. 14:154-162.

Ogden, R.W. 1972. Large deformation isotropic elasticity-on the correlation of theory and experiment for incompressible rubberlike solids. *Proc. R. Soc. London Ser. A*. 326:565-584.

Ogden, R.W. 1984. *Non-linear Elastic Deformations*. England: Ellis Horwood.

Ogden, R.W. 1986. Recent advances in the phenomenological theory of rubber elasticity. *Rubber Chemistry and Technology*. 59(3):361-383.

Shariff, M.H.B.M. 1997. Single constant strain energy function for rubberlike materials.*Plast. Rubber Process Appl.*. 26:285-290.

Shariff, M.H.B.M. & I.D. Stalker 1999. A single constant strain energy function for elastomers and its finite element implementation. In: D. Boast & V. Coveney (eds).*Finite Element Analysis of Elastomers*:41-60. London: Professional Engineering Publishing.

Tobisch, K. 1981. A three-parameter strain energy density function for filled and unfilled elastomers.*Rubber Chem. Technol.*. 54:930-939.

Treloar, L.R.G. 1944. Stress-strain data for vulcanised rubber under various types of deformation. *Trans. Faraday Soc.*. 40:59-70.

Truesdell, C.A. 1956. Das ungelöste hauptproblem der endlichen elastizitätstheorie. *Z. angew. Math. Mech.*. 36:97-103.

Tsuge, K., Arenz, R.J., Landel, S.J. & R.F. Landel 1978. Finite deformation behaviour of elastomers: dependence of strain energy density on degree of cross-linking for SBR .*Rubb. Chem Technol.*. 51:948-958.

Twizell, E.H. & R.W. Ogden 1983. Non-linear optimisation of the material constants in Ogden's stress-deformation function for incompressible isotropic elastic materials. *J. Aust. Math. Soc. Ser. B* 24:424-434.

Valanis K.C. & R.F. Landel 1967. The strain-energy function of a hyperelastic material in terms of the extension ratios. *J. Appl. Phys.*. 38:2997- 3002.

Experimental techniques

Constitutive Models for Rubber, Dorfmann & Muhr (eds) © *1999 Taylor & Francis ISBN 90 5809 113 9*

Application of flexible biaxial testing in the development of constitutive models for elastomers

C.P.Buckley
Department of Engineering Science, University of Oxford, UK

D.M.Turner
Consultant, UK

ABSTRACT: In the development of constitutive models for elastomers, there is a need for experimental means of assessing multiaxial performance of models and of validating their implementations in numerical simulations of inhomogeneous deformation. This paper proposes the use of automated biaxial tensile testing for both purposes. The approach is illustrated with experiments on latex rubber film, carried out on the Flexible Biaxial Film Tester developed at Oxford. Biaxial tests on perforated specimens, with video recording of deformation patterns, were used to assess the modelling of inhomogeneous strain. Results are used to assess the performance of a particular three-parameter constitutive model proposed recently: it is found that homogemeous and inhomogeneous biaxial response can be predicted on the basis of uniaxial test results.

1 INTRODUCTION

Recent years have seen major steps forward in the ability of Finite Element (FE) models to handle the complex patterns of material behaviour exhibited by polymers. In the case of rubbery polymers, this involves large deformations and nonlinear hyperelasticity, frequently with time-dependence superimposed. There have been several suggestions for constitutive models to describe the behaviour. The urgent need now is experimental work to establish objectively the relative merits of the various models and their FE implementations.

Further progress requires advances on two fronts. FE analysis of actual components requires knowledge of the three-dimensional constitutive response of the elastomers of interest. It is well-known that uniaxial tests alone, carried out with conventional screw-driven or servo-hydraulic testing machines, are insufficient for this purpose. Multiaxial testing is required; however this can be experimentally difficult and there is currently no suitable equipment generally available. There have been a few reports of biaxial mechanical testing of elastomers in the past. However, the literature is sparse, and a consensus has not yet emerged on which model provides the best description of multiaxial response, or on how parameters appearing in constitutive models can be obtained most conveniently in practice.

The second need is to establish benchmark multiaxial tests, with well-defined geometry and boundary conditions, for testing the accuracy of numerical continuum models incorporating elastomer constitutive models. Finally, since these materials exhibit time-dependence in their constitutive response, the multiaxial tests need to encompass a range of strain histories, encompassing those encountered in applications.

In this paper, we suggest how these needs may be met by automated biaxial straining tests on rubber sheets, combined with video imaging of the specimen. We illustrate the approach with results obtained with the Flexible Biaxial Film Tester (FBFT), developed at Oxford for biaxial testing of polymers in thin sheet form, above the glass transition. Motion on two perpendicular axes and temperature are under computer software control, providing a high degree of flexibility in choice of experimental sequence.

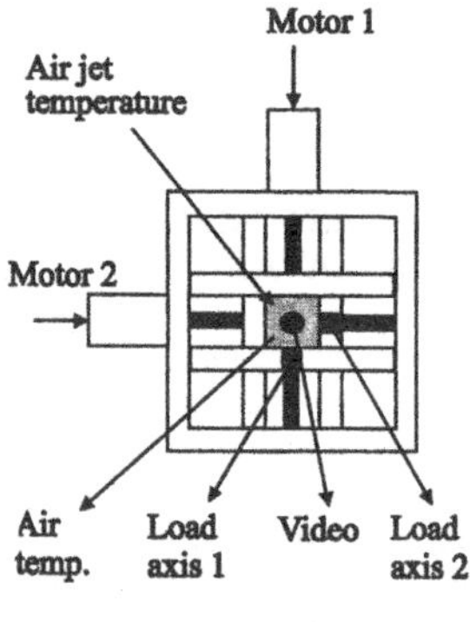

Figure 1. Schematic diagram of the Flexible Biaxial Film Tester stretching frame.

2 BIAXIAL TESTING MACHINE

The FBFT has been described elswhere (Gerlach *et al*). A schematic diagram is given below in Figure 1. The essential feature is a horizontal biaxial stretching frame, driven by two reverse-thread leadscrews. The specimen is square and supported in 24 miniature pneumatic piston grips. Each leadscrew is powered by a brushless DC servo motor providing high torque and low intertia, thereby making possible a wide range of strain-rates extending up to 40/s. The motor drive units are interfaced to a PC, from which the required motion profile is down-loaded at the start of each experiment.
If required, the specimen may also be subjected to blown air at elevated temperature, also set by the PC.

One of the centre grips on each axis contains a miniature load cell, recording tension in the specimen, and thermocouple junctions are located close to the specimen surface. Signals from load-cells and thermcoouples are captured by the PC, via a Biodata Microlink interface with IEEE-488 data bus.

For monitoring deformation in the specimen, it is marked with an ink grid and observed during the test by a video camera mounted above it. Image analysis software is then employed to obtain quantitative measurements of the deformation of the specimen during the test.

This set-up provides considerable flexibility in choice of experimental sequence of biaxial extension and temperature, under full computer control without intervention by the operator.

The FBFT has already proved a powerful tool for studying the constitutive response of amorphous thermoplastic polyester films, during biaxial stetching above the glass transition (Buckley *et al* 1996, Gerlach *et al* 1998). In the present work, we have extended its application to the biaxial stretching of elastomeric films at room temperature, with homogeneous and inhomogenous deformation. The only modification required was the use of a vacuum specimen holder, when mounting square specimens in the grips, as otherwise the low stiffness of the films caused excessive sag.

3 BIAXIAL TESTING OF ELASTOMERS – HOMOGENEOUS DEFORMATION

The FBFT provides a convenient means of measuring the constitutive response of elastomers during biaxial straining with various strain ratios. Such information is widely recognised to be essential in the development of multiaxial constitutive models, as it cannot be assumed *a priori* that a uniaxial test alone will be adequate for assessing the multiaxial performance of a model, or for obtaining model parameters. In the present work we have conducted biaxial tests on several elastomers in sheet form.

Example results are shown below for a vulcanised latex film of thickness 150 μm, subjected to an extension rate of 0.01/s on axis 1, but different extension rates on axis 2. The specimen temperature was 25±2°C. Here, we illustrate the usefulness of such data by showing their application in assessing a particular constitutive model proposed by one of the authors – see Turner *et al* 1997. This is a three parameter model (the TBf model) based on the Turner-Brennan one parameter model (successful at medium stretches up to *ca* 2 – see Turner & Brennan 1990) but with the addition of a further term in the predicted stress, to account for strain-stiffening at higher stretches. The latter term captures the inextensibility of the network chains, by representing them as nonlinear elastic filaments, and involves two additional parameters. For plane stress problems where principal stretches are λ_i ($\lambda_1, \lambda_2 \geq \lambda_3$), the equations for in-plane principal true stresses according to this model are as follows:

$$\sigma_1 = \frac{E}{(1-\nu^2)}[\lambda_1 - 1 + \nu(\lambda_2 - 1)] + \tilde{\sigma}_1 \tag{1}$$

$$\sigma_2 = \frac{E}{(1-\nu^2)}[\lambda_2 - 1 + \nu(\lambda_1 - 1)] + \tilde{\sigma}_2 \tag{2}$$

where ν is a strain-dependent Poisson's ratio:

$$\nu = \frac{(1-\lambda_3)}{(\lambda_1 + \lambda_2 - \lambda_3 - 1)} \tag{3}$$

Extra stresses $\tilde{\sigma}_1$, $\tilde{\sigma}_2$ are the resolved components of stress due to stretching of the non-linear elastic filaments, initially inclined equally to the three principal directions, described by

$$\tilde{\sigma}_i = \lambda_i \tilde{E} \frac{\tilde{e}}{1 - \dfrac{\tilde{e}}{\tilde{e}_{max}}} \cos\alpha_i \quad (i=1,2) \tag{4}$$

where $\tilde{e}$ and α_i are the extension of the filaments and their angles of inclination to the principal axes respectively

$$\tilde{e} = \sqrt{\tfrac{1}{3} I_1} - 1; \quad \cos\alpha_i = \frac{\lambda_i}{\sqrt{I_1}} \tag{5}$$

where I_1 is the first invariant of the stretch

$$I_1 = \sum_{i=1}^{3} \lambda_i^2 . \tag{6}$$

Figure 2 shows uniaxial stress-strain data for the latex film, fitted to the TBf model. As expected, the three adjustable parameters of the model allow an excellent fit to be achieved.
Tests under *biaxial* stress allowed the performance of the model to be tested. Thus, Figure 3 shows measurements of true stress on axes 1 and 2 during straining on axis 1, while axis 2 was constrained to zero strain (i.e. constant width straining). The full lines show the predictions of the TBf model, calculated using the parameters obtained independently from the fit shown in Figure 2. The fit is good: deviations are comparable with scatter in the data.

The model was further tested by carrying out tests with equal-biaxial deformation on the same material. A typical result is shown in Figure 4, together with the calculated stress as obtained from the constitutive model with the same set of parameters obtained from Figure 2. For this particular data set the fit is less good, with significant deviation from experiment at nominal strain beyond 2 – the model overestimating the degree of strain stiffening.

Experiments such as these clearly have the potential to discriminate between competing constitutive models, in their ability to simulate multiaxial response.

4 BIAXIAL TESTING OF ELASTOMERS – THE SHEET-WITH-A-HOLE EXPERIMENT

4.1 *Introduction*

The practical application of constitutive models involves their solution within a continuum model of the component of interest under prescribed loads. Often this is a Finite Element model. The need then is to validate experimentally the numerical implementation: specifically its ability to predict correctly the inhomogeneous patterns of deformation. For this purpose we have again found biaxial testing to provide a useful solution.

The test consists of subjecting a square specimen, in the form of a sheet with initial thickness t and width w, and with a central circular hole of initial diameter d, to biaxial in-plane stress, such that the principal stretches far from the hole are λ_1 and λ_2 ($w >> d >> t$). We may characterise such an experiment in terms of its far-field biaxiality ratio

$$\phi = \frac{\lambda_2 - 1}{\lambda_1 - 1} . \tag{7}$$

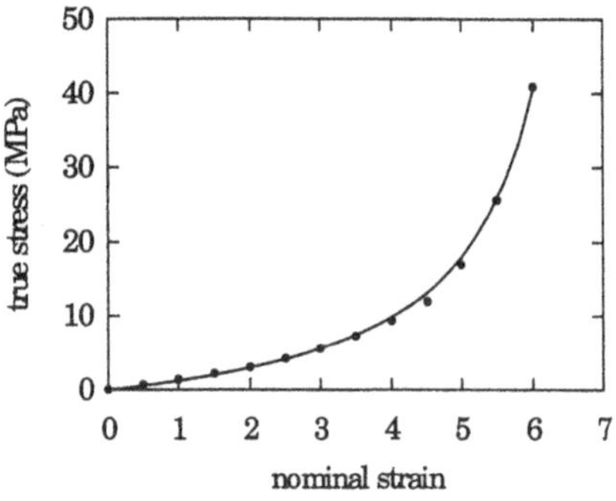

Figure 2. True stress versus nominal strain, for uniaxial straining of latex rubber film at room temperature (points), and DTf constitutive model fitted to the data (full line).

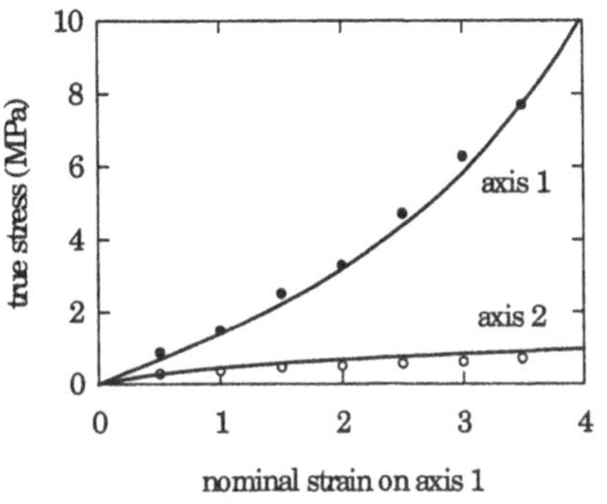

Figure 3. True stress on axes 1 and 2, during constant – width straining of latex rubber film along axis 1: experimental data (points) and predictions of the TBf constitutive model (lines), with parameters obtained from the fit in Figure 2.

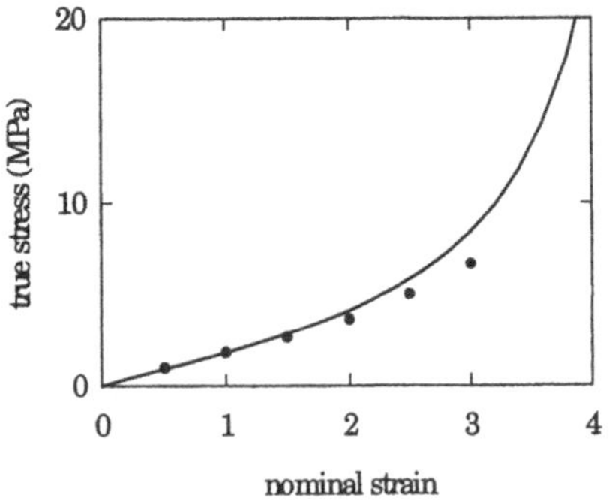

Figure 4. True stress versus nominal strain, during equal biaxial straining of latex rubber film: experimental data (points) and predictions of the TBf constitutive model, with parameters obtained from the fit in Figure 2.

This is an example of the classical solid mechanics problem of the centrally-perforated sheet under biaxial stress. It has several advantages for the present purpose.

(i) It involves a specimen of simple and reproducible shape; readily obtainable for elastomers, by casting or moulding of a thin sheet, followed by careful cutting to size, and cutting of the hole.

(ii) A single specimen is exposed to a range of stress states within one test.

(iii) The experiment has two planes of mirror symmetry, so the continuum model need model only one quarter.

(iv) The test has a well-known analytical solution in the limit of small deformation, isotropic, linear elasticity.

The special case of equal biaxial straining ($\phi = 1$), corresponds to the test employed by Rivlin and Thomas (1950), in early investigations of the Mooney-Rivlin constitutive model. Now, however, with computer-controlled testing, and convenient means of observing the specimen, it is possible to encompass a range of biaxialities and to have the entire deformation field measurable, for making the comparison with simulations.

For example, the deformation of the hole may be considered. Under load, it changes size and (if $\phi \neq 1$) elongates to a diametral width d_1 on axis 1, and d_2 on axis 2. These may be expressed as diametral stretches of the hole:

$$\lambda_1^h = \frac{d_1}{d}; \quad \lambda_2^h = \frac{d_2}{d} \tag{8}$$

For an effectively incompressible material such as an elastomer, the theory of linear elasticity predicts deformation of the hole to be given by

$$\lambda_1^h - 1 = (\lambda_1 - 1)\left(\frac{10}{3} + \frac{2\phi}{3}\right) \tag{9}$$

$$\lambda_2^h - 1 = (\lambda_2 - 1)\left(\frac{10}{3} + \frac{2}{3\phi}\right). \tag{10}$$

These equations demonstrate the well-known strain-concentrating effect of the hole, for most far-field strain states.

Another useful result from the linear theory is the condition for ensuring in-plane stresses remain positive, thereby avoiding the danger of buckling in the thin sheet specimen ($w >> t$). Around the hole, the smallest stress is the tangential tensile stress normal to the direction of maximum stretch. According to the linear theory, this will remain positive provided $\phi \geq -0.2$. Clearly, this implies the need for far-field biaxial stress, since a uniaxial stress alone would cause $\phi = -0.5$ at small strains.

4.2 *Implementation and demonstration*

In the present work, specimens of the same elastomers as mentioned above were perforated with a central circular hole, of diameter 7 mm, using a sharp punch, after first marking them with an ink grid. They were then extended on two axes with the FBFT. The video camera, mounted above the specimen, was used to record evolution of the pattern of deformation.

As expected, the hole caused localisation of the deformation. Moreover, at small deformations, the rate of growth of the hole was proportional to that of the far-field stretch, as predicted by the linear theory (equations (9) and (10)), with the predicted gradient. Thus, in the case of equal biaxial stretch of the latex sheets ($\phi = 1$) this gradient was 4 – see the dashed line in Figure 5.

For the axisymmetric case shown in Figure 5, the relation between stretch of the hole λ^h and the far-field stretch λ was also calculated numerically using a finite difference scheme, for the TBf constitutive model. Expansion of the hole diameter was imposed, and the required far-field stretch was computed by invoking equilibrium and compatability within many (typically 50) concentric rings. The results of this calculation for the latex sheet are included in Figure 5 (full line), where it can be seen to provide an excellent description of the hole growth.

The pattern of deviation from the linear prediction is worthy of note. Both the data and the TBf model

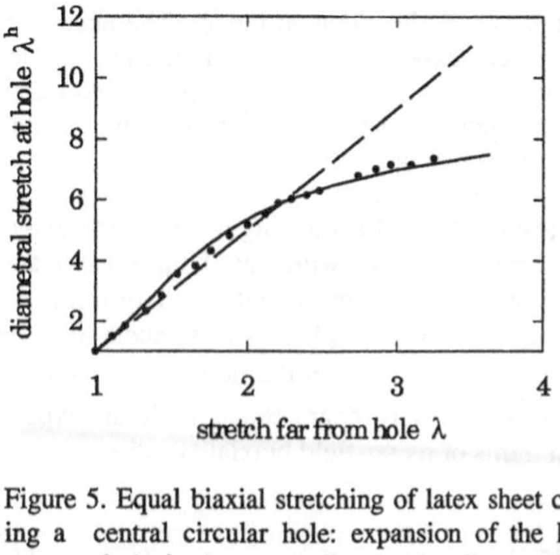

Figure 5. Equal biaxial stretching of latex sheet containing a central circular hole: expansion of the hole as measured (points), as predicted by linear elasticity (dashed line) and as predicted by the TBf constitutive model (full line).

simulation show initially a small deviation above the dashed line, but later a deviation below it. We have found this to be the general pattern for the elastomers studied (latex and polyurethane), and for thermoplastic films above the glass transition – see Buckley *et al* (1999). It reflects the shape of the uniaxial *nominal* stress versus nominal strain curve. The periphery of the hole is subject to uniaxial extension, at higher stretches than elswhere. Minimisation of the overall strain energy of the specimen therefore requires it to extend more rapidly than the linear prediction when on the strain-softening segment of the stress-strain curve (a small effect in elastomers), and more slowly when strain-stiffening.

5. CONCLUSIONS

Biaxial tensile testing, with video imaging of the specimen and the flexibility that computer-control provides, is a convenient means of assessing constitutive models for elastomers, and their application in continuum models. The stress response to homogeneous biaxial stretching allows the biaxial performance of constitutive models to be studied. In particular, it provides a means of answering the practically important question: does a model fitted to uniaxial tensile stress-strain data (relatively easily obtained) provide a good description in other stress states? For assessing the accuracy of implementations of constitutive models, the biaxial stretch of a sheet with a circular perforation provides a useful validation test.

The present work, exploiting the Oxford FBFT testing machine, has demonstrated the approach with latex films. In particular the Turner-Brennan constitutive model, with the addition of nonlinear elastic filament stresses, has been found to give reasonable predictions of biaxial response, using parameters obtained from uniaxial data. Moreover, this model gave an excellent prediction of the growth of the hole during biaxial straining of a perforated sheet.

ACKNOWLEDGEMENTS

The authors are grateful to Ms A.M.Adams for carrying out the experiments, and to the London International Group for financial support.

REFERENCES

Buckley, C.P., Jones, D.C. & Jones, D.P. 1996. Hot-drawing of poly(ethylene terephthalate) under biaxial stress: application of a three-dimensional constitutive model. *Polymer* 37: 2403-2414.

Buckley, C.P., Shirodkar, J., Dooling, P.J. & Harvie, J.L. 1999. A validation test for finite element models of polymer film drawing processes. *15th Annual Meeting of the Polymer Processing Society, 's Hertogenbosch, 1-4 June.*

Gerlach,C., Buckley, C.P. & Jones, D.P. 1998. Development of an integrated approach to the modelling of polymer film orientation processes. *Trans. I. Chem. E.* 76 Part A: 38-44.

Rivlin, R.S. & Thomas, A.G. 1950. Large elastic deformations of isotropic materials VIII Strain distribution around a hole in a sheet. *Phil. Trans. Roy. Soc.* A, 243: 289-298.

Turner, D.M. & Brennan, M. 1990. The multiaxial behaviour of rubber. *Plastics and Rubber Processing and Applications* 14: 183-188.

Turner, D.M., Boast, D. & Jarosz, R. 1997. Definitions and calculations of stress and strain in rubber. *Conf. on Finite Element Analysis of Elastomers, I.Mech.E, London, 15 October.*

Constitutive Models for Rubber, Dorfmann & Muhr (eds) © 1999 Taylor & Francis ISBN 90 5809 113 9

Bi-axial experimental techniques highlighting the limitations of a strain-energy description of rubber

H.R.Ahmadi, J.Gough & A.H.Muhr
Tun Abdul Razak Research Centre, MRPRA, Brickendonbury, UK

A.G.Thomas
Queen Mary and Westfield College, London, UK

ABSTRACT: Experimental techniques for applying simultaneous torsion and tension of a rubber cylinder, and for simultaneous simple shear in orthogonal directions of a double shear testpiece, are described and theoretical frameworks for interpreting the results are given. It is shown that the stresses associated with a small torsion superposed on a large tension cannot be modelled for filled rubber using any hyperelastic model. It is suggested that this phenomenon is related to that of energy dissipation in a strain cycle. The literature on biaxial shear of rubber shows that for both unfilled butyl and filled natural rubber energy dissipation occurs if the deformation is held at a constant magnitude but rotated through the material. It is concluded that it would be most natural to relate the dissipative stress contributions incrementally to changes in the current strain configuration.

1 INTRODUCTION

Since Rivlins development of a phenomenological theory of rubber elasticity (Rivlin, 1948) it has been conventional to model the force-deformation behaviour of rubber in terms of a strain energy function. Many different forms of the strain energy function have been proposed and are reviewed by Treloar (1975); among the most popular are the neo-Hookean, Mooney-Rivlin and Ogden forms which are often used as material models in finite element analysis. It is implicit in all such functions that rubber behaves as a perfectly elastic material although it is well known that filled rubbers especially exhibit significant departures from elasticity such as hysteresis, stress-softening and permanent set.

This paper describes two biaxial experiments which highlight limitations in the use of strain energy functions. In the first, combined torsion and uniaxial extension of a cylinder was used to demonstrate the deviations in the behaviour of a filled rubber from the predictions of large strain elasticity theory.

The second technique involves simple shear applied simultaneously in two orthogonal directions. This configuration is realised in the application of laminated rubber bearings for earthquake isolation (Dusi et al., 1999). In addition, because kinematic departures from linearity are insignificant in simple shear to infinite strains, the results are amenable to discussion in simplified, linearised, terms.

2. COMBINED TORSION AND EXTENSION OF A CYLINDER

2.1 *Theory*

Rivlin (1949; with Saunders, 1951) has shown that a particular relationship between the applied normal force and the torsional couple at small angles of twist will hold for any material whose behaviour is describable by a strain energy function W (I_1, I_2), regardless of the form of W. I_1 and I_2 are the strain invariants.

Consider an unstrained solid rubber cylinder of radius a and height ℓ_o (Figure 1). The cylinder is stretched to a length $\lambda\ell_o$, and the top surface is rotated by an angle $\theta = \psi\ell_o$ relative to the bottom surface.

The forces required to maintain this state of strain are a normal force, N, and a couple, M given by (Rivlin, 1949):-

$$M = 4\pi\psi\int_o^a r^3\left(\frac{\partial W}{\partial I_1} + \frac{1}{\lambda}\frac{\partial W}{\partial I_2}\right)dr \tag{1}$$

$$N = 4\pi\left(\lambda - \frac{1}{\lambda^2}\right)\int_o^a r\left(\frac{\partial W}{\partial I_1} + \frac{1}{\lambda}\frac{\partial W}{\partial I_2}\right)dr - 2\pi\psi^2\int_o^a r^3\left(\frac{\partial W}{\partial I_1} + \frac{2}{\lambda}\frac{\partial W}{\partial I_2}\right)dr \tag{2}$$

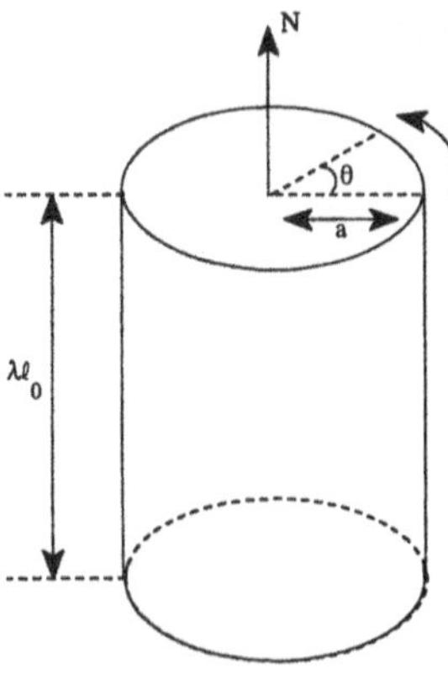

Figure 1: Cylinder subjected to combined torsion and uniaxial extension

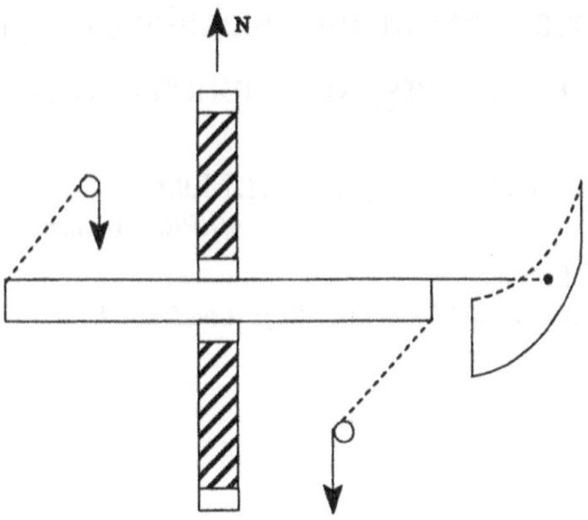

Figure 2: Schematic diagram of experimental set-up for combined torsion and extension of a cylinder

where the strain invariants are given by:-

$$I_1 = \lambda^2 + \frac{2}{\lambda} + \lambda\psi^2 r^2 \text{ and } I_2 = 2\lambda + \frac{1}{\lambda^2} + \psi^2 r^2 \qquad (3)$$

As $\psi \to 0$, terms in ψ^2 become negligible the strain invariants reduce to $I_1 = \lambda^2 + 2/\lambda$ and $I_2 = 2\lambda + 1/\lambda^2$, and are therefore independent of r. Thus, in equations (1) and (2), $\partial W/\partial I_1$ and $\partial W/\partial I_2$ may be treated as constants for integration with respect to r. Also, the second term in equation (2) vanishes. Integration of equations (1) and (2) yields:-

$$M_{\psi=0} = \pi\psi a^4\left(\frac{\partial W}{\partial I_1} + \frac{1}{\lambda}\frac{\partial W}{\partial I_2}\right) \qquad (4)$$

$$N_{\psi=0} = 2\pi a^2\left(\lambda - \frac{1}{\lambda^2}\right)\left(\frac{\partial W}{\partial I_1} + \frac{1}{\lambda}\frac{\partial W}{\partial I_2}\right) \qquad (5)$$

hence in the limit $\psi \to 0$:-

$$\frac{N}{2(\lambda - \lambda^{-2})} = \frac{M}{\psi a^2} \qquad (6)$$

Equation (6) is independent of the form of W. Thus a comparison of experiment with the predictions of equation (6) provides an indication of whether the material may be described by a strain energy function.

2.2 *Materials*

The measurements were carried out on compression moulded cylinders of two natural rubber formulations. One was essentially unfilled, the other contained 45 parts of carbon black filler. They were designated EDS19 and EDS16 respectively. The formulations are given in Table 1.

The rubbers were vulcanized to maximum rheometer torque at 140°C.

2.3 *Method*

Threaded end-pieces were bonded to the rubber cylinders during moulding to facilitate clamping. The length of each cylinder was 128mm (nominal) and its diameter was 25mm (nominal) A pair of identical cylinders were bolted to a 10kN Zwick screw-driven test machine with an aluminium torsion bar fixed between the cylinders (Figure 2). Small torsional couples were applied to the cylinders by means of weights hung from threads attached to the ends of the torsion bar and passing over pulleys. The angle of twist was calculated from the displacement of the torsion bar as measured from a suitably fixed scale chart. After each extension of the cylinders of the filled rubber, two minutes elapsed before measurements were taken in order to minimise the scatter in the measurement due to stress relaxation. As

Table 1. Rubber formulations

	unfilled	filled
natural rubber (SMRCV60)	100	100
carbon black HAF (ASTM N-330)	1	45
process oil[a]	-	4.5
zinc oxide	5	5
stearic acid	2	2
antioxidant[b]	3	3
antiozonant wax	2	2
accelerator, CBS[c]	0.6	0.6
sulphur	2.5	2.5

a low viscosity naphthenic oil
b N-(1,3-dimethylbutyl)-N'-phenyl-phenylene-diamine
c N-cyclohexylbenzothiazole-2-sulphenamide

a negligible amount of stress relaxation would occur in the unfilled rubber over the time scale of the measurements, readings were taken immediately after applying the normal load.

2.4 *Results*

For small applied couples plots of M against ψ were linear; thus it was possible to obtain a constant value for M/ψ at small ψ for each extension of the cylinders.

In order to test the validity of equation (6) $N/(2\pi a^2(\lambda - \lambda^{-2}))$ and $M/\psi\pi a^4$ were each plotted against the extension ratio, λ, in Figure 3. The ratio of these quantities has been plotted in Figure 4.

2.5 *Discussion*

For the unfilled rubber the experimental data agree adequately with the theoretical prediction from equation (6) thus supporting the use of a strain energy function as a way of describing the behaviour of rubber. However, the behaviour of the black-filled rubber shows a clear deviation from the theoretical predictions, with the torsional couple larger than the theory predicts from the stiffness in tension. This behaviour is analogous to the behaviour often observed during dynamic tests for small cycles superimposed upon larger strains where the small strain stiffness is larger than that obtained from the underlying large strain (Gregory (1984, 1985)). In the case of the present experiment, the torsional stiffness could be interpreted as the small strain stiffness, superimposed upon a larger tensile strain, for which the stiffness is lower.

Figure 3 shows that for the unfilled rubber the stiffness in both tension and torsion falls slightly with increasing extension. The stiffness of the filled rubber, however, falls steeply and then rises more slowly. For the normal force term, where the extension was rising, this is the shape expected from the stress-strain behaviour in uniaxial tension. However, it is less obvious what the effect of a tensile strain on the small angle torsional stiffness would be, and it is interesting to see that the shape of the curve is similar, leading to a roughly constant ratio at moderate strains, shown more clearly in Figure 4.

For the filled rubber both the normal stiffness and the torsional stiffness decrease very rapidly as λ increases for small values of λ. It can be seen from Figure 4 however that the change in the normal stiffness is more pronounced, thus the behaviour of the filled rubber approaches the theoretical predictions close to $\lambda = 1$. This is to be expected in view of the fact that at $\lambda = 1$, both measurements are small strain measurements. It supports the hypothesis that the disagreement with theory is a consequence of the high stiffness for small deformations, which is observed to occur whenever there is a change in the direction of straining, regardless of whether the overall state of strain is large or small.

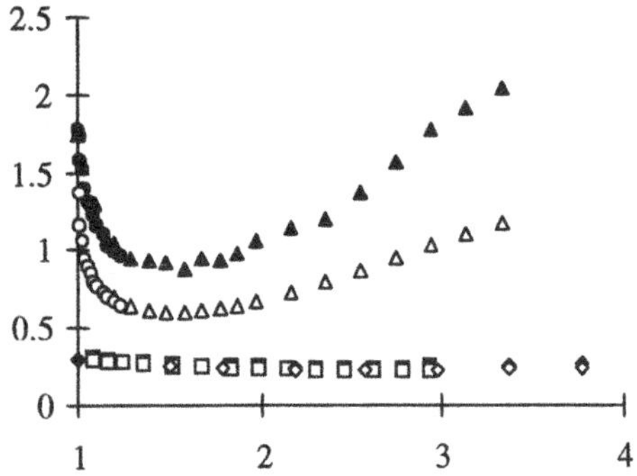

Figure 3: Comparison of the uniaxial and torsional "moduli" of cylinders in combined torsion and extension. On ordinate: black symbols, $M/\psi\pi a^4$; white symbols $N/(2\pi a^2 \ (\lambda - \lambda^{-2}))$. Squares and diamonds unfilled rubber; circles and triangles, filled rubber. (Two experiments in each case).

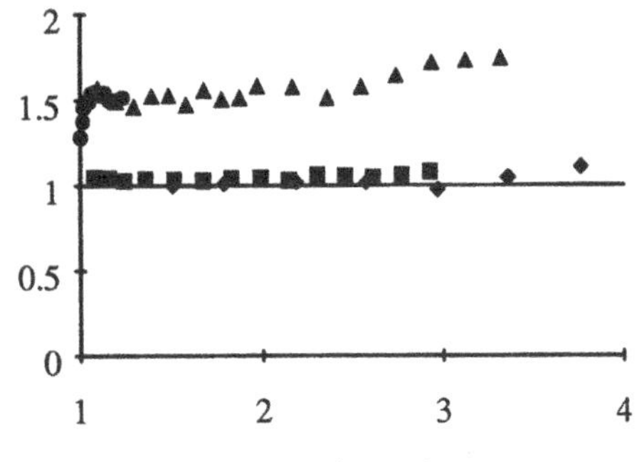

Figure 4: Comparison of experiment with the predictions of equation (6) for the combined torsion and extension of a cylinder. On ordinate: $2M(\lambda - \lambda^{-2})/N\psi a^2$. ■, ◆ unfilled rubber, ▲, ● filled rubber (two experiments in each case), — theoretical (equation 6).

These results are a clear indication of the inadequacy of the concept of an elastic strain energy function for describing the stress-strain behaviour of filled rubbers.

3. SIMPLE SHEAR IN ORTHOGONAL AXES

Gent [1960] devised a test machine which imposes a homogeneous deformation (simple shear) on a rotating double shear testpiece enabling the evaluation of the dynamic properties of the elastomer being tested. The apparatus is shown diagrammatically in Figure 5. The testpiece is a double-bonded shear consisting of two rubber cylinders R. The outer metal end pieces are secured in double ball races so that the testpiece may rotate freely about its axis. The central metal piece is surrounded by a ball race in an outer casing to which weights are applied, thus producing a vertical movement at right angles to the axis of rotation. The two rubber cylinders are thus subjected to a simple shear deformation, the direction of which is continuously rotating. The central ball race assembly moves spontaneously out of alignment, by a horizontal displacement perpendicular to the axis of rotation. This displacement is caused by stresses associated with energy dissipation in the rubber cylinders. By means of calibrated helical springs S the magnitude of the horizontal force F_s necessary to restore the central assembly to its original position in the horizontal plane could be determined. The quantities therefore measured are F_s, the vertical imposed force (F_c) and the vertical deflection (d) under this force.

It is interesting to examine the amount of the energy dissipated when the testpiece is rotated through one revolution. The energy dissipated is given by the torque $F_s d$ acting over an angle of 2π and hence the energy E dissipated per cycle of rotary shear test is given by:-

$$E = 2\pi F_s d \qquad (7)$$

The non-zero value of F_s observed during the rotary shear experiment shows that energy is dissipated when the direction of the principal stretches are changing even when their magnitudes are held constant.

In an alternative technique, using essentially the same apparatus as Figure 5, the central bearing housing may be rigidly connected to the actuator of a servohydraulic testing machine instead of being free to move horizontally. The forces F_s and F_c may be measured using a biaxial load cell mounted in line with the actuator.

The load cells measuring F_c and F_s are fixed in space but the forces can be resolved into two orthogonal directions X and Y, rotating with angular speed, ω: (Figure 6). Thus:

$$\left.\begin{aligned} F_x &= F_c\cos\theta - F_s\sin\theta \\ F_y &= F_c\sin\theta + F_s\cos\theta \\ x &= d\cos\theta \\ y &= d\sin\theta \\ \text{where } \theta &= \omega t \end{aligned}\right\} \qquad (8)$$

The above equations can therefore be used to relate shear forces in the co-ordinate system rotating with the material, having angle θ to the co-ordinate system fixed in space, providing that F_c, F_s and θ are measured.

In the case that ω is constant, Gent observed that F_c and F_s are constant so that equations (8) may usefully be rewritten as:-

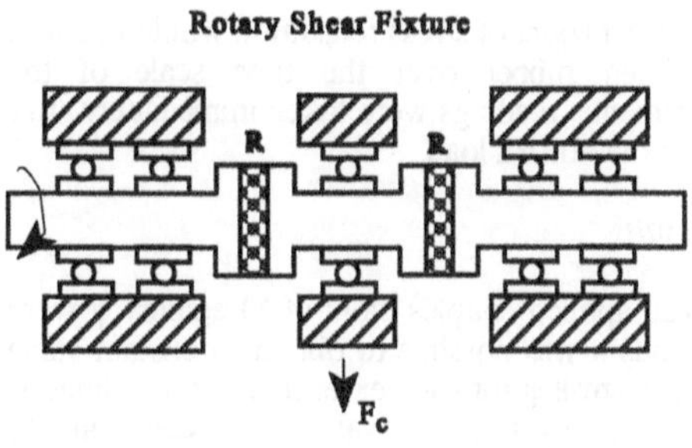

Figure 5: Apparatus for measuring the forces F_c and F_s (normal to the plane of the drawing) of a rotating shear testpiece

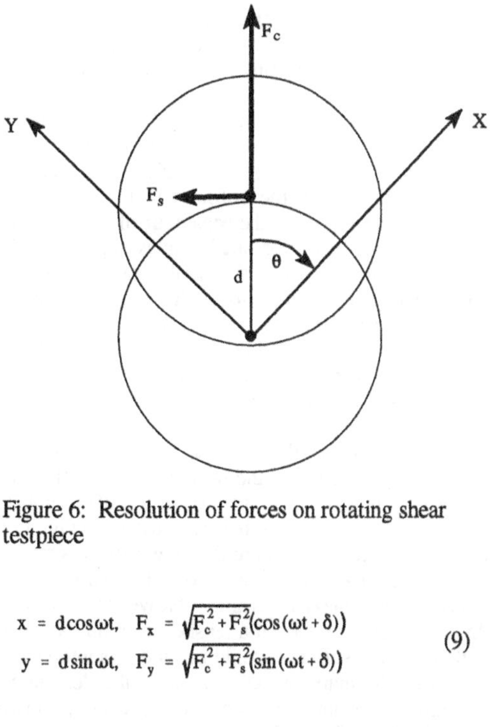

Figure 6: Resolution of forces on rotating shear testpiece

$$\begin{aligned} x &= d\cos\omega t, \quad F_x = \sqrt{F_c^2 + F_s^2}(\cos(\omega t + \delta)) \\ y &= d\sin\omega t, \quad F_y = \sqrt{F_c^2 + F_s^2}(\sin(\omega t + \delta)) \end{aligned} \qquad (9)$$

where $\tan\delta = F_s/F_c$. The energy dissipated by each of F_x and F_y in one revolution is readily shown to be $\pi d\sqrt{F_c^2 + F_s^2}\sin\delta$ so that:-

$$E_x = E_y = \pi\sqrt{F_c^2 + F_s^2}\, d\sin\delta = \pi d F_s \qquad (10)$$

Equation (10) is in agreement with equation (7), since $E = E_x + E_y$.

Working with an unfilled butyl vulcanizate, Gent found that measurements of F_s/F_c were in quite good agreement with measurements of loss factor from the decay of free vibrations. Thus $\tan\delta$ of equation (9) may be identified with the loss factor of the material, and we note that the total energy dissipated in one revolution is twice that of one cycle of a conventional single-axis sinusoidal shear test of the same testpiece with amplitude d.

In principle it would be possible to vary d and ω as functions of t in such a way that the double shear testpiece could be subjected to biaxial shear tests equivalent to those reported by Dusi et al (1999). Equations (8) could be used to calculate F_x and F_y from the measurements of F_c and F_s (as functions of time). One important experimental consideration is that the load cell for measuring F_s must be very stiff, as a lateral deflection would reduce its value and mean that, in Gent's technique, the energy dissipated in one cycle would then be $F_c e + F_s d$ where e is the lateral deflection. Thus the measured value of F_s/F_c would now underestimate the loss factor.

4. LINEAR INTERPRETATION OF BIAXIAL SHEAR

We may consider a rubber spring to be a point which can be moved in space from the origin to a position $\underline{x}$ by application of a force $\underline{F}(\underline{x}, \dot{\underline{x}}, t)$. This simple model is sufficient for discussing most rubber suspension components, for which the effects of torques associated with angular deflections of the spring are of negligible, or at least secondary, consequence. For a linear elastic spring we would have $\underline{F} = \underline{\underline{K}}\underline{x}$ where $\underline{\underline{K}}$ is a generalised stiffness.

For an isotropic two-dimensional spring, such as an earthquake isolation bearing in simple shear, we would simply have:-

$$\underline{F} = k_s \underline{x} \tag{11}$$

Suppose now the spring has hysteresis. If the nature of the hysteresis resembled an isotropic dashpot, we would have:-

$$\underline{F} = k_s \underline{x} + c\dot{\underline{x}} \tag{12}$$

so that for components x, y of $\underline{x}$:

$$F_x = k_s x + c\dot{x}$$
$$F_y = k_s y + c\dot{y}$$

Although $\underline{F}$ is no longer parallel to $\underline{x}$, there is no interaction, or coupling, between the orthogonal components. In a forced sinusoidal displacement, $x = d \sin \omega t$, $y = d \sin(\omega t + \phi)$, the work dissipated by the x-components of force in one cycle is just:-

$$\begin{aligned} E_x &= \oint F_x dx \\ &= \int_0^{2\pi/\omega} (k_s d \sin\omega t + c\omega d \cos\omega t)(d\omega \cos\omega t)\, dt \\ &= c\omega^2 d^2 \int_0^{2\pi/\omega} \cos^2\omega t\, dt \\ &= \pi c \omega d^2 \end{aligned} \tag{13}$$

Because there is no coupling, the energy dissipated in a cycle of equibiaxial shear, for example at 90° phase as in Gent's technique, would be exactly twice that given by equation 13. The result does not depend on the relative phase, since there is no interaction.

Following Dusi et al (1999) we may look in more detail at the time-dependence of the energy dissipation E(t):-

$$E(t) = \int_0^t \underline{F} \cdot d\underline{x} \tag{14}$$

Substituting $x = d \sin \omega t$, $y = d \sin(\omega t + \phi)$ into (14) gives:-

$$E(t) = \omega d \int_0^t \left[F_x \cos\omega t + F_y (\cos\omega t + \phi)\right] dt \tag{15}$$

Substituting from (12) for F_x and F_y leads to:-

$$\begin{aligned} E(t) = \frac{d^2}{4}\big[&k_s(1 - \cos(2\omega t) + \cos 2\phi - \cos 2(\omega t + \phi)) \\ &+ c\omega(4\omega t + \sin(2\omega t) + \sin 2(\omega t + \phi) - \sin 2\phi)\big] \end{aligned} \tag{16}$$

From equation (16) we see that if $\phi = \pi/2$ then $E(t) = cd^2 \omega^2 t$ so that E increases monotonically. Similarly if $\omega t = n\pi$, $E(t) = cd^2 \omega^2 t$, or $2\pi\, c\omega d^2$ per cycle $(2\pi/\omega)$ so that for other values of ϕ the overall rate of energy dissipation is the same but the work done has a periodic character, extra work to that being dissipated being alternatively supplied to and extracted from the rubber. Of course, the latter also occurs for a uniaxial shear test, but the energy dissipated between times separated by $2\pi/\omega$ is then only $\pi\, c\omega d^2$ (cf. equation 13, referring to just one axis).

The features described above are in good semi-quantitative agreement with Figure 7 of the paper by Dusi et al. (1999). However, the energy dissipated for phase angles of 30 and 15° is less than that for a phase angle of 90°.

As an alternative to the viscoelastic model embodied in equation 12, we could envisage an origin for hysteresis resembling an isotropic frictional resistance:

$$\underline{F} = k_s \underline{x} + f\dot{\underline{x}}/|\dot{\underline{x}}| \tag{17}$$

So that $F_x = k_s x + f\dot{x}/\sqrt{\dot{x}^2 + \dot{y}^2}$

$F_y = k_s y + f\dot{y}/\sqrt{\dot{x}^2 + \dot{y}^2}$

There is thus interaction, the presence of the orthogonal component effectively reducing the frictional contribution to the force.

For a uniaxial sinusoidal excitation, $x = d \sin \omega t$ the energy dissipated in one cycle is given by:

$$\begin{aligned} E &= \omega d \int_0^{\frac{2\pi}{\omega}} \left(k_s d \sin \omega t + f \operatorname{sgn}(\dot{x})\right) \cos \omega t \, dt \\ &= 4 \int_0^{\frac{\pi}{2\omega}} f \omega d \cos \omega t \, dt \\ &= 4fd \end{aligned} \tag{18}$$

For $x = d \sin \omega t$, $y = d \sin (\omega t + \phi)$ we find that:

$$\sqrt{\dot{x}^2 + \dot{y}^2} = \omega d \sqrt{1 + \tfrac{1}{2}(\cos 2\omega t + \cos 2(\omega t + \phi))} \tag{19}$$

For the special case that $\phi = \pi/2$, this term reduces to ωd and equations (17) become uncoupled. The response is now formally identical to the viscoelastic model, and making the substitution $f \equiv c\omega d$ we see from the discussion of equation (16) that $E(t) = fd\omega t$ so the total energy dissipated per cycle is $2\pi fd$. This is only $\pi/2$ times that dissipated in a uniaxial cycle to the same displacement, as given by equation (18). In the case that $\phi = 0$, $\dot{x}^2 + \dot{y}^2 = 2d^2\omega^2\cos^2\omega t$ and the equations for F_x and F_y are formally the same as that for uniaxial excitation except that f is reduced by the factor $1/\sqrt{2}$. Adding both contributions to the energy dissipation together, we see that the total energy dissipation is only $\sqrt{2}$ times that for a single axis excitation.

5. DISCUSSION

The experiments on combined torsion and extension of a rubber cylinder lead us to the conclusion that modelling filled rubber as a hyperelastic material can lead to a serious underprediction of the stresses needed to bring about small changes in strain for a state of finite strain. Similar findings have been made by Gregory (1984, 1985) for small dynamic simple shear strains superimposed on a large static shear strain, and by Busfield et al. (1997) for small pure shear strains applied to a rubber strip pre-strained in tension in the orthogonal direction. Since these extra contributions to stress for small changes in strain cannot be accounted for using elastic theory, they must originate from some dissipative mechanism in the material. It seems natural, also, to ascribe the initial high stiffness at low strains of filled rubber to the same mechanism, although this feature can be modelled, in the first strain cycle, using an appropriate "strain-energy" function (Davies et al., 1994, Gough et al., 1999). Since the effect is not very sensitive to rate several authors have proposed models for it with a frictional character (Coveney et al., 1999, Dusi et al., 1999, Ahmadi & Muhr, 1998).

Estimates of the amount of energy dissipated on deforming rubber are often based on the assumption that some fixed fraction of the maximum elastic energy supplied in a strain cycle is lost. This simplification conflicts with Gent's (1960) observation that energy is dissipated for unfilled butyl rubber even for a constant elastic energy if the strain directions are rotated, and with the similar observation of Dusi et al. for "high damping rubber" (specially formulated natural rubber with reinforcing filler). Greenwood et al. (1961) also illustrated the shortcomings of the assumption, showing experimentally that the energy loss of unfilled styrene-butadiene rubber can be much larger, for a complicated stress cycle, than it would imply. They hypothesised for an orthogonal stress combination of tension, s, and simple shear, τ, that the energy dissipated for a path on the (s, τ) plane is proportional to $\int X d\ell$ where $d\ell = \sqrt{ds^2 + d\tau^2}$ and X is a suitable length on the (s, τ) diagram, such as the distance from the centroid of the overall stress path to the element of $d\ell$. If we take s and τ to be simple shear stresses in orthogonal directions we see that this approach is not quantitatively consistent with equation (7) and the single axis loss (eg. equation (10)), since it predicts that the energy lost in one cycle of Gent's experiment will be π times that of a single axis cycle of the same amplitude, not twice. Nevertheless, the conclusion of Greenwood et al. (1961) are at least qualitatively correct. Their work is interesting not only from the point of view of the results of their experiments and analysis, but also because they were motivated by the need to explain the high energy losses in rolling and sliding friction on rubber, a subject of great practical importance for tyres.

Coveney et al. (1999) generalised their uniaxial "triboelastic" model of the dissipative process in filled rubber to three dimensions by assuming that the principal stresses are coaxial with the principal stretches. Such a model might reproduce behaviour similar to the results of section 2 above, but is clearly incapable of explaining the experimental observations of Gent (1960) and Greenwood et al (1961) since no work could be dissipated for changes in the principal stretches orthogonal to their current direction, such as if their directions are rotated.

Instead, it seems more natural to identify the dissipative contributions to the overall stresses and then to relate these contributions to the changes in the principal stretches, in a similar way to the simplified discussion in Section 4. In generalising this approach to non-linear continuum mechanics, it is probably most appropriate to relate the dissipative stresses to changes in the current configuration only since the magnitude of the departure from the original ground state is of secondary relevance. This has been found to be the case for the loss modulus of swollen elastomers, whether unfilled (Akutagawa et al., 1996) or filled (Busfield et al., 1999). It also seems to be the case for the high stiffness associated with small changes in strain for filled elastomers (Gregory, 1984, 1985). It is beyond the scope of this paper to provide detailed constitutive equations, but we would suggest using an appropriate strain energy function for calculating the elastic contributions to the total stress from the deformation relative to the ground state, and adding on frictional and viscous dissipative stresses related incrementally to changes in the current configuration.

6. CONCLUSIONS

(i) The stresses associated with combined torsion and tension of an unfilled natural rubber cylinder can be adequately modelled using an isotropic strain-energy function. However, application of such a model to natural rubber incorporating 45 parts per hundred of rubber of carbon black underpredicts the stresses associated with a small torsion superposed on a large tension by 50% or so.

(ii) When a simple shear deformation of constant magnitude is rotated through unfilled butyl or filled natural rubber the energy dissipated, per cycle, is approximately twice that dissipated in a sinusoidal uniaxial simple shear cycle of the same amplitude.

(iii) A simple linear model is given, predicting a factor of exactly 2 for the result given in (ii) above for a viscous dissipation mechanism and a factor of $\pi/2$ for a "frictional" mechanism.

(iv) Together, the results suggest that there are "dissipative" contributions to the total stress as well as the "elastic" contributions calculable from an appropriate strain energy function, and that the "dissipative" stresses depend primarily on changes in the current configuration of the material.

7. REFERENCES

Ahmadi, H.R. & Muhr, A.H. (1997) "Modelling dynamic properties of filled rubber", Plastics, Rubber & Composites Processing and Application, **26**, 451-461

Akutagawa, K., Davies, C.K.L. & Thomas, A.G., (1996) "Effect of low molar mass liquids on dynamic properties of elastomers under strain", Prog. Rubber & Plastics Tech., **12**, 174-190

Busfield, J.J.C., Ratsimba, C.H. & Thomas, A.G., (1997) "Crack growth and strain-induced anisotropy in carbon black filled natural rubber", J. Nat. Rubber Research, **12**, 131-141

Busfield, J.J.C., Deepasertkul, C., & Thomas, A.G. (1999) "Effect of liquids on the dynamic properties of carbon black filled natural rubber as a function of prestrain" Proc. European Conf. on Constitutive Models for Rubber, Vienna, Publ. Balkema

Coveney, V.A., Jamil, S.,Johnson, D.E. Keavey, M.A., Menger, C. & Williams H.T. (1999) "Implementation in FEA of a triboelastic law for dynamic behaviour of filled elastomers" in Finite Element Analysis of Elastomers, ed Boast, D. & Coveney, V.A., Prof. Eng. Publ., London

Davies, C.K.L., De, D.K. & Thomas, A.G. (1994) "Characterisation of the behaviour of rubber for engineering purposes I-Stress-strain relations" Rubber Chem. & Tech., **67**, 716-728

Dusi, A., Bettinali, F., Rebecchi, V. & Bonacina, G. (1999) "High damping rubber bearings: a simplified non-linear model with exponential constitutive law. Model description and validation through experimental activities" Proc. European Conference on Constitutive Models for Rubber, Vienna; publ. Balkema

Gent, A.N. (1960) "Simple rotary dynamic testing machine" Brit. J. Appl. Phys., **11**, 165-167

Gough, J., Muhr, A.H. & Thomas, A.G. (1998) "Material characterisation for FEA of rubber components" J. Rubber Research, **1**, 222-239

Gough, J., Gregory, I.H. & Muhr, A.H. (1999) "Determination of constitutive equations for vulcanized rubber" in Finite Element Analysis of Elastomers, ed. Boast, D. & Coveney, V.A. Prof. Eng. Publ., London

Greenwood, J.A., Minshall, H. & Tabor, D. (1961) "Hysteresis losses in rolling and sliding friction" Proc. Roy. Soc., **259A**, 480-507.

Gregory, M.J. (1984) "Measurement of rubber properties for design" Polymer Testing **4**, 211-223

Gregory, M.J. (1985) "Dynamic properties of rubber in automotive engineering" Elastomerics **117**, No.11, 19-24

Rivlin R.S. (1948) "Large elastic deformations of isotropic materials Part 1. Fundamental concepts" Phil. Trans. Roy. Soc. **240A**, 459-490

Rivlin, R.S. (1949) "Large elastic deformations of isotropic materials, part VI: Further results in the theory of torsion, shear and flexure" Phil. Trans. Roy. Soc. **243A**, 173-195

Rivlin, R.S. & Saunders D.W. (1951) "Large elastic deformations of isotropic materials. VII Experiments on the deformation of rubber" Phil. Trans. Roy. Soc. **243A**, 251-288

Treloar (1975) "The Physics of Rubber Elasticity" Third edition, Clarendon Press, Oxford

Constitutive Models for Rubber, Dorfmann & Muhr (eds) © 1999 Taylor & Francis ISBN 90 5809 113 9

The need for equi-biaxial testing to determine elastomeric material properties

R. Johannknecht & S.J. Jerrams
Rubber Research and Technology Unit, Coventry University, UK

ABSTRACT: A range of tests is employed to provide data for hyperelastic material models. It is broadly accepted that different modes of deformation are required to provide the physical constants that are input to elastomeric finite element analysis (FEA). Research conducted into the bubble inflation method raises the question of whether a single equi-biaxial test can provide adequate data to allow the determination of material parameters. It is concluded that a combination of uniaxial testing and equi-biaxial tensile testing is necessary to permit material constants to be found. The process described employs a continuous, on-line, optical observation of the bubble inflation. A special element grid and charge coupled device (CCD) cameras are used to observe gradually increasing deformation at the pole of the bubble and the relation between internal pressure, bubble height and local shell strain is evaluated. Both hydraulic and pneumatic methods and free and tube guided inflation are considered. The strain energy (density) of elastomers can be expressed by three strain invariants. The second strain invariant I_2 has greater importance in multi-axial deformation than in uniaxial deformation. Consequently, acquiring reliable biaxial data is a prerequisite for establishing a plausible material model. Tests have been conducted on hydrogenated acrylonitrile butadiene rubber (HNBR) and comparisons are made between uniaxial and equi-biaxial data. It is shown that the shape of the inflated bubble is influenced by the material. Existing conceptions, about the bubble contour achieved at high strains, are challenged. A methodology is proposed, for establishing material properties by using equi-biaxial and uniaxial deformation, in materials that exhibit high elastic strains.

1 INTRODUCTION

Finite element analysis of elastomers is complex for a number of reasons. These are:-

i) elastomers are subject to large deformations and rotations when loaded and there are associated changes in boundary conditions,
ii) the material is assumed incompressible,
iii) instabilities result from the material description
iv) multiple solutions may exist.

Even an experienced FE analyst cannot always foresee the problems that ensue from these complexities. Some are due to the material description, some to the discretisation of the model and others to a combination of both. The merits of the material models available are debated endlessly, whilst the quality of the tests to provide stress-strain data for determining parameters and the plausibility of constants obtained from regression analysis are rarely considered. A methodology has been advanced for improving hyperelastic FEA by applying data weighting and plausibility analysis (Johannknecht et al. 1999) and a test procedure for obtaining equi-biaxial stress-strain data from subjecting rubber membranes to bubble inflation is described here. Previous tests relied on stretching a rectilinear rubber membrane in two mutually perpendicular directions. Predictably localised stretching, at the points where the membranes are gripped, detract from the accuracy of results. To study strains in biomembranes, polyisoprene sheets have been inflated (Hsu et al. 1995). The material was used because of its similarity to soft tissue such as pericardia and not to provide stress-strain data for material modelling. The test specimens were inflated and deflated with a syringe pump whilst immersed in oxygenated physiologic solution in a test chamber. Others have researched the physical properties of inflated elastomeric membranes, but not to obtain material constants for FEA (Dickie and Smith 1969, Bhate and Kardos 1984, Khayat and Detouri 1995).

The deformations in this work were achieved by using both pneumatic and hydraulic media applied to free and tube guided inflation. The compound used throughout was test grade hydrogenated acrylonitrile butadiene rubber (HNBR) having a 50 Shore A hardness. Surface strains were obtained by digitising images produced using couple charged device (CCD) monochrome cameras and software written to convert bright-dark contrast to pixels and then to deformations plotted against inflation pressure.

A program is used to calculate engineering stress σ against stretch ratio λ. This information is evaluated along with uniaxial tensile test data to determine parameters for a range of elastomeric phenomenological material models. Two-term Ogden models had provided the closest correlation between FEA and physical test results after applying data weighting and plausibility analysis to data from other test modes (Johannknecht 1999). Consequently, plausible material constants using this function were derived from the uniaxial and bubble inflation tests and employed in FEA models of multi-axial deformations and other standard tests. The simulations were capable of precisely representing the diverse physical deformations in the components and tests. The inflation tests confirmed that the bubble profile assumes an elliptical form beyond a particular ratio of bubble height to test piece diameter. This ratio is dependent on the material and dimensions of the specimen. Previously it has been assumed that the bubble retains a spherical shape.

2. THE THEORY OF SHELLS APPLIED TO BUBBLE INFLATION

Bubble inflation is considered to comply with theory for thin shell structures under pressure, alternatively described as membrane theory, which assumes negligible bending stiffness. For an ideal isotropic material and an axisymmetric set-up, the bubble contour exhibits rotational symmetry and therefore the deformation at the pole of the bubble is equi-biaxial as shown in figure 1.

Pressure p is applied to inflate a thin sheet with initial thickness t_0 to a bubble shell with stretched sheet thickness t. With the measurement of pressure p and the radius of curvature r, the physical stress relation at the pole is expressed by equation (1)

$$\sigma = p \cdot \frac{r}{2t} \qquad (1)$$

3. BUBBLE INFLATION PROCEDURES

For inflation two different set-ups are possible (Bretz 1999). The simplest method is a free inflation that allows a free bubble development. The only applied strain restriction to the disc shaped specimen is the clamping at the circumference as shown in figure 2.

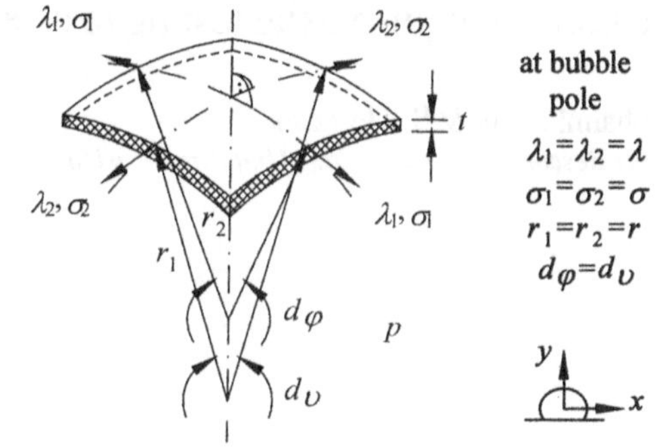

Figure 1. Bubble inflated shell element.

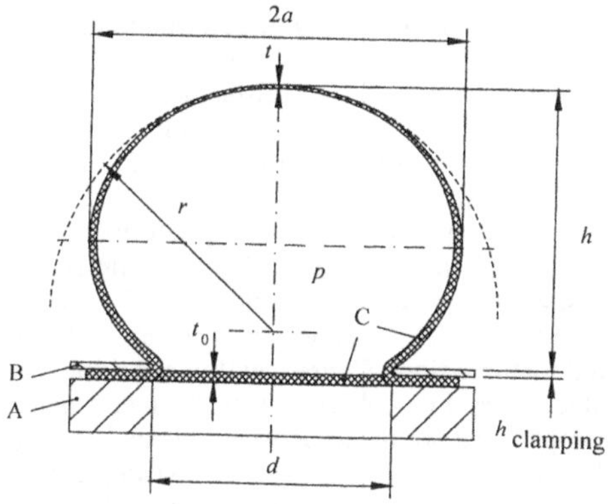

Nomenclature: A) Inflation orifice, B) Clamping orifice, C) specimen sheet.

Figure 2. Free bubble inflation

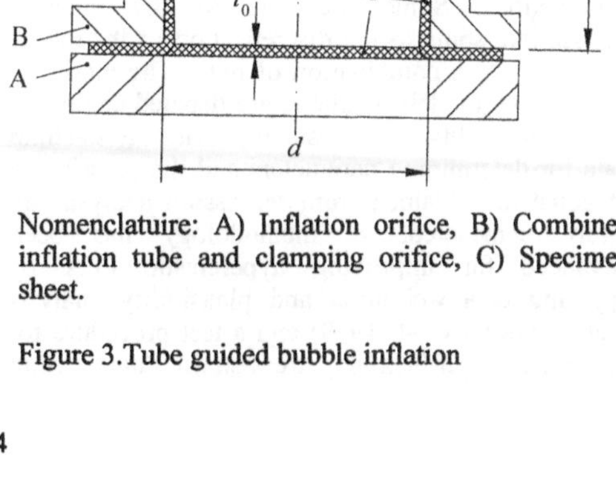

Nomenclatuire: A) Inflation orifice, B) Combined inflation tube and clamping orifice, C) Specimen sheet.

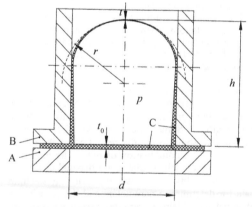

Figure 3.Tube guided bubble inflation

An alternative inflation process uses a restricted bubble development. The contour width $2a$ cannot exceed the orifice diameter d and is therefore forced to deflect towards the tube opening. This method is referred to as tube guided inflation and is shown in figure 3.

3.1 *Optical measurement of biaxial strains*

In the free inflation tests, biaxial strain at the bubble pole and the bubble contour were determined by using two monochrome CCD cameras. One camera was aligned with the bubble axis of symmetry and recorded the displacements in a grid on the surface of the bubble pole. These displacements were digitised to allow strains to be calculated in a computer analysis. The second camera was positioned normal to the axis of symmetry and simultaneously recorded bubble height and contour.

4. A COMPARISON OF RESULTS FROM UNIAXIAL AND BIAXIAL LOADING

Figures 4 and 5 show that for different rubber compounds and for any hardness there is no consistent relationship between uniaxial and equi-biaxial stress states. It is evident that material parameters calculated from a regression analysis using curves A and B, whether or not in conjunction with curve C, would be dissimilar and would result in different FEA predictions.

In figure 5, the tensile stress ratios ψ (equation (2)) show no particular master relation for different HNBR compounds over a range of hardness. Thus, stress-strain predictions that include equi-biaxial behaviour derived from a material model having parameters determined from uniaxial tensile data can not represent the actual deformation of a component subjected to multi-axial loading. Similarly, a component exhibiting predominantly uniaxial tensile or compressive strain can not be accurately modelled by parameters derived solely from equi-biaxial load test data. It is therefore essential to consider stress-strain relations for all three deformation modes to fully characterise the material's physical behaviour and carry out realistic FEA.

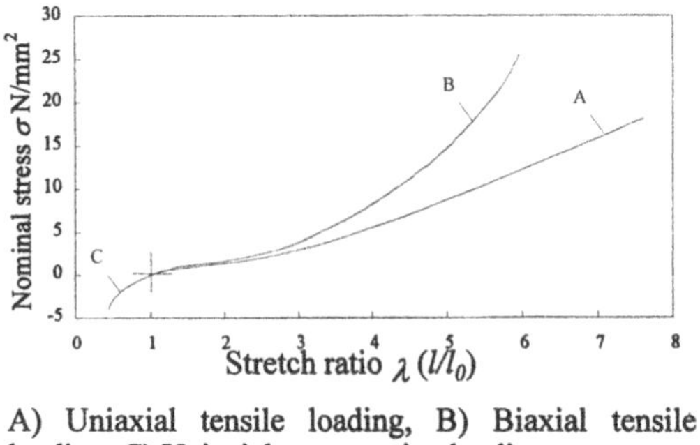

A) Uniaxial tensile loading, B) Biaxial tensile loading, C) Uniaxial compressive loading

Figure 4. Stress-strain relations for uniaxial and equi-biaxial loading (50 Shore A)

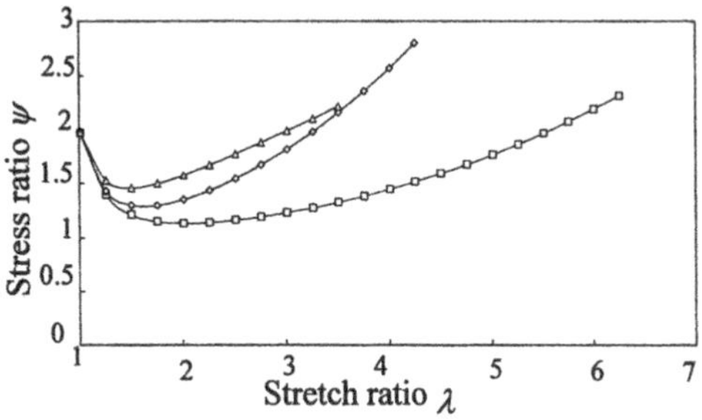

□ HNBR 50 Shore A, ◇ HNBR 60 Shore A, △ HNBR 70 Shore A.

Figure 5 Ratio of uniaxial to biaxial stress plotted against stretch ratio for three hardness'.

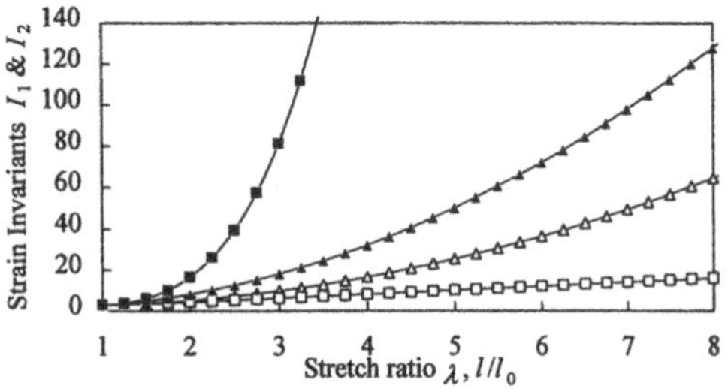

△ I_1 in uniaxial loading, □ I_2 in uniaxial loading, ▲ I_1 in equi-biaxial loading, ■ I_2 in equi-biaxial loading.

Figure 6. Disparities in strain invariants for uniaxial and equi-biaxial loading

The need to determine material constants from tests in different deformation modes is demonstrated by figure 6. The two strain invariants used in strain energy density functions for isochoric materials ($I_1 = \lambda_1^2 + \lambda_2^2 + \lambda_3^2$, $I_2 = \lambda_1^2\lambda_2^2 + \lambda_2^2\lambda_3^2 + \lambda_3^2\lambda_1^2$) are seen to vary markedly with increases in stretch ratio (λ). This variation is particularly pronounced for I_2 and substantiates that parameters taken in one deformation mode are incapable of modelling component deformations comprising multi-axial loading.

$$\psi = \frac{\sigma_{\text{equi-biaxial}}}{\sigma_{\text{uniaxial}}} \quad (2)$$

5. CONCLUSIONS & RECOMMENDATIONS

When hyperelastic material parameters are required for FEA, it is essential to include data from an equi-biaxial loading test. A bubble inflation test, using optical strain measuring techniques, provides a reliable and accurate method of obtaining this data and has none of the disadvantages inherent in stretch frames. Free bubble inflation is preferred to tube guided inflation since the bubble profile and pole height can be observed.

Inflating the bubble hydraulically is more problematic than using a pneumatic inflation process. Also it is advantageous to strain the specimen to failure and consequently provide data over the broadest strain range. The hydraulic media diminishes the effectiveness of optical techniques, is wasteful and hazardous.

To fully characterise the material behaviour it is essential to include data from uniaxial compression and tension tests. All parameters should be evaluated by a visual observation of the quality of the curve fit in regression analysis, plausibility analysis and data weighting if required.

REFERENCES

Johannknecht R., Jerrams S.J. and Clauss G. 1999. *The uncertainty of implemented curve fitting procedures in finite element software.* Finite element of elastomers. I Mech E/The Institute of Materials, Seminar (1997) and Finite Element Analysis of Elastomers. Ed. D. Boast and V. Coveny Professional Engineering Publishing (I Mech E).

Hse F.P.K., Lui A.M.C., Downs J., Rigamonti D. and Humphrey J.D. 1995. *A triplane video based experimental system for studying axisymmetrically inflated biomembranes.* IEEE Transactions on biomedical engineering. Vol 42 No. 5, pp 442-451.

Dickie R.A. and Smith T.L. 1969. *Ultimate tensile properties of elastomers VI. Strength and extensibility of a styrene-butadiene rubber vulcanizate in equal biaxial tension.* Journal of Polymer Science, Vol. 7, Part A-2, p687-707.

Bhate A.P. and Kardos J.L. 1984. *A novel technique for the determination of high frequency equi-biaxial stress-deformation behaviour of viscoelastic elastomers.* Polymer Engineering and Science, Vol. 24, No. 11.

Khayat R.E. and Detouri A. 1995. *Stretch and inflation of hyperelastic membrane as applied to blow molding.* Polymer Engineering and Science, Vol. 35, No. 23.

Johannknecht R. 1999. *The physical testing and modelling of hyperelastic materials for finite element analysis.* PhD. Dissertation, Coventry University.

Bretz H. 1998. *Aufbrau und Erprobung einer Wölbvorrichtung zur Ermittlung des hyperelastischen Verhalten elastomerer Werkstoffe unter mehrachsiger Beanspruchung.* Final year project, University of Applied Science, Heilbronn, Germany.

Hartwig K. 1997. *Simulation des Streckverblasverfahrens und Characterisierung des prozessrelevanten Materialverhaltens.* 1st Edition Aachen, Verlag Mainz, Germany.

Viscoelasticity

Constitutive Models for Rubber, Dorfmann & Muhr (eds) © 1999 Taylor & Francis ISBN 90 5809 113 9

Constitutive model for a class of hyperelastic materials with embedded rheological properties

I. Dobovšek
Faculty of Mechanical Engineering, University of Maribor, Slovenia

ABSTRACT: The paper presents a continuum mechanics approach in a mathematical formulation of a class of constitutive relations which tend to descibe mechanical as well as thermo–rheological behavior of a material with two limiting cases of consitutive behavior: thermo–hyperelastic representing the rate independent behavior and thermo–viscoelastic in a general rheological sense. In the rate–independent limit the mechanical part can be represented by appropriate strain energy functions. We consider potential functions of Murnagan's type for compressible and Signorini's type for incompressible materials. In addition to the standard kinematical variables, the time dependent part of the assumed general form of constitutive relation is postulated to depend on an internal variable similar to the reduced time scaled by a shift function as in classical thermo–viscoelasticity, but with much broader functional relationship. The corresponding expressions for stress, entropy, and dissipation inequality are derived from strict thermodynamic considerations and are given in an explicit form. Development is based on the principle of fading memory established by (Coleman 1961), (Coleman & Noll 1964) which was subsequently specialized for the class of viscoelastic materials by (Lianis 1965, 1970). A fundamental assumption of the proposed procedure is that in the expansion of the rheological time–dependent part of a free energy function small differences in histories of strain and temperature fields enter the thermodynamic potential or the free energy function only up to quadratic terms. Consequently, the derived expressions are quite amenable for further development. We discuss a thermo–rheologically simple constitutive model of a material which is described by a constitutive relation of a single–integral equation type.

1 INTRODUCTION

In recent years, the use of elastomers and rubbers has increased substantially in engineering practice where these materials form one of the most important components of a wide range of products. Considerable advances have been made in the development of mathematical theories to model the mechanical behavior of rubber–like materials. Elastomers are polymers capable of recovering substantially in size and shape after removal of the load. Natural and synthetic rubbers can recoup quite fast from large deformations. At high deformations the stress–strain relationship for these materials is non–linear and is affected by dynamic and thermal effects. A classical concept of an elastic material as a simple material, which may or may not exhibit a non–linear material response as well as a non–linear geometric response in a sense of finite deformations, is based on an assumption that the material behavior does not depend on a strain history. Such an assumption is usually augmented by the requirement of the existence of a stress potential, which is a function of deformation gradient or some other possible strain measure. Stored energy described in the form of strain energy function serves as a starting point for modeling different kinds of compressible and incompressible elastomers. Compressible elastomers are usually described by the Blatz–Ko type of constitutive relation, while the Rivlin–Saunders type of constitutive equation is generally used in the description of incompressible materials. Two derived categories from the Rivlin-Saunders class of constitutive relations, namely the neo–Hookean and the Mooney-Rivlin types of constitutive equations, have been used quite frequently in the modeling of natural rubber and other rubber like materials such as synthetic elastomers, polymers, and biological tissues. However, with the advent of new advanced materials like silicon rubber, the need for a general type of constitutive equation for more accurate description of constitutive properties of such materials has increased significantly. Materials of that kind display many features of the standard solid type hyperelastic behavior, but at the same time can exhibit a fluid-like material response with a substantial strain rate dependence. Consequently, the constitutive modeling based on the classical concept of hyperelasticity with no strain rate effects is not adequate for the description of this class of materials. To incorporate the strain rate dependence and strain memory properties in the formulation of constitutive equations, the integral hereditary operators of Volterra type are introduced in place of free material constants of a general constitutive model of the Signorini's type with the formulation based on the Almansi strain tensor. An equivalent formulation is also possible for Murnaghan's model based on the representation of specific strain energy in the form of polynomial of the Cauchy-Green strain tensor. Then, the established general 3D constitutive relation can be specialized for the specific state of stress, and free parameters, which are suitable for experimental determination of material constants, can be identified. The nonlinearity can be expressed in an explicit form for different states of stress. Since the material functions and the free constants of the model depend only on a single time-like variable, they can be directly calibrated from the experiment. Within the range of solid behaviour, materials such as plastics have mechanical properties which are extremely temperature–sensitive. It is obvious that thermal effects are significant in many instances and applications. Therefore, in what follows, we describe a strict thermodynamic consideration of thermo–rheologically simple material and propose a constitutive model which can accomodate a broad range of thermo–hyperelastic as well as thermo–viscoelastic behavior.

2 FOUNDATIONS

Consider a material body in an initial state of zero stress and strain which is referred to as a preferred natural configuration in a three dimensional Euclidean space $\mathbb{R}^3$. The material elements or particles of a continuous medium, or body $\mathcal{B}$, are denoted by the position vector $\mathbf{X} \in \mathcal{B}$ in a fixed reference configuration with the mass density ϱ_R. We define the process $\mathcal{P}$ at particle $\mathbf{X}$ and time t as an ordered set of functions

$$\mathcal{P} = \{\chi, \theta, \mathcal{E}, \eta, \boldsymbol{\sigma}, \mathbf{q}, \mathbf{b}, r\}, \tag{1}$$

where during the process $\mathcal{P}$

$\mathbf{x} = \chi(\mathbf{X}, t)$ is the motion,
$\theta = \theta(\mathbf{X}, t)$ is the absolute temperature,
$\mathcal{E} = \mathcal{E}(\mathbf{X}, t)$ is the specific internal energy per unit mass,
$\eta = \eta(\mathbf{X}, t)$ is the specific entropy per unit mass,
$\boldsymbol{\sigma} = \boldsymbol{\sigma}(\mathbf{X}, t)$ is the Cauchy stress tensor,
$\mathbf{q} = \mathbf{q}(\mathbf{X}, t)$ is the heat flux vector,
$\mathbf{b} = \mathbf{b}(\mathbf{X}, t)$ is the specific body force per unit mass,
$r = r(\mathbf{X}, t)$ is the radiant heating per unit mass.

According to the fundamental laws of continuum mechanics each process $\mathcal{P}$ is required to satisfy:
i) the balance of linear momentum

$$\frac{d}{dt}\int_{\mathcal{B}} \varrho \mathbf{v}\, dv = \int_{\partial\mathcal{B}} \boldsymbol{\sigma}\mathbf{n}\, da + \int_{\mathcal{B}} \varrho \mathbf{b}\, dv, \tag{2}$$

ii) the balance of energy

$$\frac{d}{dt}\int_{\mathcal{B}} \varrho(\mathcal{E} + \tfrac{1}{2}\mathbf{v}\cdot\mathbf{v}) dv = \int_{\mathcal{B}} \varrho(\mathbf{v}\cdot\mathbf{b} + r) dv + \int_{\partial\mathcal{B}} (\mathbf{v}\cdot\boldsymbol{\sigma}\mathbf{n} - \mathbf{q}\cdot\mathbf{n}) da, \tag{3}$$

iii) the entropy inequality

$$\frac{d}{dt}\int_{\mathcal{B}} \varrho\eta dv \geq -\int_{\partial\mathcal{B}} \frac{\mathbf{q}\cdot\mathbf{n}}{\theta} da + \int_{\mathcal{B}} \varrho\frac{r}{\theta}\, dv, \tag{4}$$

where $\mathbf{v} = \dot{\mathbf{x}} = \partial\chi(\mathbf{X}, t)/\partial t$ is the velocity of the particle $\mathbf{X}$ at time t, $\mathbf{n}(\mathbf{x}, t)$ is the outer unit normal at $\mathbf{x} \in \mathcal{B}$, and $\varrho(\mathbf{x}, t)$ is the spatial mass density function induced by the motion $\chi(\mathbf{X}, t)$ which is connected with the reference mass density by the relation

$$\varrho_R = \varrho|\det\mathbf{F}|, \quad \mathbf{F} = \mathbf{F}(\mathbf{X}, t) = \nabla\chi(\mathbf{X}, t). \tag{5}$$

$\mathbf{F}$ is the deformation gradient of $\chi(\mathbf{X}, t)$. In the sequel ∇ will denote differentiation with respect to the particle $\mathbf{X}$, while grad will denote differentiation with respect to the spatial coordinate $\mathbf{x}$.

By introducing the Helmholtz free energy $\Psi = \mathcal{E} - \theta\eta$ and assuming that the fundamental fields of the process $\mathscr{P}$ are sufficiently smooth, the global balance laws can be transformed into the corresponding local or strong form

$$\text{div}\boldsymbol{\sigma} + \varrho\mathbf{b} = \varrho\dot{\mathbf{v}}, \tag{6}$$

$$\varrho\dot{\mathcal{E}} = \boldsymbol{\sigma}\colon\mathbf{L} - \text{div}\mathbf{q} + \varrho r, \tag{7}$$

$$\varrho(\dot{\Psi} + \eta\dot{\theta}) - \boldsymbol{\sigma}\colon\mathbf{L} + \frac{\mathbf{q}\cdot\mathbf{g}}{\theta} \leq 0. \tag{8}$$

$\mathbf{L} = \text{grad}\mathbf{v}$ and $\mathbf{g} = \text{grad}\,\theta$ are the spatial gradients of velocity and temperature, respectively, and $\text{div}(\cdot)$ is the contraction of $\text{grad}(\cdot)$. The superimposed dot denotes the material time derivative of the quantity on which it operates. The inequality (8) is called the dissipation inequality.

In further development additional kinematics relations will be utilized. The polar decomposition of the deformation gradient

$$\mathbf{F} = \mathbf{R}\mathbf{U} = \mathbf{V}\mathbf{R}, \tag{9}$$

where the orthogonal tensor $\mathbf{R}$ characterizes the local rotation of the element, while $\mathbf{U}$ and $\mathbf{V}$ are called the right and left stretch tensors, respectively. By using these fields the right and left deformation tensors are defined as

$$\mathbf{C} = \mathbf{F}^T\mathbf{F} = \mathbf{U}^2, \quad \mathbf{B} = \mathbf{F}\mathbf{F}^T = \mathbf{V}^2. \tag{10}$$

The material time rate of change of the deformation defined by the velocity gradient $\mathbf{L}$ can be written as

$$\mathbf{L} = \text{grad}\mathbf{v} = \dot{\mathbf{F}}\mathbf{F}^{-1} = \mathbf{D} + \mathbf{W}, \tag{11}$$

where the symmetric part $\mathbf{D} = (\mathbf{L} + \mathbf{L}^T)/2$ and the antisymmetric part $\mathbf{W} = (\mathbf{L} - \mathbf{L}^T)/2$ are named the stretching and the spin tensor, respectively.

Sometimes it is preferable to use the current placement of a body as a natural reference configuration and define the past histories with respect to it. If $\mathbf{x}(t)$ denotes the current position of a particle, then the spatial position history of the particle can be expressed as $\mathbf{x}(\tau) = \boldsymbol{\chi}(\mathbf{x}(t), \tau)$, where $\mathbf{x}(\tau)$ denotes the location at time τ of the particle that will occupy the position $\mathbf{x}(t)$ at time t so that $\boldsymbol{\chi} = \boldsymbol{\chi}_t(\mathbf{x}, \tau)$ represents a backward path line of a particle $\tau \leq t$ presently at $\mathbf{x} = \boldsymbol{\chi}_t(\mathbf{x}, t)$. Consequently, the relative deformation gradient and the relative right Cauchy–Green deformation tensor can be defined as

$$\mathbf{F}_t(\tau) = \nabla_{\mathbf{x}}\boldsymbol{\chi}_t(\mathbf{x}, \tau), \quad F_{ij} = \partial\chi_i/\partial x_j, \tag{12}$$

$$\mathbf{C}_t(\tau) = \mathbf{F}_t^T(\tau)\mathbf{F}_t(\tau) = \mathbf{C}_t^T(\tau). \tag{13}$$

3 THE MODEL

We begin with a brief description of two fundamental models of a single integral type class of isothermal constitutive relations due to (Coleman & Noll 1961) and (Lianis 1970). The starting point in the discussion of our constitutive model is a functional relationship established by (Truesdell & Noll 1965)

$$\boldsymbol{\sigma}(t) = \mathscr{G}_{\tau=-\infty}^{t}[\mathbf{C}_t(\tau); \mathbf{B}(t)], \tag{14}$$

where $\mathscr{G}$ is an isotropic functional and $\mathbf{B}$ according to the previous definitions denotes the left Cauchy–Green strain tensor. The dependence on $\mathbf{B}$ implies the existence of a natural reference state. From the equation above a single integral type of constitutive relation for a compressible isotropic material with fading memory according to the definitions of (Truesdell & Noll 1965) can be derived which retains only linear terms in the strain history. The equation reads

$$\begin{aligned}
\boldsymbol{\sigma}(t) &= -p\mathbf{I} + \beta_1\mathbf{B} + \beta_2\mathbf{B}^2 + \\
&+ \sum_{A=0}^{2}\int_{-\infty}^{t}\chi_A(t-\tau)\times \\
&\times [\mathbf{B}^A\dot{\mathbf{C}}_t(\tau) + \dot{\mathbf{C}}_t(\tau)\mathbf{B}^A]d\tau + \\
&+ \sum_{A=0}^{2}\sum_{B=0}^{2}\int_{-\infty}^{t}\chi_{AB}(t-\tau)\mathbf{B}^A\times \\
&\times \text{Tr}[\mathbf{B}^B\dot{\mathbf{C}}_t(\tau)]d\tau.
\end{aligned} \tag{15}$$

The model consists of twelve relaxation functions $\chi_A = \chi_A(I_k^B)$ and $\chi_{AB} = \chi_{AB}(I_k^B)$, $(k = 1, 2, 3)$, as well as of pressure p, β_1, and β_2 which are functions of strain invariants

$$\begin{aligned}
I_1^B &= \text{Tr}\mathbf{B}, \quad I_3^B = \text{Det}\mathbf{B}, \\
I_2^B &= \tfrac{1}{2}(\text{Tr}^2\mathbf{B} - \text{Tr}\mathbf{B}^2).
\end{aligned} \tag{16}$$

The equation has been simplified significantly by (Lianis 1970), where by considering certain thermodynamic considerations the symmetry relation $\chi_{AB} = \chi_{BA}$ has been established. For an incompressible isotropic finite viscoelastic material the constitutive equation due to Lianis is of the form

$$\begin{aligned}
\boldsymbol{\sigma}(t) &= -p\mathbf{I} + \beta_1\mathbf{B} + \beta_2\mathbf{B}^2 + \\
&+ 2\int_{-\infty}^{t}\chi_0(t-\tau)\dot{\mathbf{C}}_t(\tau)d\tau +
\end{aligned}$$

$$+ \int_{\infty}^{t} \chi_1(t-\tau)[\mathbf{B}\dot{\mathbf{C}}_t(\tau) + \dot{\mathbf{C}}_t(\tau)\mathbf{B}]d\tau +$$

$$+ \int_{-\infty}^{t} \chi_2(t-\tau)[\mathbf{B}^2\dot{\mathbf{C}}_t(\tau) + \mathbf{C}_t(\tau)\mathbf{B}^2]d\tau +$$

$$+ \mathbf{B}\int_{-\infty}^{t} \chi_3(t-\tau) \times$$

$$\times \ \mathrm{Tr}[\mathbf{B}\dot{\mathbf{C}}_t(\tau) + \dot{\mathbf{C}}_t(\tau)\mathbf{B}]d\tau. \qquad (17)$$

In what follows we will present an extention of previously described isothermal relations into a temperature domain. The resulting relations can be considered as a special case of a generalized thermo–rheological model which is still of a single integral type. For thermo–rheologically simple materials the concept of the so–called reduced time scaled by the appropriate shift function, which is in most cases determined experimentally, has proven useful in establishing a general constitutive model. The reduced time ξ is scaled by a shift function $a > 0$ through the relation

$$\xi = \int_0^t a[\theta(z), \mathbf{C}(z), \dot{\mathbf{C}}(z), \ldots]dz. \qquad (18)$$

In addition to deformation measures defined at the beginning of the section we introduce differences in past histories $(\tau) - (t)$, i.e., $\mathbf{C}^* = \mathbf{C}(\tau) - \mathbf{C}$, $\theta^* = \theta(\tau) - \theta$ needed in further development.

The functional of Helmholtz free energy is written in the form given by Green and Rivlin as an additive decomposition

$$\Psi = \Psi^{\infty}(\mathbf{C}, \theta) +$$

$$+ \int_{-\infty}^{t} \mathcal{F}(\mathbf{C}, \theta, \mathbf{C}^*, \theta^*, \zeta) \times$$

$$\times \ a[\theta(\tau), \mathbf{C}(\tau), \ldots]d\tau, \qquad (19)$$

where Ψ^{∞} is equilibrium free energy and $\mathcal{F}(\cdot)$ sufficiently smooth function of its arguments. The second law of thermodynamics expressed with respect to the reference natural state can be written in the form of Clausius–Duhem inequality as

$$\frac{1}{2}\mathbf{T}{:}\dot{\mathbf{C}} - \varrho_R(\dot{\Psi} + \eta\dot{\theta}) - \frac{1}{\theta}\mathbf{q}\cdot\mathbf{g}_R =$$

$$= \theta\mathscr{D} - \frac{1}{\theta}\mathbf{q}\cdot\mathbf{g}_R \geq 0, \qquad (20)$$

where $\mathbf{T}$ is the second Piola–Kirchhoff symmetric stress tensor, $\mathscr{D}$ is internal dissipation, and

$$\zeta = \int_{\tau}^{t} a[\theta(z), \mathbf{C}(z), \dot{\mathbf{C}}(z), \ldots]dz, \qquad (21)$$

is the relative reduced time sometimes also called a relative pseudotime. Substituting (19) into the Clausius–Duhem inequality (20) yields a general expression for the second Piola–Kirchhoff stress tensor

$$\frac{1}{2\varrho_R}\mathbf{T} = \frac{\partial\Psi^{\infty}}{\partial\mathbf{C}} + \qquad (22)$$

$$+ \int_{-\infty}^{t} \left(\frac{\partial\mathcal{F}}{\partial\mathbf{C}} - \frac{\partial\mathcal{F}}{\partial\mathbf{C}^*}\right) a[\theta(\tau), \ldots]d\tau,$$

a general expression for specific entropy

$$\eta = -\frac{\partial\Psi^{\infty}}{\partial\mathbf{C}} \qquad (23)$$

$$- \int_{-\infty}^{t} \left(\frac{\partial\mathcal{F}}{\partial\mathbf{C}} - \frac{\partial\mathcal{F}}{\partial\mathbf{C}^*}\right) a[\theta(\tau), \ldots]d\tau,$$

and a general expression for internal dissipation

$$\mathscr{D} = -\frac{1}{\theta}a[\theta(z), \mathbf{C}(z), \dot{\mathbf{C}}, \ldots] \times$$

$$\times \int_{-\infty}^{t} \frac{\partial\mathcal{F}}{\partial\zeta} a[\theta(\tau), \ldots]d\tau. \qquad (24)$$

From (20) it follows that $\mathscr{D} \geq 0$ implies

$$\frac{\partial}{\partial\zeta}\mathcal{F}(\mathbf{C}, \theta, \mathbf{C}^*, \theta^*, \zeta) \leq 0. \qquad (25)$$

Let us consider constant histories in the time frame $-\infty < \tau \leq t$ such that

$$\mathbf{C}^*(\tau) = 0, \quad \theta^*(\tau) = 0, \quad \text{and} \qquad (26)$$

$$\lim_{\zeta\to\infty} \zeta\mathcal{F}(\mathbf{C}, \theta, 0, 0, \zeta) = 0 \qquad (27)$$

$$\zeta_0 = \int_{\tau}^{t} a[\theta(z), \mathbf{C}(z), 0, \ldots]dz, \qquad (28)$$

By considering relationships defined by (20), (24), and (25) the following restrictions on evolution of the rheological function of state $\mathcal{F}$, which are imposed by constant strain and temperature histories, are obtained

$$\mathcal{F}(\mathbf{C}, \theta, 0, 0, \zeta_0) = 0, \qquad (29)$$

$$\frac{\partial}{\partial\mathbf{C}}\mathcal{F}(\mathbf{C}, \theta, 0, 0, \zeta_0) = 0, \qquad (30)$$

$$\frac{\partial}{\partial\mathbf{C}^*}\mathcal{F}(\mathbf{C}, \theta, 0, 0, \zeta_0) = 0, \qquad (31)$$

$$\frac{\partial}{\partial\theta}\mathcal{F}(\mathbf{C}, \theta, 0, 0, \zeta_0) = 0, \qquad (32)$$

$$\frac{\partial}{\partial \theta^*}\mathcal{F}(\mathbf{C},\theta,0,0,\zeta_0) = 0, \qquad (33)$$

$$\frac{\partial}{\partial \zeta}\mathcal{F}(\mathbf{C},\theta,0,0,\zeta_0) = 0. \qquad (34)$$

According to the theory of algebraic invariants a compressible isotropic material depends on all three invariants of C, while the rheological function $\mathcal{F}$ in most general form depends also on terms with mixed invariants of C and C*:

$$\Psi^\infty = \Psi^\infty(I_k^C,\theta), \quad k=1,2,3, \qquad (35)$$

$$\mathcal{F} = \mathcal{F}(I_k^C,\mathcal{I}_k,\mathcal{J}_L,\theta,\theta^*,\zeta), \qquad (36)$$

$$L = 1,2,3,4.$$

The corresponding invariants are defined as

$$I_1^C = \mathrm{Tr}\mathbf{C}, \quad I_3^C = \mathrm{Det}\mathbf{C},$$

$$I_2^C = \tfrac{1}{2}(\mathrm{Tr}^2\mathbf{C} - \mathrm{Tr}\mathbf{C}^2), \qquad (37)$$

$$\mathcal{I}_N = \mathrm{Tr}\mathbf{C}^{*N}, \quad N=1,2,3; \qquad (38)$$

$$\mathcal{J}_1 = \mathrm{Tr}\mathbf{C}\mathbf{C}^*, \quad \mathcal{J}_2 = \mathrm{Tr}\mathbf{C}^2\mathbf{C}^*, \qquad (39)$$

$$\mathcal{J}_3 = \mathrm{Tr}\mathbf{C}\mathbf{C}^{*2}, \quad \mathcal{J}_4 = \mathrm{Tr}\mathbf{C}^2\mathbf{C}^{*2}$$

In order to simplify the general model we follow the concept described by (Coleman & Noll 1961), who assumed that in cases when differences in histories of C* and θ^* are small, both functions are contained in expansion for $\mathcal{F}$ up to the second order

$$\begin{aligned}\mathcal{F} &= \varphi_1\mathcal{I}_1^2 + \varphi_2\mathcal{I}_1\mathcal{J}_1 + \varphi_3\mathcal{I}_1\mathcal{J}_2 + \varphi_4\mathcal{I}_2 + \\ &+ \varphi_5\mathcal{J}_1\mathcal{J}_2 + \varphi_6\mathcal{J}_1^2 + \varphi_7\mathcal{J}_2^2 + \varphi_8\mathcal{J}_3 + \\ &+ \varphi_9\mathcal{J}_4 + \varphi_{10}\mathcal{I}_1\theta^* + \varphi_{11}\mathcal{J}_1\theta^* + \\ &+ \varphi_{12}\mathcal{J}_2\theta^* + \varphi_{13}\theta^{*2}, \qquad (40)\end{aligned}$$

where functions $\varphi_r = \varphi_r(I_k^C,\theta,\xi)$ depend on strain invariants, temperature, and the reduced time. Introducing (35) and (40) into (22) we find the corresponding expression for stress

$$\begin{aligned}\frac{1}{2\varrho_R}\mathbf{T} &= \alpha_0\mathbf{I} + \alpha_1\mathbf{C} + \alpha_2\mathbf{C}^2 + \qquad (41)\\ &+ \int_{-\infty}^{t}\Big[\tilde{\alpha}_0\mathbf{I} + \tilde{\alpha}_1\mathbf{C} + \tilde{\alpha}_2\mathbf{C}^2 + \\ &+ 2\varphi_4\mathbf{C}^* + \varphi_8(\mathbf{C}\mathbf{C}^* + \mathbf{C}^*\mathbf{C}) + \\ &+ \varphi_9(\mathbf{C}^2\mathbf{C}^* + \mathbf{C}^*\mathbf{C}^2)\Big]\,a d\tau,\end{aligned}$$

with characteristic coefficients due to the thermo–hyperelastic part of the relation which are defined as

$$\begin{aligned}\alpha_0 &= \frac{\partial\Psi^\infty}{\partial I_1^C} + I_1^C\frac{\partial\Psi^\infty}{\partial I_2^C},\\ \alpha_1 &= -\frac{\partial\Psi^\infty}{\partial I_2^C}, \quad \alpha_2 = \frac{1}{I_3^C}\frac{\partial\Psi^\infty}{\partial I_3^C}, \qquad (42)\end{aligned}$$

while the coefficients $\tilde{\alpha}_k(\varphi_r,\mathcal{I}_k,\mathcal{J}_L)$ from the rheological part of constitutive relation exhibit the following dependences on strain invariants and temperature

$$\begin{aligned}\tilde{\alpha}_0 &= 2\varphi_1\mathrm{Tr}\mathbf{C}^* + \varphi_2\mathrm{Tr}\mathbf{C}\mathbf{C}^* + \\ &+ \varphi_3\mathrm{Tr}\mathbf{C}^2\mathbf{C}^* + \varphi_{10}\theta^*,\\ \tilde{\alpha}_1 &= \varphi_2\mathrm{Tr}\mathbf{C}^* + 2\varphi_6\mathrm{Tr}\mathbf{C}\mathbf{C}^* + \\ &+ \varphi_5\mathrm{Tr}\mathbf{C}^2\mathbf{C}^* + \varphi_{11}\theta^*, \qquad (43)\\ \tilde{\alpha}_2 &= \varphi_3\mathrm{Tr}\mathbf{C}^* + \varphi_5\mathrm{Tr}\mathbf{C}\mathbf{C}^* + \\ &+ \varphi_7\mathrm{Tr}\mathbf{C}^2\mathbf{C}^* + \varphi_{12}\theta^*.\end{aligned}$$

Combination of (40) and (23) yields the expression for specific entropy,

$$\begin{aligned}\eta &= -\frac{\partial\Psi^\infty}{\partial\theta} + \int_{-\infty}^{t}[\varphi_{10}\mathrm{Tr}\mathbf{C}^* + \varphi_{11}\mathrm{Tr}\mathbf{C}\mathbf{C}^* + \\ &+ \varphi_{12}\mathrm{Tr}\mathbf{C}^2\mathbf{C}^* + 2\varphi_{13}\theta^*]\,a d\tau, \qquad (44)\end{aligned}$$

while the expression for internal dissipation reads

$$\mathscr{D} = -\frac{a}{\theta}\int_{-\infty}^{t}\frac{\partial}{\partial\xi}\mathcal{F}(\cdot)\,a d\tau, \qquad (45)$$

$$\mathcal{F} = \mathcal{F}(I_k^C,\mathcal{I}_k,\mathcal{J}_L,\xi,\theta,\theta^*).$$

For incompressible materials imposing the constraint $I_3^C = 1$ and considering the well–known relation between the second Piola–Kirchhoff and Cauchy stress tensors yields

$$\begin{aligned}\frac{1}{2\varrho}\boldsymbol{\sigma} &= -p\mathbf{I} + \frac{1}{2\varrho}\mathbf{F}\mathbf{T}\mathbf{F}^T = \\ &= -p\mathbf{I} + \beta_1\mathbf{B} + \beta_2\mathbf{B}^2 +\end{aligned}$$

$$+\int_{-\infty}^{t}\Big[\tilde\beta_1\mathbf{B}+\tilde\beta_2\mathbf{B}^2+ \tag{46}$$

$$+\hat\beta_0\mathcal{G}+\hat\beta_1(\mathcal{G}\mathbf{B}+\mathbf{B}\mathcal{G})+$$

$$+\hat\beta_2(\mathbf{B}^2\mathcal{G}+\mathcal{G}\mathbf{B}^2)\Big]\,ad\tau,$$

where $\mathcal{G}$ is defined as

$$\mathcal{G}=\mathbf{F}^T\mathbf{C}^{-1}(\tau)\mathbf{F}^{-1}-\mathbf{I}. \tag{47}$$

By analogous considerations as have led to the expressions (43) we could obtain explicit function relationships for the coefficients $\tilde\beta_k$ and $\hat\beta_k$, but these relations are not given here. We just indicate general dependences which are of the following form

$$\tilde\beta_k = \tilde\beta_k(\vartheta_r,\mathrm{Tr}\mathcal{G},\mathrm{Tr}\mathbf{B}\mathcal{G},\mathrm{Tr}\mathbf{B}^2\mathcal{G}),$$

$$\hat\beta_k = \hat\beta_k(\vartheta_r,\mathrm{Tr}\mathcal{G},\mathrm{Tr}\mathbf{B}\mathcal{G},\mathrm{Tr}\mathbf{B}^2\mathcal{G}), \tag{48}$$

$$\vartheta_r = \vartheta_r(I_1^B,I_2^B),\quad r=1,\ldots,10.$$

The model described so far is quite general. By assigning particular functions Ψ^∞ and $\mathcal{F}$ different models can be obtained. To illustrate the method, let us consider a generalization of a potential due to Murnaghan which can describe the thermo–hyperelastic behavior of compressible materials. The potential has the form

$$\Phi(I_1^C,I_2^C,I_3^C,\theta) = \gamma_0(\theta)+\gamma_1(\theta)(I_1^C-3)^2+$$

$$+\gamma_2(\theta)(I_2^C-3)+$$

$$+\gamma_3(\theta)(I_1^C-3)^2+$$

$$+\gamma_4(\theta)(I_1^C-3)(I_2^C-3)+$$

$$+\gamma_5(\theta)(I_3^C-1). \tag{49}$$

We assume that the material is thermo–rheologically simple (Day 1972, 1976), and that the rheological function $\mathcal{F}$ does not depend explicitly on temperature or temperature history. Apart from that, we assume that the shift function a depends on Cauchy–Green deformation tensor and temperature only, but not on any rate quantity, i.e., $a=a[\mathbf{C}(\tau),\theta(\tau)]$. Then for the single–step histories in the time frame $-\infty<\tau<\infty$ we have

$$\mathbf{C}(\tau) = \mathbf{I}+\mathbf{C}_0H(\tau), \tag{50}$$

$$\theta(\tau) = \theta_0+(\theta_1-\theta_0)H(\tau), \tag{51}$$

$$\mathbf{C}^* = \mathbf{I}-\mathbf{C}_0[H(\tau)-1], \tag{52}$$

$$\theta^* = (\theta_1-\theta_0)[H(\tau)-1], \tag{53}$$

where $H(\tau)$ is the Heaviside unit step time function, while $\mathbf{C}_0$, θ_0, and θ_1 are constant in time. Under such assumptions an appropriate choice for the free energy function may be

$$\Psi = \Phi^\infty(I_1^C,I_2^C,I_3^C,\theta)+ \tag{54}$$

$$+\int_{-\infty}^{t}\Big[\mathcal{I}_1^2[\chi_0(\zeta,\theta)+(I_1^C-3)\chi_1(\zeta,\theta)]+$$

$$+\mathcal{I}_1\mathcal{J}_1\chi_2(\zeta,\theta)+(\mathcal{I}_2-\tfrac{4}{7}\mathcal{J}_3)\chi_3(\zeta,\theta)+$$

$$+\mathcal{I}_2(I_1^C-3)\chi_4(\zeta,\theta)\Big]\,a[\mathbf{C}(\tau),\theta(\tau)]d\tau.$$

From (41) follows the expression for the symmetric Piola–Kirchhoff stress tensor

$$\frac{1}{2\varrho_R}\mathbf{T} = \alpha_0\mathbf{I}+\alpha_1\mathbf{C}+\alpha_2\mathbf{C}^2- \tag{55}$$

$$-\int_{-\infty}^{t}\Big[2\Big(\chi_0(\zeta,\theta)+$$

$$+(I_1^C-3)\chi_1(\zeta,\theta)\Big)\mathrm{Tr}\mathbf{C}^*+$$

$$+\chi_2(\zeta,\theta)\mathrm{Tr}\mathbf{C}\mathbf{C}^*\Big]\mathbf{I}+\chi_2(\zeta,\theta)(\mathrm{Tr}\mathbf{C}^*)\mathbf{C}+$$

$$+2[\chi_3(\zeta,\theta)+(I_1^C-3)\chi_4(\zeta,\theta)]\mathbf{C}^*-$$

$$-\tfrac{4}{7}\chi_3(\zeta,\theta)(\mathbf{C}\mathbf{C}^*+\mathbf{C}^*\mathbf{C})\Big]\,ad\tau.$$

Here the expressions for characteristic functions α_0,α_1, and α_2 are polynomial functions of I_k^C and γ_L^∞

$$\alpha_k=\alpha_k(I_1^C,I_2^C,I_3^C,\gamma_1^\infty,\gamma_2^\infty,\gamma_3^\infty,\gamma_4^\infty,\gamma_5^\infty). \tag{56}$$

After some tedious but straightforward algebraic manipulations one obtains

$$\alpha_0 = 2\gamma_1^\infty(\mathrm{Tr}\mathbf{C}-3)+ \tag{57}$$

$$+\gamma_3^\infty(3\mathrm{Tr}^2\mathbf{C}-17\mathrm{Tr}\mathbf{C}+27)+$$

$$+\tfrac{3}{2}\gamma_4^\infty(\mathrm{Tr}^2\mathbf{C}-2\mathrm{Tr}\mathbf{C}-\tfrac{1}{3}\mathrm{Tr}\mathbf{C}^2-2)+$$

$$+\tfrac{1}{2}\gamma_5^\infty(\mathrm{Tr}^2\mathbf{C}-\mathrm{Tr}\mathbf{C}^2),$$

$$\alpha_1 = -\gamma_2^\infty-\gamma_5^\infty-\gamma_4^\infty(\mathrm{Tr}\mathbf{C}-3),\quad \alpha_2=\gamma_5^\infty.$$

The expression for specific entropy can be determined from

$$\eta = -\frac{\partial}{\partial\theta}\Phi^{\infty}(I_1^C, I_2^C, I_3^C, \theta), \tag{58}$$

while the formal expression of a dissipation function is given as

$$\mathscr{D} = -\frac{a}{\theta}\int_{-\infty}^{t}\frac{\partial}{\partial\zeta}\mathcal{F}(I_1^C, \mathcal{I}_1, \mathcal{I}_2, \mathcal{J}_1, \mathcal{J}_2, \mathcal{J}_3, \theta, \zeta)\, a[\mathrm{C}(\tau), \theta(\tau)]d\tau. \tag{59}$$

For each specific material model we need to prescribe the particular form of rheological function of state $\mathcal{F}$ and evalute required partial derivatives. However, we refrain from such undertaking, since it is not necessary for further development.

Next we consider a theoretical development for a class of incompressible materials. An incompressible material can sustain only isochoric deformations. Consequently, every deformation of an incompessible material is subject to the internal material constraint $I_3^C = 1$ or $\mathrm{DetF} = 1$. By imposing such a constraint and using a general form of potential function for a class of incompressible materials due to Signorini, (Locket 1972), we obtain

$$\begin{aligned}\Phi(I_1^C, I_2^C, \theta) &= \gamma_0(\theta) - \tfrac{1}{2}\gamma_2(\theta)(I_2^C - 3) + \\ &+ \left[\tfrac{1}{2}\gamma_1(\theta) + \gamma_2(\theta)\right](I_1^C - 3) + \\ &+ \tfrac{1}{4}\left[\tfrac{1}{2}\gamma_3(\theta) - \gamma_2(\theta)\right](I_1^C - 3)^2.\end{aligned} \tag{60}$$

For $\gamma_3 = -2\gamma_2$ the Mooney–Rivlin type of potential is recovered, while $\gamma_2 = \gamma_3 = 0$ leads to the neo–Hookean type of material.

Under the constraint of incompressibility the free energy function reduces to

$$\begin{aligned}\Psi &= \Phi^{\infty}(I_1^C, I_2^C, \theta) - \frac{1}{8}\int_{-\infty}^{t}\Big(\Phi_1(\zeta)\mathrm{Tr}^2\mathrm{C}^{*2} + \\ &+ 2\Phi_2(\zeta)\mathrm{Tr}\mathrm{C}^{*2}\Big)a[\mathrm{C}(\tau), \theta(\tau)]d\tau,\end{aligned}$$

Following definitions in (22) and (41) we obtain expressions for the symmetric Piola–Kirchhoff stress tensor

$$\begin{aligned}\frac{1}{2\varrho_R}\mathrm{T} &= p\mathrm{C}^{-1} + \tfrac{1}{2}\gamma_2^{\infty}\mathrm{C} + \\ &+ \tfrac{1}{4}[2(\gamma_1^{\infty} - \gamma_2^{\infty}) + \gamma_3^{\infty}(I_1^C - 3)]\mathrm{I} + \\ &+ \frac{1}{4}\int_{-\infty}^{t}\Big(\Phi_1(\zeta)(\mathrm{Tr}\mathrm{C}^*)\mathrm{I} + \\ &+ 2\Phi_2(\zeta)\mathrm{C}^*\Big)a[\mathrm{C}(\tau), \theta(\tau)]d\tau,\end{aligned} \tag{61}$$

specific entropy

$$\eta = -\frac{\partial}{\partial\theta}\Phi^{\infty}(I_1^C, I_2^C, \theta), \tag{62}$$

and internal dissipation

$$\begin{aligned}\mathscr{D} &= \frac{a}{8\theta}\int_{-\infty}^{t}\Big(\mathrm{Tr}^2\mathrm{C}^*\frac{\partial\Phi_1}{\partial\zeta} + \\ &+ 2\mathrm{Tr}\mathrm{C}^{*2}\frac{\partial\Phi_2}{\partial\zeta}\Big)a[\mathrm{C}(\tau), \theta(\tau)]d\tau.\end{aligned} \tag{63}$$

The expression for the free energy Ψ at reduced time ξ on the single–step time histories (50–53) reads

$$\begin{aligned}\Psi^{\xi} &= \gamma_0^{\infty} + \tfrac{1}{2}(\mathrm{Tr}\mathrm{C}_0 - 3)(\gamma_1^{\infty} + 2\gamma_2^{\infty}) + \\ &+ \tfrac{1}{8}(\mathrm{Tr}\mathrm{C}_0 - 3)^2(2\gamma_2^{\infty} + \gamma_3^{\infty}) - \\ &- \tfrac{1}{4}(\mathrm{Tr}\mathrm{C}_0 - 3)^2\int_{\xi}^{\infty}\Phi_1(z)dz - \\ &- \tfrac{1}{4}(\mathrm{Tr}\mathrm{C}_0^2 - 2\mathrm{Tr}\mathrm{C}_0 + 3)\int_{\xi}^{\infty}\Phi_2(z)dz.\end{aligned} \tag{64}$$

The derived form of dissipative inequality, which must be fulfilled over the whole time frame $0 \leq \xi < \infty$

$$\begin{aligned}&- 3(2\mathrm{Tr}\mathrm{C}_0 - 3)[\Phi_1(\xi) + \tfrac{2}{3}\Phi_2(\xi)] + \\ &+ \mathrm{Tr}^2\mathrm{C}_0\Phi_1(\xi) + 2\mathrm{Tr}\mathrm{C}_0^2\Phi_2(\xi) \leq 0,\end{aligned} \tag{65}$$

is of a special importance, since it imposes necessary restrictions on constitutive behavior of relaxation functions $\Phi_1(\xi)$ and $\Phi_2(\xi)$ in their experimental determination in combination with different deformation tensors C_0.

4 CONCLUSIONS

The main scope of the paper was to demonstrate a theoretical development of a constitutive model of thermo–rheologically simple material using strict thermodynamic approach based on the principle of fading memory. The proposed constitutive model can accomodate a broad range of thermo–hyperelastic as well as thermo–viscoelastic material properties. In contrast to multiple integral type of constitutive relations which require a substantial experimental work in determination of multi–argument rheological functions, the single–integral type of constitutive equation is more suitable for experimental treatment. However, equations of this class are not capable of capturing an adequate constitutive behavior of the material in all situations. Nevertheless, for many practical purposes single–integral type of constitutive relations can be quite satisfactory. On the other hand, the theoretical basis and experimental data should be sophisticated enough to be able to describe a more complicated constituive response of the material. When the mechanical behavior can be represented by linear and infinitesimal strain theory, then experimental material characterization can be done quite successfully, since for linear viscoelastic behavior the mathematical theory and experimental methods are well established. However, at high deformations the stress strain relationship is nonlinear. Due to complexity of nonlinear material behavior, such a characterization is quite involved and far from being trivial. Moreover, the design procedures for nonlinear viscoelastic behavior are still not adequately developed. This is particularly true for gathering of experimental data in relation to the particular theoretical model. Somethimes the complexity of particular model exceeds capabilities of the numerical simulation for most practical purposes. Due to certain limitations in experimental determination of the data for calibration of theoretical models, the rigour of theoretical model frequently exceeds the capabilities of experimental work. Therefore, it is very important to develop constitutive models which can accomodate both extremes. By comparing the equations derived from a given model with measured experimental data, the model can be subjected to the scrutiny of experimental verification. Established general 3D constitutive relation can then be specialized for the specific state of stress, and free parameters, which are suitable for experimental determination of material constants, can be identified. Since the material functions and the free constants of the proposed model depend only on a single time–like variable, they can be directly calibrated from the experiment. Dissipative inequality, which is given in an explicit form for the particular case considered in the paper, may prove usefull in the analysis of intricacies that may be encountered in an experimental determination of the rheological function of state whose behavior is restricted by the second law of thermodynamics.

REFERENCES

Beatty, M. F. 1987. Topics in finite elasticity: Hyperelasticity of rubber, elastomers, and biological tissues—with examples. *Appl. Mech. Rev.* 40: 1699–1734.

Coleman, B. D. & W. Noll 1961. Foundations of linear visco–elasticity. *Rev. Mod. Phys.* 33: 239–249.

Coleman, B. D. 1964. Thermodynamics of materials with memory. *Arch. Rat. Mech. Analysis* 17: 3–46.

Chadwick, P. 1974. Thermo–mechanics of rubberlike materials. *Phil. Trans. R. Soc. London A* 276: 371–403.

Day, W. A. 1972. *The Thermodynamics of Simple Materials with Fading Memory.* Berlin: Springer–Verlag.

Day, W. A. 1976. Entropy and hidden variables in continuum thermodynamics. *Arch. Rat. Mech. Analysis* 17: 3–46.

Ferry, J. D. 1980. *Viscoelastic Properties of Polymers.* New York: John Wiley & Sons.

Findley, W. N., J. S. Lai & K. Onaran 1976. *Creep and Relaxation of Nonlinear Viscoelastic Materials.* Amsterdam: North Holland.

Germain, P., Q. S. Nguyen & P. Suquet 1983. Continuum thermodynamics. *J. Appl. Mech.* 50: 1010–1020.

Lianis, G. 1965. Application of irreversible thermodynamics in finite viscoelastic deformations. *J. Appl. Mech.* 32:166–173.

Lianis, G. 1970. Application of thermodynamics of viscoelastic materials with fading memory integral constitutive equations. *Int. J. Non–Linear Mech.* 5: 23–34.

Lockett, F. J. 1972. *Nonlinear viscoelastic solids.* London–New York: Academic Press.

Truesdell, C. & W. Noll 1992. *The Non–Linear Field Theories of Mechanics.* Berlin: Springer–Verlag.

Truesdell, C. 1984. *Rational Thermodynamics.* New York, Berlin: Springer–Verlag.

Ward, I. M. 1985. *Mechanical Properties of Solid Polymers.* Chichester: John Wiley & Sons.

Constitutive Models for Rubber, Dorfmann & Muhr (eds) © 1999 Taylor & Francis ISBN 90 5809 113 9

Effect of liquids on the dynamic properties of carbon black filled natural rubber as a function of pre-strain

J.J.C. Busfield, C. Deeprasertkul & A.G. Thomas
Materials Department, Queen Mary and Westfield College, London, UK

ABSTRACT: A free oscillation technique has been adopted to measure the dynamic storage and loss moduli of carbon black filled natural rubber materials. These tests are conducted with small oscillations that are superimposed on a range of tensile pre-strains. In addition, the effect of temperature on the dynamic moduli is measured as well as the effect of swelling the materials by different amounts using oil (dibutyl-adipate). It is observed that the dynamic storage and loss moduli do not depend strongly on the pre-strain at small strains and an increase in the temperature causes a dramatic reduction in both the storage and loss moduli. The filled materials are then swollen with a non volatile oil. The measured behaviour can be approximately ascribed to the combined effects of a reduction in the modulus of the rubber matrix (caused by the swelling action) and a reduction in the effective volume fraction of the filler.

1 INTRODUCTION

The dynamic behaviour of filled rubbers is of key importance in the performance of rubber engineering components. In many cases, and in particular tyres, mineral oils are incorporated into the compound during manufacture. The purpose of this is both to improve the processing behaviour and to influence the final properties of the product.

There appear to have been few scientific studies of the effect of the addition of liquids on the dynamic behaviour of filled rubbers, although there has been work by Davies et al. (1996) on the effect for unfilled rubbers. This work covered a range of liquids of widely varying viscosity values, and of rubbers of different crosslink densities. In this work, the liquids were incorporated by swelling them into an already vulcanised material. This contrasts with the procedure used in manufacture where the liquid is mixed into the raw rubber together with the other compounding ingredients. The swelling method has the considerable advantage that the degree of crosslinking can be found from measurements on the unswollen material, using rubber elasticity theory. In contrast, if the material is vulcanised after the incorporation of the liquid the stress-strain data is less reliably interpretable. Also, the liquid may, and often does, interfere with the chemistry of the crosslinking reactions. Therefore for the present work on the effect of liquids on the dynamic properties of filled rubbers, which may be regarded as an extension of that referred to above, the same swelling procedure is used. It is recognised that the usual manufacturing procedure for 'oil extension' will produce, even for identical crosslink densities, a different molecular chain configuration in the vulcanisate from that when the liquid is introduced after crosslinking, although it seems unlikely that the change in dynamic properties caused by the presence of the liquid will be much different in the two cases.

The experimental technique used is similar to that described by Davies et al (1996). A static tensile strain is applied to a strip and a small oscillation superimposed. This has relevance to the various engineering applications where an oscillation is superimposed on a static deflection. It is known that the modulus for this small oscillation is greater than that deduced from the slope of the static force-deflection relation at the appropriate deflection, and the difference is often referred to as the 'dynamic/static ratio'. This, and other effects like the 'Mullins effect' and the related stress-softening phenomenon, are associated with the complications in the stress-strain behaviour due to the presence of fillers. Because of this, care has to be taken to avoid any inadvertent pre-stressing of the test piece during the experimentation.

The technique of using a free oscillation, with measurements of frequency and logarithmic decrement to determine the dynamic properties, was adopted by Davies et al. (1996) because it gives a much more accurate means of assessing the loss modulus for low loss materials than most forced oscillation methods. It was retained for similar reasons in the present work. The method has the limitations that the frequency cannot be easily controlled, and the test amplitude is, of necessity, variable. Fortunately for these materials, the frequency dependence of their behaviour is only slight; the possible non-linearity at the small test strains is considered later.

2 EXPERIMENTAL

A plan of the apparatus is given in Figure 1 and the significant geometry is shown in Figure 2. The apparatus measures the dynamic visco-elastic properties as a function of pre-strain and temperature. The oscillating beam, with weights of independently measured moment of inertia, was supported by the knife-edge. The beam was clamped at the centre of a rubber specimen. The specimen was stretched by moving the position of the end grips and was left to reach equilibrium. The oscillation was activated remotely by switching off an electromagnet. The signal from the capacitance transducer was recorded on a chart recorder and the frequency and logarithmic decrement of the oscillation calculated.

The behaviour of linear visco-elastic materials can be represented by a complex modulus E^* (Young & Lovell 1996),

$$E^* = E' + iE''. \tag{1}$$

Here E' is the storage modulus (elastic component) and E'' is the loss modulus (dissipated energy component). In our case the definitions of E' and E'' are expressed in terms of the deformed geometry. This approach was adopted because Davies et al. (1996) showed that for unfilled rubbers the loss modulus expressed in this way was independent of the pre-extension up to quite large pre-strains (about 200%).

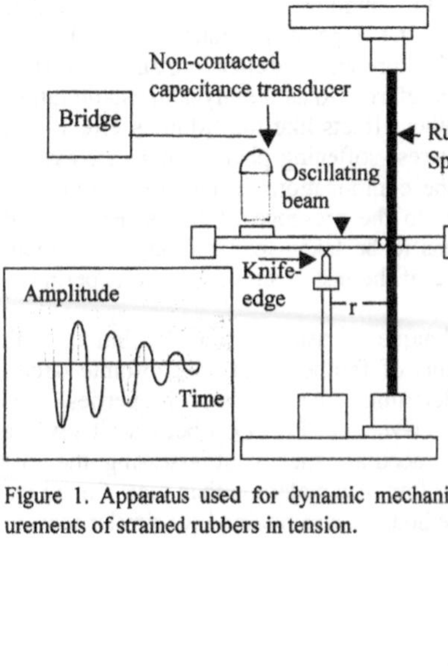

Figure 1. Apparatus used for dynamic mechanical measurements of strained rubbers in tension.

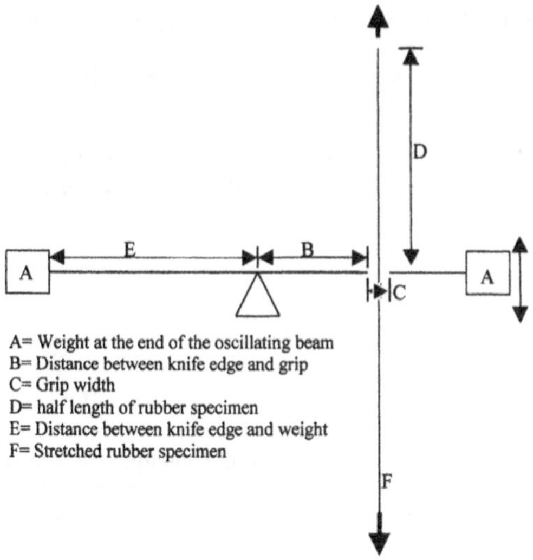

Figure 2. A schematic of the oscillating beam used for this work.

With our experimental arrangement E' and E'' are given by

$$E' = \left(\frac{I\omega^2}{2r^2}\right)\left(\frac{l_0}{a_0}\right)\lambda^2 \tag{2}$$

$$E'' = \left(\frac{\Lambda}{\pi}\right)E' \tag{3}$$

Here I is the moment of inertia of the beam, λ the extension ratio, r the distance from knife edge to the clamped rubber, ω is 2π times the frequency of oscillation, Λ its logarithmic decrement, l_0 and a_0 are the original length and cross-sectional area of the test-piece respectively and λ is its extension ratio. Akutagawa (1995) showed that it is important that the width, C, of the grip attaching the strip to the inertia beam is small compared with the knife edge-grip length. This prevents a significant distortion of the geometry on rotation of the beam.

The formulations of the rubbers used in these experiments are tabulated in Table 1. The compounded rubbers were vulcanised into 1mm thickness sheets at a temperature at 155°C for 10 minutes. From these sheets test pieces of dimensions 100x5x1mm were prepared. The tensile stress-strain tests were conducted on an Instron 1122 at a crosshead speed of 1mm/min, the load being recorded at extension ratio intervals of 0.1.

For this study dibutyl-adipate oil (DBA) was used. The viscosity of the DBA was determined by the falling ball method using the Stokes equation as 0.968±0.1 Pa.s at 16°C.

Swelling experiments were conducted. Initially the rubber samples were weighed (using an elec-

Table 1. The rubber formulations.

Rubber Ingredients	NR 0	NR 29	NR50	NR 59
Natural Rubber	100	100	100	100
Carbon Black	0	29	50	59
Anti-oxidant	3	2	3	2
Anti-ozonant	0	3	0	3
Accelerator	1.5	2.4	1.5	2.4
Sulphur	1.5	1	1.5	2.7

tronic balance) before and after immersing them in the oil. The changes in the weight of the rubber samples were monitored with respect to time. The volume of liquid absorbed, V_s, was calculated from the increase in the weight of the swollen sample. Hence, the volume fraction of rubber, V_r, in the swollen sample, referred to the rubber fraction of the material could be calculated using the relationship,

$$V_r = V_m/(V_m+V_s) \tag{4}$$

where V_r is the rubber volume fraction in swollen rubber, V_m is the volume of rubber network and V_s is the volume of absorbed liquid.

The rubber network refers to the network formed from the rubber hydrocarbon and the curative by the vulcanisation process. V_m can be calculated from the initial weights of the samples and the mix compositions, using the relationship

$$V_m = (M_0.M_m)/(M_t.\rho_{RH}). \tag{5}$$

Where M_m is the weight of rubber network (including the weight of the rubber and the sulphur), M_t is the total weight of mix formulation, M_0 is the weight of the specimen and ρ_{RH} is the density of natural rubber.

A range of swellings was attained by immersing the samples for various times, removing them from the liquid and allowing enough time for a uniform distribution to be reached. This uniform distribution time was assumed equivalent to that required for equilibrium swelling in the oil to be reached. The dynamic properties of these swollen samples were then measured as before.

3 RESULTS AND DISCUSSION

The analysis of the results as storage and loss moduli requires that for the small oscillations used the materials behave in a linearly visco-elastic manner. It is important to check this for filled rubbers as these are known to behave a non-linear manner at strains of only a few percent. This was done as follows. A plot of the deflection versus time on a chart recorder was obtained for all the tests conducted. Typical results are shown in Figure 3 for the unswollen compound containing 29 parts of carbon

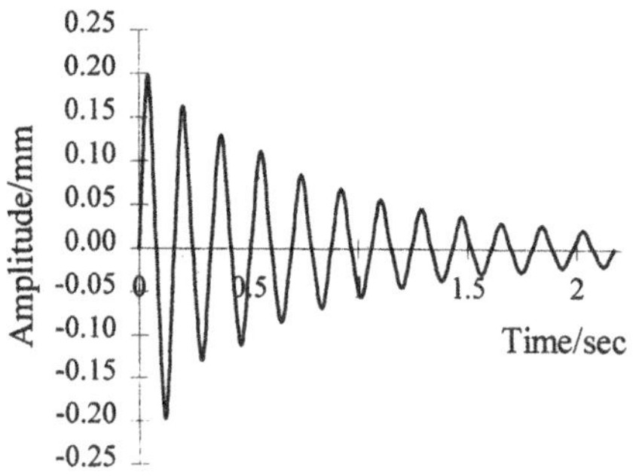

Figure 3. Oscillation amplitude plotted against time for NR29 material.

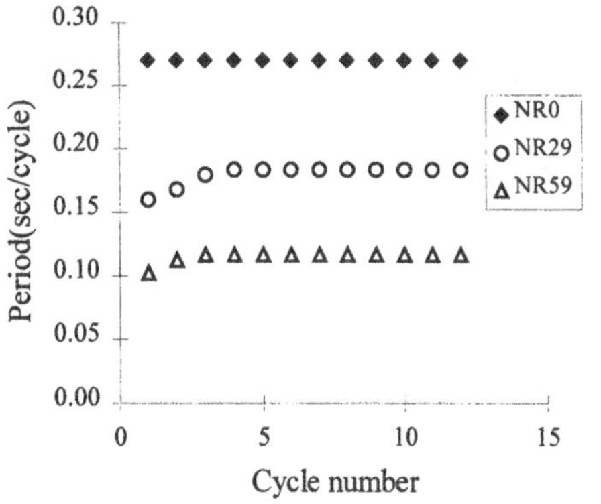

Figure 4. Period of oscillation versus the number of cycles derived from Figure 3.

black filler. From this the period, and hence frequency, of each cycle could be calculated. The results for each of the unswollen materials are shown in Figure 4. Apart from the first two cycles, where transient effects associated with the triggering process would be expected, the frequency is constant despite the continually changing amplitude shown in Figure 3. This implies that the stiffness is independent of amplitude and thus a linear analysis is valid.

With a free oscillation method as used here, it is not easy to cover a wide range of frequency and it was not intended to investigate frequency dependence in any detail in this work. It is difficult to maintain a precisely fixed frequency as the various variables are changed, and a range of about 4 to 15Hz was encountered. Some experiments were therefore done in which the frequency was intentionally changed, with the result that a frequency effect of less than about 5% in the loss modulus, was found for NR29 over this range. This is not unexpected in the light of published data on rubber viscoelastic behaviour, including those due to Davies et al. (1996), and accordingly this slight effect, being comparable to the experimental precision, was ignored.

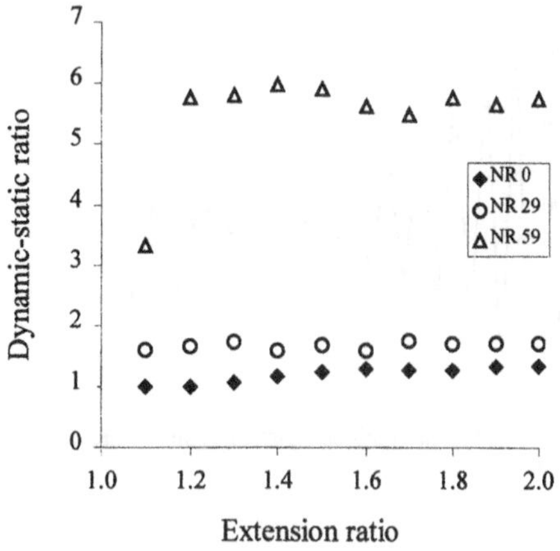

Figure 5. A comparison between the modulus measured from the tangent of the stress-strain behaviour and E' measured using the oscillating beam.

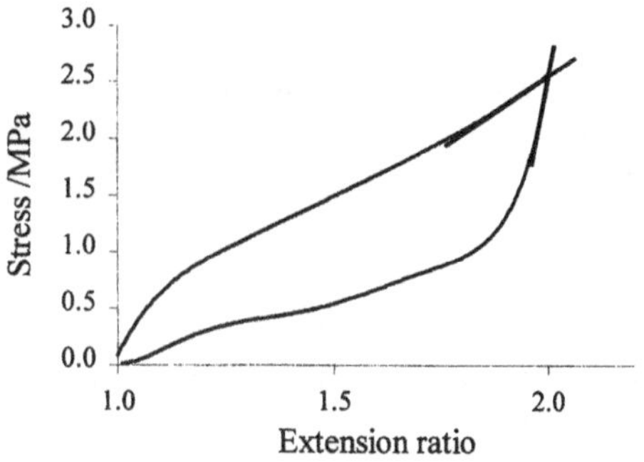

Figure 6. The extension and retraction stress versus strain relationship for the first loading cycle for NR59.

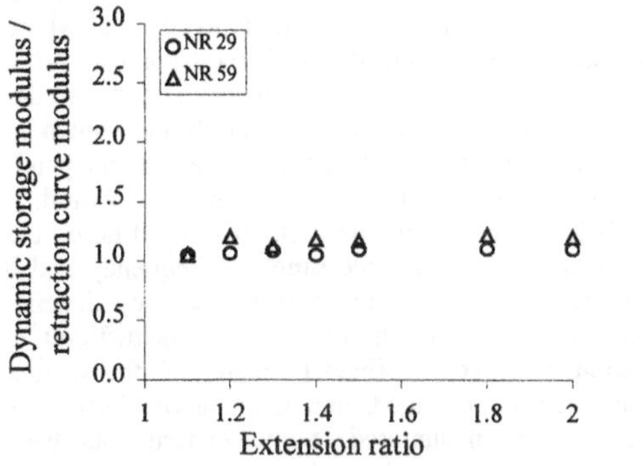

Figure 7. A comparison between modulus measured fro the tangent of the retraction stress-strain behaviour for NR29 and NR59 compared with the dynamic storage modulus, E'

For a perfectly elastic material the slope of the extension stress-strain curve would be equal to the dynamic, incremental, modulus as measured in the present experiments. For imperfectly elastic materials this would of course no longer be the case although for unfilled natural rubber vulcanizates, being highly resilient, Medalia (1978) showed that the slope and the dynamic modulus are very similar. However, Figure 5 shows that for a filled rubber the dynamic modulus is typically several times greater than the slope of the extension stress-strain curve. In engineering applications this effect is well known, and is usually referred to as the 'dynamic/static ratio'. This is a somewhat misleading term as it suggests that the effect is associated with an influence of the frequency of the oscillation. From results quoted in the previous paragraph and from other published work by Medalia (1978) it can be seen that any influence of this sort would contribute only a small fraction to this effect. It is generally thought that the slope of the retraction stress-strain curve is the determining factor in the dynamic modulus as measured here (Davey & Payne 1964).

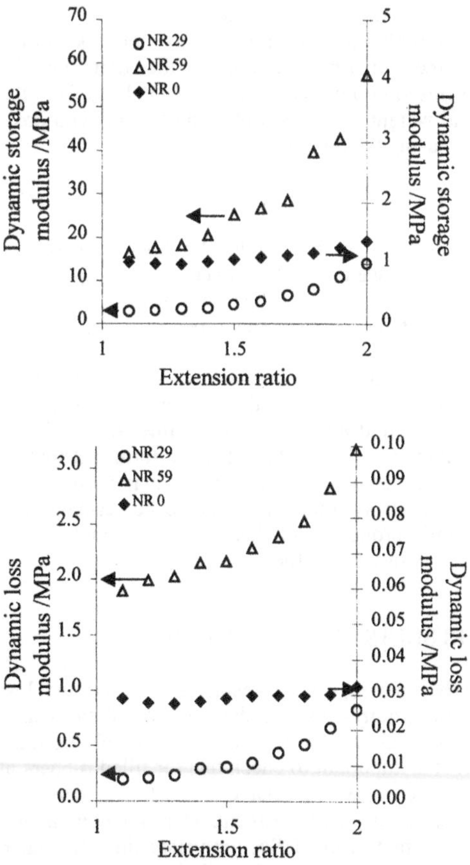

Figure 8a and 8b. The dynamic storage (a) and loss (b) moduli of NR0, NR29 and NR59 plotted as a function of pre-strain.

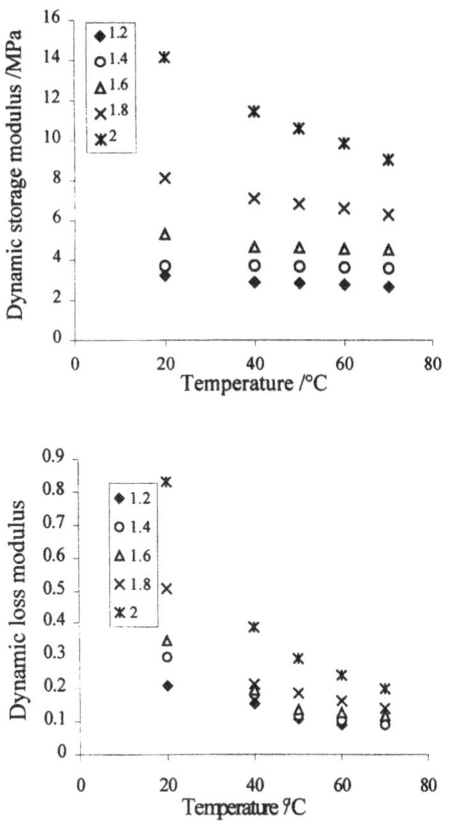

Figure 9a and 9b. The dynamic storage (a) and loss (b) moduli of NR29 plotted as a function of temperature. The different series represent the different pre-extension ratios.

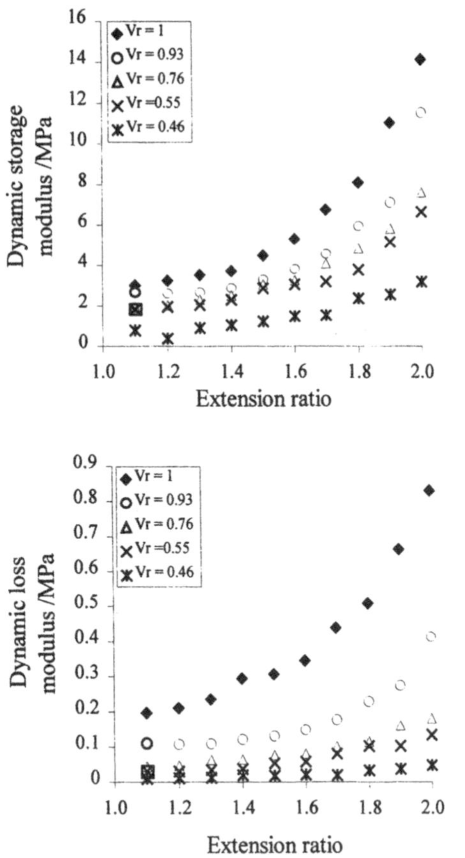

Figure 10a and 10b. The dynamic storage (a) and loss (b) moduli of NR29, swollen in DBA to different volume fractions as a function of strain.

Figure 6 shows the extension and retraction curves for NR59. The straight lines indicate the relevant extension and retraction moduli. The retraction modulus is measured in this way for a range of strains, and a comparison is made between this and the dynamic storage modulus in Figure 7. This figure shows that these two quantities are in fact closely similar. The high slope of the retraction curve at the pre-strain point is a consequence of the Mullins (1950) effect shown by filled rubbers.

The dynamic properties of the three unswollen rubbers were measured over a range of pre-strains, and the results are shown as the variation of E' and E'' in Figures 8a and 8b. The expected marked increase in both E' and E'' due to the filler is clear. There is a similar qualitative dependence on strain to that found for the unfilled material by Davies et al. (1996). There is a region at low to moderate strains where both E' and E'' are approximately constant, and for higher strains both show marked increases.

Comparison with the earlier work on unfilled materials shows that the upsweep occurs at lower strains with the present filled materials. This can be attributed to what has been termed by Mullins (19??) as the strain amplification effect. This implies that the strains in the rubber matrix are greater than that of the overall strain of the filled material, because of the presence of an inextensible filler material. So any non-linearity in the behaviour of the rubber itself will appear at a lower overall strain.

Some measurements have been done over a range of temperatures (23 to 70° C) with the results shown in Figure 9a and 9b. The storage modulus E' shows a decrease with increasing temperature, the effect being more pronounced at higher pre-strains. A similar effect is found for E''. An unfilled rubber will normally show an increase in modulus with temperature in accordance with the statistical theory of rubber-like elasticity, although viscous effects may mask this behaviour if measurements are made at too high a frequency or too low a temperature. In contrast, filled rubbers are known to show a decrease of modulus with temperature, the mechanism for which is not completely clear. One possible mechanism is

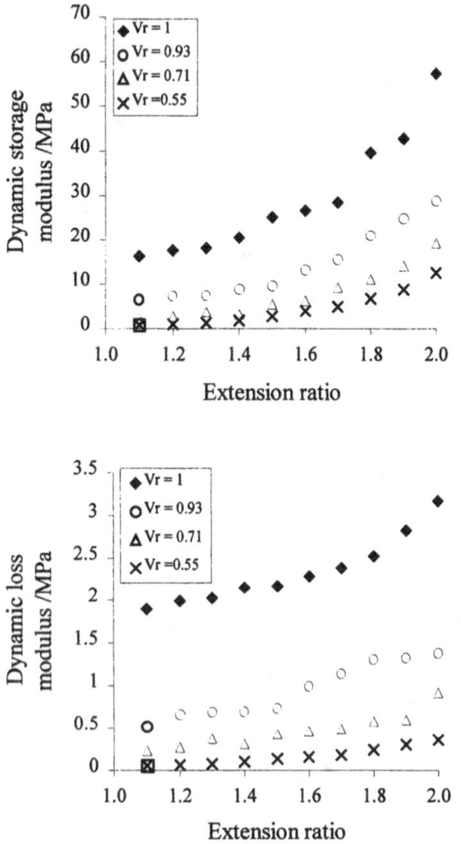

Figure 11a and 11b. The dynamic storage (a) and loss (b) moduli of NR59, swollen in DBA to different volume fractions as a function of strain.

that slippage of the rubber over the filler surface may occur more readily at high temperatures, thus reducing the stiffening effect of the filler. Another possibility is that filler aggregates, which are known to be present, may be disrupted by the application of a small stress, and the strength of these aggregates may decrease with temperature due to thermal agitation. It is not easy to distinguish between these mechanisms by purely mechanical measurements in the absence of quantitative theories. However, electrical conductivity data reported by Donnet et al. (1993) suggest that significant aggregate disruption does occur with the application of small strains of only a few percent. Thus, thermal agitation may be potent enough to give an appreciable weakening of the filler aggregate. The dissipative process determining *E''*, which will involve the breakage and reformation of the aggregate, may be expected to parallel the *E'* behaviour.

The dynamic behaviour of the swollen materials over a range of strain and degrees of swelling has

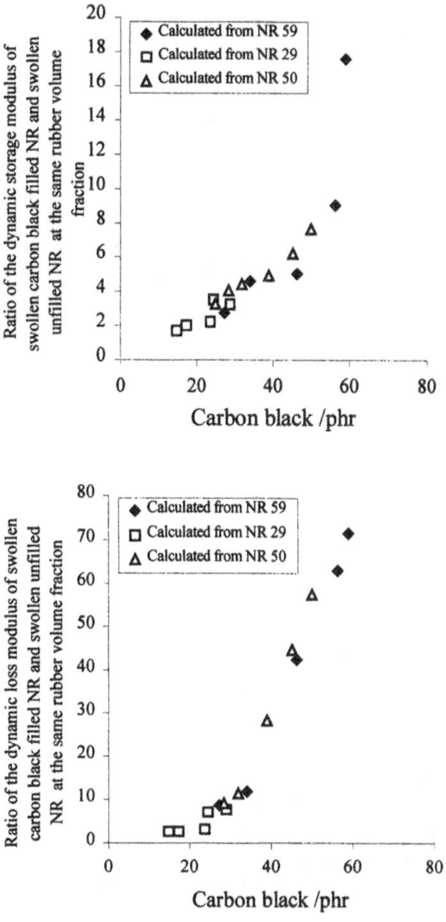

Figure 12a and 12b. The dynamic storage (a) and loss (b) moduli plotted as a function of rubber volume fraction.

also been examined; the results are shown in Figures 9 to 11. The amount of swelling is expressed as the volume fraction of rubber v_r in the rubber phase. The liquid used in the data shown is di-butyl adipate. The general behaviour is similar to that of the unfilled rubber, but the dependence of both *E'* and *E''* on swelling is much stronger, particularly for the more highly filled materials. This is perhaps to be expected, because when the material is swollen not only is the modulus of the rubber matrix decreased but also the effective volume fraction of the carbon black filler is reduced. A lower filler content will of course lead to decreases in *E'* and *E''*, as found above. It is possible to check whether a combination of these effects is sufficient to account for the observed changes.

Figures 12a, 12b, show *E'* and *E''* respectively, for the filled swollen materials (at low extensions) divided by the corresponding values for the unfilled materials at the same degree of swelling, are plotted against the volume fraction of filler, allowing for the decrease in the filler loading due to the swelling. Results for all the rubbers, with a range of both initial filler loading and degree of swelling are shown on a single graph. If the combination of the effects of swelling suggested above is correct then all the data will be consistent with a single curve. Figure 12 suggests that this is at least approximately true.

4 CONCLUSIONS

The dynamic behaviour of natural rubber vulcanisates containing several levels of filler (carbon black) loading has been studied for small oscillations superposed on a range of tensile pre-strains. The storage and loss moduli, E' and E'', do not depend strongly on pre-strain at small pre-strains, which is qualitatively similar to the behaviour of unfilled rubber (Davies et al. 1995) although the increases at higher strains occur earlier in the filled case. This is probably ascribable to the strain amplification effect usually invoked to explain the general stress-strain behaviour of filled rubbers. Increases in temperature produce substantial decreases in both E' and E''.

Swelling the materials in non-volatile solvents produces relatively large decreases in E' and E'' compared with the corresponding changes in unfilled rubbers. These decreases can be at least approximately ascribed to a combination of the softening effect of swelling on the rubber matrix and the decrease, with swelling, in the effective volume fraction of filler.

REFERENCES

Akutagawa, K. 1995. *Ph.D. Thesis*, Queen Mary and Westfield College, University of London.

Davey A.B. & Payne, A.R. 1964. *Rubber in Engineering Practice*. London: MacLaren & Son.

Davies C.K.L., Thomas A.G. & Akutagawa K. 1996. The Effect of Low Molar Mass Liquids on the Dynamic Mechanical Properties of Elastomer Under Strain, *Progress in Rubber and Plastics Technology* 12: 174.

Donnet, J-B. Bansal R.C. & Wang, M.J. 1993. *Carbon black Science and Technology*. New York: Marcel Dekker.

Medalia, A.I. 1978. Effect of Carbon black on Dynamic Properties of Rubber Vulcanizates. *Rubber Chemistry and Technology*. 51: 437.

Mullins, L. 1986. Engineering with Rubber. *Rubber Chemistry and Technology*. 59(3): G69-82

Mullins, L. 1950. Thixotropic Behaviour of Carbon Black in Rubber. *J. Phys. and Colloid Chem.* 54: 239.

Young R.J. & Lovell P.A. 1996. *Introduction to Polymers*. London: Chapman & Hall.

Constitutive Models for Rubber, Dorfmann & Muhr (eds) © 1999 Taylor & Francis ISBN 90 5809 113 9

A model of cooperative relaxation in finite viscoelasticity of amorphous polymers

Aleksey D. Drozdov
Institute for Industrial Mathematics, Beersheba, Israel

ABSTRACT: New constitutive equations are derived for the isothermal response of rubbery polymers at finite strains. An amorphous polymer is treated as an ensemble of cooperatively rearranging regions. The viscoelastic behavior is modeled as rearrangement of relaxing regions occurring at random times. The kinetics of transformation is described by the concept of thermally activated processes. Adjustable parameters in stress–strain relations are found by fitting observations for poly(methyl methacrylate) and polydienes functionalized by benzoic units in tensile relaxation tests. Fair agreement is demonstrated between experimental data and results of numerical simulation.

1 INTRODUCTION

The study is concerned with modeling the nonlinear viscoelastic response of amorphous rubbery polymers using the concept of cooperative relaxation (Adam & Gibbs 1965). Our objective is to derive stress–strain relations with finite strains using the trapping concept which takes account of random changes in energies of flow units at the instants of their rearrangement (Brawer 1984, Bässler 1987, Dyre 1987,1995, Monthus & Bouchaud 1996, Sollich 1998) and to verify the constitutive equations by comparison of experimental data in isothermal relaxation tests with results of numerical simulation.

Unlike conventional techniques that do not distinguish reformation of initial and rearranged flow units, we describe the distribution of relaxing regions by the function of three variables $\Xi(t, \tau, w)$. Stress–strain relations for an amorphous polymer are developed for an arbitrarily three-dimensional loading (previous studies have been confined to uniaxial loading) based on the laws of thermodynamics (conventional approaches employ oversimplified relationships, where the material viscosity is calculated as a reciprocal to the mean time for rearrangement).

2 REARRANGEMENT OF FLOW UNITS

Time-dependent response of rubbery polymers requires cooperative dynamics of long chains when scores of neighboring strands simultaneously change their position with respect to the bulk material (Adam & Gibbs 1965). The concept of cooperative relaxation is confirmed experimentally in tests on dipolar reorientation in viscous liquids (Dyre 1987).

A relaxing region (a domain containing the minimum number of strands that allows rearrangement) is a globule consisting of long chains, short chains and free volume clusters. Flow units are treated as mesoscopic elements which are (i) small enough to neglect their interaction and (ii) large enough to apply standard techniques in continuum mechanics to their description (Sollich 1998). Their characteristic size is amounted to several nanometers (Rizos & Ngai 1999).

The rearrangement process in the phase space is modeled as a sequence of random hops of flow units from one local equilibrium state on the energy landscape to another (Bässler 1987). The depth of the potential well where a flow unit is located at the current time t with respect to the liquid-like (reference) state (Goldstein 1969) on the energy landscape is determined by its potential energy $w > 0$ [the difference between the energy in the liquid-like state and the potential energy in the local equilibrium (Dyre 1987)].

In a stress-free medium, flow units are located at the bottom levels of their potential wells. After application of loads, relaxing regions ascend to higher energy levels, and their positions with respect to the reference state are determined by the difference $w_0 = w - u$, where u is the excess energy of a flow unit above the bottom level of its potential well. Observations show that the excess energy u grows slower than the specific mechanical energy of a relaxing region U, since a part of the energy U is spent to produce a "shoving work" on the surrounding medium (Dyre 1998) and to bring into being "irreversibility in the material structure" (Goldstein 1969). To account for these phenomena, we set

$$u = \delta U, \tag{1}$$

where $\delta \in (0, 1)$ is an adjustable parameter.

The rearrangement kinetics is described by equations similar to those for reformation of temporary networks (Drozdov 1998). Denote by $\Xi(t, \tau, w)$ the number (per unit mass) of relaxing units with potential energy w that (i) have rearranged before time τ and (ii) have not hopped into the liquid-like state within the interval (τ, t). The function $\Xi(t, \tau, w)$ entirely determines the rearrangement process: the quantity $\Xi(t, 0, w)$ is the current concentration of regions with energy w which have not rearranged in the interval $[0, t)$; the derivative

$$\frac{\partial \Xi}{\partial \tau}(t, \tau, w)\Big|_{t=\tau}$$

is the rate of hopping into traps with energy w at time τ; and the amounts

$$-\frac{\partial \Xi}{\partial t}(t, 0, w), \qquad -\frac{\partial^2 \Xi}{\partial t \partial \tau}(t, \tau, w)$$

determine the rates of changes of traps at time t for initial flow units and for regions rearranged at time τ.

The relative rates of rearrangement equal the ratios of the rates of hops to the current concentrations of flow units

$$\Gamma_0(t, w) = -\frac{1}{\Xi(t, 0, w)} \frac{\partial \Xi}{\partial t}(t, 0, w), \tag{2}$$

$$\Gamma(t, \tau, w) = -\Big[\frac{\partial \Xi}{\partial \tau}(t, \tau, w)\Big]^{-1} \frac{\partial^2 \Xi}{\partial t \partial \tau}(t, \tau, w). \tag{3}$$

According to the concept of thermally activated processes (Eyring 1936), the probability κ that a flow unit reaches the liquid-like state per unit time reads

$$\kappa = \kappa_0 \exp\Big(-\frac{w - u}{k_B \Theta}\Big),$$

where κ_0 is the rate of hops for high-energy states (close to the reference state with $w = 0$), k_B is Boltzmann's constant, and Θ is the absolute temperature. This equality together with and Eq. (1) implies that

$$\begin{aligned} \Gamma_0(t, w) &= \Gamma_*(w) \exp\big[\eta U_0(t)\big], \\ \Gamma(t, \tau, w) &= \Gamma_*(w) \exp\big[\eta U(t, \tau)\big], \end{aligned} \tag{4}$$

where $U_0(t)$ is the mechanical energy of an initial flow unit, $U(t, \tau)$ is the mechanical energy of a relaxing region rearranged at time τ, $\eta = \delta/(k_B \Theta)$ and

$$\Gamma_*(w) = \Upsilon \exp\Big(-\frac{w}{k_B \Theta}\Big), \tag{5}$$

where Υ is the maximal rate of rearrangement.

The initial condition for Eq. (2) is given by

$$\Xi(0, 0, w) = \Xi_0 p_0(w), \tag{6}$$

where $p_0(w)$ is the initial probability density of flow units and Ξ_0 is a (time-independent) number density of relaxing units (per unit mass).

Denote by $N(t)$ the number of regions (per unit mass) rearranging per unit time and by $\nu(t)$ the ratio of the number of rearranging domains (per unit time) to the total number of flow units

$$\nu(t) = \frac{N(t)}{\Xi_0}$$

The number of relaxing regions that hop into traps with energy w (per unit time and unit mass) reads

$$\frac{\partial \Xi}{\partial \tau}(t, \tau, w)\Big|_{t=\tau} = \Xi_0 \nu(t) p(t, w), \tag{7}$$

where $p(t, w)$ is the current probability to hop into a cage with energy w. Integration of Eqs. (2) and (3) with initial conditions (6) and (7) results in

$$\begin{aligned} \Xi(t, 0, w) &= \Xi_0 p_0(w) \exp\Big[-\int_0^t \Gamma_0(s, w) ds\Big], \\ \frac{\partial \Xi}{\partial \tau}(t, \tau, w) &= \Xi_0 \nu(\tau) p(\tau, w) \\ & \qquad \exp\Big[-\int_\tau^t \Gamma(s, \tau, w) ds\Big] \end{aligned} \tag{8}$$

The number density of rearranging regions with energy w is given by

$$\begin{aligned}\Xi(t,t,w) &= \Xi(t,0,w)+\int_0^t \frac{\partial \Xi}{\partial \tau}(t,\tau,w)d\tau \\ &= \Xi_0\Big\{p_0(w)\exp\Big[-\int_0^t \Gamma_0(s,w)ds\Big] \\ &\quad +\int_0^t \nu(\tau)p(\tau,w) \\ &\quad \exp\Big[-\int_\tau^t \Gamma(s,\tau,w)ds\Big]d\tau\Big\} \end{aligned} \tag{9}$$

Summing the quantities $\Xi(t,t,w)$ with various energies w and bearing in mind the conservation law

$$\int_0^\infty \Xi(t,t,w)dw = \Xi_0, \tag{10}$$

we arrive at the integral equation

$$\int_0^t C(t,\tau)\nu(\tau)d\tau = 1 - C_0(t)$$

with

$$\begin{aligned}C_0(t) &= \int_0^\infty p_0(w)\exp\Big[-\int_0^t \Gamma_0(s,w)ds\Big]dw, \\ C(t,\tau) &= \int_0^\infty p(\tau,w)\exp\Big[-\int_\tau^t \Gamma(s,\tau,w)ds\Big]dw.\end{aligned}$$

It follows from Eq. (9) that the function $\mathcal{P}(t,w) = \Xi(t,t,w)/\Xi_0$ satisfies the equality

$$\begin{aligned}\mathcal{P}(t,w) &= p_0(w)\exp\Big[-\int_0^t \Gamma_0(s,w)ds\Big] \\ &\quad +\int_0^t \nu(\tau)p(\tau,w) \\ &\quad \exp\Big[-\int_\tau^t \Gamma(s,\tau,w)ds\Big]d\tau.\end{aligned} \tag{11}$$

Neglecting the effect of mechanical factors on the rates of rearrangement,

$$\Gamma_0(t,w) = \Gamma(t,\tau,w) = \Gamma_*(w),$$

we differentiate Eq. (11) with respect to time and arrive at the energy master-equation (Dyre 1987,1995, Bouchaud 1992, Monthus & Bouchaud 1996)

$$\begin{aligned}\frac{\partial \mathcal{P}}{\partial t}(t,w) &= -\Gamma_*(w)\mathcal{P}(t,w)+\nu(t)p(t,w), \\ \mathcal{P}(0,w) &= p_0(w).\end{aligned} \tag{12}$$

In the theory of transient networks, a nonaging polymer is thought of as a medium where concentrations of all kinds of long chains are time-uniform (Drozdov 1998). In the concept of cooperative relaxation, a counterpart of this condition is the assertion that the number densities of flow units with various energies w are independent of time,

$$\mathcal{P}(t,w) = p_0(w). \tag{13}$$

We substitute expression (13) into Eq. (11), differentiate the result with respect to time, and arrive at the integral equation

$$\begin{aligned}h(t,w) &= \Gamma_0(t,w)p_0(w)\exp\Big[-\int_0^t \Gamma_0(s,w)ds\Big] \\ &\quad +\int_0^t \Gamma(t,\tau,w)h(\tau,w) \\ &\quad \exp\Big[-\int_\tau^t \Gamma(s,\tau,w)ds\Big]d\tau\end{aligned} \tag{14}$$

for the function $h(t,w) = \nu(t)p(t,w)$.

3 MECHANICAL ENERGY OF AN AMORPHOUS POLYMER

According to the transition-state theory (Goldstein 1969), rearrangement of a flow unit consists of two stages: first, it hops into the liquid-like state where stresses totally relax, and, afterward, it lands into a new potential well. This implies that the natural (stress-free) configuration of a relaxing region rearranged at instant τ coincides with the actual configuration of the bulk medium at that time.

Flow units are modeled as incompressible elastic media with the specific mechanical energy (per region)

$$U(t,\tau) = \Omega\big(I_1(t,\tau), I_2(t,\tau)\big), \tag{15}$$

where $I_k(t,\tau)$ is the kth principal invariant for the relative Finger tensor $\hat{F}(t,\tau)$ for transition from the actual configuration at time τ to the actual configuration at time t, and Ω is a prescribed function that satisfies the condition $\Omega(3,3) = 0$ which means that the strain energy density vanishes in the natural (stress-free) configuration with $I_1 = I_2 = 3$.

Neglecting the energy of interaction between rearranging regions, we calculate the specific mechanical energy of an amorphous polymer (per unit mass) at the current time t as a sum of the mechanical energies of initial flow units and domains rearranged in the interval $[0,t]$. The mass concentration of initial regions is

$$m\int_0^\infty \Xi(t,0,w)dw,$$

where m is an average mass of a flow unit, and their mechanical energy is given by

$$m\Omega\Big(I_{10}(t),I_{20}(t)\Big)\int_0^\infty \Xi(t,0,w)dw,$$

where $I_{k0}(t)$ is the kth principal invariant for the Finger tensor $\hat{F}_0(t)$ for transition from the initial configuration to the actual configuration at time t. The mass concentration of regions rearranged within the interval $[\tau,\tau+d\tau]$ is

$$md\tau\int_0^\infty \frac{\partial\Xi}{\partial\tau}(t,\tau,w)dw,$$

and their mechanical energy reads

$$md\tau\Omega\Big(I_1(t,\tau),I_2(t,\tau)\Big)\int_0^\infty \frac{\partial\Xi}{\partial\tau}(t,\tau,w)dw.$$

Summing these expressions and using Eqs. (8), we find the strain energy density of a rubbery polymer

$$\begin{aligned}\Pi(t) &= \chi\Big\{\Omega\Big(I_{10}(t),I_{20}(t)\Big)\int_0^\infty p_0(w)\\ &\exp\Big[-\int_0^t\Gamma_0(s,w)ds\Big]dw\\ &+\int_0^t\nu(\tau)\Omega\Big(I_1(t,\tau),I_2(t,\tau)\Big)d\tau\\ &\int_0^\infty p(\tau,w)\\ &\exp\Big[-\int_\tau^t\Gamma(s,\tau,w)ds\Big]dw\Big\},\end{aligned} \tag{16}$$

where $\chi=m\Xi_0$. Differentiation of Eq. (16) results in the formula (simple transformations are omitted)

$$\frac{d\Pi}{dt}(t)=\chi\Big[J_1(t)-J_2(t)\Big] \tag{17}$$

with

$$\begin{aligned}J_1(t) &= 2\hat{P}(t):\hat{D}(t),\\ J_2(t) &= \Omega\Big(I_{10}(t),I_{20}(t)\Big)\int_0^\infty\Gamma_0(t,w)p_0(w)\\ &\exp\Big[-\int_0^t\Gamma_0(s,w)ds\Big]dw\\ &+\int_0^t\nu(\tau)\Omega\Big(I_1(t,\tau),I_2(t,\tau)\Big)d\tau\\ &\int_0^\infty\Gamma(t,\tau,w)p(\tau,w)\\ &\times\exp\Big[-\int_\tau^t\Gamma(s,\tau,w)ds\Big]dw.\end{aligned} \tag{18}$$

Here $\hat{D}(t)$ is the rate-of-strain tensor and

$$\begin{aligned}\hat{P}(t) &= \Big[\psi_{10}(t)\hat{F}_0(t)+\psi_{20}(t)\hat{F}_0^2(t)\Big]\\ &\int_0^\infty p_0(w)\exp\Big[-\int_0^t\Gamma_0(s,w)ds\Big]dw\\ &+\int_0^t\nu(\tau)\Big[\psi_1(t,\tau)\hat{F}(t,\tau)\\ &+\psi_2(t,\tau)\hat{F}^2(t,\tau)\Big]d\tau\\ &\int_0^\infty p(\tau,w)\exp\Big[-\int_\tau^t\Gamma(s,\tau,w)ds\Big]dw.\end{aligned}$$

The functions $\psi_k(t,\tau)$ are given by

$$\begin{aligned}\psi_1(t,\tau) &= \frac{\partial\Omega}{\partial I_1}\Big(I_1(t,\tau),I_2(t,\tau)\Big)\\ &+I_1(t,\tau)\frac{\partial\Omega}{\partial I_2}\Big(I_1(t,\tau),I_2(t,\tau)\Big), \qquad (19)\\ \psi_2(t,\tau) &= -\frac{\partial\Omega}{\partial I_2}\Big(I_1(t,\tau),I_2(t,\tau)\Big),\end{aligned}$$

and the functions $\psi_{k0}(t)$ are determined by Eqs. (19) where I_k is replaced by I_{k0}.

4 THERMODYNAMIC POTENTIALS AND CONSTITUTIVE RELATIONS

For an incompressible medium, the first law of thermodynamics reads

$$\rho\frac{d\Phi}{dt}=\hat{s}:\hat{D}-\bar{\nabla}\cdot\bar{q}+\rho R.$$

Here $\bar{\nabla}$ is the gradient operator in the actual configuration, ρ is mass density, Φ is the specific internal energy per unit mass, $\bar{q}$ is the heat flux vector, R is the heat supply per unit mass, and $\hat{s}$ is the deviatoric part of the Cauchy stress tensor $\hat{\sigma}$,

$$\hat{\sigma}=-\breve{p}\hat{I}+\hat{s}, \tag{20}$$

where $\breve{p}$ is a pressure. The Clausius–Duhem inequality implies that

$$\rho\frac{dQ}{dt}=\rho\frac{dS}{dt}+\bar{\nabla}\cdot\Big(\frac{\bar{q}}{\Theta}\Big)-\frac{\rho R}{\Theta}\geq 0,$$

where S is the specific entropy per unit mass, and Q is the specific entropy production. Combining these equalities and taking into account that $\Phi=\Psi+S\Theta$, where Ψ is the specific (Helmholtz) free energy, we find that

$$\Theta\frac{dQ}{dt}=-S\frac{d\Theta}{dt}-\frac{d\Psi}{dt}+\frac{1}{\rho}\Big(\hat{s}:\hat{D}-\frac{1}{\Theta}\bar{q}\cdot\bar{\nabla}\Theta\Big)\geq 0. \quad (21)$$

The specific free energy Ψ equals the sum of the specific free energy Ψ_0 in the stress-free configuration at the reference temperature Θ_0, the specific free energy of thermal motion Ψ_1, and the mechanical energy Π,

$$\Psi=\Psi_0+\Psi_1+\Pi.$$

Assuming the specific heat capacity in the stress-free state

$$c=-\Theta\frac{\partial^2\Psi_1}{\partial\Theta^2}$$

to be temperature-independent, we find that

$$\Psi_1=(c-S_0)(\Theta-\Theta_0)-c\Theta\ln\frac{\Theta}{\Theta_0},$$

where S_0 is the specific entropy in the natural configuration at the reference temperature. It follows from these equalities and Eq. (17) that

$$\frac{d\Psi}{dt}=-\Big(S_0+c\ln\frac{\Theta}{\Theta_0}\Big)\frac{d\Theta}{dt}+\chi\big(J_1-J_2\big).$$

Substitution of this equality and Eq. (18) into Eq. (21) results in

$$\begin{aligned}\Theta\frac{dQ}{dt} &= -\Big(S-S_0-c\ln\frac{\Theta}{\Theta_0}\Big)\frac{d\Theta}{dt}\\ &\quad+\frac{1}{\rho}\big(\hat{s}-2\rho\chi\hat{P}\big):\hat{D}\\ &\quad+\chi J_2-\frac{1}{\rho\Theta}\bar{q}\cdot\bar{\nabla}\Theta\geq 0.\end{aligned}$$

Equating the expressions in braces to zero and using Eq. (20), we arrive at the conventional formula for the specific entropy

$$S(t)=S_0+c\ln\frac{\Theta(t)}{\Theta_0}$$

and the novel constitutive equation

$$\begin{aligned}\hat{\sigma}(t) &= -\breve{p}\hat{I}+2\rho\chi\Big\{\Big[\psi_{10}(t)\hat{F}_0(t)+\psi_{20}(t)\hat{F}_0^2(t)\Big]\\ &\quad\int_0^\infty p_0(w)\exp\Big[-\int_0^t\Gamma_0(s,w)ds\Big]dw\\ &\quad+\int_0^t\nu(\tau)\Big[\psi_1(t,\tau)\hat{F}(t,\tau)\\ &\quad+\psi_2(t,\tau)\hat{F}^2(t,\tau)\Big]d\tau\\ &\quad\int_0^\infty p(\tau,w)\\ &\quad\exp\Big[-\int_\tau^t\Gamma(s,\tau,w)ds\Big]dw\Big\}. \end{aligned}\quad (22)$$

It follows from these formulas and Eq. (18) that the Clausius–Duhem inequality (21) is satisfied for an arbitrary loading program, provided that the heat flux vector $\bar{q}$ obeys the Fourier law $\bar{q}=-K_0\bar{\nabla}\Theta$ with a positive thermal diffusivity K_0.

The stress–strain relation (22) does not belong to the class of K–BKZ constitutive relations (Bernstein et al. 1963), since it contains the functions $\nu(t)$ and $p(t,w)$ obeying Volterra equations (9) and (14) with coefficients depending on the current state of flow units through Eqs. (4). Under isothermal loading, the constitutive equations are determined by two material functions: the initial probability density $p_0(w)$ and the mechanical energy of relaxing units $\Omega(I_1,I_2)$.

5 MATERIAL FUNCTIONS IN NONLINEAR VISCOELASTICITY

Referring to the random energy model (Richert & Bässler 1990), we accept the Gaussian formula for the probability density $p_0(w)$,

$$p_0(w)=\frac{1}{\sqrt{2\pi}\Sigma}\exp\Big[-\frac{(w-W)^2}{2\Sigma^2}\Big], \quad (23)$$

where W is the mean energy of a flow unit and Σ is its standard deviation. Equation (23) has been previously employed by Dyre (1995), Monthus & Bouchaud (1996) and Sollich (1998).

We accept the Mooney–Rivlin equation for the function Ω,

$$\Omega(I_1,I_2)=c_1(I_1-3)+c_2(I_2-3), \quad (24)$$

where c_1 and c_2 are adjustable parameters. Formula (24) correctly predicts experimental data in

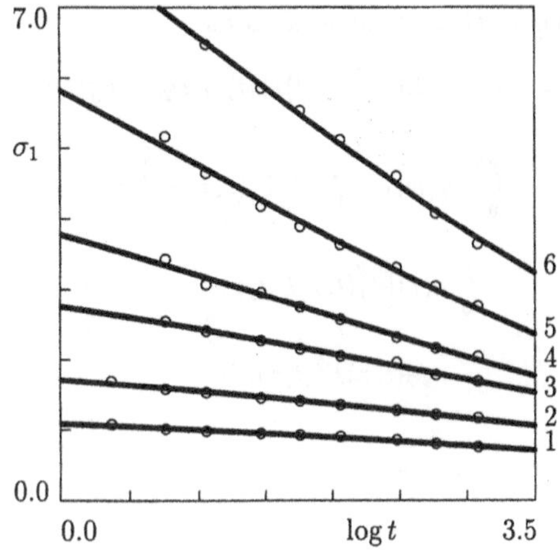

Figure 1: The longitudinal stress σ_1 MPa versus time t s for poly(methyl methacrylate). Circles: experimental data obtained by Marrucci and de Cindio (1980). Solid lines: predictions of the model with $W_* = 10.4$ and $\Sigma_* = 9.2$. Curve 1: $\lambda = 1.32$; curve 2: $\lambda = 1.59$; curve 3: $\lambda = 2.00$; curve 4: $\lambda = 2.39$; curve 5: $\lambda = 2.91$; curve 6: $\lambda = 3.64$

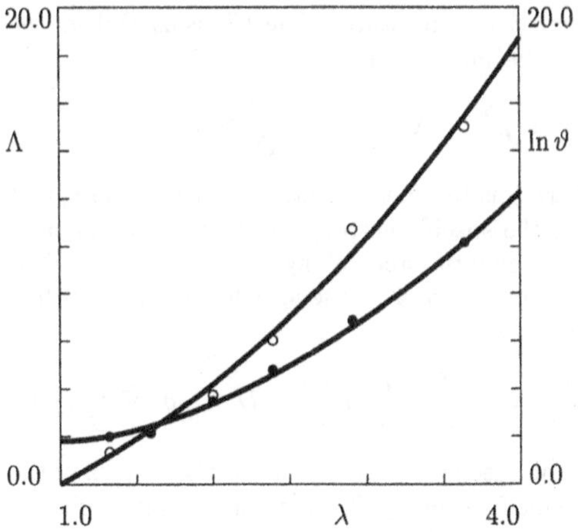

Figure 2: The parameter Λ (open circles) and the rate of rearrangement ϑ (filled circles) versus the extension ratio λ for poly(methyl methacrylate). Symbols: treatment of observations obtained by Marrucci and de Cindio (1980). Solid lines: approximations of experimental data by Eqs. (27) with $\mu_1 = 1.1809$, $\mu_2 = 0\ 0286$, $\kappa_1 = 0.7666$, $\kappa_2 = 0.0186$ and $\ln \Upsilon = 1.8062$

tensile and shear relaxation tests and in tests with constant rates of strain for crosslinked and uncrosslinked polymers (Zang et al. 1986). Substitution of expressions (23) and (24) into Eq. (22) results in

$$\begin{aligned}\hat{\sigma}(t) &= -\breve{p}(t)\hat{I} \\ &\quad + \Big[\Big(\mu_1 + \mu_2 I_{10}(t)\Big)\hat{F}_0(t) - \mu_2 \hat{F}_0^2(t)\Big] \\ &\quad \int_0^\infty p_0(w) \times \exp\Big\{-\Gamma_*(w) \\ &\quad \int_0^t \exp\Big[\kappa_1\Big(I_{10}(s) - 3\Big) \\ &\quad + \kappa_2\Big(I_{20}(s) - 3\Big)\Big] ds\Big\} dw \\ &\quad + \int_0^t \nu(\tau)\Big[\Big(\mu_1 + \mu_2 I_1(t,\tau)\Big)\hat{F}(t,\tau) \\ &\quad - \mu_2 \hat{F}^2(t,\tau)\Big] d\tau \int_0^\infty p(\tau, w) \\ &\quad \times \exp\Big\{-\Gamma_*(w) \\ &\quad \int_\tau^t \exp\Big[\kappa_1\Big(I_1(s,\tau) - 3\Big) \\ &\quad + \kappa_2\Big(I_2(s,\tau) - 3\Big)\Big] ds\Big\} dw \qquad (25)\end{aligned}$$

with $\mu_k = 2\rho\chi c_k$ and $\kappa_k = \eta c_k$. Equations (23) and (25) are determined by 7 adjustable parameters: κ_1, κ_2, μ_1, μ_2, Υ, W and Σ. To determine these constants and to validate constitutive relations, we compare experimental data in uniaxial tensile tests with results of numerical simulation.

6 UNIAXIAL TENSILE TEST

Uniaxial extension of an incompressible specimen is described by the equations

$$x_1 = \lambda(t) X_1,$$

$$x_2 = \lambda^{-\frac{1}{2}}(t) X_2,$$

$$x_3 = \lambda^{-\frac{1}{2}}(t) X_3,$$

where $\lambda(t)$ is the extension ratio and $\{X_i\}$, $\{x_i\}$ are Cartesian coordinates in the initial and actual configurations, respectively. We substitute these equalities into Eq. (25), exclude pressure $\breve{p}$ from

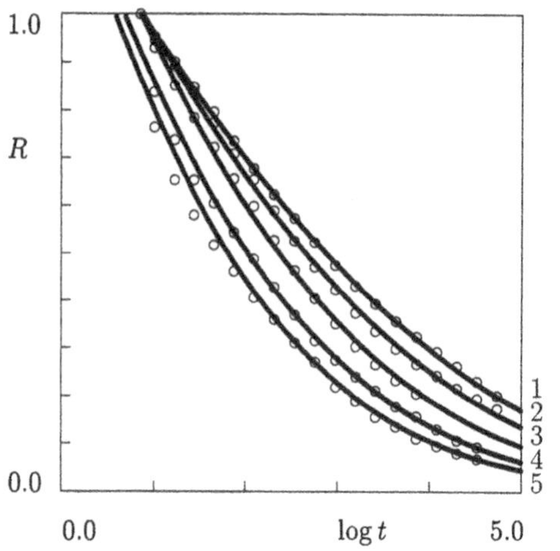

Figure 3: The ratio R versus time t s for PB–U4A–2. Circles: experimental data obtained by Dardin *et al.* (1997). Solid lines: predictions of the model with $W_* = 4.0$ and $\Sigma_* = 7.0$. Curve 1: $\lambda = 1.5$; curve 2: $\lambda = 2.0$; curve 3: $\lambda = 3.0$; curve 4: $\lambda = 5.0$; curve 5: $\lambda = 6.0$

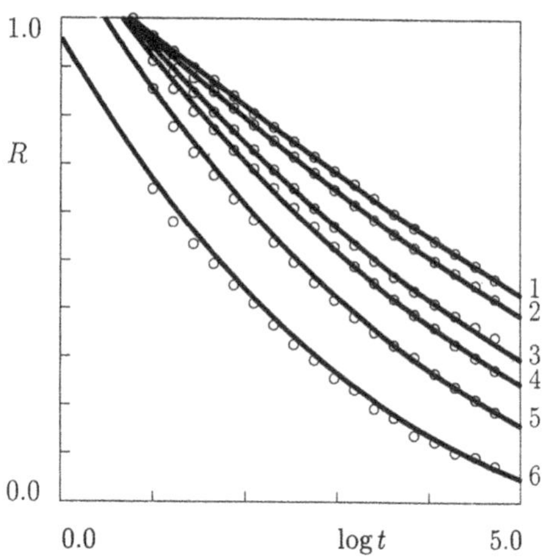

Figure 4: The ratio R versus time t s for PB–U4A–4. Circles: experimental data obtained by Dardin *et al.* (1997). Solid lines: predictions of the model with $W_* = 3.5$ and $\Sigma_* = 12.4$. Curve 1: $\lambda = 1.5$; curve 2: $\lambda = 2.0$; curve 3: $\lambda = 3.0$; curve 4: $\lambda = 4.0$; curve 5: $\lambda = 5.0$; curve 6: $\lambda = 6.0$

the boundary condition in stresses on the lateral surface of the specimen, and find the longitudinal stress σ_1. For a one-step relaxation test with the extension ratio λ,

$$\sigma_1(t,\lambda) = \frac{\Lambda(\lambda)}{\sqrt{2\pi}\Sigma_*}\int_0^\infty \exp\Big[-\Big(\frac{(z-W_*)^2}{2\Sigma_*^2} + \vartheta(\lambda)\exp(-z)t\Big)\Big]dz, \qquad (26)$$

where $z = w/(k_B\Theta)$, $W_* = W/(k_B\Theta)$, $\Sigma_* = \Sigma/(k_B\Theta)$,

$$\Lambda(\lambda) = \Big(\mu_1 + \frac{\mu_2}{\lambda}\Big)\Big(\lambda^2 - \frac{1}{\lambda}\Big),$$
$$\ln\vartheta(\lambda) = \ln\Upsilon + \Big[\kappa_1(I_1-3) + \kappa_2(I_2-3)\Big], \qquad (27)$$

and the principal invariants I_k read $I_1 = \lambda^2 + 2\lambda^{-1}$, $I_2 = \lambda^{-2} + 2\lambda$.

We begin with matching observations for poly(methyl methacrylate) at 140 °C. For a detailed description of the experimental procedure, see Marrucci & de Cindio (1980). Given W_*, Σ_* and $\vartheta(\lambda)$, the coefficient $\Lambda(\lambda)$ in Eq. (26) is found by the least-squares method. The quantities W_* and Σ_* are determined by fitting the relaxation curve at the minimum extension ratio $\lambda = 1.32$ by the steepest descent technique. The parameter $\vartheta(\lambda)$ is found to ensure the best approximation of the relaxation curve at the extension ratio λ.

Observations in tensile relaxation tests are plotted in Figure 1 together with the results of numerical simulation. The quantities $\Lambda(\lambda)$ and $\vartheta(\lambda)$ are depicted in Figure 2 together with their approximations by Eqs. (27). The coefficients μ_k, κ_k and $\ln\Upsilon$ are determined by a modification of the least-squares method which accounts for the proportionality of μ_k and κ_k. Figures 1 and 2 demonstrate fair agreement between experimental data and predictions of the model.

To ascertain that the model adequately predicts experimental data for other rubbery polymers, we fit observations for two kinds of polybutadiene modified by benzoic acid. A detailed description of specimens and the experimental procedure is provided in Dardin et al. (1997), where the ratios of the longitudinal stress $\sigma_1(t)$ to that at the reference time t_0 (when the required extension ratio λ is reached in tests) are plotted as functions of time,

$$R(t) = \frac{\sigma_1(t)}{\sigma_1(t_0)}.$$

Since the dependence $t_0(\lambda)$ is not provided by Dardin et al. (1997), the function $\Lambda(\lambda)$ cannot be determined uniquely. To reduce the number of ad-

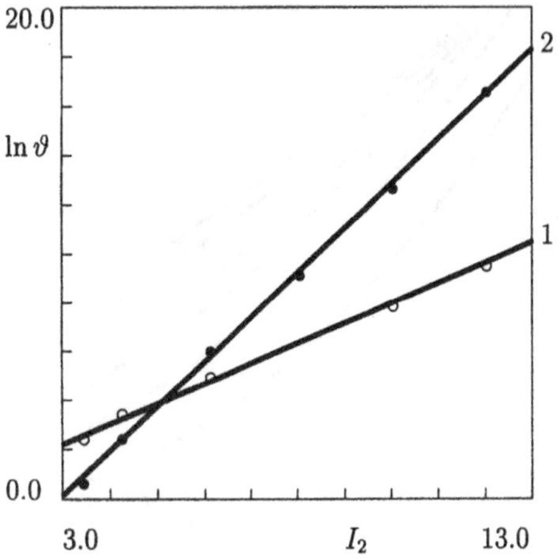

Figure 5: The rate of rearrangement ϑ versus the second principal invariant I_2 of the Finger tensor. Symbols: treatment of observations obtained by Dardin *et al.* (1997). Solid lines: approximations of experimental data by Eq. (27). Curve 1: PB–U4A–2, $\ln \Upsilon = 2.1955$, $\kappa_2 = 0.8270$; curve 2: PB–U4A–4, $\ln \Upsilon = 0.1301$, $\kappa_2 = 1.8174$

justable parameters, we suppose that $\mu_1 = 0$ and $\kappa_1 = 0$, which means that the strain energy density Ω depends on the second invariant I_2 of the Finger tensor only.

Observations for polydienes PB–U4A–2 and PB–U4A–4 at room temperature are depicted in Figures 3 and 4 together with predictions of the model. The rate of rearrangement ϑ and its approximation by Eq. (27) are plotted in Figure 5. These figures demonstrate an acceptable agreement between observations and results of numerical simulation.

7 CONCLUSIONS

Constitutive equations have been derived for the viscoelastic response of amorphous rubbery polymers at finite strains. The model is based on the theory of cooperative relaxation in a version of the concept of traps. A polymer is treated as an ensemble of flow units rearranged at random times. The rate of rearrangement is determined by the theory of thermally activated processes. The relative height of a potential barrier for rearrangement is determined by the current energy w of a flow unit and some part u of its mechanical energy.

Stress–strain relations for a viscoelastic material are developed using the laws of thermodynamics. These equations generalize the K–BKZ constitutive equations in finite viscoelasticity into which they turn when (i) mechanical factors do not affect the rates of hops and (ii) flow units are trapped in their cages.

Adjustable parameters are found by fitting experimental data for poly(methyl methacrylate) and polydienes functionalized by benzoic units in tensile relaxation tests. Fair agreement is demonstrated between observations and results of numerical simulation.

ACKNOWLEDGEMENT

The research was supported by the Israeli Ministry of Science through grants 9641–1–96 and 1202–1–98.

REFERENCES

Adam, G. and Gibbs, J.H. 1965. On the temperature dependence of cooperative relaxation properties in glass-forming liquids. *J. Chem. Phys.* 43: 139–146.

Bässler, H. 1987. Viscous flow in supercooled liquids analyzed in terms of transport theory for random media with energetic disorder. *Phys. Rev. Lett.* 58: 767–770.

Bernstein, B., Kearsley, E.A. and Zapas, L.J. 1963. A study of stress relaxation with finite strains. *Trans. Soc. Rheol.* 7: 391–410.

Bouchaud, J.P. 1992. Weak ergodicity breaking and aging in disordered systems. *J. Phys. I France* 2: 1705–1713.

Brawer, S.A. 1984. Theory of relaxation in viscous liquids and glasses. *J. Chem. Phys.* 81: 954–975.

Dardin, A., Spiess, H.W., Stadler, R. and Samulski, E.T. 1997. Relaxation of stress and orientation in thermo-reversible networks. *Polymer Gels Networks* 5: 37–54.

Dyre, J.C. 1987. Master-equation approach to the glass transition. *Phys. Rev. Lett.* 58: 792–795.

Dyre, J.C. 1995. Energy master equation: a low-temperature approximation to Bässler's random-walk model. *Phys. Rev. B* 51: 12276–12294.

Dyre, J.C. 1998. Source of non-Arrhenius average relaxation time in glass-forming liquids. *J. Non-Cryst. Solids* 235–237: 142–149.

Drozdov, A.D. 1998. *Mechanics of viscoelastic solids.* Chichester: Wiley.

Eyring. H. 1936. Viscosity, plasticity, and diffusion as examples of absolute reaction rates. *J. Chem. Phys.* 4: 283–291.

Goldstein, M. 1969. Viscous liquids and the glass transition: a potential energy barrier picture. *J. Chem. Phys.* 51: 3728–3739.

Marrucci, G. and de Cindio, B. 1980. The stress relaxation of molten PMMA at large deformations and its theoretical interpretation. *Rheol. Acta* 19: 68–75.

Monthus, C. and Bouchaud, J.-P. 1996. Models of traps and glass phenomenology. *J. Phys. A* 29: 3847–3869.

Richert, R. and Bässler, H. 1990. Dynamics of supercooled melts treated in terms of the random walk concept. *J. Phys.: Condens. Matter* 2: 2273–2288.

Rizos, A.K. and Ngai, K.L. 1999. Experimental determination of the cooperative length scale of a glass-forming liquid near the glass transition temperature. *Phys. Rev. E* 59: 612–617.

Sollich, P. 1998. Rheological constitutive equation for a model of soft glassy materials. *Phys. Rev. E* 58: 738–759.

Zang, Y.H., Muller, R. and Froelich, D. 1986. Interpretation of the rheological behaviour in elongation of uncrosslinked polystyrene melts in terms of the Mooney–Rivlin equation. *Polymer* 27: 61–65.

Tyres and friction

Constitutive Models for Rubber, Dorfmann & Muhr (eds) © 1999 Taylor & Francis ISBN 90 5809 113 9

A generalized orthotropic hyperelastic constitutive model for reinforced rubber-like materials

M. Itskov & Y. Başar
Institute for Statics and Dynamics, Ruhr-University Bochum, Germany

ABSTRACT: In the present paper, a new orthotropic hyperelastic constitutive model is proposed for a numerical simulation of reinforced rubber-like structures. The model presents a non-linear extension of the orthotropic St.Venant-Kirchhoff material and is described in each principal material direction by an arbitrary isotropic tensor function coupled with the corresponding structural tensor. In the special case of isotropy the model proposed can be reduced to the Valanis-Landel hypothesis or the well-known Ogden material and can therefore be considered as their generalization to orthotropic media. Constitutive relations and the elastic modulus of the model presented are expressed in terms of eigenvalue-bases of the right Cauchy-Green tensor **C**. For the analysis of shells this model is then coupled with a six (five in incompressible case) parametric shell kinematics able to deal with large strains as well as finite rotations. The application of the model is finally illustrated by a number of numerical examples.

1 INTRODUCTION

Fabric, cord or fibre reinforced rubber structures such as pneumatic membranes, automobile tires, hydraulic hoses and many others are characterized by strongly anisotropic material properties and can simultaneously undergo large elastic deformations. The elastic constitutive relations of these structures can be derived on the basis of an orthotropic hyperelastic material model. Such a constitutive model is proposed in the present paper.

The model is obtained as a non-linear extension of the orthotropic St.Venant-Kirchhoff material. To this end, the Green-Lagrange strain tensor **E** is substituted in the expression of the strain energy function by three arbitrary isotropic tensor functions; each of them is coupled with one of the structural tensors. In the special case of isotropy the constitutive law proposed can be reduced to the Valanis-Landel hypothesis or to the well-known Ogden model and can therefore be considered as their generalization to orthotropic materials.

For isotropic hyperelastic materials, the Ogden model (Ogden 1984) with a strain energy function formulated in terms of principal stretches has been shown to be of grate accuracy in spite of a relatively complicated numerical realization (see e.g. Başar & Itskov 1998). This model permits to consider an arbitrary number of material constants and polynomal powers, demonstrates a good agreement with experimental results at large strains and involves, finally, many other material laws as special cases. For this reason, the anisotropic generalization of the Ogden model enables to consider various orthotropic hyperelastic material models within a unified concept.

Constitutive relations and the elastic modulus of the model presented are expressed in terms of eigenvalue-bases of the right Cauchy-Green tensor and obtained for the cases of distinct and coinciding eigenvalues as well. The spectral decomposition of **C** can be obtained by means of the well-known closed formula (Betten 1984, Ting 1985, Morman 1986) without explicit solving the eigenvalue problem.

For the analysis of shells this model is then coupled with a six (five in incompressible case) parametric shell kinematics able to deal with large strains as well as finite rotations. The incompressibility condition is satisfied by eliminating the transverse normal strains C_{33} and the stretching parameter at the element level. For numerical simulations we use a four-node isoparametric finite shell element, where transverse shear stains are interpolated according to the assumed strain concept. Finally, we present a number of numerical examples to illustrate the application of the constitutive model proposed and to demonstrate the efficiency of the developed finite element.

2 BASIC NOTATIONS

Throughout the paper we use the absolute tensor notations and stick to the tensor algebra and differentiation rules given by Itskov (1999). Some of important notations and definitions to be used in the further derivations are recorded below.

A second-order tensor (a bold capital letter):

$$\mathbf{A} = \mathrm{A}_{ij}\mathbf{g}^i \otimes \mathbf{g}^j = \mathrm{A}^{ij}\mathbf{g}_i \otimes \mathbf{g}_j = \mathrm{A}^{i}_{\bullet j}\mathbf{g}_i \otimes \mathbf{g}^j = \ldots \quad (2.1)$$

The second-order identity tensor:

$$\mathbf{I} = \mathbf{g}_i \otimes \mathbf{g}^i . \quad (2.2)$$

A fourth-order tensor (a bold italics capital letter):

$$\boldsymbol{D} = D_{ijkl}\mathbf{g}^i \otimes \mathbf{g}^j \otimes \mathbf{g}^k \otimes \mathbf{g}^l = D^{ijkl}\mathbf{g}_i \otimes \mathbf{g}_j \otimes \mathbf{g}_k \otimes \mathbf{g}_l$$

$$= D^{i\ \ k}_{\bullet j \bullet l}\mathbf{g}_i \otimes \mathbf{g}^j \otimes \mathbf{g}_k \otimes \mathbf{g}^l = \ldots \quad (2.3)$$

The fourth-order identity tensor:

$$\boldsymbol{I} = \mathbf{I} \otimes \mathbf{I} = \mathbf{g}_i \otimes \mathbf{g}^i \otimes \mathbf{g}_j \otimes \mathbf{g}^j . \quad (2.4)$$

The simple contraction of arbitrary order tensors:

$$\mathbf{A}\boldsymbol{D} = \left(\mathrm{A}^{i}_{\bullet j}\mathbf{g}_i \otimes \mathbf{g}^j\right)\left(D^{k\ \ m}_{\bullet l \bullet n}\mathbf{g}_k \otimes \mathbf{g}^l \otimes \mathbf{g}_m \otimes \mathbf{g}^n\right)$$

$$= \mathrm{A}^{i}_{\bullet j} D^{j\ \ m}_{\bullet l \bullet n}\mathbf{g}_i \otimes \mathbf{g}^l \otimes \mathbf{g}_m \otimes \mathbf{g}^n . \quad (2.5)$$

Tensor products of two second-order tensors:

$$\boldsymbol{D} = \mathbf{A} \otimes \mathbf{B} = \left(\mathrm{A}^{i}_{\bullet j}\mathbf{g}_i \otimes \mathbf{g}^j\right) \otimes \left(\mathrm{B}^{k}_{\bullet l}\mathbf{g}_k \otimes \mathbf{g}^l\right)$$

$$= \mathrm{A}^{i}_{\bullet j}\mathrm{B}^{k}_{\bullet l}\mathbf{g}_i \otimes \mathbf{g}^j \otimes \mathbf{g}_k \otimes \mathbf{g}^l , \quad (2.6)$$

$$\boldsymbol{D} = \mathbf{A} \times \mathbf{B} = \left(\mathrm{A}^{i}_{\bullet j}\mathbf{g}_i \otimes \mathbf{g}^j\right) \times \left(\mathrm{B}^{k}_{\bullet l}\mathbf{g}_k \otimes \mathbf{g}^l\right)$$

$$= \mathrm{A}^{i}_{\bullet j}\mathrm{B}^{k}_{\bullet l}\mathbf{g}_i \otimes \mathbf{g}_k \otimes \mathbf{g}^l \otimes \mathbf{g}^j . \quad (2.7)$$

Double contractions of a fourth-order tensor with a second-order one (4:2) and (2:4):

$$\boldsymbol{D} : \mathbf{A} = \left(D^{i\ \ k}_{\bullet j \bullet l}\mathbf{g}_i \otimes \mathbf{g}^j \otimes \mathbf{g}_k \otimes \mathbf{g}^l\right) : \left(\mathrm{A}^{m}_{\bullet n}\mathbf{g}_m \otimes \mathbf{g}^n\right)$$

$$= D^{i\ \ k}_{\bullet j \bullet l}\mathrm{A}^{j}_{\bullet k}\,\mathbf{g}_i \otimes \mathbf{g}^l , \quad (2.8)$$

$$\mathbf{A} : \boldsymbol{D} = \left(\mathrm{A}_{m\bullet}^{\ \ n}\mathbf{g}^m \otimes \mathbf{g}_n\right) : \left(D^{i\ \ k}_{\bullet j \bullet l}\mathbf{g}_i \otimes \mathbf{g}^j \otimes \mathbf{g}_k \otimes \mathbf{g}^l\right)$$

$$= D^{i\ \ k}_{\bullet j \bullet l}\mathrm{A}_{i\bullet}^{\ \ l}\,\mathbf{g}^j \otimes \mathbf{g}_k . \quad (2.9)$$

The partial derivative of a second-order tensor with respect to another second-order one (2,2):

$$\mathbf{B},_{\mathbf{A}} = \frac{\partial\left(\mathrm{B}^{i}_{\bullet j}\mathbf{G}_i \otimes \mathbf{G}^j\right)}{\partial\left(\mathrm{A}^{k}_{\bullet l}\mathbf{g}_k \otimes \mathbf{g}^l\right)} = \frac{\partial \mathrm{B}^{i}_{\bullet j}}{\partial \mathrm{A}^{k}_{\bullet l}}\mathbf{G}_i \otimes \mathbf{g}^k \otimes \mathbf{g}_l \otimes \mathbf{G}^j \quad (2.10)$$

and with respect to its symmetric part:

$$\mathbf{B},_{\mathrm{sym}\mathbf{A}} = \left(\mathbf{B},_{\mathbf{A}}\right)' \quad \text{with} \quad \mathrm{sym}\mathbf{A} = \tfrac{1}{2}\left(\mathbf{A} + \mathbf{A}^{\mathrm{T}}\right), \quad (2.11)$$

where the operation $(\ldots)'$ is defined for a fourth-order tensor by

$$(\boldsymbol{E})' = \left(E^{i\ \ j}_{\bullet k \bullet l}\mathbf{g}_i \otimes \mathbf{g}^k \otimes \mathbf{g}_j \otimes \mathbf{g}^l\right)'$$

$$= \frac{1}{2} E^{i\ \ j}_{\bullet k \bullet l}(\mathbf{g}_i \otimes \mathbf{g}^k \otimes \mathbf{g}_j \otimes \mathbf{g}^l$$

$$+ \mathbf{g}_i \otimes \mathbf{g}_j \otimes \mathbf{g}^k \otimes \mathbf{g}^l) . \quad (2.12)$$

To complete this Section we present some important algebraic tensor identities:

$$\mathbf{A} \otimes \mathbf{B} : \mathbf{C} = \mathbf{ACB} , \qquad \mathbf{C} : \mathbf{A} \otimes \mathbf{B} = \mathbf{A}^{\mathrm{T}}\mathbf{C}\mathbf{B}^{\mathrm{T}} , \quad (2.13)$$

$$(\mathbf{A} \times \mathbf{B}) : \mathbf{C} = (\mathbf{B} : \mathbf{C})\mathbf{A} , \quad \mathbf{C} : (\mathbf{A} \times \mathbf{B}) = (\mathbf{C} : \mathbf{A})\mathbf{B} \quad (2.14)$$

and differentiation rules emanating from (2.5-2.7) and (2.10):

$$(\mathbf{AB}),_{\mathbf{C}} = \mathbf{A},_{\mathbf{C}}\mathbf{B} + \mathbf{AB},_{\mathbf{C}} , \quad (2.14)$$

$$(\alpha\mathbf{A}),_{\mathbf{B}} = \mathbf{A} \times \alpha,_{\mathbf{B}} + \alpha\mathbf{A},_{\mathbf{B}} , \quad (2.15)$$

$$(\mathbf{A} : \mathbf{B}),_{\mathbf{C}} = \mathbf{A} : \mathbf{B},_{\mathbf{C}} + \mathbf{B} : \mathbf{A},_{\mathbf{C}} , \quad (2.16)$$

where α denotes a scalar function.

3 MATERIAL MODEL

In an arbitrary point of an undeformed body we consider a set of orthogonal unit base vectors $\mathbf{l}_i = \mathbf{l}^i$ $(i = 1, 2, 3)$ coinciding with the principal material directions. Geometrically, the material orthotropy can be characterized through the symmetry group $\mathcal{G}$, which conserves the *structural tensors*

$$\mathbf{L}_i = \mathbf{l}_i \otimes \mathbf{l}_i , \quad (i = 1, 2, 3) \quad (3.1)$$

unchanged by any orthogonal transformation $\mathbf{Q} = \mathbf{Q}^{-\mathrm{T}}$ belonging to $\mathcal{G}$, thus:

$$\mathbf{L}_i = \mathbf{Q}^{\mathrm{T}}\mathbf{L}_i\mathbf{Q} , \quad (i = 1, 2, 3) \quad \forall \mathbf{Q} \in \mathcal{G} . \quad (3.2)$$

The mechanical properties of an orthotropic material can be described by means of the following nine *engineering elastic constants:*

E_1, E_2, E_3	Young's moduli referred to the principal material directions $\mathbf{l}_i$,
$G_{ij} = G_{ji}\ (i \neq j)$	Lame's shear moduli referred to the planes $\mathbf{l}_i\mathbf{l}_j$,

$$\nu_{ij} = \nu_{ji}\frac{E_j}{E_i} \quad (i \neq j) \quad \text{Poisson's ratios}$$

and the material parameter a_{ij} $(i,j = 1,2,3)$:

$$a_{ii} = E_i \frac{1-\nu_{jk}\nu_{kj}}{\Delta}, \quad \text{(no sum. over } i\text{)}$$

$$a_{ij} = a_{ji} = E_i \frac{\nu_{ij} + \nu_{kj}\nu_{ik}}{\Delta} \qquad (k \neq i \neq j, \;\; i,j = 1,2,3)$$

with $\Delta = 1 - \nu_{12}\nu_{21} - \nu_{23}\nu_{32} - \nu_{31}\nu_{13} - 2\nu_{21}\nu_{32}\nu_{13}$. (3.3)

If the material under consideration is incompressible the number of independent elastic constants reduces to six: E_i and G_{ij} $(i \neq j, \;\; i,j = 1, 2, 3)$. The other material parameters can be expressed in this case by:

$$a_{ii} = \frac{1}{D}\left(\frac{2}{3E_j} + \frac{2}{3E_k} - \frac{1}{3E_i}\right),$$

$$a_{ij} = \frac{1}{D}\left(\frac{1}{6E_i} + \frac{1}{6E_j} - \frac{5}{6E_k}\right),$$

$$\nu_{ij} = \frac{1}{2}E_j\left(\frac{1}{E_i} + \frac{1}{E_j} - \frac{1}{E_k}\right), \quad (k \neq i \neq j, i,j = 1,2,3)$$

$$D = \frac{3}{4}\left(\frac{2}{E_1E_2} + \frac{2}{E_2E_3} + \frac{2}{E_3E_1} - \frac{1}{E_1^2} - \frac{1}{E_2^2} - \frac{1}{E_3^2}\right). \tag{3.4}$$

As a strain measure of an orthotropic hyperelastic material we introduced the so-called oriented strain tensors

$$\mathbf{\Theta}_i(\mathbf{C}) = \mathbf{\Omega}_i(\mathbf{C})\mathbf{L}_i, \quad (i = 1,2,3) \tag{3.5}$$

where $\mathbf{\Omega}_i(\mathbf{C})$ denotes isotropic tensor-valued tensor functions of the right Cauchy-Green tensor $\mathbf{C}$. The strain energy function (per unit undeformed volume) of an orthotropic hyperelastic material is assumed to be of the form:

$$W(\mathbf{C}) = \frac{1}{2}\sum_{i,j}^{3} a_{ij}\mathrm{tr}\mathbf{\Theta}_i\mathrm{tr}\mathbf{\Theta}_j + \sum_{i,j\neq i}^{3} G_{ij}\mathrm{tr}\left(\mathbf{\Theta}_i\mathbf{\Theta}_j\right), \tag{3.6}$$

which corresponds to the St.Venant-Kirchhoff orthotropic model at infinitesimal strains.

For incompressible materials the tensor $\mathbf{C}$ is subjected in (3.6) to the constraint:

$$\mathrm{III}_{\mathbf{C}} = \det\mathbf{C} = 1, \tag{3.7}$$

which can be automatically satisfied using the Lagrange multiplier method. Accordingly, the strain energy function can be given by

$$\psi(\mathbf{C}) = W(\mathbf{C}) + p\left(\mathrm{III}_{\mathbf{C}} - 1\right) \tag{3.8}$$

with p as a Lagrange multiplier. For a compressible hyperelastic material the energy function $\psi(\mathbf{C})$ can be formulated in the form of the decoupled volumetric/isochoric response:

$$\psi(\mathbf{C}) = W(\overline{\mathbf{C}}) + U(J), \quad J = \mathrm{III}_{\mathbf{C}}^{1/2}, \quad \overline{\mathbf{C}} = J^{-2/3}\mathbf{C}, \tag{3.9}$$

where $U(J)$ denotes a scalar function describing the volumetric part of the strain energy.

Keeping in mind that the functions $\mathbf{\Omega}_i(\mathbf{C})$ are isotropic it can be easily verified that the strain parameters $\mathrm{tr}\mathbf{\Theta}_i$ and $\mathrm{tr}\left(\mathbf{\Theta}_i\mathbf{\Theta}_j\right)$ appearing in (3.6) present invariants within the symmetry group $\mathcal{L}$:

$$\mathrm{tr}\left[\mathbf{\Omega}_i\left(\mathbf{Q}^{\mathrm{T}}\mathbf{C}\mathbf{Q}\right)\mathbf{L}_i\right] = \mathrm{tr}\left[\mathbf{Q}^{\mathrm{T}}\mathbf{\Omega}_i(\mathbf{C})\mathbf{Q}\mathbf{L}_i\right]$$

$$= \mathbf{\Omega}_i(\mathbf{C}) : \left(\mathbf{Q}\mathbf{L}_i\mathbf{Q}^{\mathrm{T}}\right) = \mathrm{tr}\mathbf{\Theta}_i, \tag{3.10}$$

$$\mathrm{tr}\left[\mathbf{\Omega}_i\left(\mathbf{Q}^{\mathrm{T}}\mathbf{C}\mathbf{Q}\right)\mathbf{L}_i\mathbf{\Omega}_j\left(\mathbf{Q}^{\mathrm{T}}\mathbf{C}\mathbf{Q}\right)\mathbf{L}_j\right]$$

$$= \mathrm{tr}\left[\mathbf{Q}^{\mathrm{T}}\mathbf{\Omega}_i(\mathbf{C})\mathbf{Q}\mathbf{L}_i\mathbf{Q}^{\mathrm{T}}\mathbf{\Omega}_j(\mathbf{C})\mathbf{Q}\mathbf{L}_j\right]$$

$$= \mathbf{\Omega}_i(\mathbf{C})\mathbf{L}_i\mathbf{\Omega}_j(\mathbf{C}) : \left(\mathbf{Q}\mathbf{L}_j\mathbf{Q}^{\mathrm{T}}\right)$$

$$= \mathrm{tr}\left(\mathbf{\Theta}_i\mathbf{\Theta}_j\right) \qquad \forall\mathbf{Q} \in \mathcal{L}. \tag{3.11}$$

Introducing some additional non-symmetric structural tensors

$$\mathbf{L}_{ij} = \mathbf{l}_i \otimes \mathbf{l}_j, \qquad \mathbf{L}_{ij}^{\mathrm{T}} = \mathbf{L}_{ji} \quad (i \neq j = 1, 2, 3) \tag{3.12}$$

the strain invariants $\mathrm{tr}\left(\mathbf{\Theta}_i\mathbf{\Theta}_j\right)$ can be alternatively expressed by

$$\mathrm{tr}\left(\mathbf{\Theta}_i\mathbf{\Theta}_j\right) = \left(\mathbf{\Omega}_i\mathbf{l}_i \otimes \mathbf{l}_i\mathbf{\Omega}_j\mathbf{l}_j \otimes \mathbf{l}_j\right) : \mathbf{I} = \left(\mathbf{l}_i\mathbf{\Omega}_j\mathbf{l}_j\right)\left(\mathbf{l}_j\mathbf{\Omega}_i\mathbf{l}_i\right)$$

$$= \mathrm{tr}\left(\mathbf{\Omega}_i\mathbf{L}_{ji}\right)\mathrm{tr}\left(\mathbf{\Omega}_j\mathbf{L}_{ij}\right), \tag{3.13}$$

which will be used to simplify the further derivation of constitute relations and elastic moduli.

Using for the introduced invariants the abbreviations

$$\mathrm{I}_i = \mathrm{tr}\left(\mathbf{\Omega}_i\mathbf{L}_i\right), \tag{3.14}$$

$$\mathrm{I}_{ij} = \mathrm{tr}\left(\mathbf{\Omega}_i\mathbf{L}_{ji}\right) \;\; (i \neq j = 1, 2, 3) \tag{3.15}$$

the strain energy function (3.6) can be now written as

$$W(\mathbf{C}) = \frac{1}{2}\sum_{i,j}^{3} a_{ij}\mathrm{I}_i\mathrm{I}_j + \sum_{i,j\neq i}^{3} G_{ij}\mathrm{I}_{ij}\mathrm{I}_{ji}. \tag{3.16}$$

At infinitesimal strains the proposed constitutive model (3.6) has to match the St.Venant-Kirchhoff orthotropic material described by the energy function (see e.g. Başar et al. 1999)

$$W(\mathbf{C})=\frac{1}{2}\sum_{i,j}^{3}a_{ij}\,\mathrm{tr}(\mathbf{EL}_i)\,\mathrm{tr}(\mathbf{EL}_j)$$

$$+\sum_{i,j\neq i}^{3}G_{ij}\,\mathrm{tr}(\mathbf{EL}_i\mathbf{EL}_j). \qquad (3.17)$$

This imposes the following conditions on the functions $\mathbf{\Omega}_i(\mathbf{C})$:

$$\mathbf{\Omega}_i(\mathbf{I})=\mathbf{0}, \quad [\mathbf{\Omega}_i(\mathbf{C})]_{,\mathbf{C}}\big|_{\mathbf{C}=\mathbf{I}}=\mathbf{E}_{,\mathbf{C}}=\frac{1}{2}\mathbf{I}'. \qquad (3.18)$$

In the further derivation we assume that $\mathbf{\Omega}_i(\mathbf{C})$ belongs to a special class of isotropic tensor functions, which can be expanded in power series:

$$\mathbf{\Omega}(\mathbf{C})=\alpha_0\mathbf{I}+\alpha_1(\mathbf{C}-\beta\mathbf{I})+\alpha_2(\mathbf{C}-\beta\mathbf{I})^2+\ldots \qquad (3.19)$$

Herein α_i and β are scalar-valued constants. Such functions are referred here to as analytical tensor functions.

Now, using the spectral decomposition of $\mathbf{C}$ in terms of the eigenvector bases $\mathbf{M}_k=\mathbf{n}_k\otimes\mathbf{n}_k$ (k = 1, 2, 3):

$$\mathbf{C}=\sum_{k=1}^{3}\Lambda_k\mathbf{M}_k=\sum_{k=1}^{3}\lambda_k^2\mathbf{M}_k, \qquad (3.20)$$

with $\lambda_k=\sqrt{\Lambda_k}$ denoting the principal stretches, the functions $\mathbf{\Omega}_i(\mathbf{C})$ and $\mathbf{\Theta}_i(\mathbf{C})$ can be expressed by

$$\mathbf{\Omega}_i(\mathbf{C})=\sum_{k=1}^{3}\overline{\omega}_i(\Lambda_k)\mathbf{M}_k, \qquad (3.21)$$

$$\mathbf{\Theta}_i(\mathbf{C})=\sum_{k=1}^{3}\overline{\omega}_i(\Lambda_k)\mathbf{M}_k\mathbf{L}_i \qquad (3.22)$$

in terms of the so-called diagonal functions $\overline{\omega}_i(\Lambda)$ $=\omega_i(\lambda)$:

$$\overline{\omega}(\Lambda)=\alpha_0+\alpha_1(\Lambda-\beta)+\alpha_2(\Lambda-\beta)^2+\ldots \qquad (3.23)$$

Attention is now focused on the special case, where the principal directions of $\mathbf{C}$ and the principal material directions coincide $\mathbf{n}_i=\mathbf{l}_i$ $(i=1,2,3)$. Considering (3.6) and (3.22) we then obtain:

$$W(\mathbf{C})=\frac{1}{2}\sum_{i,j}^{3}a_{ij}\,\omega_i(\lambda_i)\,\omega_j(\lambda_j) \qquad (3.24)$$

demonstrating clearly that the material properties in each orthotropic direction are described by an individual scalar function of the corresponding stretching.

It can be observed, that for the special case of isotropy, where

$$a_{11}=a_{22}=a_{33}=\frac{1-\nu}{(1+\nu)(1-2\nu)}E=\lambda+2G,$$

$$a_{12}=a_{23}=a_{31}=\frac{\nu}{(1+\nu)(1-2\nu)}E=\lambda,$$

$$G_{12}=G_{23}=G_{31}=\frac{E}{2(1+\nu)}=G,$$

$$\mathbf{\Omega}_1(\mathbf{C})=\mathbf{\Omega}_2(\mathbf{C})=\mathbf{\Omega}_3(\mathbf{C})=\mathbf{\Omega}(\mathbf{C}) \quad\Rightarrow$$

$$\overline{\omega}_1(\Lambda)=\overline{\omega}_2(\Lambda)=\overline{\omega}_3(\Lambda)=\overline{\omega}(\Lambda), \qquad (3.25)$$

the strain energy function (3.6) reduces to

$$W(\mathbf{C})=\lambda\frac{1}{2}[\mathrm{tr}\mathbf{\Omega}(\mathbf{C})]^2+G\,\mathrm{tr}\big[\mathbf{\Omega}(\mathbf{C})^2\big]. \qquad (3.26)$$

If, additionally, the material under consideration is assumed to be incompressible and $\lambda=0$, the strain energy function (3.26) takes a form of the Valanis-Landel hypothesis (Valanis & Landel 1967):

$$W(\mathbf{C})=G\,\mathrm{tr}\big[\mathbf{\Omega}(\mathbf{C})^2\big]=G\sum_{k=1}^{3}\omega^2(\lambda_k) \qquad (3.27)$$

involving the well-known Ogden model (Ogden 1984) as a special case. Thus, we see that the material model given by (3.6) presents the generalization of the Valanis-Landel hypothesis to the case of orthotropy. It can be also regarded as the non-linear extension of the St.Venant-Kirchhoff orthotropic material.

Similar to the Ogden model, the functions $\mathbf{\Omega}_i(\mathbf{C})$ and the corresponding diagonal ones $\overline{\omega}_i(\Lambda)=\omega_i(\lambda)$ will be approximated by the following power series:

$$\mathbf{\Omega}_i(\mathbf{C})=\sum_r\frac{\mu_{ri}}{\alpha_{ri}}\left(\mathbf{C}^{\alpha_{ri}/2}-\mathbf{I}\right), \qquad (3.28)$$

$$\omega_i(\lambda)=\sum_r\frac{\mu_{ri}}{\alpha_{ri}}\left(\lambda^{\alpha_{ri}}-1\right), \qquad (3.29)$$

where μ_{ri} and α_{ri} present the Ogden-type material constants associated with the orthotropic direction i (i = 1, 2, 3). The above functions automatically satisfy the first condition in (3.18), while the second one requires, that

$$\sum_r\mu_{ri}=1. \qquad (3.30)$$

The unknown material parameters μ_{ri} and α_{ri} can be evaluated from an experiment, where the loading axes coincide with the principal material directions. In this case, the physical Cauchy stresses are given (for an incompressible material) by:

$$\sigma^{<ii>} = \lambda_i \frac{\partial W}{\partial \lambda_i} + p\,, \qquad (i = 1, 2, 3) \qquad (3.31)$$

which leads under consideration of (3.24) to:

$$\sigma^{<ii>} = \lambda_i \omega_i'(\lambda_i) \sum_j^3 a_{ij} \omega_j(\lambda_j) + p\,. \qquad (3.32)$$

The number of unknown coefficients μ_{r_i} and α_{r_i} in series representation (3.29) should correspond to the number of points $\sigma^{<ii>} = \sigma^{<ii>}(\lambda_1, \lambda_2, \lambda_3)$ available from the experimental study. Inserting these results into (3.32) leads to a nonlinear equation system, which can then be solved for μ_{r_i} and α_{r_i} e.g. by means of the Newton-Raphson method.

4 CONSTITUTIVE RELATIONS AND ELASTIC MODULI

The stress-strain relations for the material model under consideration can be obtained on the basis of the Doyle-Ericksen formula (Doyle & Ericksen 1967):

$$\mathbf{S} = 2\frac{\partial \psi}{\partial \mathbf{C}}\,. \qquad (4.1)$$

For incompressible materials described by (3.8) this relation yields under consideration of (3.14-3.16):

$$\mathbf{S} = 2\sum_{i,j}^3 a_{ij} \mathrm{I}_i \mathrm{I}_j{,}_{\mathbf{C}} + 4\sum_{i,j\neq i}^3 G_{ij} \mathrm{I}_{ij} \mathrm{I}_{ji}{,}_{\mathbf{C}} + 2p\mathrm{III}_{\mathbf{C}} \mathbf{C}^{-1} \qquad (4.2)$$

with

$$\mathrm{I}_i{,}_{\mathbf{C}} = \mathbf{L}_i : [\boldsymbol{\Omega}_i(\mathbf{C})]_{,\mathbf{C}}\,, \quad \mathrm{I}_{ij}{,}_{\mathbf{C}} = \mathbf{L}_{ij} : [\boldsymbol{\Omega}_i(\mathbf{C})]_{,\mathbf{C}}\,. \qquad (4.3)$$

For the derivative of the function $\boldsymbol{\Omega}_i(\mathbf{C})$ with respect its argument involved in (4.3), we use the closed-formula solution by Itskov (1999) in terms of the eigenvector bases of $\mathbf{C}$

$$\boldsymbol{\Omega}_i(\mathbf{C})_{,\mathbf{C}} = \sum_a^3 \overline{\omega}_i'(\Lambda_a) \mathbf{M}_a \otimes \mathbf{M}_a + \sum_{a,b\neq a}^3 \frac{\overline{\omega}_i(\Lambda_a) - \overline{\omega}_i(\Lambda_b)}{\Lambda_a - \Lambda_b} (\mathbf{M}_a \otimes \mathbf{M}_b)'\,. \qquad (4.4)$$

Inserting (4.4) into (4.3) we receive the expression of the derivative $\mathrm{I}_i{,}_{\mathbf{C}}$ appearing in the constitutive relation (4.2) in the case of

(i) distinct eigenvalues $\Lambda_1 \neq \Lambda_2 \neq \Lambda_3$:

$$\mathrm{I}_i{,}_{\mathbf{C}} = \sum_{a=1}^3 \overline{\omega}_i'(\Lambda_a) \mathbf{M}_a \mathbf{L}_i \mathbf{M}_a + \sum_{a,b\neq a}^3 \frac{\overline{\omega}_i(\Lambda_a) - \overline{\omega}_i(\Lambda_b)}{\Lambda_a - \Lambda_b} \mathbf{M}_a \mathbf{L}_i \mathbf{M}_b\,. \qquad (4.5)$$

For coinciding eigenvalues one can obtain considering a limit case $\Lambda_a - \Lambda_b = \Delta \to 0$ in (4.5):

(ii) double coalescence of eigenvalues: $\Lambda_a \neq \Lambda_b = \Lambda_c = \Lambda$, $\mathbf{C} = \Lambda_a \mathbf{M}_a + \Lambda(\mathbf{I} - \mathbf{M}_a)$,

$$\mathrm{I}_i{,}_{\mathbf{C}} = \overline{\omega}_i'(\Lambda_a) \mathbf{M}_a \mathbf{L}_i \mathbf{M}_a + \overline{\omega}_i'(\Lambda)(\mathbf{I} - \mathbf{M}_a)\mathbf{L}_i(\mathbf{I} - \mathbf{M}_a) + 2\frac{\overline{\omega}_i(\Lambda_a) - \overline{\omega}_i(\Lambda)}{\Lambda_a - \Lambda} \mathrm{sym}[\mathbf{M}_a \mathbf{L}_i (\mathbf{I} - \mathbf{M}_a)]\,, \qquad (4.6)$$

(iii) triple coalescence of eigenvalues: $\Lambda_a = \Lambda_b = \Lambda_c = \Lambda$, $\mathbf{C} = \Lambda \mathbf{I}$,

$$\mathrm{I}_i{,}_{\mathbf{C}} = \overline{\omega}_i'(\Lambda) \mathbf{L}_i\,. \qquad (4.7)$$

The elastic modulus can be formulated in a material description starting from the definition (Simo & Marsden 1984):

$$\mathbb{C} = 4\frac{\partial^2 \psi}{\partial \mathbf{C} \partial \mathbf{C}} = 2\frac{\partial \mathbf{S}}{\partial \mathbf{C}}\,. \qquad (4.8)$$

Inserting (4.2) into (4.8) delivers:

$$\begin{aligned}\mathbb{C} = {} & 4\sum_{i,j}^3 a_{ij} \left(\mathrm{I}_j{,}_{\mathbf{C}} \times \mathrm{I}_i{,}_{\mathbf{C}} + \mathrm{I}_i \mathrm{I}_j{,}_{\mathbf{CC}}\right) \\ & + 8\sum_{i,j\neq i}^3 G_{ij} \left(\mathrm{I}_{ij}{,}_{\mathbf{C}} \times \mathrm{I}_{ji}{,}_{\mathbf{C}} + \mathrm{I}_{ij} \mathrm{I}_{ji}{,}_{\mathbf{CC}}\right) \\ & + 4p\left[\mathbf{C}^{-1} \times \mathbf{C}^{-1} - \left(\mathbf{C}^{-1} \otimes \mathbf{C}^{-1}\right)'\right] \\ & + 4\left(\mathbf{C}^{-1} \times p{,}_{\mathbf{C}} + p{,}_{\mathbf{C}} \times \mathbf{C}^{-1}\right).\end{aligned} \qquad (4.9)$$

By considering (4.5) and using tensor differentiation rules (see Itskov 1999) the closed-form expressions for the derivatives $\mathrm{I}_{ji}{,}_{\mathbf{C}}$, $\mathrm{I}_i{,}_{\mathbf{CC}}$ and $\mathrm{I}_{ji}{,}_{\mathbf{CC}}$ presented in (4.9) can be obtained for the cases of distinct as well as coinciding eigenvalues.

Considering the relations (4.2) and (4.9) it can be seen that the strain energy function proposed satisfies the following necessary conditions (see also Almeida & Spilker 1998):

1. Existence of a natural state $\mathbf{S}(\mathbf{C} = \mathbf{I}) = \mathbf{0}$,

2. Positive definiteness of the stiffness matrix in the natural state

if the restrictions (3.18) imposed on the functions $\boldsymbol{\Omega}_i(\mathbf{C})$ are fulfilled.

5 SHELL KINEMATICS AND INCOMPRESSIBILITY CONDITION

The constitutive model proposed is coupled with a shell kinematics to be presented in this section (for details see Başar & Ding, 1997, Başar & Itskov 1998). The shell kinematics involves six (five in incompressible case) independent parameters and is based on the assumption that the position vector x* related to the arbitrary point P* of the deformed shell continuum is described by a linear expression

$$\mathbf{x}^{*}(\theta^{i}) = \mathbf{x}(\theta^{\alpha}) + \theta^{3}\lambda\mathbf{d}(\theta^{\alpha}) \tag{5.1}$$

with respect to the thickness coordinate θ^3. Herein the director **d** is supposed to be a unit vector such that

$$\mathbf{d}\cdot\mathbf{d} = 1 \quad \rightarrow \quad \mathbf{d},_{\alpha}\cdot\mathbf{d} = 0\,. \tag{5.2}$$

Through the multiplicative decomposition of the second term in (5.1) (Simo et al. 1990) the stretching parameter λ is decoupled from the director **d** subjected in view of (5.2) to a pure rotation. In the present study the rotation of the director will be parametrized with respect to the global Cartesian frame by means of the Euler angles φ_α.

The differentiation of (5.1) with respect to the curvilinear coordinates θ^i yields the base vectors related to P*:

$$\mathbf{g}_\alpha = \mathbf{a}_\alpha + \theta^3(\lambda,_\alpha \mathbf{d} + \lambda\mathbf{d},_\alpha), \qquad \mathbf{g}_3 = \lambda\mathbf{d}\,, \tag{5.3}$$

where $\mathbf{a}_\alpha = \mathbf{x}(\theta^\alpha),_\alpha$ presents the base vectors of the deformed midsurface.

For the definition of 2D strain variables we use the right Cauchy-Green tensor given in terms of the deformation gradient $\mathbf{F} = \mathbf{g}_i \otimes \mathbf{G}^i$ by:

$$\mathbf{C} = \mathbf{F}^{\mathrm{T}}\mathbf{F} = \mathbf{g}_i\cdot\mathbf{g}_j\mathbf{G}^i\otimes\mathbf{G}^j = g_{ij}\mathbf{G}^i\otimes\mathbf{G}^j$$

$$= C_{ij}\mathbf{G}^i\otimes\mathbf{G}^j\,. \tag{5.4}$$

Introducing the base vectors (5.3) into (5.4) the components of **C** are obtained in the form:

$$C_{ij} = \begin{bmatrix} C_{\alpha\beta} & C_{\alpha 3} \\ C_{3\alpha} & C_{33} \end{bmatrix} = \begin{bmatrix} \overset{0}{C}_{\alpha\beta} + \theta^3 \overset{1}{C}_{\alpha\beta} & \overset{0}{C}_{\alpha 3} \\ \overset{0}{C}_{3\alpha} & \overset{0}{C}_{33} \end{bmatrix}, \tag{5.5}$$

where the following kinematic relations hold for 2D strains denoted by $\overset{n}{(\dots)}$, $n = 0,1$:

tangential strains:

$$\overset{0}{C}_{\alpha\beta} = \mathbf{a}_\alpha\cdot\mathbf{a}_\beta\,, \tag{5.6}$$

$$\overset{1}{C}_{\alpha\beta} = \lambda(\mathbf{d},_\alpha\cdot\mathbf{a}_\beta + \mathbf{d},_\beta\cdot\mathbf{a}_\alpha) + \mathbf{d}\cdot(\lambda,_\alpha\mathbf{a}_\beta + \lambda,_\beta\mathbf{a}_\alpha), \tag{5.7}$$

transverse shear strains:

$$\overset{0}{C}_{\alpha 3} = \lambda\mathbf{a}_\alpha\cdot\mathbf{d}\,, \tag{5.8}$$

transverse normal strains:

$$\overset{0}{C}_{33} = \lambda^2\,. \tag{5.9}$$

Note that in accordance with the approximation level of the base vectors (5.3) the quadratic and linear terms are neglected in the expressions of tangential $C_{\alpha\beta}$ and transverse shear $C_{3\alpha}$ strains, respectively.

Now, we focus the attention on the incompressibility condition (3.7), which can be rewritten in the form:

$$\mathrm{III_C} = e_{ijk}C^i_1 C^j_2 C^k_3 = 1 + R + C^3_3 S = 1\,. \tag{5.10}$$

It is automatically satisfied if the transverse normal strain C^3_3 is determined by means of the relation:

$$C^3_3 = -\frac{R}{S} \tag{5.11}$$

with the abbreviations:

$$R = e_{ij\alpha}C^i_1 C^j_2 C^\alpha_3 - 1, \quad S = e_{\alpha\beta 3}C^\alpha_1 C^\beta_2\,. \tag{5.12}$$

The approximation for C^3_3 (5.11) should be consistent with (5.9), which requires that

$$\lambda = \frac{[\mathbf{A}_1\,\mathbf{A}_2\,\mathbf{A}_3]}{[\mathbf{a}_1\,\mathbf{a}_2\,\mathbf{d}]}\,. \tag{5.13}$$

Herein $\mathbf{A}_i$ denotes the base vectors related to the midsurface of the undeformed shell. The condition (5.13) is considered through a numerical elimination of the stretching parameter at the element level.

6 FINITE ELEMENT FORMULATION AND NUMERICAL EXAMPLES

Numerical implementations have been carried out using a four-node isoparametric finite shell element. The main characteristic features of the finite element formulation are summarised below.

- Independent kinematic variables **x** and $\mathbf{d}(\varphi_\alpha)$ are interpolated by the standard bilinear polynomials.
- To avoid shear locking the assumed strain concept (Dvorkin & Bathe 1984) is used for the interpolation of constant transverse shear deformations $\overset{0}{C}_{\alpha 3}$.

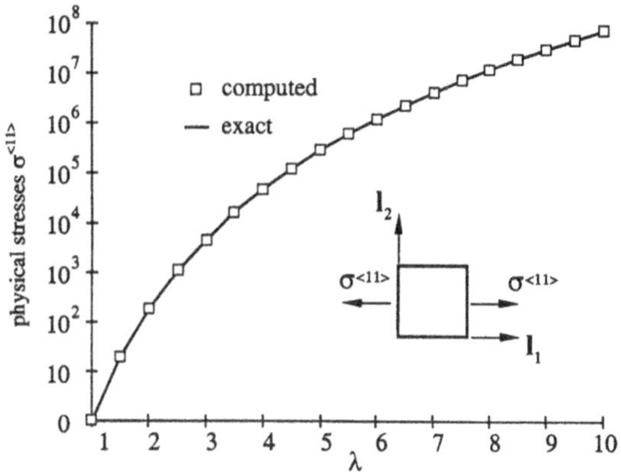

Material data

$E_1 = 10.0$,

$E_2 = E_3 = 1.0$,

$G_{12} = G_{13} = 2.2$

$\mu_{1i} = 0.8$, $\alpha_{1i} = 4.0$,

$\mu_{2i} = -0.1$, $\alpha_{2i} = 1.0$,

$\mu_{3i} = 0.3$, $\alpha_{3i} = -2.8$

$(i = 1, 2, 3)$

λ	$\sigma^{<11>}$
1	0.0
2	191.49
3	$4.763 \cdot 10^3$
4	$4.710 \cdot 10^4$
5	$2.795 \cdot 10^5$
6	$1.199 \cdot 10^6$
7	$4.111 \cdot 10^6$
8	$1.196 \cdot 10^7$
9	$3.066 \cdot 10^7$
10	$7.120 \cdot 10^7$

Figure 1. Simple tension test of a rectangular sheet.

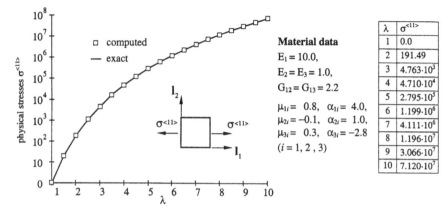

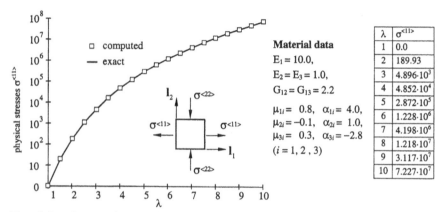

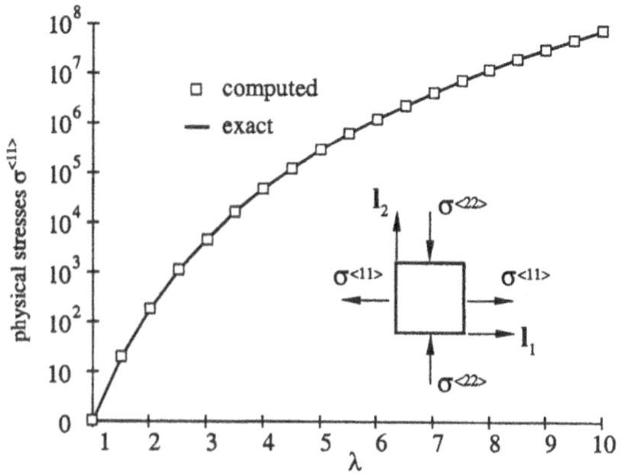

Material data

$E_1 = 10.0$,

$E_2 = E_3 = 1.0$,

$G_{12} = G_{13} = 2.2$

$\mu_{1i} = 0.8$, $\alpha_{1i} = 4.0$,

$\mu_{2i} = -0.1$, $\alpha_{2i} = 1.0$,

$\mu_{3i} = 0.3$, $\alpha_{3i} = -2.8$

$(i = 1, 2, 3)$

λ	$\sigma^{<11>}$
1	0.0
2	189.93
3	$4.896 \cdot 10^3$
4	$4.852 \cdot 10^4$
5	$2.872 \cdot 10^5$
6	$1.228 \cdot 10^6$
7	$4.198 \cdot 10^6$
8	$1.218 \cdot 10^7$
9	$3.117 \cdot 10^7$
10	$7.227 \cdot 10^7$

Figure 2. Pure shear test of a rectangular sheet.

- The 3D material model is directly enforced in the finite element formulation through numerical thickness integration.

In the following we consider a number of numerical examples demonstrating the effectiveness and accuracy of the finite element developed.

6.1 *Homogeneous deformation of a rectangular sheet*

In this test example a rectangular sheet made of transversely isotropic material is subjected to the simple tension and pure shear deformation (Fig. 1-2), in which the principal material axes and principal direction of the deformation coincide. In this situation the corresponding analytical solution can be constructed and then compared with numerical results. Using (3.23) and (3.31) we have for these loading cases:

1. Simple tension: $\lambda_1 = \lambda, \ \lambda_2 = \lambda_3 = \frac{1}{\sqrt{\lambda}},$

$$W(\mathbf{C}) = \frac{1}{2} a_{11} [\omega_1(\lambda_1) - \omega_2(\lambda_2)]^2, \tag{6.1}$$

$$\sigma^{<11>}(\lambda) = a_{11} [\omega_1(\lambda) - \omega_2(\lambda^{-1/2})] \left[\lambda \omega_1'(\lambda) + \frac{1}{2\sqrt{\lambda}} \omega_2'(\lambda^{-1/2}) \right],$$

$$\sigma^{<22>} = \sigma^{<33>} = 0. \tag{6.2}$$

2. Pure shear: $\lambda_1 = \frac{1}{\lambda_2} = \lambda, \ \lambda_3 = 1,$

$$W(\mathbf{C}) = \frac{1}{2} a_{11} [\omega_1^2(\lambda_1) - \omega_1(\lambda_1)\omega_2(\lambda_2)] + \frac{1}{2} a_{22} \omega_2^2(\lambda_2), \tag{6.3}$$

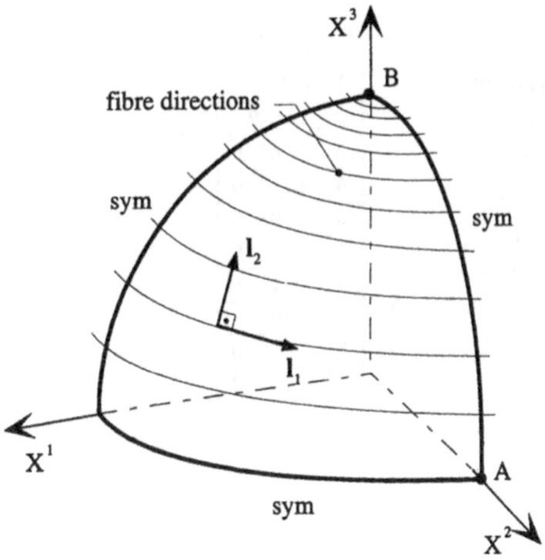

Geometry
R = 10, h=0.5

Material data
$E_1 = 10.0,\ E_2 = E_3 = 1.0,\ G_{12} = G_{13} = 2.2$
$\mu_{1i} = 0.8,\ \mu_{2i} = -0.1,\ \mu_{3i} = 0.3,$
$\alpha_{1i} = 4.0,\ \alpha_{2i} = 1.0,\ \alpha_{3i} = -2.8\ (i = 1, 2, 3)$

Load
internal hydrostatic pressure p

Figure 3. Inflation of a spherical balloon made of a transversely isotropic hyperelastic material.

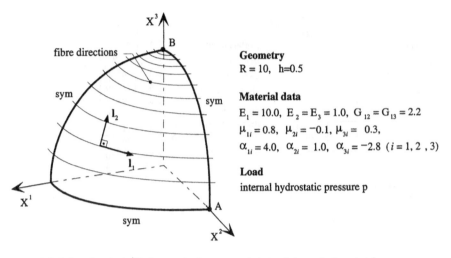

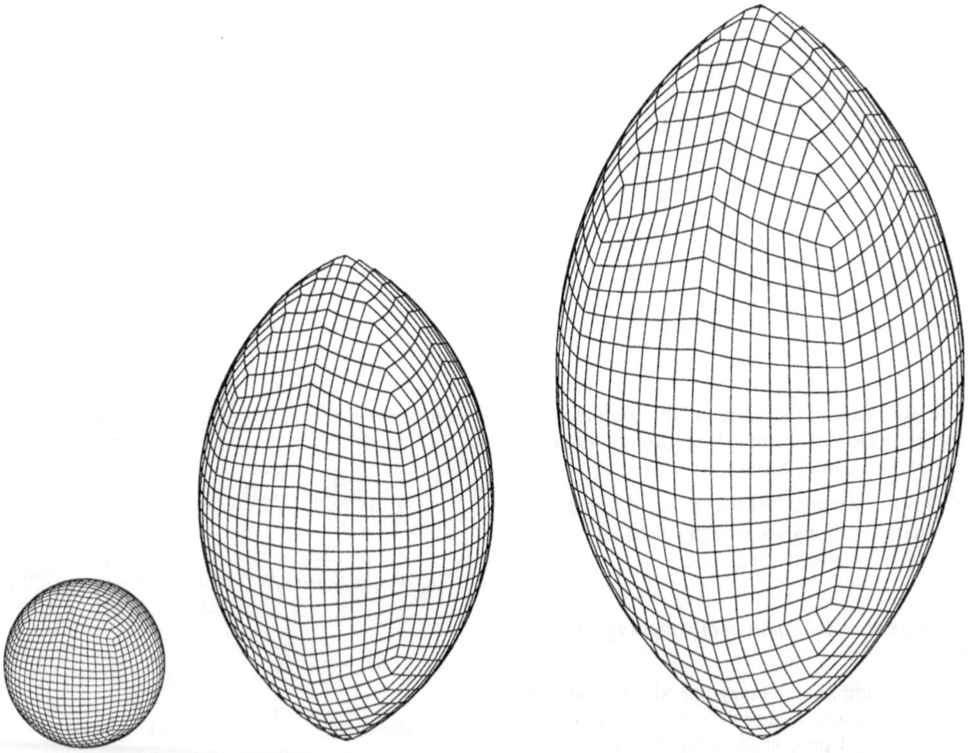

Initial configuration | Deformed configuration (p = 2) | Deformed configuration (p = 40)

Figure 4. Inflation of a balloon made of a transversely isotropic material: initial and deformed configurations.

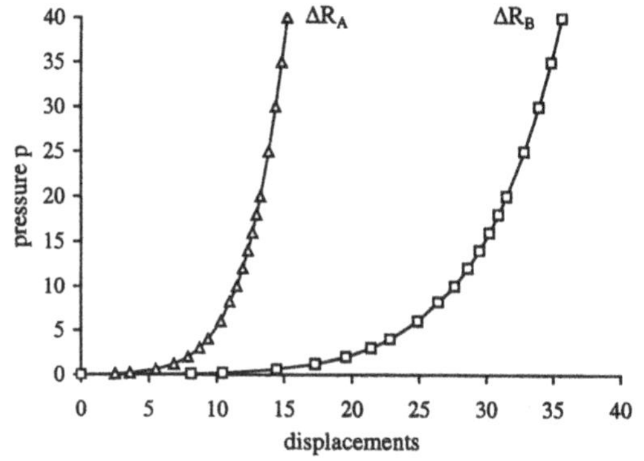

Figure 5.Inflation of a balloon made of a transversely isotropic material: load-displacement diagrams.

$$\sigma^{<11>}(\lambda)=a_{11}\lambda\left[\omega_1(\lambda)\omega_1'(\lambda)-\frac{1}{2}\omega_1'(\lambda)\omega_2(\lambda^{-1})\right],$$

$$\sigma^{<22>}(\lambda)=\frac{1}{\lambda}\left[a_{22}\omega_2(\lambda^{-1})\omega_2'(\lambda^{-1})-\frac{1}{2}a_{11}\omega_1(\lambda)\omega_2'(\lambda^{-1})\right],$$

$$\sigma^{<33>}=0. \tag{6.4}$$

For both cases the numerical results are in full agreement with the analytical solution.

6.2 *Inflation of a spherical balloon*

A spherical balloon made of transversely isotropic hyperelastic material is submitted to the internal hydrostatic pressure (Fig.3). The reinforcing fibres are arranged in the circumferential direction such that the problem under consideration is axisymmetric. To ensure an adequate quality of the finite element discretization in the vicinity of the pole we have analyzed an octant of the structure. Through the transversal isotropy of the material the initially spherical shell takes by loading the form of a rugby ball (Fig. 4). The material anisotropy also becomes apparent from the load-displacement diagrams plotted in Fig. 5 for two different points of the balloon. Large strains and finite rotations involved in this example should be emphasized as well.

7 CONCLUSION

For an adequate numerical simulation of reinforced rubber structures we have proposed a new orthotropic hyperelastic constitutive model. In each principal orthotropic direction the material properties are described through an individual isotropic tensor function of the strain tensor coupled with the corresponding structural tensor. Similar to the Ogden model these isotropic tensor functions are approximated by power series. Due to arbitrary number of polynomial coefficients and power, the model can be easily matched to experimental results. At infinitesimal strains the model proposed coincides with the orthotropic St.Venant-Kirchhoff material. Thus the condition of existence of the natural state and the requirement of the positive definiteness of the elasticity tensor related to this state are automatically fulfilled.

Using the algebra and differentiation rules of fourth and second order tensors (Itskov 1999) the constitutive relations and elastic moduli are derived in terms of eigenvalue-bases of the right Cauchy-Green tensor and obtained for the cases of distinct and coinciding eigenvalues. The numerical examples presented have demonstrated the effectiveness of the finite shell element build on the basis of the proposed constitutive model.

REFERENCES

Almeida, E.S. & R.L. Spilker 1998. Finite element formulations for hyperelastic transversely isotropic biphasic soft tissues. *Comput. Methods Appl. Mech. Engrg.* 151: 513-538.

Başar, Y. & Y. Ding 1997. Shear Deformation Models for Large-Strain Shell Analysis. *Int. J. Solids & Structures* 34(14): 1687-1708.

Başar, Y. & M. Itskov 1998. Finite Element Formulation of the Ogden Material Model with Application to Rubber-Like Shells. *Int. J. Numer. Meth. Engng.* 42: 1279-1305.

Başar, Y., M. Itskov & A. Eckstein 1999. Composite Laminates Nonlinear Interlaminar Stress Analysis by Multi-Layer Shell Elements. *Comp. Meth. Appl. Mech. Engrg.* (in press).

Betten, J. 1984. Interpolations Methods for Tensor Functions. In Avula, X.J.R. et al. (eds.), *Mathematical Modeling in Science and Technology*, Rergamon Press, New York.

Doyle, T.C. & J.L. Ericksen 1956. Nonlinear elasticity. In *Advances in Applied Mechanics* IV, Academic Press, New York.

Dvorkin, E.N & K.-J. Bathe 1984. A continuum mechanics based four-node shell element for general nonlinear analysis. *Engng. Comput.*1: 77-88.

Itskov, M. 1999. On the theory of fourth-order tensors and their application in computational mechanics. *Comp. Meth. Appl. Mech. Engrg.* (submitted).

Morman, K.N. 1986. The generalized strain measure with application to non homogeneous deformations in rubber-like solids. *J. Appl. Mech.* 53: 726-728.

Ogden, R.W. 1984. *Non-Linear Elastic Deformations*. Ellis Horwood series in mathematics and its applications, Chichester.

Simo, J.C. & J.E. Marsden 1984. On the Rotated Stress Tensor and the Material Version of the Doyle-Ericksen Formula. *Arch. Rat. Mech. Anal.* 86: 213-231.

Simo, J.C., M.S. Rifai & D.D. Fox 1990. On a stress resultant geometrically exact shell model. Part IV: variable thickness with through-the-thickness stretching. *Comp. Meth. Appl. Mech. Engrg.* 81: 91-126.

Ting, T.C.T. 1985. Determination of $\mathbf{C}^{1/2}$, $\mathbf{C}^{-1/2}$ and More General Isotropic Tensor Functions of **C**. *Journal of Elasticity* 15: 319-323.

Valanis, K.C. & R.F. Landel 1967. The Strain-Energy Function of a Hyperelastic Material in Terms of the Extension Ratios. *J. Appl. Phys.* 38: 2997-3002.

Constitutive Models for Rubber, Dorfmann & Muhr (eds) © 1999 Taylor & Francis ISBN 90 5809 113 9

Modelling rolling friction of rubber for prediction of tyre behaviour

V. Dorsch, A. Becker & L. Vossen
Continental AG, Research and Process Development, Hannover, Germany

ABSTRACT: When simulating the behaviour of a rolling tyre, modelling the rubber friction between tyre tread and road surface is crucial. Friction tests were done with different loads and velocities. From the analysis of the experimental results a phenomenological friction law for the friction coefficient is proposed. Finite Element computations of the experiments show the validity of this model. A main application of this friction model are simulations of the rolling tyre to predict wear and tire characteristics. Thus an understanding of mechanical processes in the contact patch enables the tyre engineer to optimise tyre design with the aid of simulation.

1 FRICTIONAL BEHAVIOUR OF RUBBER

For tyre design and tread compound layout understanding the frictional behaviour of rubber is very important. Due to the polymer nature of rubber and its nonlinear viscoelastic behaviour also rubber friction is viscose and nonlinear. Kummer & Meyer (1967) introduced a unified theory of rubber friction which explains the influence of the sliding velocity by the viscoelastic frequency dependent behaviour in a microscopic scale. Being nonlinear elastic, the normal pressure also influences rubber friction (Grosch & Schallamach 1969).

In spite of these well known frictional models most numerical Finite Element computations of rubber parts still use the simple Coulomb's law:

$$F_t = \mu \cdot F_n \tag{1}$$

F_t denotes the tangential force. In case of sliding it results from multiplying the normal force F_n with the constant sliding frictional coefficient μ.

To predict complex attributes like wear and characteristics of rolling tires using this simple frictional law is not sufficient in situations of high sliding.

2 FRICTION TESTS

At Continental AG several friction tests with different specimens but same rubber compounds and attributes of the contact partner were analysed. For this investigation all experiments were done under dry friction conditions.

2.1 *Grosch Abrader test*

This test was originally designed for abrasion analysis (Grosch & Schallamach 1969). Figure 1 shows a view of the test equipment.

The test piece, a small rubber wheel with 80mm diameter, is pressed by the wheel's load on a grinding wheel which is propelled by a motor. Due to friction the test wheel is driven by the rotating disk. The slip angle can be adjusted by turning the measurement axle. Thus, the test wheel is loaded similarly to a tyre: there is a wheel load, the difference between the driving and tyre direction is described

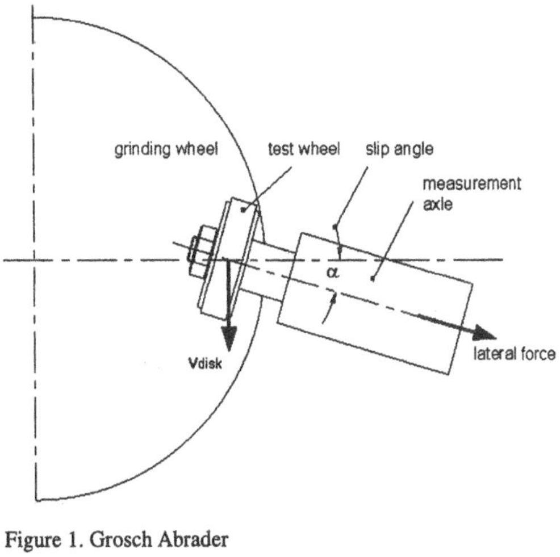

Figure 1. Grosch Abrader

by the slip angle so that a side force in lateral direction from the wheel occurs.

2.2 *Linear tester*

This test was done by Eberhardsteiner et al. (1998). A test specimen made of the same compound as for the Grosch Abrader is pressed on similar abrasive paper (so-called safety walk) and then accelerated to the specified sliding velocity. During a length of approximately 0.2 m the tangential force is measured at constant speed. Measuring speed, load, contact area and tangential force the frictional coefficient can be analysed in dependence of slip velocity and pressure.

2.3 *Rotating disk test*

These tests were carried out by Haupert (unpubl.). The test specimen is pressed onto a rotating disk covered with safety walk. With the measured load and tangential force and the rotating speed of the disk the frictional coefficient can be analysed.

2.4 *Experimental results*

The overall range of the frictional coefficient for the different tests, compounds and loading conditions varied from 0.6 to 1.8. The level of the values depends on the type of compound used, e.g. for winter or summer tyres. The pressure was between 1 bar and 10 bar, slip velocity occurred between 0.0005 m/s and 10 m/s. All experiments showed a non-linear dependence of the frictional coefficient μ from pressure and slip velocity. Figure 2 shows a typical result for μ from the Grosch Abrader measurements.

3 GENERAL FRICTION LAW

The results of the experiments show that a precise friction law is important for simulating rubber parts with friction. Here we use a phenomenological approach and establish a model with the frictional coefficient being a function of slip velocity and pressure.

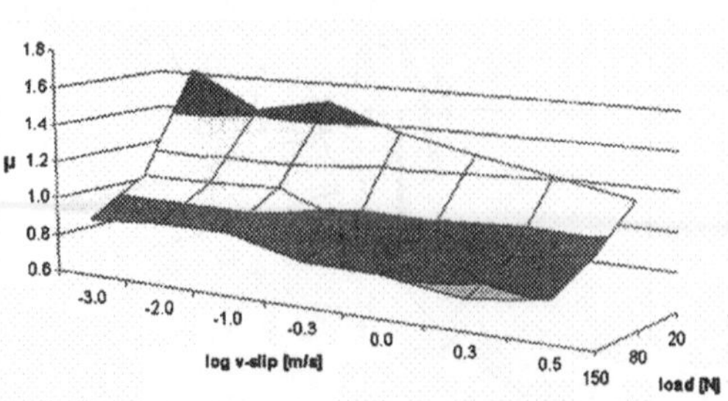

Figure 2. Frictional coefficient at different slip velocities and pressure values

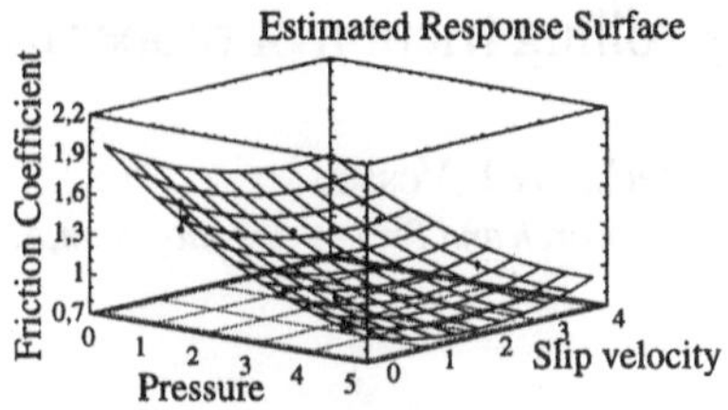

Figure 3. Frictional function for μ

As the Grosch test has a most similar loading condition to a tyre, only its results were used for the proposal of this function. Figure 3 shows the close correlation of this functional approach to the measurements.

This frictional law can now be implemented into a Finite Element program for tyre simulations. For each type of compound the parameters of the frictional function have to be handled as material parameters of the combination of rubber compound and type of ground.

4 FINITE ELEMENT FORMULATION FOR ROLLING TYRES

Here only a brief outline is given. The interested reader may confer Becker & Seifert (1997) for more details.

4.1 *Rolling kinematics*

A variational setting based on Hamilton's principle is used to involve the acceleration terms for the steady state rolling tyre. Within the present approach the position of a material point x(t) in the current configuration is a mapping φ of a material point X(t) of the reference configuration at a representative time t. The coordinate system is based in the tyre centre.

$$x(X,t) = \phi(X(t)) \tag{2}$$

Applying the chain rule leads to the material derivative involving the deformation gradient F.

$$\dot{x}(t) = GRAD_X\, x \cdot \dot{X}(t) = F \cdot \dot{X}(t) \tag{3}$$

Rigid body dynamics at an arbitrary time t is described with respect to reference coordinates ξ and the help of an orthogonal tensor R(t).

With R(t), X(t) = R(t) ξ and the orthogonal tensor $\Omega = RR^T$ the variational formulation needed for the Finite Element framework yields

$$\delta\dot{x} = \delta F \cdot \Omega(t) \cdot X(t) \tag{4}$$

The equations of motion are transferred within standard variational calculus in combination with the information of steady state rolling and Gauss' theorem. The resulting expression reads

$$\delta H(x,\delta x)=\int_V\left(\rho\ddot{x}\delta\dot{x}+f\delta x-2\frac{\partial W}{\partial C}\cdot\delta E\right)dV=0 \quad (5)$$

which is the continuum mechanical equivalent to Hamilton's principle of analytical mechanics. W, C and δE denote the strain energy, the right Cauchy Green tensor and the first variation of the Green Lagrange strain tensor. Furthermore ρ and f are associated mass density and the applied external forces. Based on this variational expression straight forward discretization of δE and $\delta\dot{x}$ gives the resulting Finite Element formulation.

4.2 *Rolling frictional contact*

Within the framework of a perturbed Lagrangian functional the residual terms are

$$G(x,\delta x)=\delta H(x,\delta x)+f_n\delta g_n+f_t\delta g_t=0 \quad (6)$$

where the contributions concerning the standard displacement part for the rolling formulation $\delta H(x,\delta x)$, the normal $f_n\delta g_n$ and tangential part $f_t\delta g_t$ of the contact forces are separated. The normal contact force $f_n = \lambda_n g_n$ depends on the nonlinear gap g_n and the penalty parameter λ_n. For this normal contribution a standard numerical treatment concerning linearisation within the Finite Element algorithm is performed. Regarding the tangential contact force a regularised Coulomb friction law is used which reads

$$f_t=-\mu\cdot\tanh\left(\frac{|v_s|}{\omega}\cdot\lambda_t\right)\cdot\frac{v_s}{|v_s|}\cdot\lambda_n g_n \quad (7)$$

where f_t is the contact force vector, g_t the nonlinear tangential gap with penalty parameter λ_t, μ the tangential friction coefficient, v_s the slip velocity and ω the associated angular velocity of the tyre. The tangential frictional coefficient μ will be a function of the slip velocity and the pressure according the general frictional law from paragraph 3. The factor $|v_s|/\omega$ can be viewed as a regularized slip velocity with the term tanh() as an abitrary chosen smoothing function (Zheng, unpubl.).

The slip velocity is computed as follows

$$v|_s=\dot{x}=F\cdot\Omega\cdot X|_{footprint}+v|_{axis}-v|_{ground} \quad (8)$$

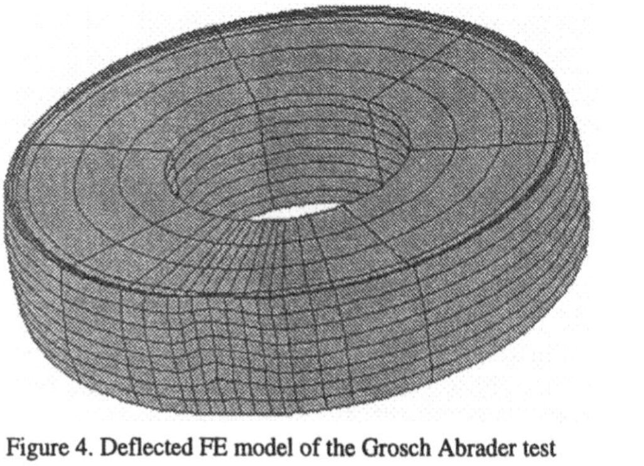

Figure 4. Deflected FE model of the Grosch Abrader test

5 VALIDATION OF THE FINITE ELEMENT MODEL OF THE GROSCH ABRADER TEST

For validation of the frictional law the Grosch Abrader test was simulated by Finite Element computations. The rolling kinematics and the frictional rolling contact described in the last paragraph are implemented in the in-house program GENFEP of Continental AG. Figure 4 shows the Finite Element structure of the test wheel deflected by the loading conditions of the experiment.

A comparison of measured and computed contact patch areas under static load show a good correlation of the model (Figs 5, 6).

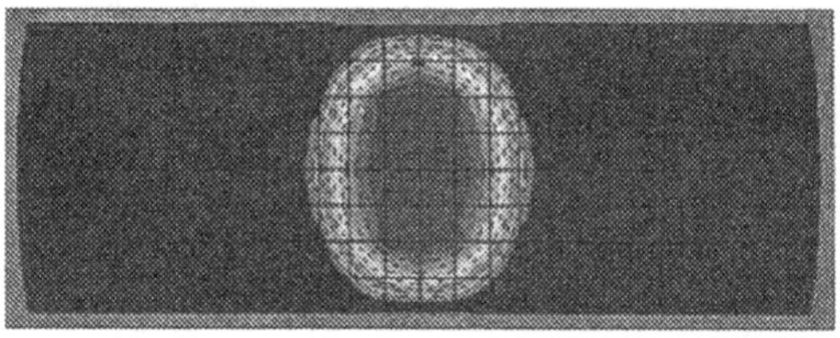

Figure 5. Computed pressure distribution in the footprint

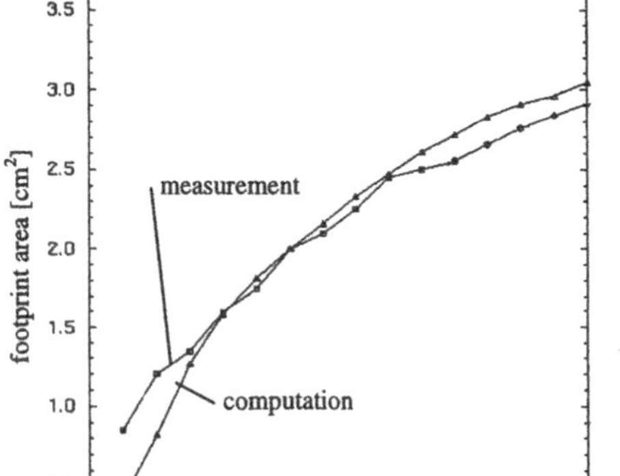

Figure 6. Computed and measured area of contact patch

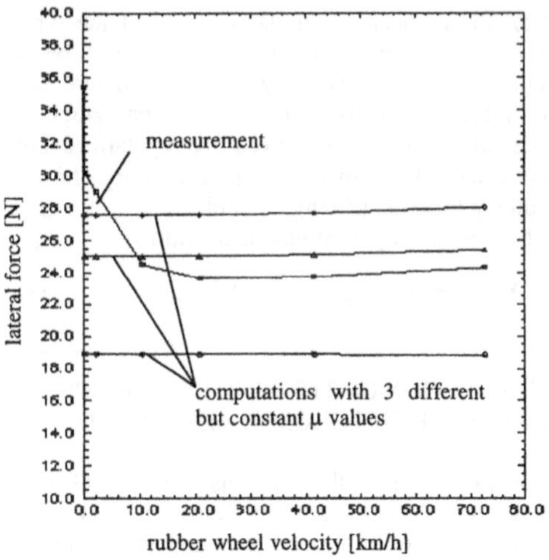

Figure 7. Measured and computed lateral forces at different speeds using constant μ values

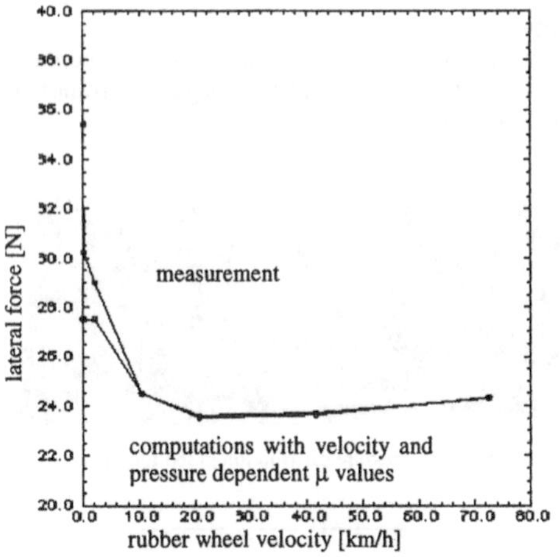

Figure 8. Measured and computed lateral forces at different speeds using velocity dependent μ values

Figure 7 shows the measured lateral force of the test wheel at different speeds. Three results of computations using a constant frictional coefficient are also plotted. It can clearly be seen, that a constant frictional coefficient cannot represent the behaviour of the rubber wheel. Figure 8 shows the good correlation of the lateral force when using a velocity and pressure dependent frictional coefficient.

The pressure distribution in the footprint of the rolling model (Fig. 9) shows a maximum which has moved to the inner side of the wheel compared to the static loading case (Fig. 5).

A pressure measurement, which could only be done at another slip angle and another wheel load, shows the same effect (Fig. 10).

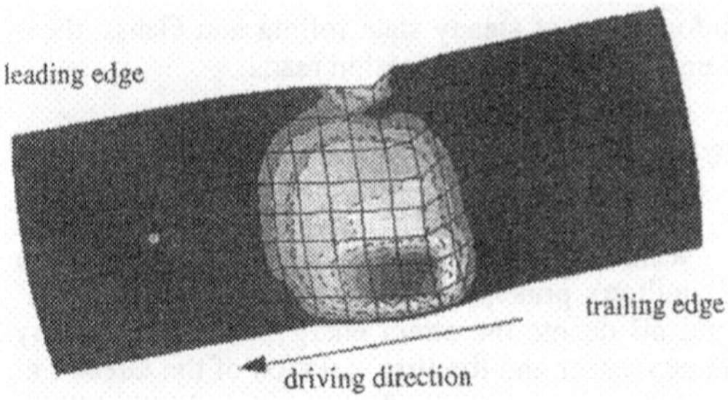

Figure 9. Computed pressure distribution of the rolling test wheel with 10^0 slip angle

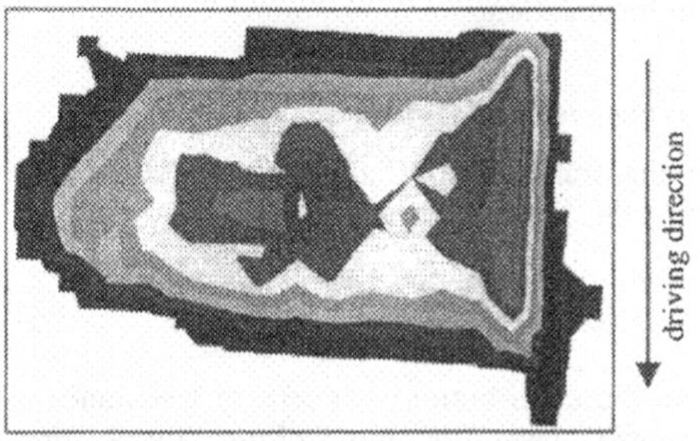

Figure 10. Measured pressure distribution of the rolling test wheel with 3^0 slip angle

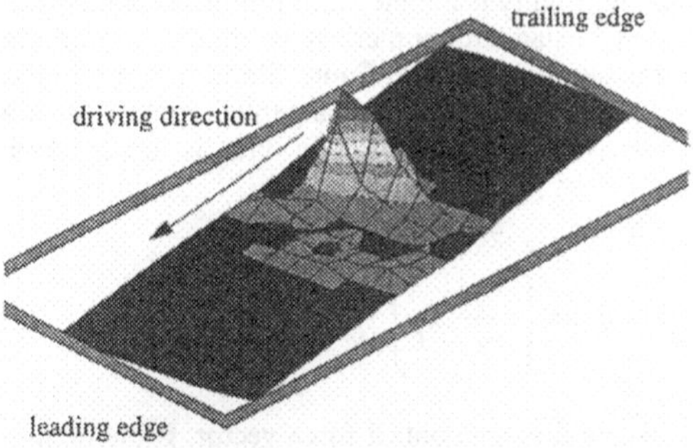

Figure 11. Computed slip velocity distribution of the rolling test wheel at 10^0 slip angle

Looking on the slip velocity distribution in driving direction the so called snap-out effect can be seen: In the leading edge when the material points touch the ground there is stick contact first. Further rolling then causes an increasing lateral deformation of the footprint. Once the maximum tangential contact force is reached the material snaps back. Thus we see a clear peak of slip velocity at the trailing edge of the rubber wheel in figure 11.

All the preceding results give a strong confidence into the simulation model, which encourages us to use this phenomenological frictional law for tyre computations.

6 FINITE ELEMENT TIRE COMPUTATIONS

Understanding and predicting the processes in the footprint of a rolling tyre is necessary for tyre design. An important application of simulation is prediction of wear and tire characteristics. The goal of such computations is on the one hand to reduce costly highway tests and test tyre building and on the other hand to understand the mechanics and thus to be able to optimize tire design.

A typical 3D FE tire model has between 24,000 and 75, 000 degrees of freedom. A hybrid element formulation is used and a composite material model describes the cord rubber layers. Ground is modelled as rigid surface. Figure 12 shows the deflected mesh of the loaded tyre.

Figure 13 shows the already mentioned snap out effect, a peak in the slip velocity distribution at the trailing edge of a tyre rolling at 80 km/h with a driving moment of 150 Nm.

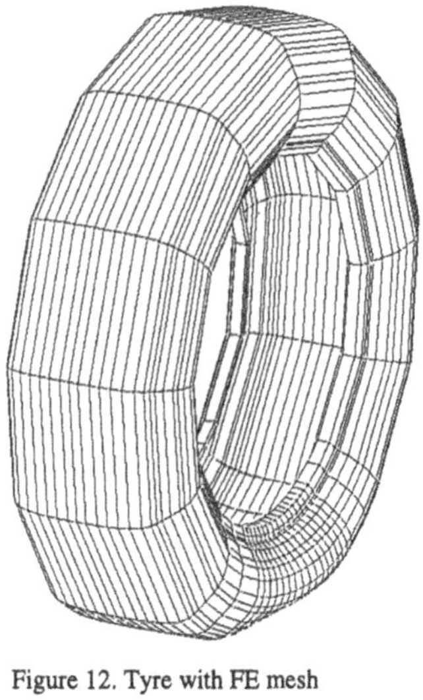

Figure 12. Tyre with FE mesh

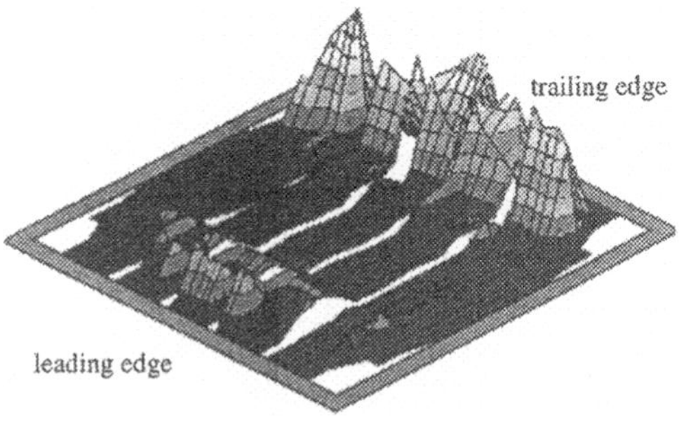

Figure 13. Circumferential slip velocity distribution of a rolling tire without camber angle

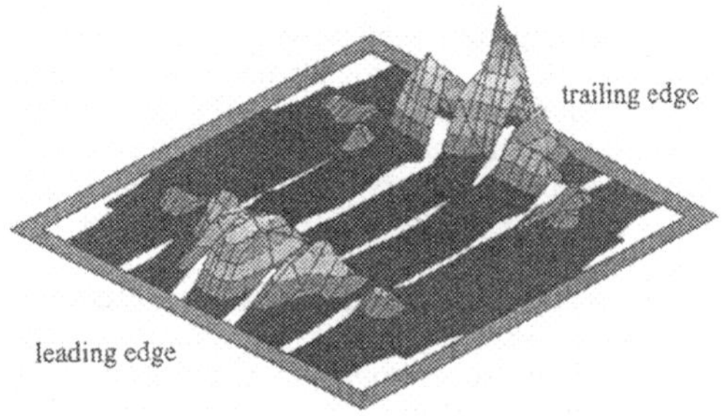

Figure 14. Frictional energy distribution of a rolling tire

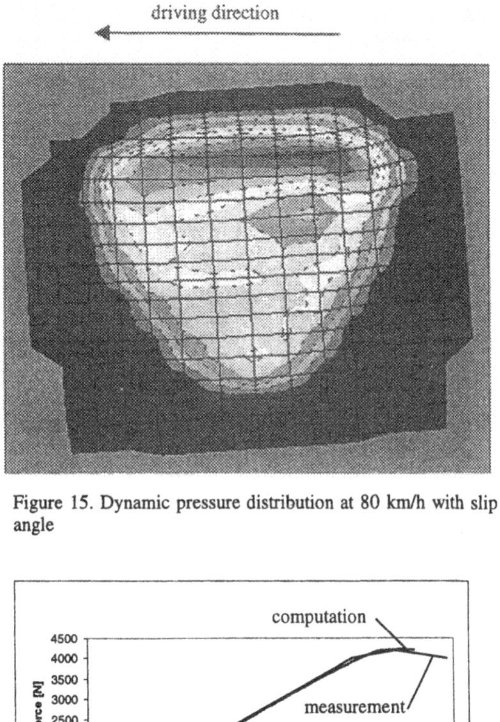

Figure 15. Dynamic pressure distribution at 80 km/h with slip angle

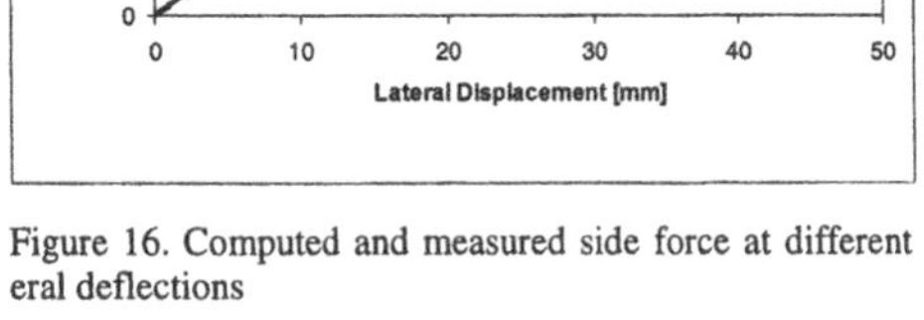

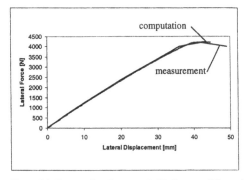

Figure 16. Computed and measured side force at different lateral deflections

The frictional energy distribution which is the product of tangential frictional forces and slip velocities indicates wear (Becker & Seifert 1997). Thus the peak of the frictional energy distribution shown in figure 14 hints for larger wear in the center.

Another important application is the prediction of static and steady state rolling tire characteristics. Vehicle handling behaviour strongly depends on those characteristic curves of its tyres. The pressure distribution of a rolling tyre running with slip angle shows a shift of the pressure peak to the outer side (Fig. 15).

Computations of the static lateral tyre characteristics show good correlation with measurements (Fig. 16). These simulations help to understand how tyre construction influences those characteristics and how to optimize tyre design.

Thus the whole process from tyre design to tyre characteristics to vehicle handling can be simulated in the computer.

REFERENCES

Grosch, K.A & Schallamach, A. 1969. The load dependence of laboratory abrasion and tyre wear. *Kautschuk und Gummi* 22(6): 280-292.

Kummer, H.W. & Meyer, W.E. 1967. *ATZ* 69: 245.

Becker, A. & Seifert, B. 1997. Simulation of wear with a FE tyre model using a steady state rolling formulation. In M.H. Aliabadi & A. Samartin (ed.), *Computational methods in contact mechanics*; *Proc. 3rd intern. conf., Madrid, 1997.* Southampton: Computational Mechanics Publications: 119-128.

Eberhardsteiner, J. & Fidi, W. & Liederer, W. Experimentelle Bestimmung der adhäsiven Reibeigenschaften von Gummiproben auf ebenen Oberflächen. Kautschuk, Gummi, Kunststoffe 5(11): 773-781.

Haupert, F. 1998. Reibversuche an einer Gummimischung mittels Stift auf Scheibe-Anordnung zur Ermittlung der Reibwerte unter verschiedenen Prüfbedingungen. *Internal report*, Institut für Verbundwerkstoffe GmbH, Kaiserslautern.

Zheng, D. 1995. Frictional rolling contact. *Internal report*, Continental AG.

Constitutive Models for Rubber, Dorfmann & Muhr (eds) © 1999 Taylor & Francis ISBN 90 5809 113 9

Experimental characterisation of friction for FEA modelling for elastomers

J.R.Daley & D.Lam
Forsheda Limited, Dowty Engineered Seals, Tewkesbury, UK

D.J.Weale & M.V.Mercy
University of Birmingham, UK

ABSTRACT: In the Non-linear Finite Element Analysis of elastomers, friction coefficients remain one of the largest unknowns within the modelling process. Even given precise material models, geometry and loading and boundary conditions, without an accurate friction function, errors are built back into the calculations. The possible value and type of friction function and the corresponding range of computerised results imply an exact solution may only be estimated, and the designer returns to engineering judgement on the limit of the solution. Through a combined project between Forsheda and The University of Birmingham, the use of modular experimental friction tests has lead to friction functions for elastomers under various assembly conditions being produced. Correlation between the experimental and theoretical FEA results is discussed and limitations in the testing procedures highlighted. Extending the friction models to more realistic sealing designs and operational environments is discussed and the accuracy of employing such measures theorised.

1 INTRODUCTION

Understanding of friction in elastomeric seals is required for the accurate analysis of seals in application. With the advent of non-linear finite element analysis the stress strain relationships within the elastomer has been investigated and several methods of characteristic based on strain energy functions have been developed (e.g. Ogden 1972, Rivlin & Sanders 1951).

At Forsheda, the employment of non-linear finite element analysis plays a vital role in the design development of new seals. FEA has been regarded as a tool for reducing lead-times and production costs, through design optimisation prior to seal manufacture. It is therefore essential that FEA results accurately predict the seal behaviour in operation, before any engineering judgement may be made. At Forsheda, this judgement is accompanied with over twelve years experience in using non-linear FEA. Currently within Forsheda, the non-linear software MARC is utilised in facilities across UK, Sweden, Italy and USA.

Within the available FEA packages, friction functions are required in either simple coefficient form or more complex functions. Although the value of μ is vital to certain results, for example those predicting assembly loads, research into a workable function for these packages has not been published widely. The major concern exists where there is no single correct method or function, which could be generically used for all FEA models of rubber sealing applications. Such applications require reduced friction forces (Schleth et al 1999, Daley & Mays 1999b).

The work conducted by Forsheda in collaboration with The University of Birmingham, attempts to shed light in determining a friction function which closely correlates with physical experimental tests.

2 FRICTION THEORY

The basic theory (Coulomb 1785) of friction being a direct relationship between axial and normal force (1) has to be extended when considering the micro level of friction theories.

$$F = \mu N \qquad (1)$$

where F = axial force; μ = friction coefficient; and N = normal force.

The friction may be classed as static and dynamic or running friction. The static element being the breakout friction level, where no relative movement actually takes place between the two mating surfaces. The dynamic friction draws in elastohydrodynamic functions into play.

There are two distinctive friction models, which can be incorporated into elastomeric contact analysis while using MARC. These two models are Coulomb Friction and stick-slip Friction.

According to MARC (MARC 1997), Coulomb friction is represented as a non-linear model dependent on both the normal force and relative velocity, and can be expressed as:

$$f_t = -\mu . f_n . t \tag{2}$$

where f_t = tangential frictional force; μ = friction coefficient; f_n = normal reaction; and t = tangential vector in the direction of the relative velocity.

A modified model, Coulomb for Rolling, can also be introduced within the MARC (MARC 1997) analyses. This friction model becomes invaluable where the velocity of an approaching contact body greatly differs from the relative velocity at the point of contact, as in the case for rolling problems.

The stick-slip model simply assumes that there is a degree of adhesion, which an external tangential force would be required to overcome before sliding can take place:

$$\|f_t\| \leq \alpha . \mu . f_n \tag{3}$$

$$\|f_t\| > \alpha . \mu . f_n \tag{4}$$

where f_t = tangential frictional force; μ = friction coefficient; f_n = normal reaction; and α = overshoot parameter.

MARC utilises equation (3) in the sticking mode and equation (4) in the slipping case. This can be explained by considering the interaction of the elastomeric contact with a smooth rigid surface, which moves from rest. Initially there is a component of static friction, which would cause the elastomeric surface to stick to the rigid body. As the rigid body continues to travel, frictional forces would increase to a threshold where the elastomer looses contact of the rigid body. This slippage gives rise to a dynamic friction coefficient, which is generally less than static values.

Elastohydrodynamic theories have been proposed (Dowson 1968) for elastic bodies and their use in the accurate numerical simulation of said mechanisms.

$$H = 1.40 \frac{G^{0.74} U^{0.74}}{W'^{0.074}} \tag{5}$$

where H = film thickness; G = material constant; W = radial contact load; and U = velocity constant.

The running friction therefore will depend on the normal load and also on the velocity in the application. There is no indication in (5) as to the level of lubricant or indeed any properties associated with the lubricant, such as viscosity. Given a lubricated case and the material used is consistent, equation (5), reduces to the film thickness being a function of the velocity. The higher the velocity the thicker the film thickness and hence a reduction (up to a limit) in friction.

3 TESTING

A test program was set up at the University of Birmingham to measure and record the breakout and running friction function from axial movement of "O" ring seals. The methodology behind this was through functional testing using a simple bore and gland housing. The tests were to investigate material, speed and lubricant parameters and their effect on the axial load required to move the seals.

The test fixture was used in the assimilation of loading data. The rig was a piston and cylinder for assessing axial load required to assemble and move the seals in a housing, shown in Figure 1. The piston had two grooves machined on it to allow two "O" rings to be tested at once, this increased the stability of the rig, and allowed some self-centering due to the seals. A Lloyd (Birmingham University) and an Instron tensile test machine (Forsheda) were both used to apply and measure the loads and displacements. Tests were carried out at 25°C and 100°C, with speeds ranging from 20mm per min to 300mm per min, under dry conditions or lubricated with engine oil, and with two nitrile rubbers of 70 and 90 IRHD hardness.

The tests involved firstly assembling the "O" rings onto the grooves of the piston, only seals of the same hardness were used together. The piston was then inserted into the cylinder with the axial load recorded using the extensometry equipment.

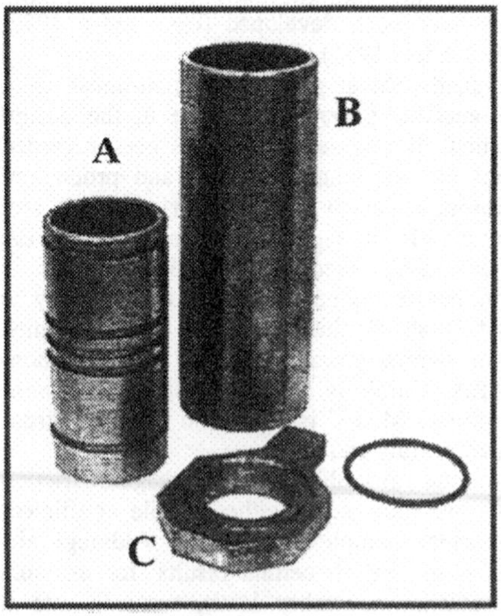

Figure 1. "O" ring loading rig consisting a piston A, bore B, and axial compression rig C.

From the experiment described above, the axial load and displacement for a given seal and conditions could be found, however the normal load and thus the friction coefficient could not easily be measured. A simple second rig C (Figure 1), was constructed to try to gain a measurement of the normal load. The rig allowed the compression of a seal in an axial groove in a flat plate constructed to simulate the radial groove. By analogy between the level of compression in rig C to the interference in the radial application the normal load may be estimated.

With the utilisation of different friction functions in the calculations, FEA simulation would attempt to reproduce results that closely match those from the physical tests.

4 NON-LINEAR FINITE ELEMENT ANALYSIS

In conjunction with the experimental work, non-linear FEA was carried out at Forsheda where the systems tested at Birmingham were modelled and analysed.

2D axisymmetric models were created of the “O” ring seals using nominal dimensions. Dimensions of the groove and housing were measured from the experimental rig and used as input to the program. Material properties of the nitrile elastomers were obtained through tensile and compression tests of samples, (Daley 1995, Daley & Mays 1999a). The ensuing results were fitted with a 2 term Ogden strain density function using the MARC software.

Figure 2 shows the graphical output from the Coulomb FEA analyses. The figure shows three positions of the two "O" rings during the assembly process (centerline of the seals being below the seals), radial Cauchy stress is plotted on the FEA mesh.

The FEA models were created for a static / quasi-static situation. In assembling an “O” ring, although the axial movement may be relatively large between the gland and the housing, the movement of material within the seal is low (assuming no rolling) and therefore dynamic material properties of the elastomer have been assumed not to be vital.

This FEA model was run with varying friction functions to simulate the insertion test with μ of 0.1, 0.2, 0.3 and 0.4. The friction models Coulomb, Coulomb for Rolling and stick-slip from the MARC software were employed.

The results from the FEA models were compared alongside those obtained from experimental work in order to generalise friction functions for the tests.

These generalised functions would then be fed back to be incorporated into FEA models of similar sealing products for design optimisation.

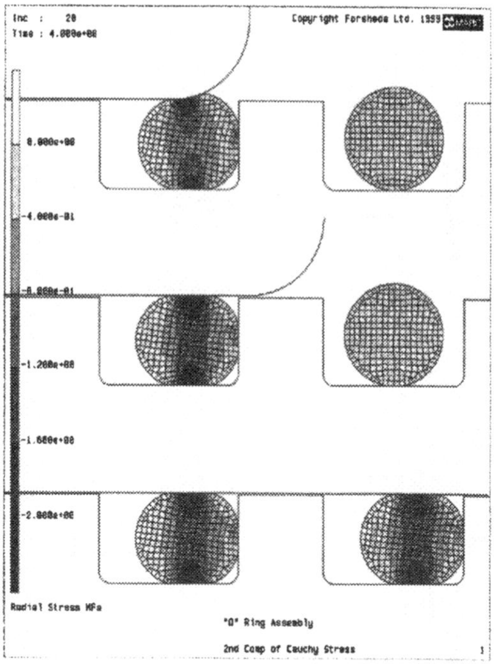

Figure 2. Diagram showing FEA simulation of the "O" ring assembly.

5 RESULTS

5.1 *Experimental Results*

Graphs of the experimental results are shown in Figures 3 to 6. Figures 3 and 4 show the axial experimental load on the assembly of the second seal (the first having already been inserted). Figure 3 shows the loading for one seal and the effect of speed of assembly and the effects of lubrication. Figure 3 is simplified and re-plotted as Figure 4 for the dry and lubricated cases with an assembly speed of 40mm/min.

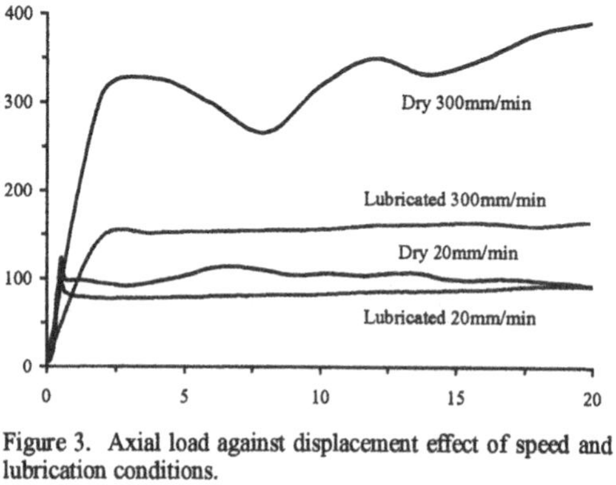

Figure 3. Axial load against displacement effect of speed and lubrication conditions.

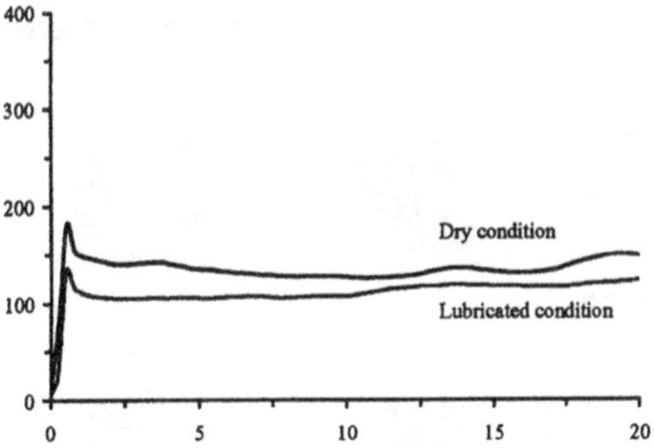

Figure 4. Axial load for the dry and lubricated conditions at 40mm/min.

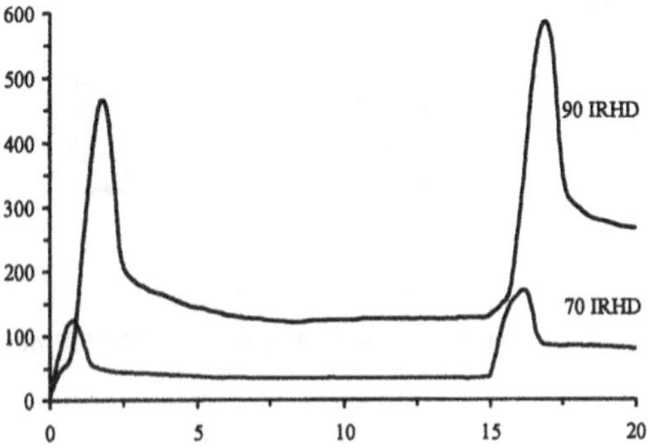

Figure 5. Axial load variation for 70 and 90 IRHD seals.

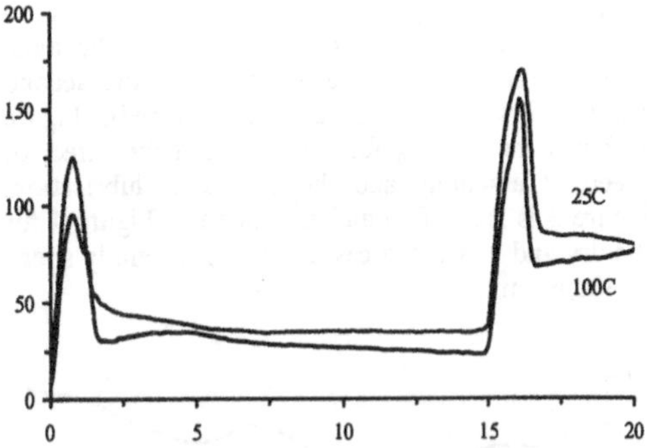

Figure 6. Axial load for 70IRHD seal at two temperatures.

The effect of lubrication was a reduction of the friction coefficient, as anticipated. The level of variation in the results was increased with the introduction of the lubricant, this is probably due to the fact lubricated friction is dependant on several layers of boundary lubrication and the effect of elastohydrodynamics, which is a whole area of research in itself.

Extending the testing, the assembly loading for two seals was measured. Figure 5 and 6 indicate the axial loading comparing the material hardness (Figure 5) and temperature (Figure 6). The figures show for the experiment the two distinct areas of breakout friction and running friction.

5.2 *Finite element analysis results*

Figures 7, 8 and 9 show the comparison between the FEA models with the various friction algorithms and the experimental data. All the figures are considering the 70 hard nitrile with the experimental insertion speed of 40mm/min. The three figures show inconsistencies between the FEA and experimental. The ratio between the FEA and experimental for the four key stages are shown in Table 1.

Table 1 shows the FEA results compared to the experimental for each part of the insertion. Table 2 puts these results relative to the axial running load of the first seal. Table 3 gives the coefficient of friction required for each of the FEA models to generate the correct value for the axial running load of the first seal, i.e. tuning the friction models for this result.

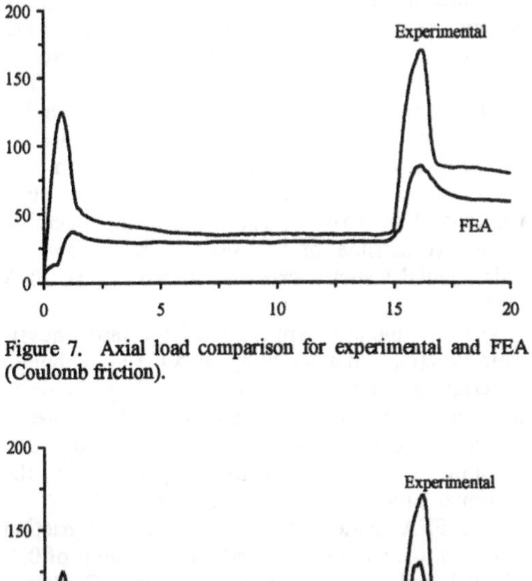

Figure 7. Axial load comparison for experimental and FEA (Coulomb friction).

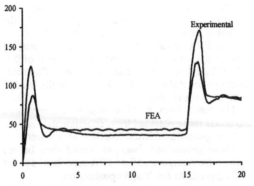

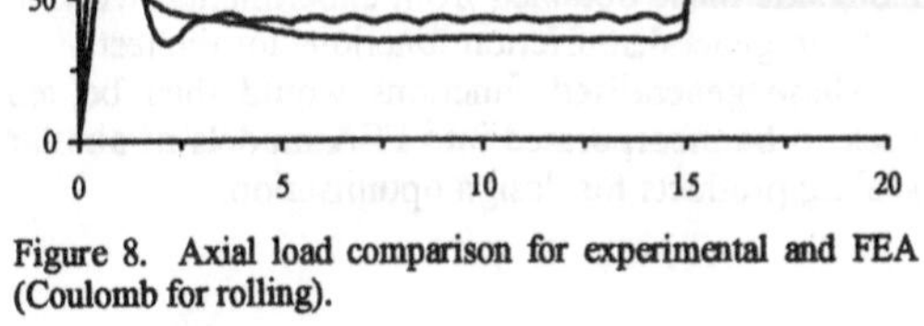

Figure 8. Axial load comparison for experimental and FEA (Coulomb for rolling).

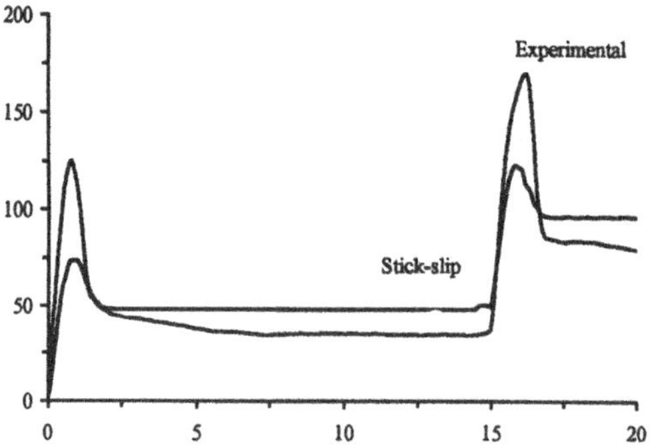

Figure 9. Axial load comparison of experimental and FEA for stick-slip.

Table 1. Ratio of FEA loads to experimental loads for key stages of test.

	1st Peak	Sliding	2nd Peak	Sliding
Experiment	1.00	1.00	1.00	1.00
Coulomb*	0.31	0.84	0.51	0.75
Coulomb Rolling*	0.69	1.21	0.76	1.07
Stick-Slip+	0.59	1.37	0.73	1.23

* Using equivalent 0.20 friction coefficients.
+ Using equivalent 0.20 running and 1.05 overshoot.

Table 2. Ratio of loads at key stages to load during first sliding portion.

	1st Peak	Sliding	2nd Peak	Sliding
Experiment	3.55	1.00	4.83	2.23
Coulomb*	1.45	1.00	2.90	2.00
Coulomb Rolling*	2.02	1.00	3.04	1.98
Stick-Slip+	1.53	1.00	2.55	2.01

* Using equivalent 0.20 friction coefficients.
+ Using equivalent 0.20 running and 1.05 overshoot.

Table 3. Apparent coefficient friction values required.

	Mean	1st Peak	Running	2nd Peak	Running
Coulomb	0.39	0.65	0.24	0.39	0.27
Coulomb Rolling	0.23	0.29	0.17	0.26	0.19
Stick-Slip	0.23	0.34	0.15	0.27	0.16

Using these tuned friction values the FEA models were re-run. With a coefficient of Coulomb friction of 0.24 a good correlation with experimental results for the running condition was obtained, as seen in Figure 10. Conversely the model showed very poor correlation with the insertion loads. During insertion the peak load was seen to fall and the shape of the peak to be substantially modified, as seen in Figure 7.

Additional FEA models were run with stick-slip coefficients of μ equal to 0.155 and value of α of 3.575 and 7.000, these results are shown in Figure 10.

The corresponding results from FEA for the peaks and running segments are shown in Table 4, for the various analyses with the Coulomb and stick-slip friction models.

6 DISCUSSION OF RESULTS

The results show that the FEA friction models Coulomb for rolling and stick-slip did give axial loading results of the same form as the experimental. Several specific features were seen in each of the FEA friction algorithms utilised.

The Coulomb model (Figure 7), gave a low prediction for the assembly load of the first seal, but a fairly realistic value and form for the running friction (in a lubricated example). Once tuned for the first running section, the Coulomb model gave a good result in both the first (4% difference) and second (7% difference) running sectors. Accurate values for the peak loads were not found however, with the first peak diminishing totally.

The Coulomb for rolling gave a closer value for the peak assembly loads, but on the running friction gave a variable result around a fixed mean. This form of running friction is closer to the unlubricated case, but the actual value of the mean is close to the running friction value found in the lubricated case. In the transition from the peak load to the running

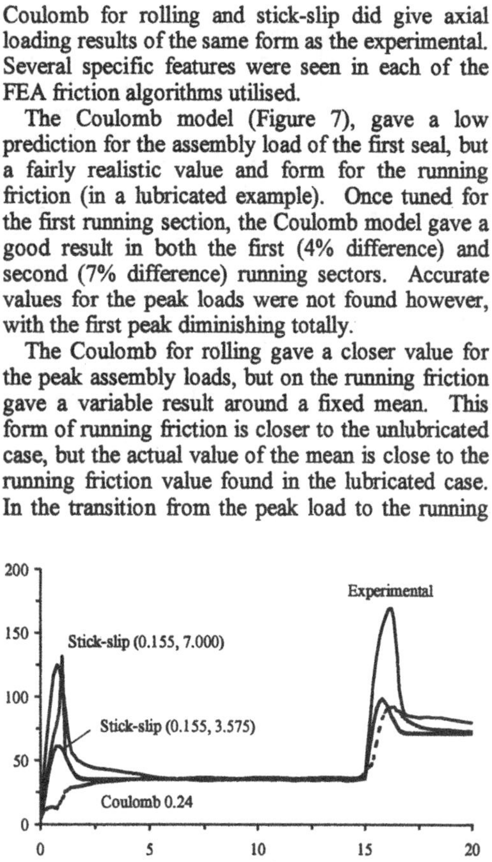

Figure 10. Axial load against displacement for FEA stick-slip and Coulomb models compared to experimental.

Table 4. Ratio of loads at key stages to load during first sliding portion.

	1st Peak	Sliding	2nd Peak	Sliding
Experiment	1.00	1.00	1.00	1.00
Coulomb∇	-	1.04	0.54	0.93
Stick-Slip*	0.31	0.84	0.51	0.75
Stick-Slip#	0.49	1.02	0.58	0.92
Stick-Slip+	1.06	1.00	0.58	0.91

∇ Using equivalent 0.24 friction coefficients.
* Using coefficients of 0.20 running, 1.05 overshoot.
# Using coefficients of 0.155 running, 3.575 overshoot.
+ Using coefficients of 0.155 running, 7.000 overshoot.

condition it may be observed the Coulomb for rolling model undershot the running value, which was not seen in the experimental or other FEA friction model results. The Coulomb for rolling model gave the greatest distortion in the FEA mesh resulting in the model undergoing large rotation, as may be seen in Figure 11.

The stick-slip model gave closer results for the peak loads compared to the coulomb friction. The stick-slip running friction gave a very smooth result (more inline with lubricated) with very little variation.

An examination of the FEA results indicated that the stick-slip model should be the closest to the experimental. The running friction with both seals inserted was approximately twice (2.23) as large as for the single seal, which is consistent with any model based on Coulomb friction. The second insertion load is roughly the insertion load of the first seal plus its subsequent running friction. This would be consistent with the stick-slip case assuming that slip conditions apply to the first seal whilst stick conditions apply to the second seal when it first contacts the cylinder. It may be calculated that the coefficient of friction for slip conditions is between 0.15 and 0.16, whilst the overshoot factor (α) should be between 3.55 and 3.60 (3.575 average) see Tables 1, 2 and 3. The results from the FEA showed that the level of the running friction was in line with the experimental, the peak value (where the

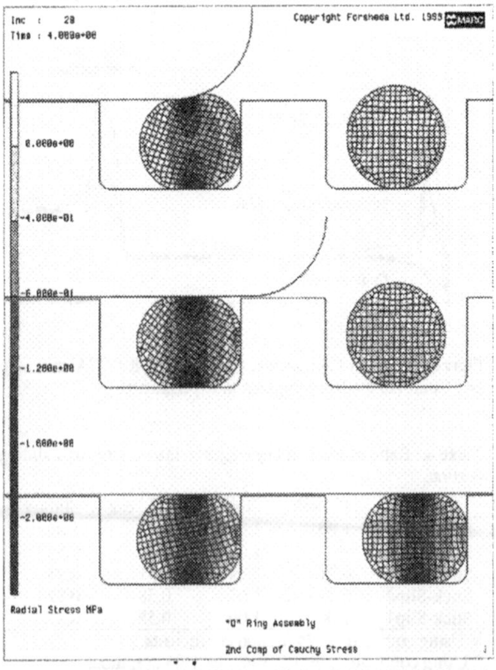

Figure 11. Mesh distortion under Coulomb for rolling friction

sticking occurred) was less than the experimental. Rerunning with an overshoot value of 7.000 gave a peak more in line with the first experimental peak and subsequent running load, but resulted in an underestimate for the second peak (Table 4). The second peak load values for the stick-slip models (ran with overshoot values of 3.575 and 7.000), are seen to be similar.

7 CONCLUSIONS

In the application the experimental tests were carried out, the results were variable. The basic results indicated that friction tended to increase with speed, temperature and hardness. These results were found through experimental methods and confirmed with the FEA technique.

The closest FEA model to the experimental was dependent on the conditions the seals were subject to, but initially the Coulomb for rolling model showed the greatest correlation in the absolute values, but large mesh distortion was observed. It was originally thought the stick-slip model would be the closest to the experimental results, and indeed gave a good correlation for the first assembly loading profile. However the second profile was not precisely predicted, this may be due to the sticking phenomenon only being applied to the first insertion peak. Using different values of overshoot did not result in a noticeable variation in the second peak, indicating that in this case the second assembly being modelled incurs minimum stick (peaks independent of α). More work to assess all the friction models, and in particular the stick-slip model is required. There may be an ultimate requirement to tailor the most suitable friction function for the final application, or as shown here tune the friction coefficient dependant on which part of the application is of most importance.

It is envisaged a mixture of the available friction functions would have to be used to accurately represent the complete loading values and form of the experimental test.

REFERENCES

Coulomb, C. 1785. Théorie des machines simples, *Mémoires de Mathématique et de Physique, Academie des Sciences,* vol. 10.

Daley, J. R. 1995. The material modelling of rubber. *Proceedings of Finite Element Analysis of Elastomers, IMechE conference,* London, 15th October 1995.

Daley, J. R. & Mays, S. 1999a. The complexity of material modelling in the design optimisation of elastomeric seals. In *Finite Element Analysis of Elastomers,* D Boast and VA Coveney (Eds), Professional Engineering Publishing, London, pp. 119-128.

Daley, J. R. & Mays, S. 1999b. The modelling of environmental conditions and the interaction on polymer seals in the fluid power industry. *Proceedings of the 11th International Sealing Conference,* 3rd-4th May, Dresden, pp. 84-96.

Dowson, D. 1968. Elastohydrodynamics. *Proceedings of the Institution of Mechanical Engineers*, Vol. 182, Pt 3A, 1967-68, pp. 151-167.

MARC 1997. Volume A, theory and user information. MARC Analysis Research Corporation.

Ogden, R.W. 1972. Large deformation isotropic elasticity. On the correlation of the theory and experiment for compressible rubber like materials. *Proceedings of the Royal Society*, London, A328, pp. 567-583.

Rivlin, R. S. & Sanders, D. W. 1951. Large elastic deformations of isotropic materials, part 7, experiments on the Deformation of rubber. *Phil. Transactions of the Royal Society*, A243, pp. 251-288.

Schleth, A., Münzenmaier, A., Muth, A. & Post, P. 1999. Use and benefits of numerical seal analysis in everyday practise of a pneumatics manufacturer. *Proceedings of the 11th International Sealing Conference*, 3rd-4th May, Dresden, pp. 31-44.

Constitutive Models for Rubber, Dorfmann & Muhr (eds) © 1999 Taylor & Francis ISBN 90 5809 113 9

Physical parameters strain energy function for rubberlike materials

M. H. B. M. Shariff
School of Computing and Mathematics, University of Teesside, Middlesbrough, UK

ABSTRACT: A strain energy function with physical parameters for rubberlike materials is proposed. Constitutive relations for the parameters are given. Sensitivity analysis on the least squares normal equations is done together with a method to obtain robust normal equations. The predicted results compare well with published experimental data.

1 INTRODUCTION

We assume that a rubberlike material is isotropic relative to a ground (stress free) state and that its isothermal mechanical properties can be represented in terms of an elastic strain energy function $S(\lambda_1, \lambda_2, \lambda_3)$, where λ_1, λ_2 and λ_3 are principal stretches. Due to isotropy, $S(\lambda_1, \lambda_2, \lambda_3)$, measured per unit volume in the ground state, must be symmetric with respect to interchange of any two of λ_1, λ_2 and λ_3. In this paper we only consider the Valanis-Landel (1967) separable form of strain energy function, i.e.,

$$S(\lambda_1, \lambda_2, \lambda_3) = r(\lambda_1) + r(\lambda_2) + r(\lambda_3) . \quad (1)$$

For moderate strains the above separable form of strain energy density function has been used by many authors to describe many types of rubberlike materials. Only incompressible materials are considered here and the incompressibility constraint

$$\lambda_1\lambda_2\lambda_3 = 1 \quad (2)$$

is adopted. In view of the above equation it is convenient to eliminate λ_3 in favour of λ_1 and λ_2 and the strain energy function takes the form,

$$W(\lambda_1, \lambda_2) = S(\lambda_1, \lambda_2, 1/(\lambda_1\lambda_2))$$

$$= r(\lambda_1) + r(\lambda_2) + r(\frac{1}{\lambda_1\lambda_2}) , \quad (3)$$

noting the symmetry,

$$W(\lambda_1, \lambda_2) = W(\lambda_2, \lambda_1) . \quad (4)$$

Many good strain energy functions (in the sense that they can represent a wide range of incompressible rubberlike materials and a wide range of strains) of the form given by equation (1) have been developed in the past. The parameters of many of these functions, however, do not have physical interpretations. When the values of these non-physical parameters are obtained via, e.g., a curve fitting method, it is often difficult to validate them using experimental stress or rate of stress values. This reason, amongst others, leads us to propose, in Section 2, a physical parameter strain energy function. The function is linear in its parameters and hence their values can be obtained from a positive definite system of linear equations when the least squares method is used to calculate them.

2 STRAIN ENERGY FUNCTION

Let σ_i $(i = 1, 2, 3)$ be the principal components of the Cauchy (true) stress. In an incompressible material they are related to the function r given in equation (1) by the relation

$$\sigma_i = \lambda_i \frac{\partial W}{\partial \lambda_i} - p = \lambda_i r'(\lambda_i) - p \quad (i = 1, 2, 3), \quad (5)$$

where the prime signifies differentiation with respect to the argument for the function in question and p is the arbitrary hydrostatic stress required because of the incompressibility constraint (2). It follows that

$$\sigma_1 - \sigma_3 = \lambda_1 r'(\lambda_1) - \frac{1}{\lambda_1\lambda_2} r'(\frac{1}{\lambda_1\lambda_2}) \tag{6}$$

and

$$\sigma_1 - \sigma_2 = \lambda_1 \frac{\partial W}{\partial \lambda_1} - \lambda_2 \frac{\partial W}{\partial \lambda_2} = \lambda_1 r'(\lambda_1) - \lambda_2 r'(\lambda_2)\,. \tag{7}$$

We note that equations (6) and (7) are unaffected by any superposed hydrostatic stress. To be consistent with the classical linear theory of incompressible isotropic elasticity, appropriate for infinitesimal deformations, we must have the relations

$$r(1) = 0, \quad r''(1) = \frac{2E}{3}, \tag{8}$$

where E is the Young's modulus. Equation (7) depends only on λ_1 and λ_2 and this is sufficient to determine r (and hence W and S) from experiments in which the principal stretches, λ_1 and λ_2 are varied independently. For a biaxial deformation we let $\sigma_b(\lambda_1, \lambda_2) = \sigma_1 - \sigma_2$ and define a biaxial modulus B as:

$$B(\lambda_1, \lambda_2) = \frac{\partial \sigma_b(\lambda_1, \lambda_2)}{\partial \lambda_1}. \tag{9}$$

We note that if a material can be characterised by the separable Valanis- Landel (1967) form then B is independent of λ_2. In ground-state

$$B(1,1) = \frac{2E}{3}. \tag{10}$$

Without loss of generality we restrict the function r so that $r'(1) = 0$. This simplifies equation (7) for $\lambda_2 = 1$, i.e.,

$$\sigma_1 - \sigma_2 = \lambda_1 r'(\lambda_1). \tag{11}$$

For simplicity we let $f(\lambda) = \lambda r'(\lambda)$. Hence the above equation becomes

$$\sigma_1 - \sigma_2 = f(\lambda_1) - f(\lambda_2)\,. \tag{12}$$

In order that the parameters of the proposed strain energy function to have physical interpretations we approximate $f(\lambda)$ using Hermite interpolation functions, i.e. ,

$$f(\lambda) = \beta_0 \hat{\phi}_0 + \sum_{i=-n, i\neq 0}^{m} \alpha_i \phi_i(\lambda) + \sum_{i=-n, i\neq 0}^{m} \beta_i \hat{\phi}_i, \tag{13}$$

where α_i and β_i are the parameters of the strain energy function, $\beta_0 = \frac{2E}{3}$, $\hat{\phi}_0(1) = 0$, $\hat{\phi}'_0(1) = 1$ and, ϕ_i and $\hat{\phi}_i$ have the properties

$$\phi_i(y_j) = \delta_{ij}, \quad \phi'_i(y_j) = 0$$

$$\hat{\phi}_i(y_j) = 0, \quad \hat{\phi}'_i(y_j) = \delta_{ij}\,, \tag{14}$$

where $i, j = -n, \ldots, -1, 1, 2, \ldots, m$, δ_{ij} is the Kronecker delta and $y_j \in (0, \infty)$ are points (Carey and Oden (1983)) where the values of α_j and β_j are defined. The function ϕ_0 takes the form

$$\phi_0(\lambda) = \frac{-y_{-1}}{8}(\zeta+1)^2(\zeta-1), \quad \lambda \in [y_{-1}, 0]\,,$$

$$\zeta = \frac{2\lambda - y_{-1}}{-y_{-1}},$$

$$\phi_0(\lambda) = \frac{y_1}{8}(\zeta+1)(\zeta-1)^2, \quad \lambda \in [0, y_1]\,,$$

$$\zeta = \frac{2\lambda - y_1}{y_1}.$$

Within an element $\Omega_e = [y_j, y_{j+1}]$ $(j = -n, \ldots, -1, 1, 2, \ldots, m-1)$ the interpolation functions take the form

$$\psi_1^e(\lambda) = \frac{(\zeta-1)^2(\zeta+2)}{4},$$

$$\psi_2^e(\lambda) = \frac{(\zeta+1)^2(2-\zeta)}{4},$$

$$\hat{\psi}_1^e(\lambda) = h_e \frac{(\zeta-1)^2(\zeta+1)}{8},$$

$$\hat{\psi}_2^e(\lambda) = h_e \frac{(\zeta+1)^2(\zeta-1)}{8},$$

$$h_e = y_{j+1} - y_j\,,$$

$$\zeta = \frac{2\lambda - y_j - y_{j+1}}{y_{j+1} - y_j}$$

and, e.g., $\phi_j = \psi_2^e \cup \psi_1^{e+1}$ for the appropriate element Ω_e. Since details of Hermite interpolation functions can be found in Carey and Oden (1983) we avoid giving an extensive account of them in the present paper. Note that $f \in C^1$ and this implies that $W \in C^2$. This restriction in continuity does not cause problems in finite element calculations since most calculations only require at the most the second derivative of W. The properties of the functions ϕ_j and $\hat{\phi}_j$ ensure that

$$\alpha_j = \sigma_b(y_j, 1)\,,$$

$$\beta_j = B(y_j, \lambda_2) = \frac{\partial \sigma_b(y_j, \lambda_2)}{\partial \lambda_1}. \tag{15}$$

Hence α_j and β_j are physical parameters that can be measured directly from experiments.

3 CONSTITUTIVE INEQUALITIES

To ensure physically reasonable responses the strain energy density function must satisfy certain restrictions (Truesdell 1956). For example, a class of inequalities was proposed by Hill (1970) for examination of isotropic elasticity and he has shown that one and only one of the inequalities (referred to as Hill's inequality) admits incompressibility. For incompressible solids Hill's inequality asserts that the scalar product $tr(\hat{\mathbf{T}}\mathbf{E})$ (where tr denotes the trace of second order tensor) is positive at all strains and for arbitrary non-zero strain-rates. The notations $\mathbf{T}$, $\mathbf{E}$ and $\hat{\mathbf{T}}$ represent the Cauchy stress, the Eulerian strain-rate and the rigid-body derivative (the rate of change on axes rotating rigidly with the local body spin) of the Cauchy stress respectively. It is shown in Ogden (1972) that, relative to the principal axes of the Eulerian strain ellipsoid, the component form of the scalar product $tr(\hat{\mathbf{T}}\mathbf{E})$ is expressible as $\hat{\sigma}_{ij}\epsilon_{ij}$(from now onwards summation is implied by repeated suffixes unless otherwise stated), where $\hat{\sigma}_{ij}$ denote the components of the rigid-body derivative of the Cauchy stress on the axes of the Eulerian strain ellipsoid and ϵ_{ij} are components of the Eulerian strain-rate on the same axes. Then by the use of equations (5), (6) and (7) and taking note $\frac{\dot{\lambda}_i}{\lambda_i} = \epsilon_{ii}$ we obtain, following Ogden (1972),

$$\hat{\sigma}_{ii} \equiv \dot{\sigma}_i = \lambda_i f'(\lambda_i)\epsilon_{ii} - \dot{p} \quad (i = 1,2,3)$$
$$(i \textit{ not summed}), \tag{16}$$

and

$$\hat{\sigma}_{ij} = \frac{{\lambda_i}^2 + {\lambda_j}^2}{\lambda_i^2 - \lambda_j^2}(f(\lambda_i) - f(\lambda_j))\epsilon_{ij} \quad (i \neq j),$$
$$(i,j \textit{ not summed}), \tag{17}$$

where the superposed dot denotes the material time derivative. It follows that the scalar product $tr(\hat{\mathbf{T}}\mathbf{E})$ is expressible as (taking note that $\epsilon_{ii} = 0$)

$$\hat{\sigma}_{ij}\epsilon_{ij} \equiv \lambda_i f'(\lambda_i)\epsilon_{ii}^2 + \frac{{\lambda_i}^2 + {\lambda_j}^2}{\lambda_i^2 - \lambda_j^2}(f(\lambda_i) - f(\lambda_j))\epsilon_{ij}^2 \tag{18}$$

In the second group of the right-hand side of (18) the summation over i and j is restricted to $i \neq j$ and $\lambda_i \neq \lambda_j$. For the expression (18) to be positive it is necessary that $f'(\lambda_j) > 0$ and $f(\lambda_i) > f(\lambda_j)$ if $\lambda_i > \lambda_j$, for $i,j = 1,2,3$. Hence a necessary condition for the expression (18) to be positive is that

$$\beta_0 > 0, \;\; \beta_j > 0 \text{ and } \alpha_j > \alpha_k, \;\; j > k \;\; ,$$

where $j,k = -n,\ldots,-1,1,\ldots,m$.

4 LEAST SQUARES METHOD

In the least squares method (without weighting) the value of the vector parameter $\boldsymbol{v}$ is obtained from the problem **(I)**:

$$\min_{\boldsymbol{v}} E(\boldsymbol{v}) = \sum_k [f_k - F(x_k, \boldsymbol{v})]^2 \,,$$

where $\{f_k\}$ is the data, $\{x_k\}$ is the set of sample points and the functional form of F depends on the strain energy function used in the method. When the least squares method is applied to the proposed strain energy function the values of α_j and β_j are obtained from a system of linear equations, called the normal equations, of the form

$$\boldsymbol{A}\boldsymbol{v} = \boldsymbol{b} \,,$$

where
$\boldsymbol{v} = [\boldsymbol{\alpha}, \boldsymbol{\beta}]^T$, $\boldsymbol{\alpha} = [\alpha_{-n}, \ldots, \alpha_{-1}, \alpha_1, \ldots, \alpha_m]^T$ and $\boldsymbol{\beta} = [\beta_{-n}, \ldots, \beta_{-1}, \beta_0, \beta_1, \ldots, \beta_m]^T$, The elements of positive definite matrix $\boldsymbol{A}$ and $\boldsymbol{b}$ depend on the experimental data. E.g. if we apply the least squares method on a biaxial experiment in which $\lambda_2 = 1$, we have

$$\boldsymbol{A} = \left(\sum_k \phi_i(x_k)\phi_j(x_k)\right)$$

and

$$\boldsymbol{b} = \left(\sum_k f_k \phi_j(x_k)\right) \,,$$

where x_k are the experimental values of the principal stretch λ_1, f_k are the exprimental values of σ_b, $(i,j = -n, \ldots, -1, 1, \ldots, m, m+1, \ldots, 2m+n+1)$ and $\phi_l = \hat{\phi}_k (k = -n, \ldots, -1, 0, 1, \ldots, m) (l = m+1, \ldots, 2m+n+1)$.

4.1 Sensitivity

The sensitivity of the solution $\boldsymbol{v}$ to the data $\boldsymbol{b}$ is described by the relation (Conte & de Boor 1965)

$$\frac{\|\delta \boldsymbol{v}\|}{\|\boldsymbol{v}\|} \leq C(\boldsymbol{A}) \frac{\|\delta \boldsymbol{b}\|}{\|\boldsymbol{b}\|} \,,$$

where $C(\boldsymbol{A}) = \|\boldsymbol{A}\| \|\boldsymbol{A}^{-1}\|$ is the condition number of $\boldsymbol{A}$ and $\|.\|$ is an appropriate norm. If the condition number is large a small relative change in the data $\boldsymbol{b}$ gives a large relative change in the parameter $\boldsymbol{v}$. For certain set of sample points the matrix $\boldsymbol{A}$ is quite often ill-conditioned (large condition number), enough so that a straightforward application of a solution algorithm produces unreliable results. We note that this ill-conditioned behaviour can also be observed for many types of (more than one linear parameters) strain energy functions, such as the Mooney-Rivlin type and in this case we generally lack confidence in the values of the parameters obtained. We note that the condition number corresponding to a linear strain energy function depends on the type of strain energy function and the set of sample points. In the case of non-linear parameters strain energy functions, such as Ogden's (1972) and Tobisch (1981), the normal equations are nonlinear (and often ill-conditioned) with respect to the parameters and the sensitivity of the system depends on the condition number of the Hessian of E given in problem **(I)**. This depends not only on the type of strain energy function and the set of sample points, but on the set of experimental stress values which depend on the type of material.

Alternative to problem **(I)** we can also obtained the value of $\boldsymbol{v}$ via the problem **(II)**

$$\min_{\boldsymbol{v}} E(\boldsymbol{v}) = \int_a^b [t(\lambda) - F(\lambda, \boldsymbol{v})]^2 d\lambda\,,$$

where $a \leq \lambda \leq b$. The function t is of the form

$$t(\lambda) = \sum_{i=1}^{P} \gamma_i \chi_i(\lambda)\,,$$

where γ_i is obtained via problem **(I)**. The bases χ are orthogonal with respect to the sample points x_k, i.e.,

$$\sum_k \chi_i(x_k)\chi_j(x_k) = 0, \quad i \neq j$$

$$\neq 0, \quad i = j.$$

$$(i, j = 1, \ldots, P)$$

Hence the matrix $\boldsymbol{A}$ of the normal equations is diagonal which is well condition. Alternatively, t is obtained using an existing (good) strain energy function in which their parameters are obtained via the least squares problem **(I)** , taking note that we are only interested in the 'visual' fitting of t to the data and not the values of the parameters of the strain energy function used in the method. If we apply the least squares method **(II)** on a biaxial experiment in which $\lambda_2 = 1$, we have, for the proposed form of strain energy ,

$$\boldsymbol{A} = \left(\int_a^b \phi_i(\lambda)\phi_j(\lambda) d\lambda \right)$$

and

$$\boldsymbol{b} = \left(\int_a^b t(\lambda)\phi_i(\lambda) d\lambda \right)$$

We note that positive definite "mass" matrix $\boldsymbol{A} = \left(\int_a^b \phi_i(\lambda)\phi_j(\lambda) d\lambda \right)$ is well conditioned and hence the value of the parameter vector $\boldsymbol{v}$ is not sensitive to the data $\boldsymbol{b}$. This overcomes the near singular problem sometimes inherited by problem **(I)**.

4.2 Relation

In uniaxial, pure shear and equibiaxial extensions the first principle component of the Cauchy stress can be simply written as

$$\sigma_1 = f(\lambda) - f(\frac{1}{\lambda^l}), \;\; l = \frac{1}{2}, 1, 2, \tag{19}$$

where $\lambda = \lambda_1 > 1$, and $l = \frac{1}{2}, 1$ and 2 correspond to a uniaxial extension, a pure shear deformation and an equibiaxial extension respectively. For a biaxial deformation we consider the stress

$$\sigma_b = \sigma_1 - \sigma_2 = f(\lambda) - f(\lambda_2). \tag{20}$$

In view of equations (19) and (20) it is interesting to note that if a material can be characterised by a separable form (as expressed in equation (1)) the stress σ_1 in a uniaxial or a pure shear or an equibiaxial deformation is related to the biaxial stress σ_b in the following manner:

For any fixed λ_2 at $\lambda = \lambda_o > 1$

$$\sigma_1(\lambda_o) = \sigma_b(\lambda_o) - \sigma_b(\frac{1}{\lambda_o^l}). \tag{21}$$

In view of the above we wish to investigate, under certain conditions, whether theoretically we could obtain, via the least squares method, the same values for the parameters in biaxial, simple tension, pure shear and equibiaxial deformations. We do this by assuming that a ground-state biaxial (or Young's) modulus is known and modifying the functions for $\hat{\phi}_{-j}$ and $\phi_{-j}\,, (j = 1, 2, \ldots, m)$. Within an element $\Omega_e = [y_j, y_{j+1}]\,(j = -m, \ldots, -1)$ the modified interpolation functions take the form

$$\psi_1^e(\lambda) = \frac{(\zeta - 1)^2(\zeta + 2)}{4}\,,$$

$$\psi_2^e(\lambda) = \frac{(\zeta + 1)^2(2 - \zeta)}{4}\,,$$

$$\hat{\psi}_1^e(\lambda) = h_e \frac{(\zeta - 1)^2(\zeta + 1)}{8}\,,$$

$$\hat{\psi}_2^e(\lambda) = h_e \frac{(\zeta + 1)^2(\zeta - 1)}{8}\,,$$

$$h_e = y_{j+1} - y_j\,,$$

$$\zeta = \frac{2\eta - y_j - y_{j+1}}{y_{j+1} - y_j}\,,$$

$$\eta = \frac{1}{\lambda^t}\,.$$

In this case, however, $\beta_j = -\dfrac{B(y_j, \lambda_2) y_j^{l+1}}{l}$ $(j = -m, \dots, -1)$, which is a negative scaled value of the biaxial modulus B.

First consider a biaxial deformation for which $\lambda_2 = 1$ and select the sample points $x_{-k} = \dfrac{1}{x_k^t}$, $x_k > 1\,(k = 1, \dots, s)$, for $\lambda_1 < 1$ and x_k for $\lambda > 1$. We also select for the element Ω_e, $y_{-k} = \dfrac{1}{y_k}$ and $y_k > 1\,(k = 1, \dots, m)$. After some manipulations and taking note that the modified functions have the properties:

$$\phi_j(x_{-k}) = \phi_{-j}(x_k) = 0\,,$$

$$\hat{\phi}_j(x_{-k}) = \hat{\phi}_{-j}(x_k) = 0$$

$$\phi_j(x_k) = \phi_{-j}(x_{-k})\,,$$

$$\hat{\phi}_j(x_k) = \hat{\phi}_{-j}(x_{-k})\,,$$

$$j = 1, \dots, m\,,$$

$$k = 1, \dots, s\,, \tag{22}$$

the least squares method gives

$$\begin{pmatrix} \boldsymbol{D} & \boldsymbol{E} \\ \boldsymbol{E} & \boldsymbol{F} \end{pmatrix} \begin{pmatrix} \boldsymbol{\alpha}^- \\ \boldsymbol{\beta}^- \end{pmatrix} = \begin{pmatrix} \boldsymbol{b}_1^- \\ \boldsymbol{b}_2^- \end{pmatrix} \tag{23}$$

and

$$\begin{pmatrix} \boldsymbol{D} & \boldsymbol{E} \\ \boldsymbol{E} & \boldsymbol{F} \end{pmatrix} \begin{pmatrix} \boldsymbol{\alpha}^+ \\ \boldsymbol{\beta}^+ \end{pmatrix} = \begin{pmatrix} \boldsymbol{b}_1^+ \\ \boldsymbol{b}_2^+ \end{pmatrix}, \tag{24}$$

where $\boldsymbol{\alpha}^- = [\alpha_{-1}, \dots, \alpha_{-m}]^T$, $\boldsymbol{\alpha}^+ = [\alpha_1, \dots, \alpha_m]^T$, $\boldsymbol{\beta}^- = [\beta_{-1}, \dots, \beta_{-m}]^T$, $\boldsymbol{\beta}^+ = [\beta_1, \dots, \beta_m]^T$,

$$\boldsymbol{b}_1^- = (\sum_{k=1}^{s} \sigma_b(x_{-k})\phi_{-j}(x_{-k}) -$$

$$\sum_{k=1}^{s} \hat{\phi}_0(x_{-k})\phi_{-j}(x_{-k}))\,,$$

$$\boldsymbol{b}_2^- = (\sum_{k=1}^{s} \sigma_b(x_{-k})\hat{\phi}_{-j}(x_{-k}) -$$

$$\sum_{k=1}^{s} \hat{\phi}_0(x_{-k})\hat{\phi}_{-j}(x_{-k}))\,,$$

$$\boldsymbol{b}_1^+ = \left(\sum_{k=1}^{s} \sigma_b(x_k)\phi_j(x_k) - \sum_{k=1}^{s} \hat{\phi}_0(x_k)\phi_j(x_k)\right),$$

$$\boldsymbol{b}_2^+ = \left(\sum_{k=1}^{s} \sigma_b(x_k)\hat{\phi}_j(x_k) - \sum_{k=1}^{s} \hat{\phi}_0(x_k)\phi_j(x_k)\right),$$

$$(j = 1, \dots, m)$$

The vector value of the parameter $\boldsymbol{v}$ is obtained by solving equations (23) and (24) independently.

We next apply the least square method on a simple tension or a pure shear or an equibiaxial deformation. After some manipulation and in view of equation (22) the least squares method gives

$$\begin{pmatrix} \boldsymbol{D} & -\boldsymbol{D} & \boldsymbol{E} & -\boldsymbol{E} \\ \boldsymbol{E} & -\boldsymbol{E} & \boldsymbol{F} & -\boldsymbol{F} \end{pmatrix} \begin{pmatrix} \boldsymbol{\alpha}^- \\ \boldsymbol{\alpha}^+ \\ \boldsymbol{\beta}^- \\ \boldsymbol{\beta}^+ \end{pmatrix} = \begin{pmatrix} \boldsymbol{b}_1^- - \boldsymbol{b}_1^+ \\ \boldsymbol{b}_2^- - \boldsymbol{b}_2^+ \end{pmatrix}.$$

The above equation does not have a unique solution. However, one of the solutions is the same as that of equations (23) and (24) for the biaxial deformation and this completes our investigation.

5 COMPARISON WITH PUBLISHED EXPERIMENTAL DATA

We first compare our theoretical results with the biaxial data of Jones & Treloar (1975). The value of the vector parameter $\boldsymbol{v}$ is obtained via problem **(I)** using only the set of data when $\lambda_2 = 1$. Four elements are used as shown below:

$$\Omega_1 = [0.3497, 0.5181]\,,\quad \Omega_2 = [0.5181, 1.000]\,,$$

$$\Omega_3 = [1.000, 1.930]\,,\quad \Omega_4 = [1.930, 2.86]. \tag{25}$$

From Figure 1 we see that the theoretical curves agree well with the five sets of the experimental data even though we apply the least squares method on only one set of data. In this case

$$\boldsymbol{v} = [-.6004, -.3961, 1.043, 2.574,$$

$$0.9977, 0.7580, 0.9582, 1.339, 1.735]^T.$$

Recently Shariff (1999) has developed a linear parameters strain emnergy function which characterises a variety of different rubbers. In his work he let the function $\hat{f}(\equiv f)$ to have the form:

$$\hat{f}(\lambda) = E \sum_{i=0}^{n} \mu_i \bar{\phi}_i(\lambda), \tag{26}$$

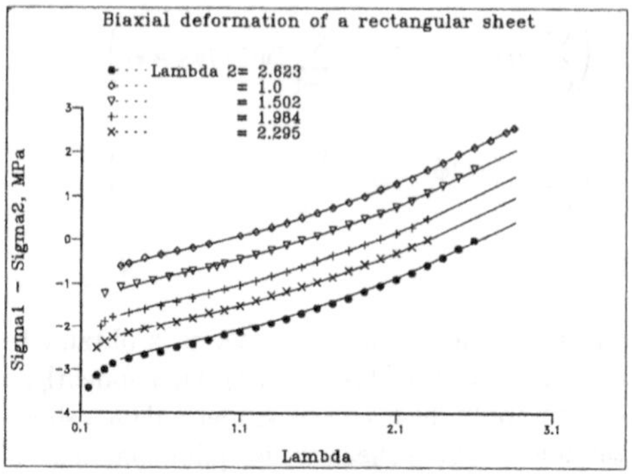

Figure 1. Comparison of theoretical curves with the data of Jones & Treloar (1975): Biaxial deformation of a rectangular sheet. Lambda = λ_1, Sigma1 = σ_1, Sigma2 = σ_2. $\sigma_1 - \sigma_2$ v. λ_1 at fixed values of λ_2

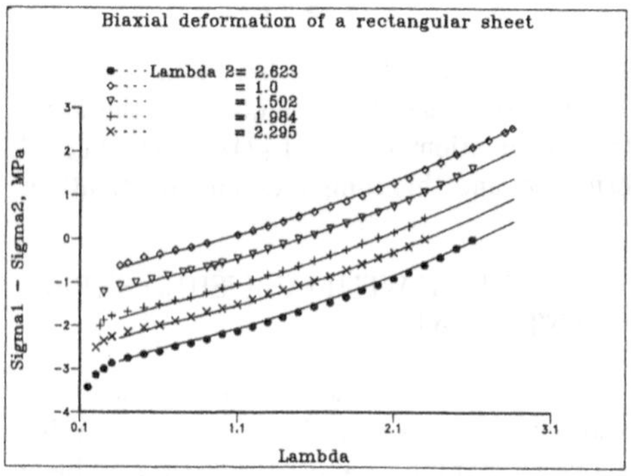

Figure 2. Comparison of theoretical curves with the data of Jones & Treloar (1975): Biaxial deformation of a rectangular sheet. Lambda = λ_1, Sigma1 = σ_1, Sigma2 = σ_2. $\sigma_1 - \sigma_2$ v. λ_1 at fixed values of λ_2

where $\mu_0 = 1$, $\mu_i (i = 1, \ldots, n)$ are parameters and

$$\bar{\phi}_0(\lambda) = \frac{2ln(\lambda)}{3}$$

$$\bar{\phi}_1(\lambda) = e^{(1-\lambda)} + \lambda - 2$$

$$\bar{\phi}_2(\lambda) = e^{(\lambda-1)} - \lambda$$

$$\bar{\phi}_3(\lambda) = \frac{(\lambda-1)^3}{\lambda^{3.6}}$$

$$\bar{\phi}_j(\lambda) = (\lambda-1)^{j-1}, \quad j = 4, 5, \ldots, n.$$

We test this function by letting

$$\alpha_j = \hat{f}(y_j) \quad (j = -n, \ldots, -1, 1, \ldots, m),$$

$$\beta_j = \hat{f}'(y_j) \quad (j = -n, \ldots, -1, 0, 1, \ldots, m). \qquad (27)$$

Using the same four elements given by equation (25) we depict, in Figure 2, the theoretical curves using the above values of α_j and β_j, and $\hat{f}(\lambda) = 1.4421(\bar{\phi}_0 + 0.880941\bar{\phi}_1 + 0.0544614\bar{\phi}_2)$ (Shariff 1999). We see that, for

$$\boldsymbol{v} = [-.6589, -.4500, 1.092, 2.58,$$

$$1.548, 1.039, 0.9614, 1.388, 1.835]^T,$$

the theory agrees well with the experiment.

To further justify our proposed form strain energy function we compare our results with the data of Treloar (1944) using the value of $\boldsymbol{v}$ obtained via equation (27) and using $\hat{f}(\lambda) = 11.0009(\bar{\phi}_0 + 1.37334\bar{\phi}_1 + 0.0471163\bar{\phi}_2 + 0.841383 \times 10^{-04}\bar{\phi}_3)$ (Shariff 1999). For uniaxial and pure shear extensions we use six elements as shown in the following:

$$\Omega_1 = [0.049, 0.1], \quad \Omega_2 = [0.1, 0.3],$$

$$\Omega_3 = [0.3, 1.0], \quad \Omega_4 = [1.0, 4.5],$$

$$\Omega_5 = [4.5, 6.0], \quad \Omega_6 = [6.0, 7.6]. \qquad (28)$$

In this case

$$\boldsymbol{v} = [-53.6448, -10.9599, -4.01201, 64.0892,$$

$$147.490, 476.676, 3292.17, 156.640,$$

$$9.26412, 7.33393, 32.9277, 92.6359,$$

$$396.550]^T.$$

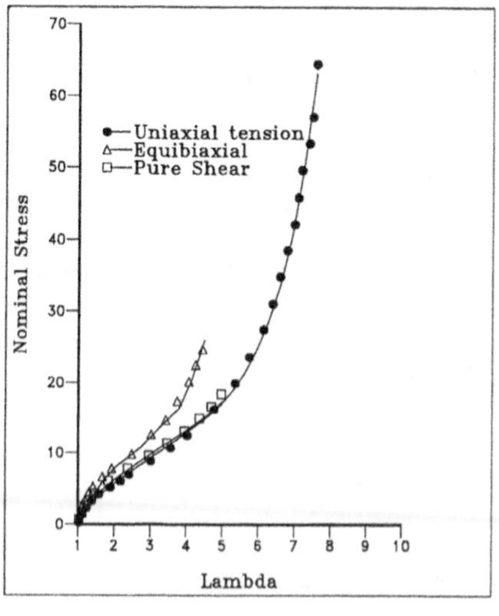

Figure 3. Comparison of theoretical curves with the data of Treloar (1944). Lambda = λ_1, Nominal stress = $\frac{\sigma_1}{\lambda_1}$

However, for the biaxial extension we note that, for six elements, we have to evaluate the parameter α_{-3} and β_{-3} at $y_{-3} = \frac{1}{y_3^2} = 0.049$. The gradient at this point is very steep and extra care is taken in selecting the elements. An ad hoc choice of elements for the biaxial extension is

$$\Omega_1 = [0.049, 0.09], \ \Omega_2 = [0.09, 0.21],$$
$$\Omega_3 = [0.21, 1.0], \ \Omega_4 = [1.0, 2.17],$$
$$\Omega_5 = [2.17, 3.33], \ \Omega_6 = [3.33, 4.5]. \quad (29)$$

In this case

$$\boldsymbol{v} = [-53.6448, -12.8785, -5.19941, 13.4845,$$
$$33.9875, 64.0892, 3292.17, 234.436,$$
$$19.0916, 7.33393, 14.9504, 20.6494,$$
$$32.9277]^T$$

Note that the gradient $\beta_{-3} = 3292.17$. The theoretical curves agree well with the data of Treloar (1944) as shown in Figure 3.

6 CONCLUSIONS

A strain energy function with physical parameters is developed and a method to obtain robust normal equations for the parameters is given. The values of the parameters can easily be checked directly from experiments. The predicted results agree well with published experimental data. The extent of the proposed strain energy function applicability to other rubbers needs to be assessed by comparing it with experimental data of a much wider class of rubberlike materials.

REFERENCES

Carey, G.F. & J.T. Oden 1983. *Finite Elements: A Second Course, Vol.2.* New Jersey, USA:Prentice-Hall.

Conte, S.D. & C. de Boor 1965. *Elementary Numerical Analysis.* Japan:McGraw-Hill Kogkusa

Hill, R. 1970. Constitutive inequalities for isotropic elastic solids under finite strain. *Proc. R. Soc. Lond..* A 314:457-472.

Jones, D.F. & L.R.G. Treloar 1975. The properties of rubber in pure homogeneous strain. *J. Phys. D..* 8:1285-1304.

Ogden, R.W. 1972. Large deformation isotropic elasticity - on the correlation of theory and experiment for incompressible rubberlike solids. *Proc. R. Soc. London, Ser. A.* 326:565-584.

Shariff, M.M.B.M. 1999. A strain energy function for filled and unfilled elastomers M.H.B.M. -this proceedings

Tobisch, K. 1981. A three-parameter strain energy density function for filled and unfilled elastomers. *Rubber Chem. Technol..* 54:930-939.

Treloar, L.R.G. 1944. Stress-strain data for vulcanised rubber under various types of deformation. *Trans. Faraday Soc..* 40:59-70.

Truesdell, C.A. 1956. Das ungeloste hauptproblem der endlichen elastizitätstheorie. *Z. angew. Math. Mech..* 36:97-103.

Valanis K.C. & R.F. Landel 1967. The strain-energy function of a hyperelastic material in terms of the extension ratios. *J. Appl. Phys..* 38:2997-3002.

Constitutive Models for Rubber, Dorfmann & Muhr (eds) © 1999 Taylor & Francis ISBN 90 5809 113 9

Experimental and numerical investigation of the friction behavior of rubber blocks on concrete and ice surfaces

T. Huemer, J. Eberhardsteiner, W. N. Liu & H. A. Mang
Institute for Strength of Materials, Vienna University of Technology, Austria

ABSTRACT: Up till now most on the work of the development of tires has been done empirically. The construction and testing of prototypes of tires is a rather expensive and time-consuming procedure. Therefore, in the last years, great effort and progress were made in the numerical simulation of the mechanical behavior of tires by the Finite Element Method (FEM). In addition to appropriate element technologies and suitable material models for realistic numerical simulations, it is important to capture frictional effects of rubber correctly. Application of the Coulomb friction law was shown to be inadequate. This contribution is focused on the experimental and numerical investigation of the frictional behavior of rubber blocks on concrete and ice surfaces.

The experimental investigation is performed with a special testing device (linear friction tester, LFT). There, rubber blocks are moved along a straight line on plane surfaces, i. e. made of ice or concrete. Friction coefficients are determined under different conditions for sliding velocity, pressure of the rubber blocks on the surface, and environmental temperature. The testing device and the testing scheme as well as selected results will be presented.

For numerical simulations, a friction law dependent on normal pressure and sliding velocity is used. The model parameters are calibrated by using different experimental results. The implementation of the friction law for 2D contact analyses in several multi-purpose finite element codes is illustrated by means of examples.

1 INTRODUCTION

One of the main objectives of the development of tires is the optimization of the load transfer capacity between the rubber tread of a tire and the surface of a road. The transfer of load is performed by a sliding friction process. The rubber blocks of the tread slide with different velocities and values of normal pressure over the road surface. This mechanism depends mainly on the surface itself, the dynamic driving conditions and the environmental temperature.

For a better understanding of the mechanism and for determination of the dependence of the traction forces on the normal pressure and the sliding velocity a comprehensive experimental investigation was carried out. For that purpose, a device for testing the frictional behavior of rubber blocks on different surfaces under various conditions was developed (Section 2).

The improvement of the performance of tires is still mainly done empirically by means of time-consuming and expensive test series which are also strongly influenced by the atmospheric conditions. In order to increase the application of numerical simulations within the process of tire-development, suitable material models and contact algorithms were developed. For the optimization of the load transfer mechanism, the frictional behavior of rubber blocks on different surfaces has to be captured correctly. Therefore, a friction law based on experimental results is necessary for a realistic numerical simulation (Section 3).

2 EXPERIMENTAL INVESTIGATION

2.1 Experimental Set-up

For the experimental investigation a linear friction tester (LFT) developed by Continental AG, Germany and the Institute for Strength of Materials at the Vienna University of Technology was used (Eberhardsteiner, Fidi, and Liederer 1998). With this testing device the friction coefficient μ of rubber blocks under various conditions can be measured.

Figure 1 shows a front view of the testing device. It mainly consists of a linear drive, which is attached to a stiff steel frame construction made of C-sections. A servo amplifier with a high positioning accuracy controls the linear drive. The loading unit and the bracket for the test specimen, as well as systems for load and displacement measuring are mounted on the leading sledge of the linear drive.

The load is applied by means of a pneumatic cylinder. With a load measuring bolt three load components L_1, L_2, L_3 (Figure 2) are measured. The vertical displacement u_V is obtained by a LVDT position transducer. The horizontal displacement u_H is recorded by pulse counting of the linear drive. The recording of all experimental quantities and the control of the movement of the test specimen is done by means of a digital measuring amplifier.

Three independent experimental quantities are obtained from the measured quantities: F_H, F_V, e (Figure 2). F_H denotes the horizontal load component in the interface between the test specimen and the contact surface. F_V is the vertically applied load, and e is the eccentricity of F_V relative to the center of the width of the test specimen (Figure 2). The friction coefficient follows as

$$\mu = \frac{F_H}{F_V} \, . \tag{1}$$

The experimental procedure is shown in Figure 3 and can be summerized as follows:

- acceleration of the test specimen to the wanted sliding velocity,
- touch-down and vertical loading of the specimen,
- movement of the loaded specimen over the sliding distance of 100 to 200 mm, and
- lift-up and retardation of the specimen.

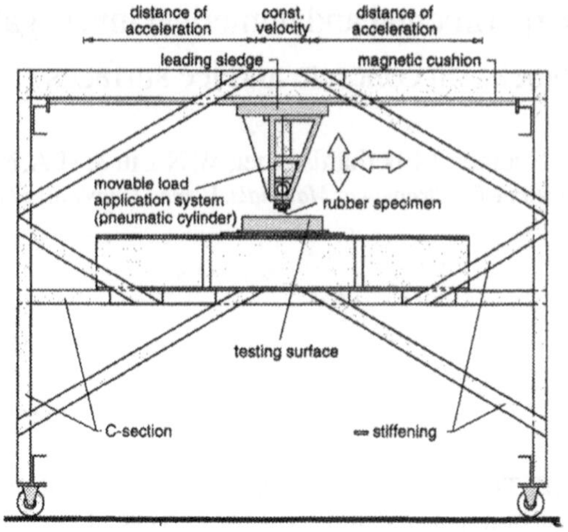

Figure 1: Front view of the testing device

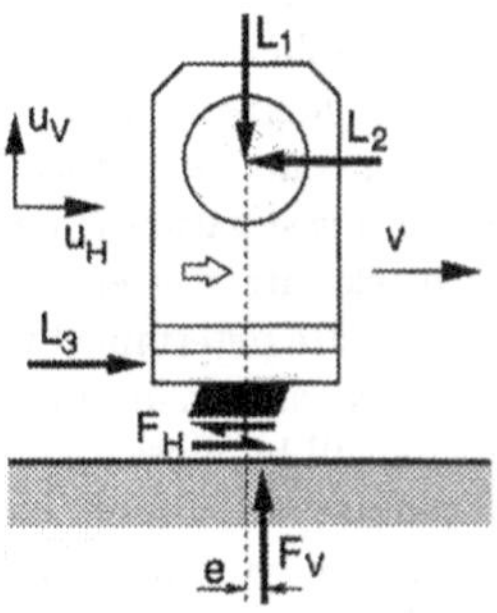

Figure 2: Measuring quantities

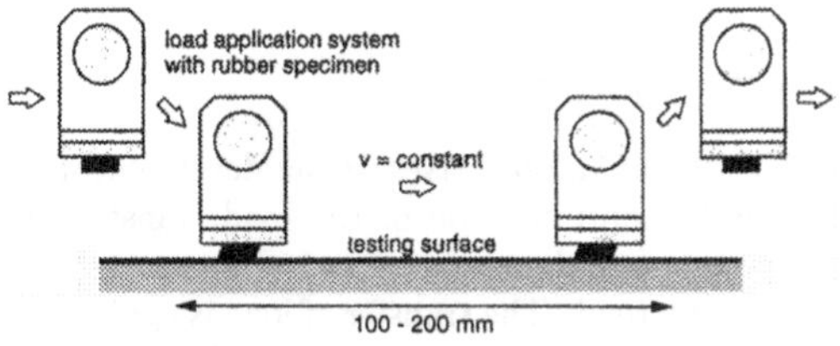

Figure 3: Experimental procedure

With the LFT equipment, the experimental data base concerning the measurement of the sliding load transfer of rubber treads on plane surfaces, such as wet and dry asphalt, concrete and ice, will be significantly extended.

The nominal vertical pressure on rubber blocks with a contact area of about 20 cm^2 can be varied from 0.5 to 5 bar, the sliding velocity from 0.1 to 1000 mm/s. Because the LFT is located in a climate chamber, the experiments can be

Figure 4: Test specimen and loading unit during experiment

performed at environmental temperatures ranging from −15° C to +35° C.

The photograph (Figure 4) shows the loading unit with a test specimen during an experiment.

2.2 Experimental Results

Experiments were performed under varying conditions: sliding velocity v = 1, 10, 100, and 1000 mm/s, nominal pressure p = 1, 2, 3, 4, and 5 bar, on ice and wet concrete surfaces. Figure 5 shows the friction coefficient related to the vertical load F_V for different types of loading cylinders. The results show good repeatability for the different set-ups.

The dependence of the friction coefficient μ on the vertical load and the relative sliding velocity is illustrated in Figure 6. The maximum of the friction coefficient for the test specimen on an ice surface can be observed at a sliding velocity of about 100 to 300 mm/s and under a nominal pressure of 2 bar.

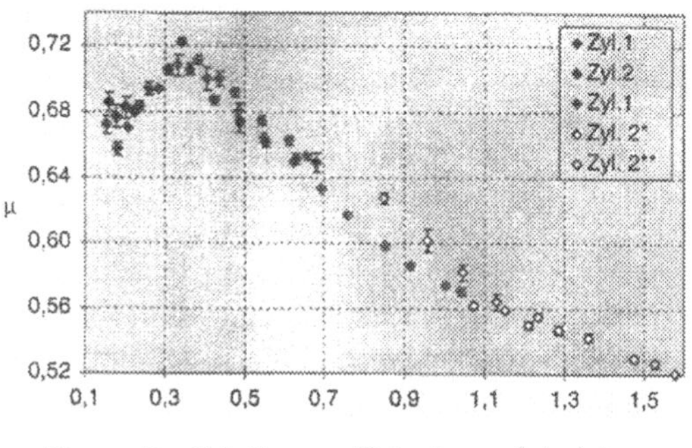

Figure 5: Friction coefficient μ related to the vertical load F_V for a soft rubber compound; sliding velocity 100 mm/s

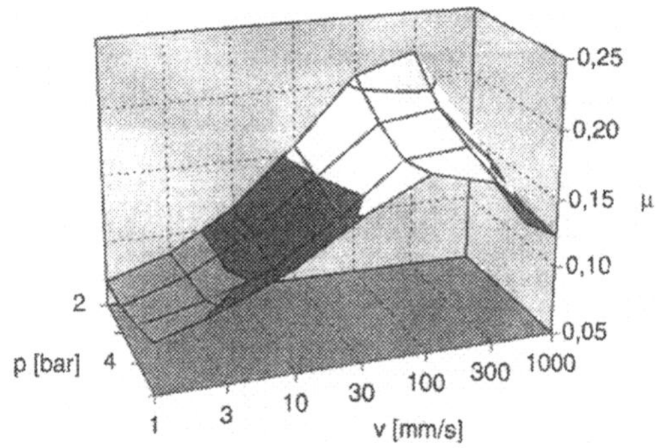

Figure 6: Friction coefficient μ as a function of vertical load (nominal pressure) and sliding velocity

The results obtained from the experiments are a basis for numerical investigations. The parameters of the friction law used are calibrated according to the results of the experiments, i.e. on concrete surfaces, by means of a least-square fitting procedure.

3 NUMERICAL INVESTIGATION

3.1 Finite Element Model

The Finite Element (FE) Method was employed for the numerical investigation. The test specimens (Figure 7a) were discretized as a 2D plane strain model (Figure 7b). The geometric shape of the FE model is derived from averaged dimensions of several test specimens.

Two commercial, multipurpose FE codes were used for the analyses, ABAQUS 5.8 (Hibbitt, Karlsson & Sorensen, Inc. 1998) and MARC K6.2 (MARC Analysis Research Corporation 1996), respectively (in alphabetical order). For the latter, a contact algorithm including frictional effects developed by (Payer, Meschke, and Mang 1998) was used, while for the former the built-in contact facilities were used.

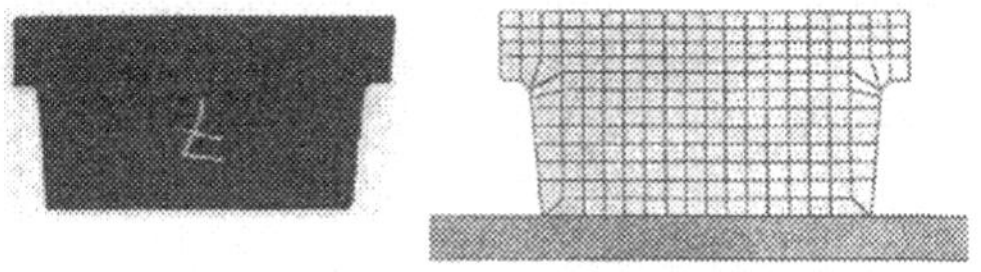

(a) (b)

Figure 7: (a) Test specimen (sample #7), (b) FE discretization

For both FE codes, an Ogden material model for rubber compounds was used with the following strain energy functions:
for ABAQUS:

$$W = \sum_{n=1}^{3} \frac{2\mu_n}{\alpha_n^2} J^{-\frac{1}{3}}(\lambda_1^{\alpha_n} + \lambda_2^{\alpha_n} + \lambda_3^{\alpha_n} - 3) \;, \qquad (2)$$

and for MARC:

$$W = \sum_{n=1}^{3} \frac{\mu_n}{\alpha_n} J^{-\frac{1}{3}}(\lambda_1^{\alpha_n} + \lambda_2^{\alpha_n} + \lambda_3^{\alpha_n} - 3) \;. \qquad (3)$$

λ_1, λ_2, and λ_3 are the principle stretches, J is the Jacobian determinant, α_i and μ_i are material constants. These constants were obtained by curve fitting procedures applied to the data from uniaxial tension/compression tests for the investigated rubber compounds. Rubber was treated as fully incompressible ($J = 1$). A hybrid finite element formulation was used: element type CPE4H for ABAQUS and element type 80 for MARC.

3.2 Friction Law

From the results of the experimental investigations it can be concluded that the classical Coulomb friction law does not adequately represent the real friction behavior of rubber blocks on different surfaces. Therefore, a friction law dependent on normal pressure, sliding velocity and environmental temperature was used in both FE codes. It is based on the pressure dependent modified Coulomb friction law proposed by (Wriggers, Vu Van, and Stein 1990)

$$f(p,\tau) = -(\alpha|p|^n + \beta|p|) + |\tau| \leq 0 \;. \qquad (4)$$

Here, τ denotes the traction shear stress, α, β and n are parameters of the interface law, which depend on the surface properties.

The implementation of the friction law for 2D contact analyses was carried out by means of a return mapping algorithm in the same way as for plasticity material models. Thus, the analogeous condition to the friction law (4) is the yield condition. As common in the theory of plasticity the tangential sliding velocity is split into an elastic part ($\dot{g}_T^{EL}$) and an inelastic part ($\dot{g}_T^{IN}$):

$$\dot{g}_T = \dot{g}_T^{EL} + \dot{g}_T^{IN} \qquad (5)$$

In order to prevent violation of the second law of thermo-dynamics a non-associated slip rule for the inelastic part of the tangential velocity

$$\dot{g}_N^{IN} = 0; \quad \dot{g}_T^{IN} = \dot{\lambda}\frac{\partial f}{\partial \tau} = \dot{\lambda}\;\mathrm{sign}(\tau) \qquad (6)$$

has to be used for the formulation of the friction law. Here, $\dot{\lambda}$ is the consistency parameter. This non-associated slip rule leads to a non-symmetric tangent coefficient matrix.

With the kinematic split (5), a slip condition (e.g. 4), and the slip rule (6), thc return mapping algorithm can be summarized as (Wriggers 1995)

1. evaluation of the elastic predictor,
2. back projection to slip surface if slip condition is violated, and
3. determination of the consistent linearization.

In order to obtain a quadratic rate of convergence near the solution the third step is unconditionally necessary in the global Newton iteration scheme.

3.3 Numerical Results

The results obtained by ABAQUS and MARC were very similar. Therefore, numerical results from both FE codes are shown. The source of the results is indicated.

Table 1 shows the average effective friction coefficient μ_{eff} for two rubber compounds (hard, soft) on both ice and wet concrete surfaces, respectively. The further parameters of these examples are: sliding velocity $v = 100$ mm/s and a constant vertical nominal pressure $p = 4$ bar.

The deformed test specimen and the stresses normal to the contact surface in a stationary sliding state obtained by MARC are shown in Figure 8. The according pressure distributions are illustrated in Figure 9 obtained by ABAQUS.

4 CONCLUSIONS

For the development of numerical algorithms like friction laws or, more general, material models, a well established experimental basis is necessary. The developed LFT gives the possibility of measuring the frictional behavior of rubber blocks on different surfaces under various conditions. The experimental investigation carried out yielded a considerable amount of results, extending the experimental base of reliable data significantly.

These experimental data form the basis for a developed interface law. Depending on normal pressure and sliding velocity the frictional bahavior of rubber on different surfaces can be captured real-

Table 1: Average friction coefficient μ_{eff} for a soft and a hard rubber compound

μ_{eff}	on ice	on wet concrete
soft	0.150	0.514
hard	0.105	0.667

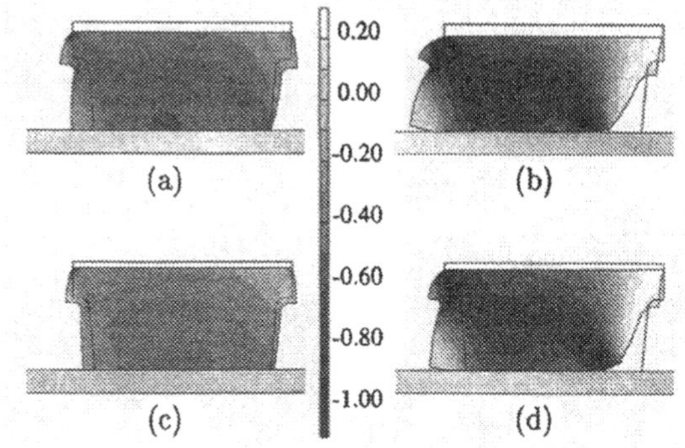

Figure 8: FE analysis of friction tests; sliding velocity 100 mm/s, nominal pressure 4 bar, distribution of the stresses normal to the contact area [N/mm^2], stationary sliding state, obtained by MARC: (a) soft rubber compound on ice surface, (b) soft rubber compound on wet concrete surface, (c) hard rubber compound on ice surface, (d) hard rubber compound on wet concrete surface.

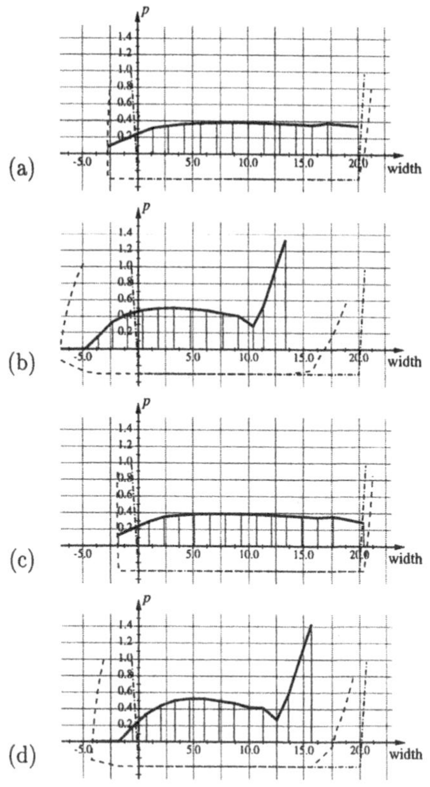

Figure 9: Distribution of the contact pressure in [N/mm^2], deformed (dashed line) and undeformed (dash-dotted line) contact area; sliding velocity 100 mm/s, nominal pressure 4 bar, stationary sliding state, obtained by ABAQUS: (a) soft rubber compound on ice surface, (b) soft rubber compound on wet concrete surface, (c) hard rubber compound on ice surface, (d) hard rubber compound on wet concrete surface.

istically. The implementation of the friction law in two commercial FE codes was done analogous to the implementation of a plasticity material model. Thus, a useful tool for numerical prediction of the load transfer between the rubber tread and the road surface is available.

Future extensions of the numerical simulations will be done by using more complex geometric shapes of the rubber blocks. Using a wider range of data from the experimental investigation will improve the quality of prediction of numerical analyses. Generalisation of the friction law for 3D contact analysis will give the opportunity for a widespread application within the development process of tires.

ACKNOWLEDGEMENTS

The authors wish to express their thanks to the Continental AG, Hannover, Germany for its permission to print this publication.

REFERENCES

Eberhardsteiner, J., W. Fidi, and W. Liederer (1998). Experimentelle Bestimmung der adhäsiven Reibeigenschaften von Gummiproben auf ebenen Oberflächen. *Kautschuk Gummi Kunststoffe 51*(11), 773–781. In German.

Hibbitt, Karlsson & Sorensen, Inc. (1998). *ABAQUS/Standard User's Manual, Version 5.8*. Pawtucket, RI 02860-4847: Hibbitt, Karlsson & Sorensen, Inc. Multi-Purpose Finite Element Package.

MARC Analysis Research Corporation (1996). *MARC Reference Manual, K6.2*. Palo Alto, CA 94306: MARC Analysis Research Corporation. Multi-Purpose Finite Element Package.

Payer, H.-J., G. Meschke, and H. Mang (1998). Geometrisch nichtlineare 2D FE-Kontaktanalysen kantiger, deformierbarer Körper. *Zeitschrift für Angewandte Mathematik und Mechanik 78*(2), 657–658. In German.

Wriggers, P. (1995). Finite element algorithms for contact problems. *Archives of Computational Methods in Engineering 2*(4), 1–49.

Wriggers, P., T. Vu Van, and E. Stein (1990). Finite element formulation of large deformation impact-contact problems with friction. *Computers & Structures 37*(3), 319–331.

Constitutive Models for Rubber, Dorfmann & Muhr (eds) © 1999 Taylor & Francis ISBN 90 5809 113 9

Material characterisation of tire structure used in explicit time integration of differential equations of the rolling process

F. Böhm
Berlin University of Technology, Germany

ABSTRACT: These characteristics are due to friction in the material, linear proportional damping and plastic effects. In case of cords, rayon to steel, an additional effects of pre-stress by inner pressure of the tire exists, so that friction between the filaments increases hysteretic behaviour. The volume constance is only of interest for holding the thickness of the sheets constant. For friction of the tread with the road the theory of Meyer and Kummer is used. Sticking is simulated by destructible chain elements, which are reintroduced when contact begins again. The time integration of tire rolling dynamics is done by using particle system approach. Parameters for material characteristics are stemming from experiments guided from continuum-mechanics ideas.

1 INTRODUCTION

The radial tire is made of high anisotropic material and consist of cords, ropes (rayon, nylon) and cables (steel), coated by rubber. It acts as a rolling airspring and the pretension forces in carcass and belt cords are only little changed by deflection. The coating rubber is mainly stressed by shear and compression. Tension in rubber may only exist between the cords and at the singularity of the belt edge. The energy of deformation can therefore be simplified for fast time integration. It is defined as a anisotropic membrane deformation energy and small parts of Timoshenko bending of the sheets and small Bernoulli bending of the rubber of tread and carcass. The bending stiffness of cords is neglected, Böhm (1996)

The dissipation of the cords consist of friction of the filaments and the rubber between it, combined damping forces of the coating rubber of the sheets, plastic dissipation forces of the tread and the inner liner. There is also some damping caused by liquid parts of the rubber compound. The tread produces in contact with road also external friction forces but which are not dissipative. The time integration of this system needs generally some rheologic ideas to formulate elastic and inelastic forces.

2 TIME INTEGRATION

Time integration of a single string is done using discrete masses and connecting them by massless but elastic and dissipative elements of rope material. The integration method is the predictor - one step corrector method and is semi-explizite, that means there is no corrector iteration. The remaining error is corrected in the next time step. So the time step Δt has to be smaller than 1/10 of shortest oscillation of the discrete system in time

$$\Delta t \le \frac{1}{10} 2\pi \sqrt{\frac{m_i}{4C_i}}, \tag{1}$$

this means it fulfils a strong shannon-criteron, Böhm (1993). The cable material produces material friction proportional to the prestress, see Figure 1. In case of compression the elastic reaction is very small. To get a smooth change from increasing length Δl to decreasing length and for the reverse change a creep behaviour is modelled by the following evolution equations:

$$\begin{aligned}
&\dot{\Delta l}_t > 0 \; and \; \dot{\Delta l}_{t-\Delta t} > 0 : F_t = G1_t - \frac{1}{2}\left(G1_{t-\Delta t} - F_{t-\Delta t}\right)\\
&\dot{\Delta l}_t < 0 \; and \; \dot{\Delta l}_{t-\Delta t} > 0 : F_t = F_{t-\Delta t} - \frac{1}{5}\left(F_{t-\Delta t} - G2_t\right)\\
&\dot{\Delta l}_t < 0 \; and \; \dot{\Delta l}_{t-\Delta t} < 0 : F_t = G2_t - \frac{1}{2}\left(G2_{t-\Delta t} - F_{t-\Delta t}\right)\\
&\dot{\Delta l}_t > 0 \; and \; \dot{\Delta l}_{t-\Delta t} < 0 : F_t = F_{t-\Delta t} + \frac{1}{5}\left(G1_{t-\Delta t} - F_{t-\Delta t}\right)
\end{aligned} \tag{2}$$

In case of a symmetric material reaction when a time step $\Delta t = T_{min}/10$ and $\Delta l_t > 0$ is used one gets also a quite good but not really smooth creep behaviour, see Figure 2. To improve this in the highest frequency range shorter time steps are necessary. Sharp unilateral constraints for contact can be avoided by

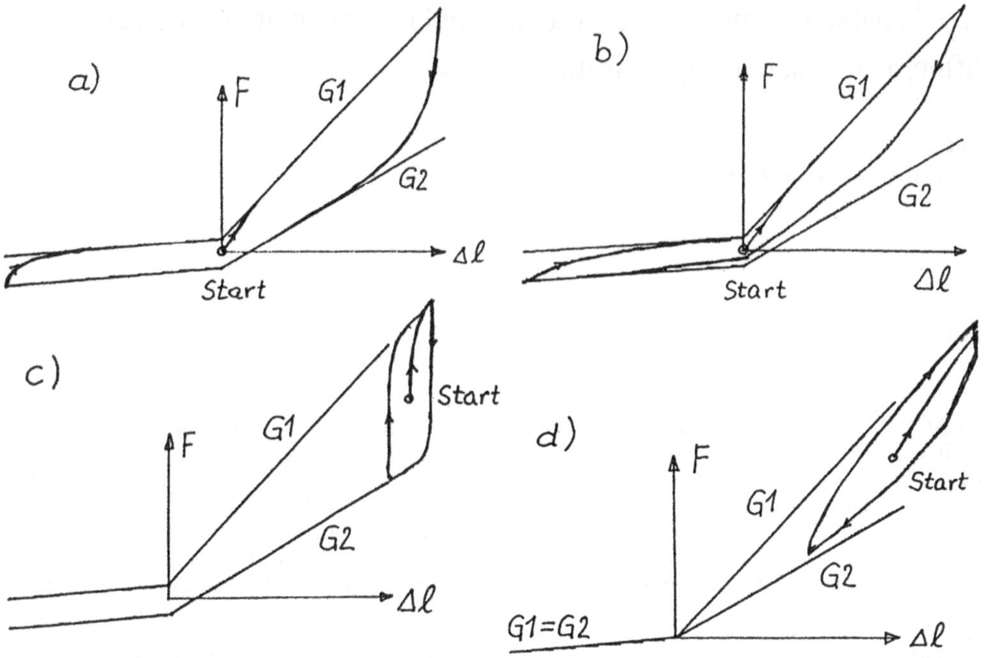

Figure 1. Hysteresis of belt. a.) 40 Steps b.) 10 Steps c.) 10 Steps d.) 10 Steps for one oszillation ± A.

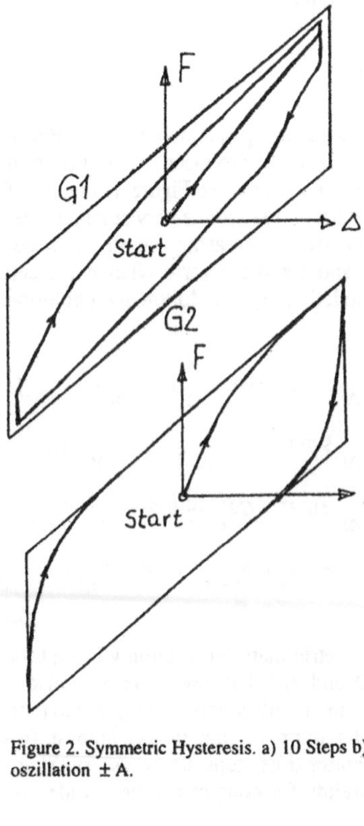

Figure 2. Symmetric Hysteresis. a) 10 Steps b) 40 Steps for one oszillation ± A.

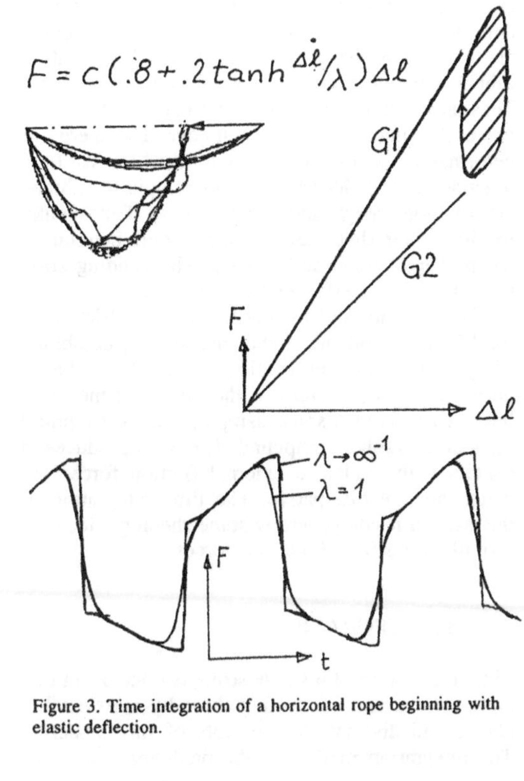

Figure 3. Time integration of a horizontal rope beginning with elastic deflection.

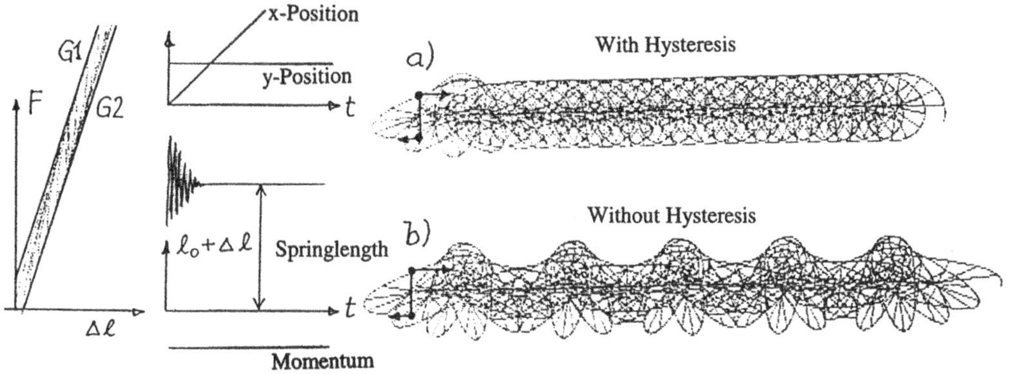

Figure 4. a) Time integration: Start of two masses with different velocity and suddenly connected by a hysteretic rod coming to stationary rotation. Linear decrease of oszillation of the spring length. X- position linear, Y-Position constant, Momentum constant. b.) Connection by pure elastic rod. No stationary rotation.

massless sensor points. For nonlinear analytic behaviour it is useful to separate between low rotational frequencies and high eigenfrequencies. The enclosed area is the dissipated work done by the friction forces between the filaments of the rope. The factors in (2) provide also a good creep behaviour when a smaller time step is used. It is not necessary to use finer time steps, this is shown in *first example* when integrating a hanging rope Figure 3. The mechanical rope model consists of 17 masspoints in horizontal positions starting with vertical gravity load and then the right end of the rope is shift to left, so that a secondary dynamic process of vertical deflection begins. After some time an equilibrium for the elastic rope is reached.

The *second example* shows the beginning of the rotation of two suddenly connected masses with friction and creep. The start of the system is horizontal but with different velocity. In Figure 4 typical creep behaviour of the connecting force, the linear reduction of radial oscillation and the evolution of a rigid body motion is presented. This is due to the famous ideas of *I. Prigogine*, who stated that a stationary or stable dynamic state can only be reached by dissipation. Because of the fact that nonlinear systems may have several fixpoints, stable or unstable, in Figure 4 the case without dissipation is shown underneath. Pure elastic systems do not have stationary states.

Only frictional material reactions (Hysteresis) lead to linear decreasing oscillations and after diminishing the material acts as a rigid body. In case of rubber the carbon black is responsible for this behaviour and in case of cords the surface friction is it. For instance in laboratory vibrations tests of Pkw-tires one needs shaker with more then 40 N force amplitude.

3 TO REACH A STATIONARY STATE OF ROLLING TIRE

Stable rotation of a rolling 3D-tire by dissipation is computed in the next *(third) example*. Here exists the engineering condition that the rolling resistance should be minimal. Such tires are not able to withstand high rolling velocities, see Figure 5. The excitation stems numerically from the unilateral contact of the discrete masses. Plastic contact is used. Plastic compound parts in tread rubber can be simulated using a limiting shear strain γ_{PL} and a memory shear strain γ_0:

$$
\begin{aligned}
&\tau = C(\gamma_t - \gamma_0)\ IF\\
&|\tau_t| > C\gamma_{PL} : THEN\ \tau_t = C\gamma_{PL}\,\mathrm{sgn}(\gamma_t - \gamma_0)\\
&\dot\gamma_t < 0\ and\ \dot\gamma_{t-\Delta t} > 0 : \gamma_0 = \gamma_t - \gamma_{PL}\\
&\dot\gamma_t > 0\ and\ \dot\gamma_{t-\Delta t} < 0 : \gamma_0 = \gamma_t + \gamma_{PL}\ ENDIF
\end{aligned}
\quad , \qquad (3)
$$

A nonlinear part exists for high tire loads so the elastic slip forces of the profil elements are seriously reduced by compression strain ε_d. Even local wrinkling may occur. Together with $C_1=(1-2\varepsilon_d)$ one gets

$$\tau_{1t} = C_1\left(\gamma_t - \frac{1}{2}\gamma_t|\gamma_t|\right). \qquad (4)$$

A liquid part of the compound produces damping:

$$\tau_{2t} = k\dot\gamma_t. \qquad (5)$$

All parts together show a realistic behaviour of the shear stress Figure 6, Böhm 1988

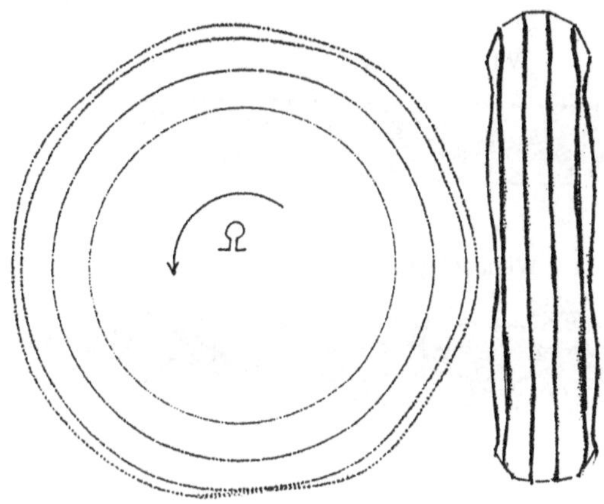

Figure 5. Time integration of a rolling tire with v=50 m/s and 3 cm deflection. Time step is $\Delta t=10^{-5}$ sec, coming to nearly stationary state.

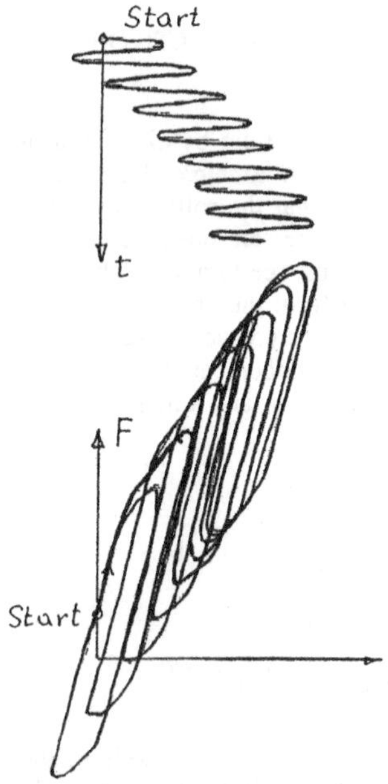

Figure 6. Elastic, plastic and damping reactions in tread compound.

4 CONTACT FRICTION AND HEAT ON SURFACE

At least the friction forces between tire and road are simulated using the idea that sticking should be a sort of molecular elastic connection of the two surfaces. Therefore the contact of one profile point produce an elastic restoring force to the tangential relative movement at begin of contact. Tangential

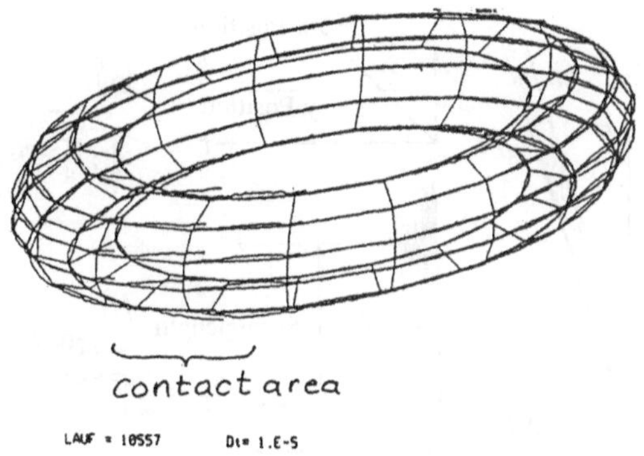

Figure 7. Time integration of a rolling tire with 5° slip angle. Traces of points in contact are shown.

springs are created in this moment not stronger than given in equation (1). When the resultant tangential stress became bigger than $u_{stick}p_z$ sliding occurs. The sliding forces are due to Meyer-Kummer-Theory, Meyer & Kummer (1962). Traces of contacting mass points are shown in Figure 7.

Thermographic video monitoring shows the distribution of dissipation. The steel-cords of the belt distribute the heating of the rubber sheets equally into the tread. In vibration test one sees the dissipation in the side-walls. Testing a tire rolling with lateral slip on a glass-plate one sees the different friction work along the with of the tread caused by different contact length. An interesting effect comes from different vulcanisation. The upper part of the moulding apparatus is more cooled than the basic part. So the tire has significant different hysteresis in both parts and there is a sharp borderline at tire equator to be seen.

REFERENCE

Böhm, F. 1996. Dynamic rolling process of tıres as layered structures. Ninth International Conference on the Mechanics of Composite Materials (Riga, October 1995).

Böhm, F. 1993. Reifenmodell für hochfrequente Rollvorgänge auf kurzwelligen Fahrbahnen. VDI-Tagung, Hannover 1993, VDI-Berichte Nr. 1088, VDI-Verlag Düsseldorf 1993, S.65-81.

Böhm, F. 1988. Einfluß der Laufflächeneigenschaften von Gürtelreifen auf den instationären Rollkontakt. Kautschuk und Gummi - Kunststoffe 4 (1988), 359-365.

Meyer, W.E. & Kummer, H.W. 1962. Mechanism of Force Transmission Between Tyre and Road. SAE Reprint, National Automobil Week, Detroit 1962.

Softening phenomena

Constitutive Models for Rubber, Dorfmann & Muhr (eds) © 1999 Taylor & Francis ISBN 90 5809 113 9

A realistic elastic damage model for rubber

L. Nasdala, M. Kaliske & H. Rothert
Institut für Statik, Universität Hannover, Germany

A. Becker
Continental AG, Tire Research, Computational Mechanics, Hannover, Germany

ABSTRACT: An elastic damage model is introduced which is well suited for an efficient and realistic description of the softening behaviour of rubber material. Stiffness reduction as a result of static or cyclic loading is determined quantitatively using damage functions. Besides, special pressure functions are introduced in order to take into account that softening effects are more distinct under compression than under tension. The influence of temperature on the stiffness of rubber is considered equivalently by temperature functions. In a finite element analysis, the footprint of a new tire and a tire in service are computed to confirm the reliability, capability and robustness of the proposed material model.

1 INTRODUCTION

In Fig. 1 a typical cyclic tension test is presented. The rubber specimen investigated depicts a reduction of stiffness and dissipative behaviour which is observed as hystereses. Viscoelastic and elastoplastic formulations can be used for the computation of the rate-dependent and rate-independent damping effects. Damage models are suitable for the characterisation of the stiffness reduction. A lot of different approaches worth mentioning have been developed so far in order to describe these nonlinear effects. For example, Kaliske and Rothert (1999) present combined viscoelastic and elastoplastic damage models for a complete description of all phenomena.

A decision on the applicability of the different formulations depends on the following criteria:

- The number of unknown material parameters has to be identified by the given experiments. For example, if an anisotropic damage model is applied, multiaxial tests are necessary.

- The chosen material formulation must be able to describe the relevant characteristics of the actual application. An elastoplastic approach can not be used to simulate a relaxation or creep test. Viscoelastic formulations should be preferred.

  On the other hand, investigations have shown that an elastoplastic approach is better suited for the simulation of cyclic tension tests at low velocities than a viscoelastic model, because the dissipative behaviour is nearly rate-independent.

- The material formulation should be numerical efficient. Viscoelastic models are time-dependent. Therefore, viscoelastic computations need much more time steps than rate-independent elastoplastic approaches.

In tire computations a combination of different models is used. For the simulation of rolling resistance, inelastic models have to be applied in order to represent the dissipative properties.

Here, an elastic damage formulation is introduced which is suited for the investigation of mod-

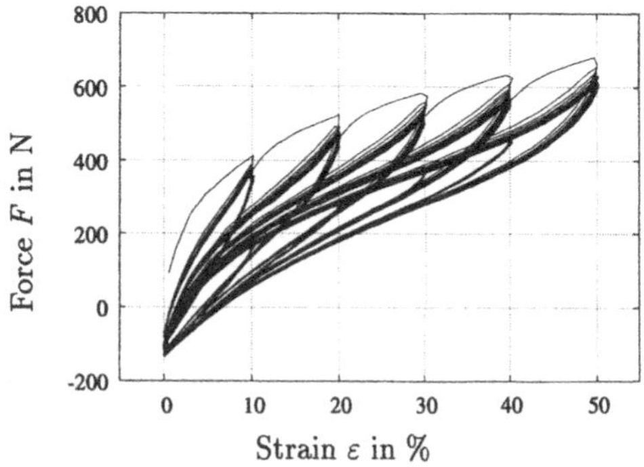

Figure 1: Measurement of a cyclic tension test

ifications of the tire contour while in service. For this topic, it is more important to describe the reduction of the stiffness than to compute the damping behaviour. Besides, a pure elastic damage model needs less material parameters than an inelastic damage formulation and it is numerically efficient.

2 DAMAGE FORMULATIONS

The expression *damage* is ambiguous in the field of mechanics. In fracture mechanics damage describes macro cracks and delaminations. However, in this paper damage is understood as a reduction of the material's stiffness as a result of micro defects, see e. g. Mullins and Tobin (1965). This effect has been experimentally observed and phenomenologically described by Kachanov (1958) for the first time. He introduced a damage variable d which has no physical motivation. It possesses the values $d = 0$ for undamaged up to 1 for fully damaged material.

For a detailed discussion of different damage models it is referred to Kachanov (1986), Chaboche (1987), Lemaitre (1992), Schieße (1994), Altenbach, Altenbach, and Zolochevsky (1995), Krajcinovic (1996).

2.1 *Basic damage model*

On the basis of the multiplicative split of the deformation gradient

$$\underline{\mathbf{F}} = J^{\frac{1}{3}}\,\overline{\underline{\mathbf{F}}} \tag{1}$$

which goes back to Flory (1961), the potential

$$\Psi = U + W^{\mathrm{d}} \tag{2}$$

is splitted into a volumetric and an isochoric part. Simo, Taylor, and Pister (1985) motivate this approach by the fact that rubber material is nearly incompressible and inelastic deformations concentrate on the isochoric part of the deformation. The volumetric part $U = U(J)$ is a function of the determinant of the deformation gradient und for nearly incompressibility $J \approx 1$. The nominal isochoric potential

$$W^{\mathrm{d}} = (1-d)\,W \tag{3}$$

depends on the scalar damage parameter $d \in [0,1]$ and the effective isochoric potential W. The effective isochoric potential $W = W(\mathrm{I}_{\overline{\mathbf{b}}}, \mathrm{II}_{\overline{\mathbf{b}}})$ can be expressed in terms of the invariants $\mathrm{I}_{\overline{\mathbf{b}}}, \mathrm{II}_{\overline{\mathbf{b}}}$ of the isochoric parts of the left Cauchy-Green Tensor

$$\overline{\underline{\mathbf{b}}} = \overline{\underline{\mathbf{F}}}\,\overline{\underline{\mathbf{F}}}^{\mathrm{T}} = J^{-\frac{2}{3}}\,\underline{\mathbf{F}}\,\underline{\mathbf{F}}^{\mathrm{T}} \tag{4}$$

After some standard manipulations (see Miehe (1995)) we get an analogous structure for the dependence of the nominal isochoric Kirchhoff stress tensor

$$\underline{\boldsymbol{\tau}}^{\mathrm{d}}_{\mathrm{iso}} = (1-d)\,\underline{\boldsymbol{\tau}}^{\mathrm{e}}_{\mathrm{iso}} \tag{5}$$

from the effective, that means elastic Kirchhoff tensor

$$\underline{\boldsymbol{\tau}}^{\mathrm{e}}_{\mathrm{iso}} = 2\,\frac{\partial W}{\partial \underline{\mathbf{g}}} \tag{6}$$

where

$$\underline{\mathbf{g}} = g_{ij}\,\mathbf{g}^i \otimes \mathbf{g}^j \tag{7}$$

is the metric tensor in the current configuration $\mathcal{B}_t$. The damage variable

$$d = d^{\mathrm{ex}}_{\alpha}(\alpha) + d^{\mathrm{ex}}_{\beta}(\beta) \tag{8}$$

consists of a discontinuous part

$$d^{\mathrm{ex}}_{\alpha}(\alpha) = d^{\infty}_{\alpha}\left[1 - \exp\left(-\frac{\alpha}{\eta_\alpha}\right)\right] \tag{9}$$

where

$$\alpha(t) = \max_{s\in[0,t]} W(s) \tag{10}$$

or expressed numerically for one time step

$$\alpha^{n+1} = \begin{cases} W^{n+1} & , \quad W^{n+1} > \alpha^n \\ \alpha^n & , \quad \text{otherwise} \end{cases} \tag{11}$$

and a continuous part

$$d^{\mathrm{ex}}_{\beta}(\beta) = d^{\infty}_{\beta}\left[1 - \exp\left(-\frac{\beta}{\eta_\beta}\right)\right] \tag{12}$$

where

$$\beta(t) = \int_0^t \left|\frac{\partial W(s)}{\partial s}\right| ds \tag{13}$$

or in an algorithmic formulation

$$\beta^{n+1} = \beta^n + |W^{n+1} - W^n| \quad . \tag{14}$$

The shape of the exponential functions $d^{\mathrm{ex}}_{\alpha}(\alpha)$ and $d^{\mathrm{ex}}_{\beta}(\beta)$ can be fitted to experimental data by the damage parameters d^{∞}_{α}, d^{∞}_{β}, η^{α} and η^{β}.

With the effective (elastic) isochoric tangential material tensor

$$\underline{\underline{\mathbf{c}}}^{\mathrm{e}}_{\mathrm{iso}} = 2\,\frac{\partial \underline{\boldsymbol{\tau}}^{\mathrm{e}}_{\mathrm{iso}}}{\partial \underline{\mathbf{g}}} \quad , \tag{15}$$

the chain rule

$$\frac{\partial d}{\partial \underline{\mathbf{g}}} = \frac{\partial d}{\partial W}\,\frac{\partial W}{\partial \underline{\mathbf{g}}} \tag{16}$$

and Eq. (5), the nominal isochoric tangential material tensor is given as

$$\begin{aligned} \underline{\underline{\mathfrak{c}}}^{\mathrm{d}}_{\mathrm{iso}} &= 2\,\frac{\partial \underline{\boldsymbol{\tau}}^{\mathrm{d}}_{\mathrm{iso}}}{\partial \underline{\mathbf{g}}} \\ &= 2\,(1-d)\,\frac{\partial \underline{\boldsymbol{\tau}}^{\mathrm{e}}_{\mathrm{iso}}}{\partial \underline{\mathbf{g}}} - 2\,\underline{\boldsymbol{\tau}}^{\mathrm{e}}_{\mathrm{iso}} \otimes \frac{\partial d}{\partial \underline{\mathbf{g}}} \\ &= (1-d)\,\underline{\underline{\mathfrak{c}}}^{\mathrm{e}}_{\mathrm{iso}} - \frac{\partial d}{\partial W}\,\underline{\boldsymbol{\tau}}^{\mathrm{e}}_{\mathrm{iso}} \otimes \underline{\boldsymbol{\tau}}^{\mathrm{e}}_{\mathrm{iso}} \;. \end{aligned} \tag{17}$$

The derivative of the damage variable d with respect to the effective isochoric potential W follows by chain rule

$$\frac{\partial d}{\partial W} = \frac{\partial d_\alpha}{\partial \alpha}\,\frac{\partial \alpha}{\partial W} + \frac{\partial d_\beta}{\partial \beta}\,\frac{\partial \beta}{\partial W} \tag{18}$$

with the partial derivatives

$$d'_\alpha = \frac{\partial d_\alpha}{\partial \alpha} = \frac{d^\infty_\alpha}{\eta_\alpha}\,\exp\left(-\frac{\alpha}{\eta_\alpha}\right) \quad , \tag{19}$$

$$d'_\beta = \frac{\partial d_\beta}{\partial \beta} = \frac{d^\infty_\beta}{\eta_\beta}\,\exp\left(-\frac{\beta}{\eta_\beta}\right) \quad . \tag{20}$$

The algorithmic realization is given by

$$\frac{\partial \alpha}{\partial W} = \begin{cases} 1 & , \quad \alpha^{n+1} = W^{n+1} > \alpha^n = W^n \\ 0 & , \quad \text{otherwise} \end{cases} \tag{21}$$

and

$$\begin{aligned} \frac{\partial \beta}{\partial W} &= \frac{\beta^{n+1} - \beta^n}{W^{n+1} - W^n} = \\ &= \frac{|W^{n+1} - W^n|}{W^{n+1} - W^n} = \operatorname{sign}(W^{n+1} - W^n) \;. \end{aligned} \tag{22}$$

The accompanying algorithmic tangential tensor $\underline{\underline{\mathfrak{c}}}^{\mathrm{d}}_{\mathrm{iso}}$ provides a quadratic convergence of the nonlinear finite element solution in a Newton-Raphson procedure.

2.2 *Cubic splines as damage function*

Investigations have shown that the behaviour of rubber can be described merely in a qualitative way by the exponential damage functions $d^{\mathrm{ex}}_\alpha(\alpha)$ and $d^{\mathrm{ex}}_\beta(\beta)$.

In order to verify this statement we try to compute the tension test of Fig. 1 with cubic splines

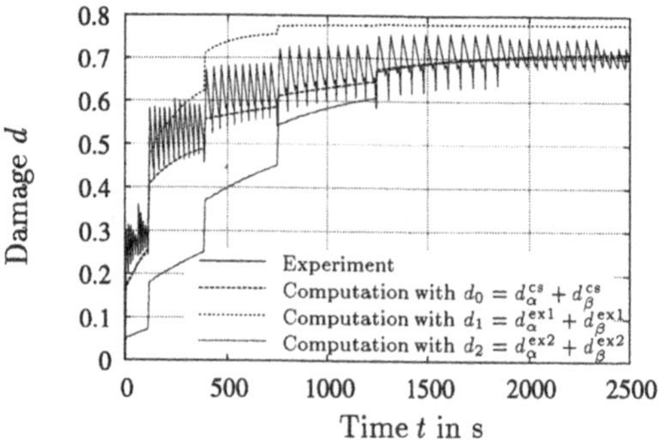

Figure 2: Fitting the damage functions

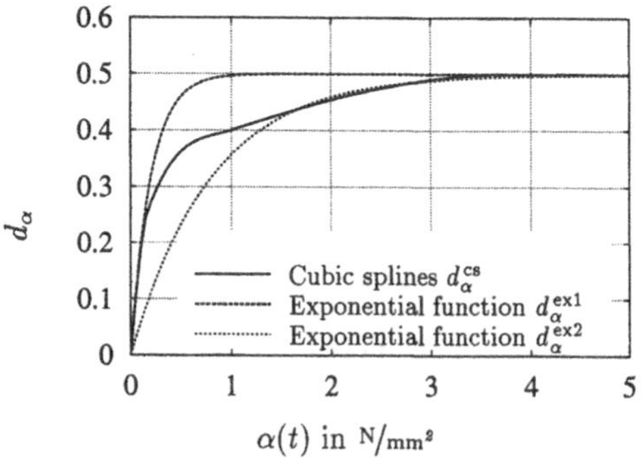

Figure 3: Discontinuous damage functions

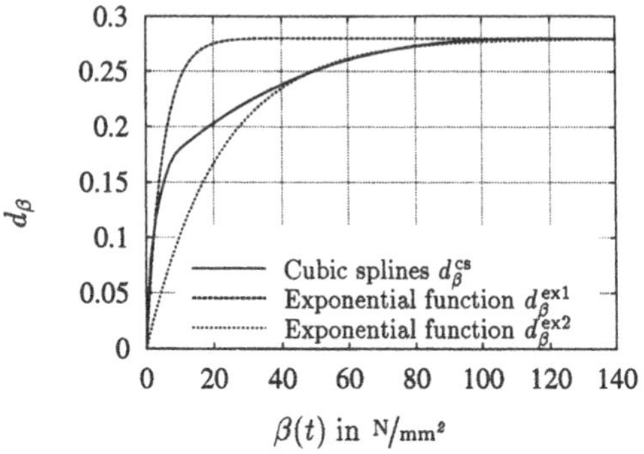

Figure 4: Continuous damage functions

$d_0 = d^{cs}_\alpha + d^{cs}_\beta$ on the one hand and with two exponential damage functions $d_1 = d^{ex1}_\alpha + d^{ex1}_\beta$ and $d_2 = d^{ex2}_\alpha + d^{ex2}_\beta$ on the other hand. First, the specimen is loaded in 10 cycles at strain amplitudes of 10%, then in 10 cycles at 20%, and so on up to 50% strain amplitude. Afterwards, the strain amplitude is reduced again.

The fitting of the material parameters is carried out as follows:

- First, a maximum amount of the elastic stiffness has to be assumed. Here, we use Mooney-Rivlin material with the two material parameters C_1 and C_2.

- We have to take into account that the force yields negative values during the experiment at small strains. Since even the reduced elastic stiffness can not be negative, in every case we have to cut all those parts in an extension test where the measured force is negative and the accompanying strain is positive. Here, we neglect the measured data with a value of the force less than 300N.

- With these modified data we can determine the damage variable d. In Fig. 2 the measured damage variable is depicted versus time. The curve is oscillating due to the hysteresis of the load-strain-curve (see Fig. 1).

- Every change in the strain amplitude causes a strong increase of the damage variable. This increase can be used to determine the discontinuous damage functions d_α which are depicted in Fig. 3.

- The remaining increase of the damage at an actual strain amplitude can be used to fit the continuous damage functions d_β which are shown in Fig. 4.

It can be seen in Fig. 2, 3 and 4. that the shape of the cubic splines is quite different from the exponential functions. The exponential functions d_α^{ex1} and d_β^{ex1} are similar to the cubic splines d_α^{cs} and d_β^{cs} just at the beginning and the exponential functions d_α^{ex2} and d_β^{ex2} match to the cubic splines just at the end.

In order to stress the reliability of the cubic splines $d_0 = d_\alpha^{\text{cs}} + d_\beta^{\text{cs}}$, the result of the simulation is shown in Fig. 5. In view of the fact that we assume an elastic damage model, the curve matches very well the measured one (see Fig. 1).

In Fig. 6 the result of a computation using the exponential damage function $d_1 = d_\alpha^{\text{ex1}} + d_\beta^{\text{ex1}}$ is given. It can be seen that during the beginning of the experiment the damage increase is too high. At the end of the experiment, there is no further increase of the damage.

In Fig. 7 the exponential damage function $d_2 = d_\alpha^{\text{ex2}} + d_\beta^{\text{ex2}}$ is used. At the beginning of the experiment, there is only a small increase of the damage, that means an underestimated reduction of the stiffness. At the end, the stiffness reduction is so high that the computed force is even decreasing when the strain varies from 40% to 50% for the first time. This softening effect causes numerical problems.

Finally, we can point out that by cubic splines the experiments can be determined more exactly.

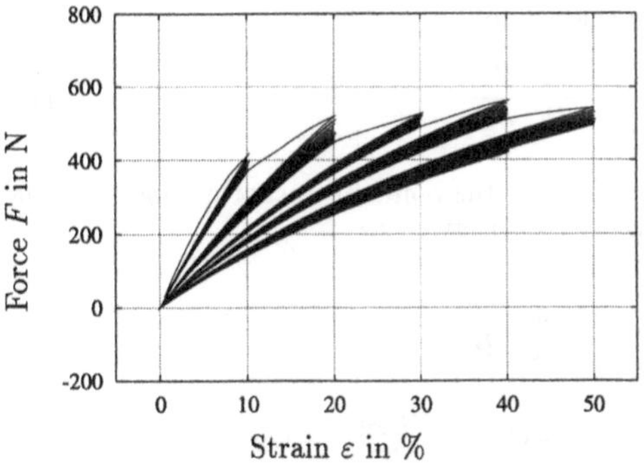

Figure 5: Computation with cubic splines $d_0 = d_\alpha^{\text{cs}} + d_\beta^{\text{cs}}$

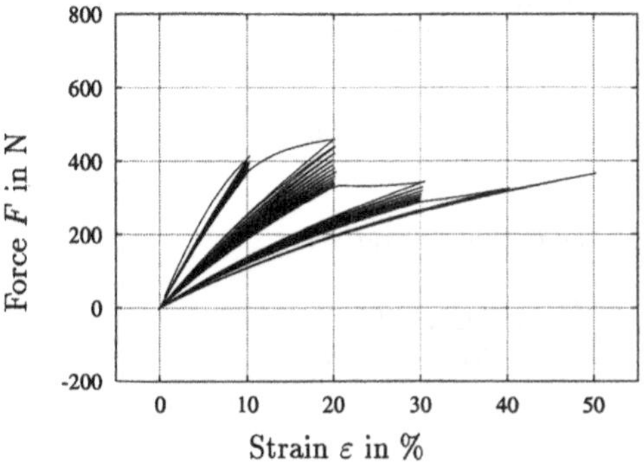

Figure 6: Computation with exponential function $d_1 = d_\alpha^{\text{ex1}} + d_\beta^{\text{ex1}}$

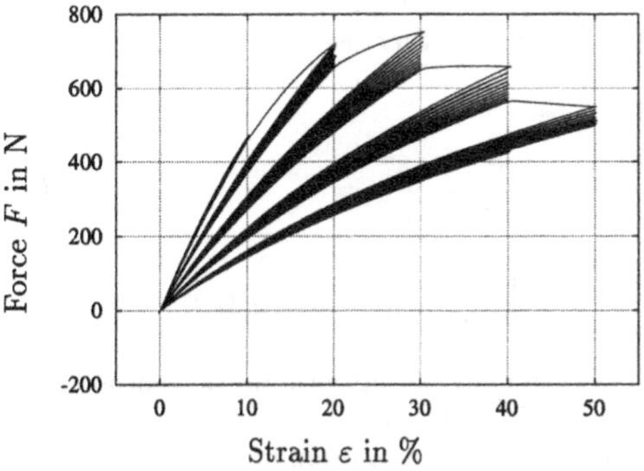

Figure 7: Computation with exponential function $d_2 = d_\alpha^{\text{ex2}} + d_\beta^{\text{ex2}}$

2.3 *Damage by tension or compression*

Another problem can be seen when we compute a cyclic compression test with material parameters coming from the cyclic tension test. In Fig. 8 the measurement of a typical cyclic compression test is shown. If we assume the same material data set as for the tension experiment, a behaviour of the simulation is obtained which is too stiff, see Fig. 9.

Not only elastomers yield this dependence of the load direction (tension or compression) on the amount of damage. In particular, this effect is well known for concrete. There are a lot of publications concerning this load direction dependence of materials. For example Oliver, Cervera, Oller, and Lubliner (1990) and Suanno (1995) introduce a scalar damage parameter as a function of the principal strains to consider the increase of micro defects as a result of tension. The damage parameter is activated when at least one of the principal strains has a positive sign. Naumenko (1996) develops on the basis of the work from Saanouni, Chaboche, and Lesne (1989) and Lemaitre and Chaboche (1990) a direction dependent damage criterium for metals.

Besides the fact, that finite strains have to be taken into account, there are the following three reasons why a new model is introduced.

- In contrast to concrete, for rubber the stiffness reduction as a result of pressure is higher than a stiffness reduction by tension. The sign of the principal strains is not an objective measure for pressure. Tension and compression tests have both, negative and positive principal strains.
- Rubber is nearly incompressible. Therefore, damage is taken into account as a function of the effective isochoric potential W. But the invariants $\mathrm{I}_{\bar{\underline{b}}}$, $\mathrm{II}_{\bar{\underline{b}}}$ or alternativly the isochoric principal strains $\bar{\lambda}_1, \bar{\lambda}_2, \bar{\lambda}_3$ do not give a hint of the direction of the loading.
- A strict distinction of damage or non-damage leads to numerical instabilities. For simple geometries no numerical problem may occur. But our computations have shown that for threedimensional tire computations no convergence is achieved.

For the given reasons the following damage function

$$\zeta(\vartheta) = \begin{cases} \widehat{\zeta} & , \ \vartheta \le -\widehat{\vartheta} \\ \frac{\widehat{\zeta}+1}{2} + \frac{\widehat{\zeta}-1}{2} \sin\left(\pi + \frac{\pi\vartheta}{2\widehat{\vartheta}}\right) & , \ -\widehat{\vartheta} < \vartheta < \widehat{\vartheta} \\ 1 & , \ \vartheta \ge \widehat{\vartheta} \end{cases} \tag{23}$$

with the pressure variable

$$\vartheta = \vartheta(J) = J - 1 \tag{24}$$

is introduced. The pressure function has the limits $\zeta = 1$ for tension and $\widehat{\zeta}$ in case of pressure. For $-\widehat{\vartheta} < \vartheta < +\widehat{\vartheta}$ the pressure function is sinusoidal. Thus, the iterative solution of the nonlinear system of equations of the finite element method has a smooth behaviour. A step function would lead to numerical problems.

The damage variable ϑ is a function of the determinant of the deformation gradient $J = \sqrt{\mathrm{III}_{\underline{b}}}$. But, as mentioned before, there is no alternative, since neither the first and the second invariant $\mathrm{I}_{\bar{\underline{b}}}$, $\mathrm{II}_{\bar{\underline{b}}}$ of the isochoric part of the left Cauchy Green strain tensor nor other variables can measure the load direction, e. g. tension or pressure at a uniaxial tension test.

It has been found out, that it is sufficient to take into account the direction dependence of damage at the discontinuous part d_α. The continuous part d_β is left unchanged. Instead of the discontious damage variable $\alpha(t)$ from Eq. (10) the modified damage variable

$$\alpha^*(t) = \max_{s\in[0,t]} \left[\zeta(\vartheta(J(s)))\, W(s))\right] \tag{25}$$

is used. Now, the (modified) damage variable is not just a function of the isochoric potential W, but also a function of the pressure function ζ, which considers the load direction. This yields the cubic splines

$$d_\alpha(\alpha) \longrightarrow d_{\alpha^*}(\alpha^*) \tag{26}$$

as modified damage function.

For the algorithmic part of the presented damage formulation the damage variables

$$\alpha^{*,n+1} = \begin{cases} \zeta(\vartheta^{n+1})\, W^{n+1} & , \ \zeta(\vartheta^{n+1})\, W^{n+1} > \alpha^{*,n} \\ \alpha^{*,n} & , \ \text{otherwise} \end{cases} \tag{27}$$

at time t_{n+1} or for the accompanying load step, respectively, are necessary. The results for $\alpha^{*,n+1}$ and ϑ^{n+1} have to be stored in a data base for the next load step.

As can seen in Fig. 10, the results are much better taking into account the pressure function. For example, without pressure function the minimum value of the force is -780 N. The use of the pressure function ζ yields a force of -510 N which is much closer to the measured value of -490 N.

2.4 *Damage due to temperature*

To take into account that an increase of the temperature results in a stiffness reduction, temperature damage $d_{\mathrm{T}}(T)$ is introduced. For the function $d_{\mathrm{T}}(T)$ a parabolic shape is applied. The temperature for each integration point is assumed to be constant.

Finally, the total damage variable

$$d = d_\alpha^{\mathrm{cs}}(\alpha^*) + d_\beta^{\mathrm{cs}}(\beta) + d_{\mathrm{T}}(T) \tag{28}$$

composed of three different parts.

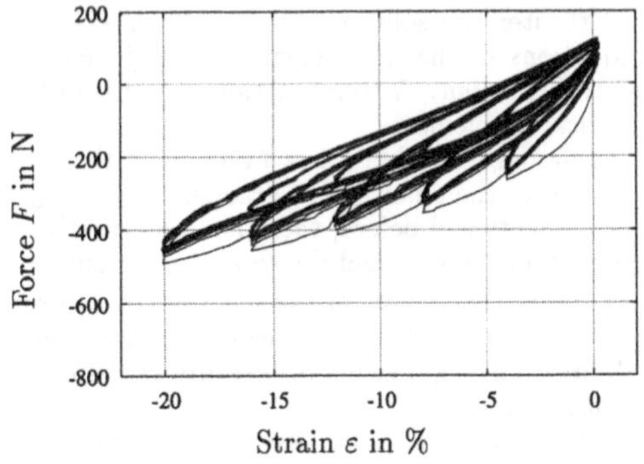

Figure 8: Measurement of a cyclic compression test

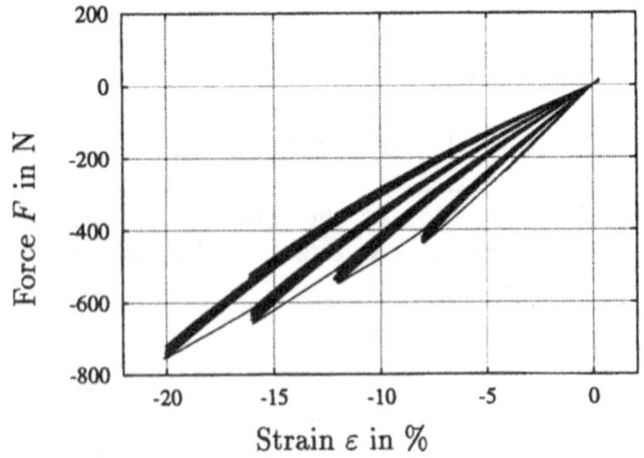

Figure 9: Computation without pressure function

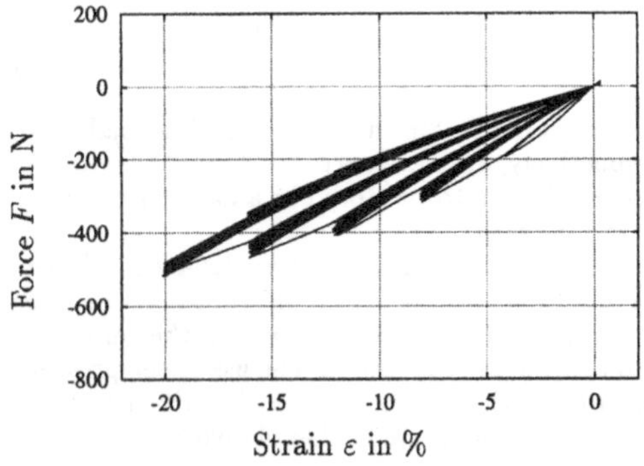

Figure 10: Computation with pressure function

2.5 *Consistent linearization*

In comparison to the linearization of the basic model the only change is the derivative of the damage variable d with respect to the metric tensor $\underline{g}$ (see Eq. (16)). It has to be taken into account that the new damage variable d is not just a function of the effective isochoric energie W, but it depends because of the introduction of the determinant of the deformation gradient J on the volumetric part of the deformation, too. The temperature has no influence on the consistent linearization, because it is chosen to be constant. With the partial derivative

$$\frac{\partial J}{\partial \underline{g}} = \frac{\partial J}{\partial \underline{b}}\,\underline{b} = \frac{J}{2}\,\underline{b}^{-1}\,\underline{b} = \frac{J}{2}\,\underline{1} \tag{29}$$

we get

$$\frac{\partial d}{\partial \underline{g}} = \frac{\partial d_{\alpha^*}}{\partial \underline{g}} + \frac{\partial d_{\beta}}{\partial \underline{g}} \tag{30}$$

where

$$\frac{\partial d_{\alpha^*}}{\partial \underline{g}} = \frac{\partial d_{\alpha^*}}{\partial \alpha^*}\,\frac{\partial \alpha^*}{\partial \underline{g}} \tag{31}$$

and

$$\frac{\partial d_{\beta}}{\partial \underline{g}} = \frac{\partial d_{\beta}}{\partial \beta}\,\frac{\partial \beta}{\partial W}\,\frac{\partial W}{\partial \underline{g}} \tag{32}$$

The derivative of the nominal isochoric Kirchhoff stress tensor $\underline{\tau}^{d}_{iso}$ with respect to the metric tensor $\underline{g}$ yields the nominal isochoric tangential material tensor

$$\underline{\underline{c}}^{d}_{iso} = 2\,\frac{\partial \underline{\tau}^{d}_{iso}}{\partial \underline{g}} = 2\,(1-d)\,\frac{\partial \underline{\tau}^{e}_{iso}}{\partial \underline{g}} - 2\,\underline{\tau}^{e}_{iso} \otimes \frac{\partial d}{\partial \underline{g}} \tag{33}$$

which is required for the Newton-Raphson iteration.

3 NUMERICAL EXAMPLE

For a validation of the proposed material formulation tire footprints of a new tire and a tire in service are computed. Fig. 11 shows the chosen finite element discretization. The temperature distribution is determined in a 2d thermomechanical computation (see Dehnert and Volk (1992), Becker, Dorsch, Kaliske, and Rothert (1998)) and used as input parameter for a 3d simulation.

In Fig. 12, the pressure distribution in the footprint of a new tire and a tire in service are compared. It can be seen that the footprint is enlarged and that the pressure distribution appears smoother. Besides, a growth of the tire at the edge of the belt can be simulated. This effect can be observed especially for truck tires.

4 CONCLUSION

An elastic damage formulation has been introduced which consists of three parts. For continuous and discontinuous damage cubic splines are proposed for a quantitative description of experiments. For temperature damage a parabolic approach has been chosen. A consistent linearization of the proposed damage model yields the tangential material tensor which is necessary for a quadratic convergence of a Newton-Raphson procedure. The algorithmic formulation is robust and yields good results compared to experiments.

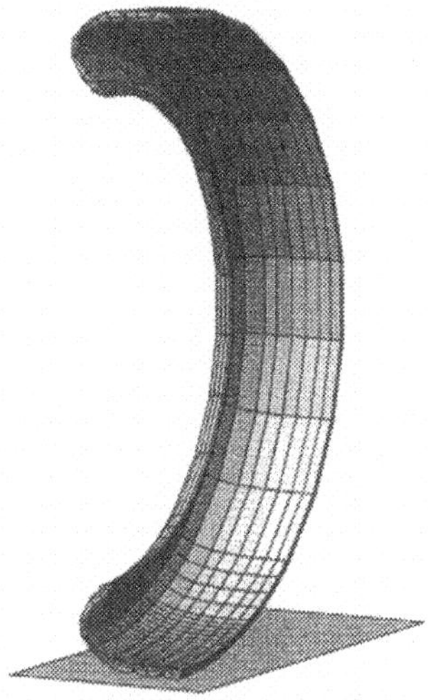

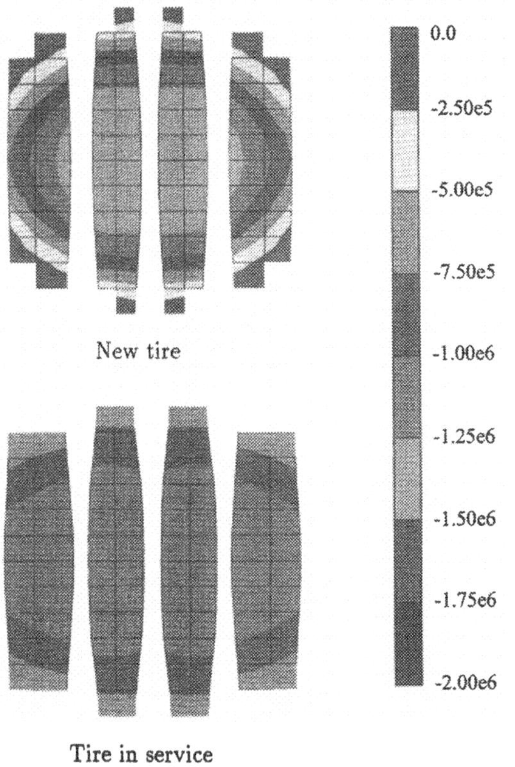

Figure 12: Computed footprints of a truck tire

The shown elastic damage model is suited for an extension to viscoelastic or elastoplastic parts. These combined material models are introduced by Kaliske and Rothert (1999).

REFERENCES

Altenbach, H., J. Altenbach, & A. Zolochevsky (1995). *Erweiterte Deformationsmodelle und Versagenskriterien in der Werkstoffmechanik.* Stuttgart: Deutscher Verlag für Grundstoffindustrie.

Becker, A., V. Dorsch, M. Kaliske, & H. Rothert (1998). A material model for simulating the hysteretic behavior of filled rubber for rolling tires. *Tire Science and Technology 26*, 132–148.

Chaboche, J. L. (1987). Continuum damage mechanics: Present state and future trends. *Nuclear Engineering and Design 105*, 19–33.

Dehnert, J. & H. Volk (1992). Prediction of the temperature distribution in rolling tires using the finite element method. *Kautschuk, Gummi, Kunststoffe 45*, 47–51.

Flory, P. J. (1961). Thermodynamic relations for high elastic materials. *Transactions of the Faraday Society 57*, 829–838.

Kachanov, L. M. (1958). Time of rupture process under creep conditions. *TVZ Acad. Nauk S.S.R., Otd. Techn. Nauk 8*, 26–31.

Kachanov, L. M. (1986). *Introduction to continuum damage mechanics.* Martinus Nijhoff Publishers, Dordrecht.

Kaliske, M. & H. Rothert (1999). Viscoelastic and elastoplastic damage formulations. In *Proceedings of the First European Conference on Constitutive Models for Rubber.* Wien, Österreich.

Krajcinovic, D. (1996). *Damage Mechanics.* Elsevier.

Lemaitre, J. (1992). *A Course on Damage Mechanics.* Springer-Verlag.

Lemaitre, J. & J. L. Chaboche (1990). *Mechanics of solid materials.* Cambridge University Press.

Miehe, C. (1995). Discontinuous and continuous damage evolution in Ogden-type large-strain elastic materials. *European Journal of Mechanics. A/Solids 14*, 697–724.

Mullins, L. & N. R. Tobin (1965). Stress softening in rubber vulcanizates. part i. use of a strain amplification factor to describe the elastic of filler-reinforced vulcanized rubber. *Journal of Applied Polymer Science 9*, 2933–3009.

Naumenko, K. (1996). *Modellierung und Berechnung der Langzeitfestigkeit dünnwandiger Flächentragwerke unter Einbeziehung von Werkstoffkriechen und Schädigung.* Ph. D. thesis, Fakultät für Maschinenbau der Universität Magdeburg.

Oliver, J., M. Cervera, S. Oller, & J. Lubliner (1990). Isotropic damage models and smeared crackanalysis of concrete. In N. Bicanic and H. Mang (Eds.), *Computer Aided Analysis and Design of Concrete Structures*, pp. 945–957. Pineridge Presse.

Saanouni, K., J. L. Chaboche, & P. M. Lesne (1989). On the creep crack-growth prediction by a non local damage formulation. *European Journal of Mechanics. A/Solids 8*, 437–459.

Schieße, P. (1994). *Ein Beitrag zur Berechnung des Deformationsverhaltens anisotrop geschädigter Kontinua unter Berücksichtigung der thermoplastischen Kopplung*. Ph. D. thesis, Institut für Mechanik, Ruhr-Universität Bochum.

Simo, J. C., R. L. Taylor, & K. S. Pister (1985). Variational and projection methods for the volume constraint in finite deformation elastoplasticity. *Computer Methods in Applied Mechanics and Engineering 51*, 177–208.

Suanno, R. L. M. (1995). *Ein dreidimensionales Simulationsmodell für Stahlbeton mit Plastizität und Schädigung*. Ph. D. thesis, Institut für Baustatik, Universität Stuttgart.

Constitutive Models for Rubber, Dorfmann & Muhr (eds) © 1999 Taylor & Francis ISBN 90 5809 113 9

Viscoelastic and elastoplastic damage formulations

M. Kaliske & H. Rothert
Institut für Statik, Universität Hannover, Germany

ABSTRACT: Loading of filled rubber material results in a modification of the material's stiffness. The elastic properties are reduced due to changes on the molecular scale. These softening phenomena of elastomers can be described by a continuum damage model. The goal of the present paper is twofold. Firstly, the constitutive formulations are introduced. Moreover, computational aspects are stressed. The material models are given together with their numerical algorithms and the accompanying tangential material tensors, which play a crucial role when solving the discretized nonlinear initial boundary value problem. Therefore, the shown approaches are presented in a format ready for a finite element implementation.

1 INTRODUCTION

An experimental study of carbon-black filled rubber material at cyclic loading shows nonlinear elastic and inelastic material characteristics. The latter properties yield hysteresis loops which stand for the dissipated energy per cycle of deformation. On the one hand, rubber material depicts time- and frequency-dependency. On the other hand, rate-independent dissipative behaviour is also observed due to amplitude-dependent plastic changes of the material's microstructure.

The hysteretic behaviour is not stationary from the beginning, but it converges after some time against a constant load-displacement hysteresis. A strong reduction of the stiffness, called Mullins-effect, is found at first loading. This phenomenon is caused by rupture of molecular links. A continuum damage model, as proposed by Miehe (1995) for elastic characteristics, may be used to model this behaviour. The strain energy is assumed to be the thermodynamical driving force for the formulation of a continuous and a discontinuous damage approach.

In order to obtain a realistic and complete material model the nonlinear elastic damage formulation is not sufficient and it has to be extended to account for inelastic features. Therefore, a viscoelastic description (see e. g. Kaliske, Rothert (1997)) may be appropriate for the modelling depending on the problem to be investigated. The paper presents a combined characterization of the mentioned aspects which results in a finite viscoelastic damage approach. From the theoretical formulation a clear structure of the algorithmic representation can be derived. A second approach derives the rate-independent properties of rubber by a rheological structure that arranges elastoplastic material elements in parallel (see Kaliske, Rothert (1998) for details). The elastoplastic model is also developed with respect to the damage formulation in order to yield a combined phenomenological description.

2 ELASTIC DAMAGE APPROACH

The molecular structure of a carbon-black or silicate filled rubber consists of the filler particles and long polymer chains. The two components and the polymeric macromolecules themselves are linked together through the vulcanization process. If this filled network is subjected to loading, the response is partly elastic and partly inelastic. Moreover, some of the links between filler and polymer

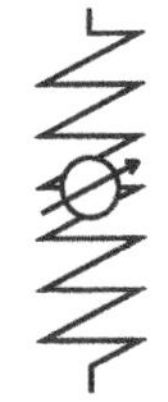

Figure 1: Elastic Hooke-element with damage

chain break and, therefore, a reduction of the material stiffness is observed. For example, at first loading of the elastomer a strong stiffness modification is found which is the so-called Mullins-effect (Mullins (1969)). This phenomenon is macroscopically measured as load-depending softening. Subsequent reloadings converge against stationary characteristics after some further load cycles.

The hyperelastic behaviour is described by a strain energy function Ψ being either a phenomenological term or a physically based model. The properties of filled rubber material exhibit different properties with respect to volumetric and deviatoric deformations. This observation is taken into account by applying an additive split of the strain energy function. The first part $\hat{U}(J)$ is a function of the volume-change J evaluated as determinant of the deformation gradient $J = \det \underline{\mathbf{F}}$. It describes the hydrostatic pressure. The second Term $\hat{W}_0(\underline{\bar{\mathbf{b}}})$ is given by the volume-preserving part $\underline{\bar{\mathbf{b}}} = J^{-\frac{2}{3}} \, \underline{\mathbf{F}} \, \underline{\mathbf{F}}^T$ of the left Cauchy-Green tensor $\underline{\mathbf{b}}$. In the subsequent derivations merely the isochoric contributions are of interest and, therefore, are given explictly. An appropriate model for the volumetric component has to be added.

The deformation induced micro-defects are characterized by

$$\Psi = \hat{U}(J) + (1-d) \, \hat{W}_0(\underline{\bar{\mathbf{b}}}), \tag{1}$$

a phenomenological approach of continuum damage mechanics. In the constitutive potential function Ψ damage, i. e. softening, is defined by the scalar damage variable $d \in [0,1]$. Usually, an isotropic scalar model is employed for rubber material although some anisotropic features are found. The value of the scalar quantity is $d = 0$ for an undamaged material and $d = 1$ when it is fully damaged. In order to represent the material model symbolically, a pointer, which stands for load depending modifications, is attached to the elastic spring, i. e. the so-called Hooke-element (Figure 1). Subsequently, this basic rheological constituent will be used in complex models. This phenomenological concept for softening materials goes back to Chaboche (1981), (1988), Lemaitre (1984) and Kachanov (1986) among others.

Departing from Equation (1) standard arguments of continuum mechanics yield the nominal isochoric Kirchhoff stress tensor

$$\underline{\tau}_{iso} = (1-d) \, \underline{\tau}_{0,iso} \tag{2}$$

relative to the current configuration analogously to the structure of the potential function using the effective stresses $\underline{\tau}_{0,iso}$ and the damage variables

$$d = \hat{d}_\alpha(\alpha) + \hat{d}_\beta(\beta). \tag{3}$$

Miehe (1995) proposes this composition of the damage phenomena. A discontinuous part $\hat{d}_\alpha(\alpha)$ where

$$\hat{\alpha}(t) = \max_{s \in [0,t]} \hat{W}_0(s) \tag{4}$$

and a continuous portion $\hat{d}_\beta(\beta)$ where

$$\hat{\beta}(t) = \int_0^t |\dot{\hat{W}}_0(s)| \, ds \tag{5}$$

include different aspects of deformation induced softening (see Miehe (1995) for details). In this formulation the effective strain energy $\hat{W}_0(\underline{\bar{\mathbf{b}}})$ is the thermodynamic force that drives the damage evolution. Equation (4) leads to an increase in damage as a function of maximum strain energy reached at the material point during the whole load history. In contrast, Equation (5) formulates the evolution of softening as function of the change of strain energy. Thus, it gives an increase in damage even for unloading.

The damage functions $\hat{d}_\alpha(\alpha)$ and $\hat{d}_\beta(\beta)$ need to be specified in order to define the damage evolution. In Miehe (1995) a simple exponential expression with a saturation form is employed. This model is convenient for qualitative studies, yields a symmetric tangent operator and will also be used in this paper (see Equation (10)). A more flexible and realistic prediction of $\hat{d}_\alpha(\alpha)$ and $\hat{d}_\beta(\beta)$ is obtained when the damage evolution is defined by cubic splines calibrated with respect to experimentally determined material characteristics (see Nasdala, Kaliske, Rothert & Becker (1999) for details). Additionally, Nasdala et al. (1999) propose to separate a compression and an extension deformation resulting in a realistic model with a non-symmetric material tensor.

3 VISCOELASTIC DAMAGE APPROACH

Experimental investigations of filled rubber material exhibit frequency-dependent stiffness and

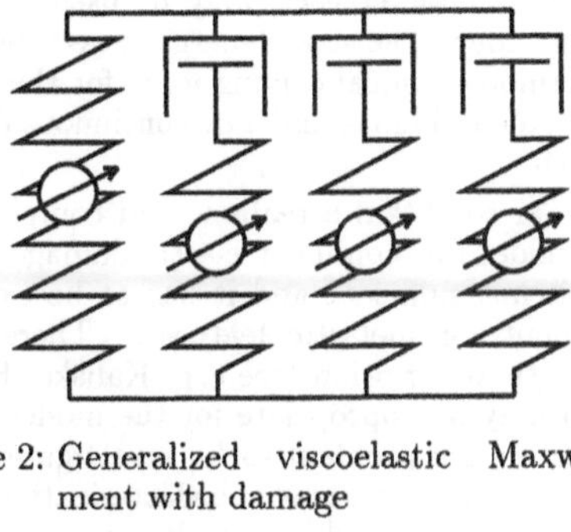

Figure 2: Generalized viscoelastic Maxwell-element with damage

damping properties. Therefore, the nonlinear elastic damage model is extended by a viscoelastic material description. Simo (1987) and Govindjee & Simo (1992) combine viscoelastic and damage formulations similar to this publication. This finite linear viscoelastic damage approach has a clear structure and it is easy to implement into a finite element program. Miehe & Keck (1999) address the same topic with a nonlinear viscoelastic approach.

3.1 *General formulation*

The generalized Maxwell-element (Figure (2)) is used as symbolic viscoelastic rheological structure and it is extended by damage to yield

$$\mathrm{DEV}\underline{\mathbf{S}} = (1-d)\,\mathrm{DEV}\underline{\mathbf{S}}_0 + \sum_{i=1}^{N} \underline{\mathbf{H}}_i \qquad (6)$$

an additive composition of the second Piola Kirchhoff stress tensor. The isochoric stresses in the reference configuration $\mathrm{DEV}\underline{\mathbf{S}}$, where

$$\mathrm{DEV}\,(\underline{\bullet}) := (\underline{\bullet}) - \frac{1}{3}\,[\underline{\mathbf{C}} : (\underline{\bullet})]\,\underline{\mathbf{C}}^{-1} \qquad (7)$$

and $\underline{\mathbf{C}}$ is the right Cauchy Green tensor, consists of the nominal elastic stress tensor $(1-d)\,\mathrm{DEV}\underline{\mathbf{S}}_0$ and the viscoelastic internal stress variables $\underline{\mathbf{H}}_i$. These quantities are described by the linear rate-equation

$$\dot{\underline{\mathbf{H}}}_i + \frac{1}{\tau_i}\underline{\mathbf{H}}_i = \gamma_i\,(1-d)\,\mathrm{DEV}\,\dot{\underline{\mathbf{S}}}_0. \qquad (8)$$

The viscosity of one Maxwell-element i is defined by the relaxation time τ_i. The material parameter γ_i gives the contribution of this rheological element to the total stiffness of the material. Starting from the rate-equation (8), the analytical hereditary integral

$$\underline{\mathbf{H}}_i = \gamma_i \int_0^t \exp\left(-\frac{t-s}{\tau_i}\right) \times \frac{\partial}{\partial s}\left((1-d)\,\mathrm{DEV}\,\hat{\underline{\mathbf{S}}}_0(s)\right)\,ds \qquad (9)$$

of the stress history is derived. This exact integral for the update of the internal stress variables $\underline{\mathbf{H}}_i$ has to be evaluated. For a detailed discussion of the shown viscoelastic model we refer to Kaliske & Rothert (1997) and the literature therein.

In this publication we choose for simplicity a convenient exponential saturation function

$$d = d_\alpha^\infty\left[1-\exp\left(-\frac{\alpha}{\eta_\alpha}\right)\right] + d_\beta^\infty\left[1-\exp\left(-\frac{\beta}{\eta_\beta}\right)\right] \qquad (10)$$

for both components of damage (Equation (3)) as mentioned before. The sum of the softening parameters is bounded $(d_\alpha^\infty + d_\beta^\infty) \in [0,1]$ by the maximum of possible damage, and η_α, η_β characterize the saturation properties. An algorithmic formulation of the discontinuous approach yields an update of the internal variables

$$\alpha^{n+1} = \begin{cases} W_0^{n+1} & : \text{ for } W_0^{n+1} - \alpha^n > 0 \\ \alpha^n & : \text{ else} \end{cases} \qquad (11)$$

and, equivalently, for continuous damage

$$\beta^{n+1} = \beta^n + |W_0^{n+1} - W_0^n| \qquad (12)$$

at current time t^{n+1} (see also Miehe (1995)). The results W_0^n, α^n, β^n at time t^n are taken from a data base where they are stored for one time step $\Delta t = t^{n+1} - t^n$.

Analytical integration and the use of a linear discretization of the nominal stress rate at time increment $\Delta t = t^{n+1} - t^n$ lead to the recursive formula

$$\begin{aligned} \mathrm{DEV}\underline{\mathbf{S}}^{n+1} &= (1-d^{n+1})\mathrm{DEV}\underline{\mathbf{S}}_0^{n+1} \\ &+ \sum_{i=1}^{N}\left\{\exp\left(-\frac{\Delta t}{\tau_i}\right)\underline{\mathbf{H}}_i^n + \gamma_i\,\frac{1-\exp(-\frac{\Delta t}{\tau_i})}{\frac{\Delta t}{\tau_i}} \times \left[(1-d^{n+1})\mathrm{DEV}\,\underline{\mathbf{S}}_0^{n+1} - (1-d^n)\mathrm{DEV}\,\underline{\mathbf{S}}_0^n\right]\right\} \end{aligned} \qquad (13)$$

which is an approximate evaluation of the convolution integral (9). It is a very efficient update of the stress history. The damage variables and results of the stress quantities at the last converged time step t^n $\sum_{i=1}^{N}\underline{\mathbf{H}}_i^n$, $\mathrm{DEV}\underline{\mathbf{S}}_0^n$ are employed in this formulation. The stress tensor $\mathrm{DEV}\underline{\mathbf{S}}^{n+1}$ is transformed by a push-forward operation

$$\underline{\boldsymbol{\tau}}_{iso}^{n+1} = \underline{\mathbf{F}}^{n+1}\,(\mathrm{DEV}\,\underline{\mathbf{S}}^{n+1})\,\underline{\mathbf{F}}^{n+1,T} \qquad (14)$$

with the deformation gradient $\underline{\mathbf{F}}^{n+1}$ into the current configuration.

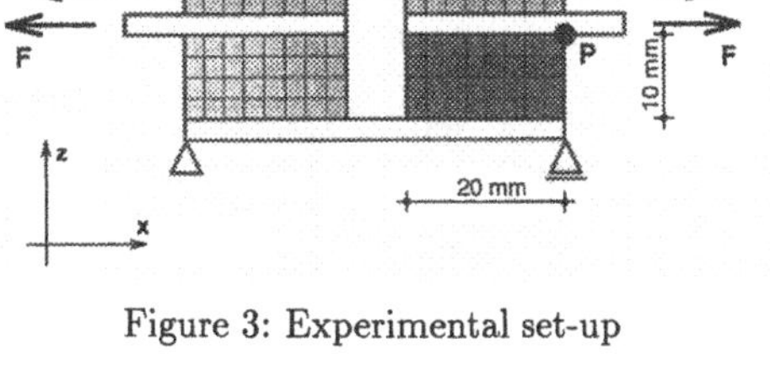

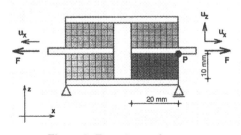

Figure 3: Experimental set-up

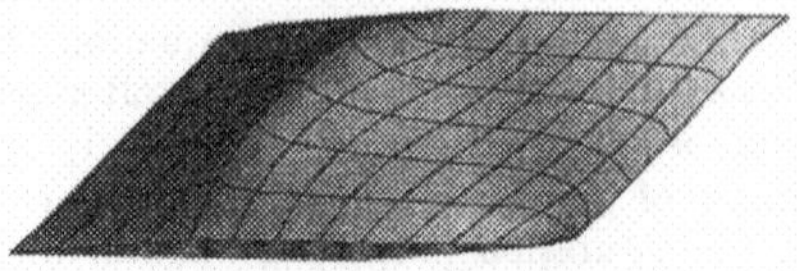

Figure 4: Deformed configuration

The iterative Newton procedure to solve a nonlinear finite element problem requires the determination of the consistent tangential material operator. It can be derived analytically for the given material formulation. The algorithmic material tensor which is developed consistently with the stress computation (Equation (13)) reads

$$\underline{\underline{\mathbf{a}}}_{iso}^{n+1} = (1 - d^{n+1})\, \mathcal{G}\; \underline{\underline{\mathbf{a}}}_{0,iso}^{n+1} - \frac{\partial d^{n+1}}{\partial W_0^{n+1}}\, \mathcal{G}\; \underline{\tau}_{0,iso}^{n+1} \otimes \underline{\tau}_{0,iso}^{n+1} \qquad (15)$$

where $\mathcal{G}$ is a constant scalar factor for constant Δt

$$\mathcal{G} = 1 + \sum_{i=1}^{N} \gamma_i \, \frac{1 - \exp\left(-\frac{\Delta t}{\tau_i}\right)}{\frac{\Delta t}{\tau_i}} \qquad (16)$$

and

$$\frac{\partial d^{n+1}}{\partial W_0^{n+1}} = \begin{cases} d'_\alpha + d'_\beta \,\mathrm{sign}[\Delta W] & : \text{ for } W_0^{n+1} - \alpha^n > 0 \\ d'_\beta \,\mathrm{sign}[\Delta W] & : \text{ else} \end{cases} \qquad (17)$$

using $\Delta W = W_0^{n+1} - W_0^n$. The presented formulation and the algorithmic discretization of finite viscoelastic damage is very efficient to implement into a finite element code.

3.2 *Numerical example*

For an illustration a numerical fe-test was carried out on the basis of a standard nonhomogeneous shear experiment (Figure 3). We make use of the symmetries of the specimen and model just a part of it. The rubber specimen is discretized by 192 mixed Q2/P1-elements according to Simo & Taylor (1991). The inner steel members are horizontally extended in a displacement controlled simulation. We measure the force and the displacements at the marked position P. At maximum displacements the configuration is strongly deformed as can be seen from Figure 4.

The result of the simulations is summarized by Figures 5 – 6. Here, force F is given as a function of horizontal and vertical displacements of point P for the viscoelastic (vis) and the combined viscoelastic damage approach (vis+d_α, vis+d_β). The viscoelastic results form a hysteresis which stands for the dissipated energy per cycle of deformation.

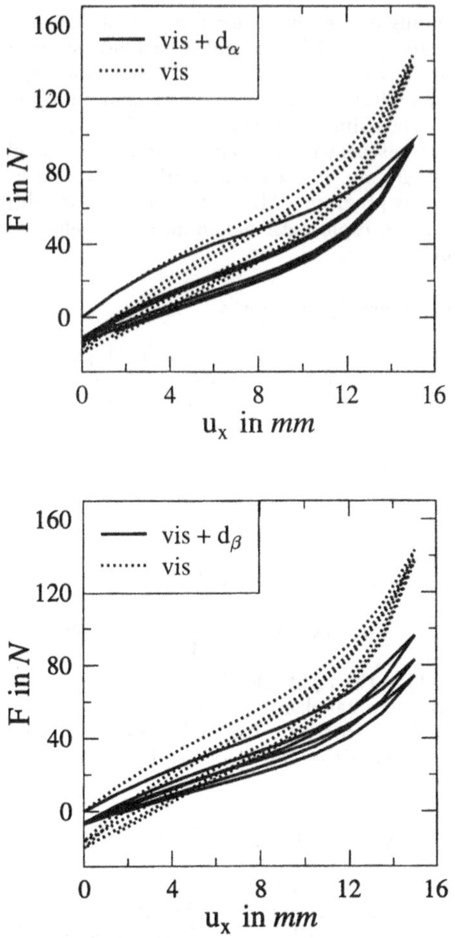

Figure 5: Horizontal displacements

All three cycles give nearly the same result, the process is quasi stationary. The stiffness of the rubber material is reduced during the first loading using the extended model with discontinuous softening. Subsequently, no further increase in softening takes place and all load-displacement loops are identical. The situation is different employing continuous softening associated with variable β. The stiffness reduction process occurs at all cycles because loading is continuously changed.

4 ELASTOPLASTIC DAMAGE APPROACH

An investigation of filled rubber at very low strainrates reveals that the material is not frequency-dependent any more, but still a significant amount of damping is found. In this range of loading material properties are amplitude-dependent.

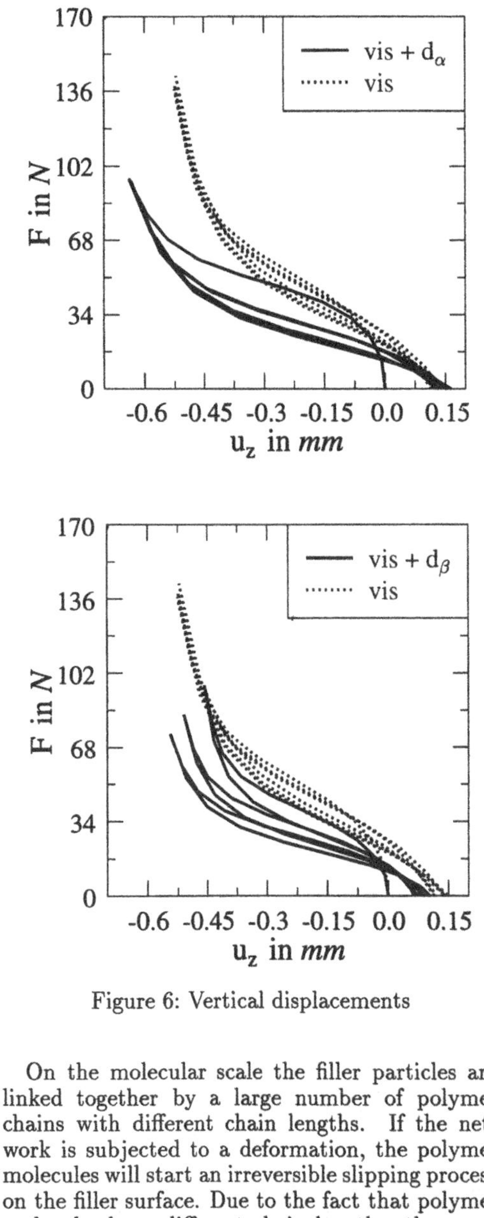

Figure 6: Vertical displacements

On the molecular scale the filler particles are linked together by a large number of polymer chains with different chain lengths. If the network is subjected to a deformation, the polymer molecules will start an irreversible slipping process on the filler surface. Due to the fact that polymer molecules have different chain lengths, the onset of sliding is found at different macroscopic states of deformation. Additionally, filler particles are deformed plastically when the elastic limit is exceeded and the filler-filler contact is also subjected to plastic slipping. Moreover, the elastoplastic process is accompanied by rupture of links between filler particles and polymer chains. For the subsequent model it is assumed that the deformation process is either elastic damage or plastic sliding. The ideal plastic phenomena do not increase the softening of the elastic properties.

These micromechanical considerations motivate the introduction of a generalized Prandtl-model (Figure 7) with softening characteristics in its elastic parts and the application to rubber material. The constitutive approach derived here is represented symbolically by arranging a finite number of elastoplastic Prandtl-elements in parallel with an elastic spring. The frictional material elements have ideal-plastic characteristics. Choosing different yield limits results in a formulation which is capable of simulating the described plastic slip process at different macroscopic strain levels as well as the plastic deformation of the filler particles. Reese & Wriggers (1997), Lion (1997) and Miehe & Keck (1999) address this class of constitutive models among others.

4.1 *General formulation*

The constitutive model is developed from a standard elastoplastic approach as for example described by Miehe & Stein (1992) and Simo (1992). The strain energy function Ψ (Equation (1)) and the yield criterion Φ

$$\mathbb{E}_{\underline{\tau}} := \{\underline{\tau} \in \mathbb{R}^6 \mid \hat{\Phi}(\underline{\tau}) \le 0\} \qquad (18)$$

are fundamental constitutive functions. Employing standard arguments of thermodynamics we get the constitutive relations for the stresses and the reduced dissipation inequality

$$\mathcal{D}^p := \underline{\tau} : \left[-\frac{1}{2}\, \mathcal{L}_v(\underline{b}^e)\, \underline{b}^{e,-1} \right] \ge 0 \qquad (19)$$

where $\mathcal{L}_v(\underline{b}^e)$ is the Lie-derivative of the elastic right Cauchy-Green tensor $\underline{b}^e$. The use of the principle of maximum dissipation leads to the flow rule

$$-\frac{1}{2}\, \mathcal{L}_v(\underline{b}^e)\, \underline{b}^{e,-1} = \lambda\, \frac{\partial \Phi}{\partial \underline{\tau}} \qquad (20)$$

and loading and unloading contitions

$$\hat{\Phi}(\underline{\tau}) \le 0, \quad \lambda \ge 0, \quad \lambda\, \hat{\Phi}(\underline{\tau}) = 0. \qquad (21)$$

According to the generalized Prandtl-element (Figure 7) the material response is determined from the superposition of a finite number of separate rheological elements. Therefore, the nominal isochoric Kirchhoff stress tensor

$$\underline{\tau}_{iso} = (1-d)\,\mathrm{dev}\underline{\tau}_0 + \sum_{j=1}^{M}(1-d_j)\,\delta_j\,\mathrm{dev}\underline{\tau}_{0,j} \qquad (22)$$

is composed of the stress response of the j material elements where δ_j defines the contribution of element j. The elastoplastic damage formulation has to be evaluated for every single element. The flow rule is expressed in effective quantities

$$-\frac{1}{2}\,\mathcal{L}_v(\overline{\underline{\mathbf{b}}}^e_j)\,\overline{\underline{\mathbf{b}}}^{e,-1}_j = \lambda_j\,\frac{\partial\,\Phi_j}{\partial\,\underline{\boldsymbol{\tau}}_j} = \frac{\lambda_j}{(1-d_j)}\,\frac{\partial\,\Phi_j}{\partial\,\underline{\boldsymbol{\tau}}_{0,j}} = \lambda_{0,j}\,\frac{\partial\,\Phi_j}{\partial\,\underline{\boldsymbol{\tau}}_{0,j}} \tag{23}$$

as well as the yield criterion of ideal von Mises plasticity

$$\Phi_j = \|\mathrm{dev}\,\underline{\boldsymbol{\tau}}_{0,j}\| - \sqrt{\frac{2}{3}}\,\sigma_{y,j} \le 0 \tag{24}$$

is used with effective deviatoric Kirchhoff stresses which gives the coupling with the damage formulation. The scalar damage variables

$$d_j = \hat{d}_\alpha(\alpha) + \hat{d}_\beta(\beta) \tag{25}$$

are determined with respect to the elastic state of deformation according to Equation (4) and Equation (5). Fundamentals of this class of elastoplastic damage models are given by Simo & Ju (1987) and Steinmann, Miehe & Stein (1994).

The algorithmic representation of the shown formulation is based on an elastic-plastic-damage operator split. In the following description of the three parts of the algorithm index j for every Prandlt-element is dropped for simplicity. It has to be applied to each material element.

4.2 *Elastic predictor*

The elastoplastic phase of the algorithm is carried out analogously to standard procedures as described for example in Miehe & Stein (1992).

The model is subsequently derived for the specific logarithmic potential function

$$W_0 = \sum_{i=1}^{3} \frac{\mu}{4}\,\ln^2 \overline{\underline{\mathbf{b}}}^e \tag{26}$$

which yields a structure identical for small and finite strains. The first step is to compute the trial state of deformation $\overline{\underline{\mathbf{b}}}^{trial,n+1}$ in order to give

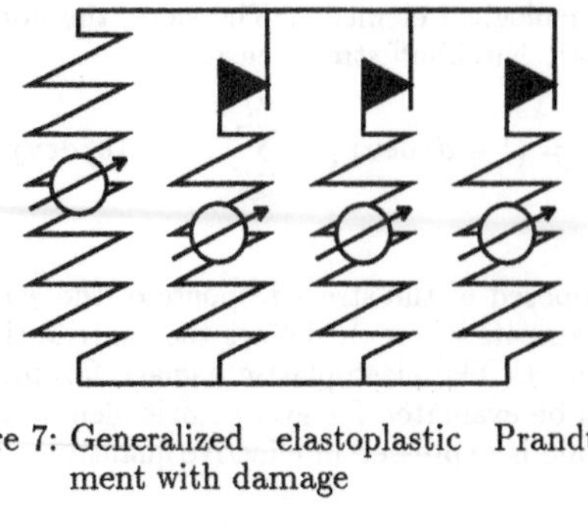

Figure 7: Generalized elastoplastic Prandtl-element with damage

$$\mathrm{dev}\,\underline{\boldsymbol{\tau}}_0^{trial,n+1} = \mu \ln \overline{\underline{\mathbf{b}}}^{trial,n+1} \tag{27}$$

the effective stress predictor.

4.3 *Plastic corrector*

In case that the consistency condition is not satisfied and $\Phi^{n+1} > 0$, a return projection

$$\mathrm{dev}\,\underline{\boldsymbol{\tau}}_0^{n+1} = \mathrm{dev}\,\underline{\boldsymbol{\tau}}_0^{trial,n+1} - 2\mu\,\Delta\lambda_0^{n+1}\,\frac{\partial\,\Phi^{n+1}}{\partial\,\underline{\boldsymbol{\tau}}_0^{n+1}} \tag{28}$$

of the effective stress tensor back to the yield surface is accomplished. The consistency parameter $\Delta\lambda_0^{n+1} = \Delta\lambda^{n+1}/(1-d^{n+1})$ has to be used as an effective quantity. For ideal von Mises plasticity the discrete yield criterion

$$\|\mathrm{dev}\underline{\boldsymbol{\tau}}_0^{trial,n+1}\| - 2\mu\,\frac{\Delta\lambda^{n+1}}{(1-d^{n+1})} - \sqrt{\frac{2}{3}}\,\sigma_y = 0 \tag{29}$$

is evaluated analytically to give finally the update of the effective stress tensor

$$\mathrm{dev}\underline{\boldsymbol{\tau}}_0^{n+1} = \sqrt{\frac{2}{3}}\,\frac{\sigma_y}{\|\mathrm{dev}\underline{\boldsymbol{\tau}}_0^{trial,n+1}\|}\,\mathrm{dev}\underline{\boldsymbol{\tau}}_0^{trial,n+1} = \beta^{n+1}\,\mathrm{dev}\underline{\boldsymbol{\tau}}_0^{trial,n+1} \tag{30}$$

where

$$\beta^{n+1} = \sqrt{\frac{2}{3}}\,\sigma_y\,\|\mathrm{dev}\underline{\boldsymbol{\tau}}_0^{trial,n+1}\|^{-1}. \tag{31}$$

The final step of the plastic correction phase is

$$\overline{\underline{\mathbf{b}}}^{e,n+1} = \exp\left(-2\,\Delta\lambda_0^{n+1}\,\frac{\partial\,\Phi^{n+1}}{\partial\,\underline{\boldsymbol{\tau}}_0^{n+1}}\right)\overline{\underline{\mathbf{b}}}^{trial,n+1} \tag{32}$$

the determination of the elastic state of deformation.

4.4 *Damage corrector*

When the elastic portion of the deformation is known, the isochoric strain energy function W_0 can be evaluated in order to compute the damage quantities

$$d^{n+1} = \hat{d}_\alpha(\alpha^{n+1}) + \hat{d}_\beta(\beta^{n+1}). \tag{33}$$

The elastic stiffness and, subsequently, the stresses are reduced

$$\mathrm{dev}\underline{\boldsymbol{\tau}}^{n+1} = (1-d^{n+1})\,\beta^{n+1}\,\mathrm{dev}\underline{\boldsymbol{\tau}}_0^{trial,n+1} \tag{34}$$

by this third step. The damage corrector phase yields the nominal stress update by applying the damage variable to the effective stress tensor. The final step requires certain history dependent quantities W_0^{n+1}, α^{n+1}, β^{n+1}, $\underline{\mathbf{b}}^{e,n+1}$ which are stored for one time step in a data base.

4.5 *Algorithmic tangent operator*

The algorthmic tangent operator for the shown stress update procedure is obtained from the general formula

$$\underline{\underline{\mathbf{a}}}^{n+1}_{iso} = 2\, \underline{\mathbf{b}}^{trial,n+1} \frac{\partial\, \mathrm{dev}\underline{\tau}^{n+1}}{\partial\, \underline{\mathbf{b}}^{trial,n+1}}\, \underline{\mathbf{b}}^{trial,n+1} \qquad (35)$$

which can be derived analytically. The material tensor is computed from

$$\begin{aligned} \underline{\underline{\mathbf{a}}}^{n+1}_{iso} &= (1 - d^{n+1})\, \beta^{n+1}\, \underline{\underline{\mathbf{a}}}^{trial,n+1}_{0,iso} \\ &- \mathcal{A}\ \mathrm{dev}\underline{\tau}_0^{trial,n+1} \otimes \mathrm{dev}\underline{\tau}_0^{trial,n+1} \\ &+ \mathcal{B}\ \mathrm{dev}\underline{\tau}_0^{trial,n+1} \otimes \mathrm{dev}\underline{\tau}_0^{trial,n+1} \end{aligned} \qquad (36)$$

where $\underline{\underline{\mathbf{a}}}^{trial,n+1}_{0,iso}$ is the isochoric fourth order elasticity tensor for the trial state and $\mathrm{dev}\underline{\tau}_0^{trial,n+1}$ is the accompanying deviatoric stress tensor. Factor $\mathcal{A}$ and $\mathcal{B}$ are determined according to

$$\mathcal{A} = 2\mu\,(1 - d^{n+1})\,\beta^{n+1}\, \|\mathrm{dev}\underline{\tau}_0^{trial,n+1}\|^{-2} \qquad (37)$$

and

$$\mathcal{B} = \frac{\partial\, d^{n+1}}{\partial\, W_0^{n+1}} \left(\frac{4\,\mu\ d^{n+1}\, W_0^{n+1}}{\|\mathrm{dev}\underline{\tau}_0^{trial,n+1}\|^2} - (\beta^{n+1})^3 \right). \qquad (38)$$

The general formulation comprises several special cases. The approach reduces to the elastic model for

$$d^{n+1} = 0.0, \quad \beta^{n+1} = 1.0,$$

$$\mathcal{A} = 0.0, \quad \mathcal{B} = 0.0. \qquad (39)$$

An elastic damage description is obtained using

$$d^{n+1} > 0.0, \quad \beta^{n+1} = 1.0,$$

$$\mathcal{A} = 0.0, \quad \mathcal{B} = \frac{\partial\, d^{n+1}}{\partial\, W_0^{n+1}} \qquad (40)$$

and elastoplasticity is retained for the parameters

$$d^{n+1} = 0.0, \quad \beta^{n+1} < 1.0,$$

$$\mathcal{A} = \frac{2\mu\ \beta^{n+1}}{\|\mathrm{dev}\underline{\tau}_0^{trial,n+1}\|^2}, \quad \mathcal{B} = 0.0. \qquad (41)$$

The shown consistent tangent modulus preserves the quadratic rate of convergence of the Newton iteration.

4.6 *Numerical example*

The cyclic nonhomogeneous shear test (see Section 3.2) is also applied in combination with the elastoplastic damage model. The plots (Figure 8, Figure 9) show the horizontal force F as function of horizontal and vertical displacements. The broken curves (pla) stand for the pure elastoplastic Prandtl-element, while the solid lines are obtained in combination with either of the softening models (pla+d_α, pla+d_β). The discontinuous softening approach yields hysteresis loops with a flatter

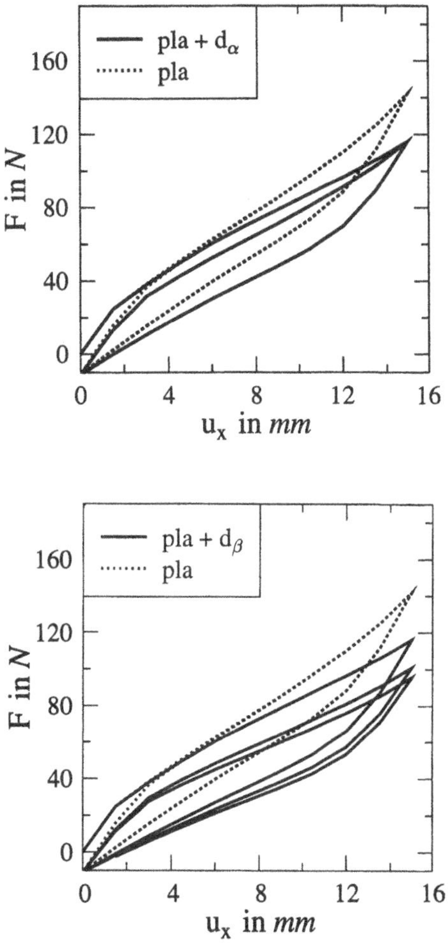

Figure 8: Horizontal displacements

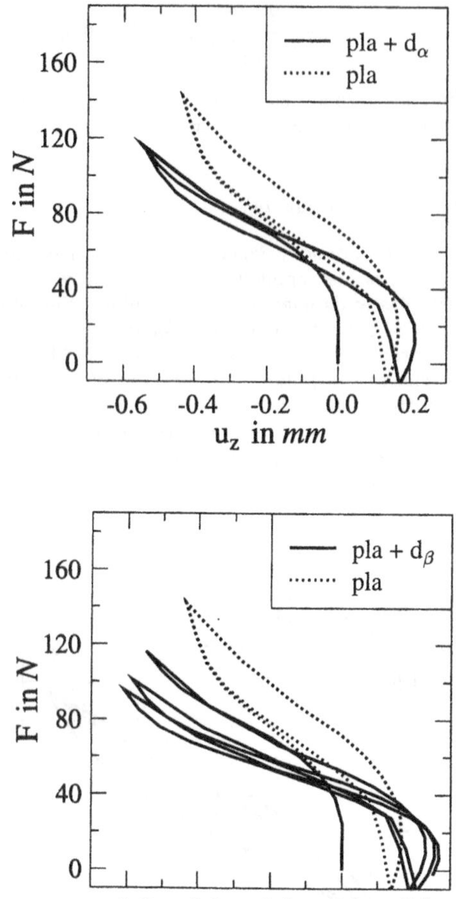

Figure 9: Vertical displacements

slope compared to the elastoplastic model. Softening occurs only at first loading. This is in contrast to continuous damage where the rubber stiffness is reduced during all parts of the numerical experiment. The important difference compared to the viscoelastic formulation (Section 3.2) is the fact that the stiffness and the damping phenomena here are rate-independent and measured even at very low strain rates.

5 CONCLUSIONS

The constitutive behaviour of elastomeric material is very complex even if the isothermal situation is considered. The constitutive models for a characterization of the material properties should be constructed from simple constituents in order to achieve a variable complexity of the formulation depending on the problem of interest. Basic rheological elements are introduced in the presentation. Main focus is on the coupling of viscoelasticity and elastoplasticity with softening, i. e. the reduction of the elastic stiffness. The approaches shown are valid for moderate strain at about 50 – 100%. Very efficient fe-implementations are presented which require no local iteration to solve constitutive equations. Therefore, the constitutive models are computationally inexpensive.

REFERENCES

Chaboche, J. (1981). Continuous damage mechanics. a tool to describe phenomena before crack initiation. *Nuclear Engineering Design 64*, 233–247.

Chaboche, J. (1988). Continuous damage mechanics. Part I: General concepts. Part II: Damage growth, crack initiation, and crack growth. *Journal of Applied Mechnics 55*, 59–72.

Govindjee, S. & J. Simo (1992). Mullins' effect and the strain amplitude dependence of the storage modulus. *International Journal of Solids and Structures 29*, 1737–1751.

Kachanov, L. (1986). *Introduction to Continuous Damage Mechanics.* Martinus Nijhoff Publishers, Dordrecht, Netherlands.

Kaliske, M. & H. Rothert (1997). Formulation and implementation of three-dimensional viscoelasticity at small and finite strains. *Computational Mechanics 19*, 228–239.

Kaliske, M. & H. Rothert (1998). Constitutive approach to rate-independent properties of filled elastomers. *International Journal of Solids and Structures 35*, 2057–2071.

Lemaitre, J. (1984). How to use damage mechanics. *Nuclear Engineering Design 80*, 233–245.

Lion, A. (1997). On the large deformation behaviour of reinforced rubber at different temperatures. *Journal of the Mechanics and Physics of Solids, in press 45*, 1805–1834.

Miehe, C. (1995). Discontinuous and continuous damage evolution in Ogden-type large strain elastic materials. *European Journal of Mechanics, A/Solids 14*, 697–720.

Miehe, C. & J. Keck (1999). Superimposed finite elastic-viscoelastic-plastoelastic stress response with damage in filled rubbery polymers. Experiments, modelling and algorithmic implementation. *Journal of the Mechanics and Physics of Solids, in press.*

Miehe, C. & E. Stein (1992). A canonical model

of multiplicative elasto-plasticity. Formulation and aspects of the numerical implementation. *European Journal of Mechanics, A/Solids 99*, 25–43.

Mullins, L. (1969). Softening of rubber by deformation. *Softening of Rubber by Deformation 42*, 339–362.

Nasdala, L., M. Kaliske, H. Rothert, & A. Becker (1999). A realistic elastic damage model for rubber. In *Proceedings of the First European Conference on Constitutive Models for Rubber.* Wien, Österreich.

Reese, S. & P. Wriggers (1997). A material model for rubber-like polymers exhibiting plastic deformation: Computational aspects and comparison with experimental results. *Computer Methods in Applied Mechanics and Engineering 148*, 279–298.

Simo, J. (1987). On a fully three-dimensional finite-strain viscoelastic damage model: Formulation and computational aspects. *Computer Methods in Applied Mechanics and Engineering 60*, 153–173.

Simo, J. (1992). Algorithms for static and dynamic multiplicative plasticity that preserve the classical return mapping schemes of the infinitesimal theory. *Computer Methods in Applied Mechanics and Engineering 99*, 61–112.

Simo, J. & J. Ju (1987). Strain- and stress-based continuum damage models. I. Formulation. II. Computational aspects. *International Journal for Solids and Structures 23*, 821–869.

Simo, J. & R. Taylor (1991). Quasi-incompressible finite elasticity in principal stretches. Continuum basis and numerical algorithms. *Computer Methods in Applied Mechanics and Engineering 85*, 273–310.

Steinmann, P., C. Miehe, & E. Stein (1994). Comparison of different finite deformation inelastic damage models within multiplicative elastoplasticity for ductile materials. *Computational Mechanics 13*, 458–474.

Constitutive Models for Rubber, Dorfmann & Muhr (eds) © 1999 Taylor & Francis ISBN 90 5809 113 9

A non-Gaussian network alteration model

L.J. Ernst & E.G. Septanika
Delft University of Technology, Netherlands

ABSTRACT: For filled vulcanised rubber materials, the time dependent behaviour for long-duration loading situations is described by extending Tobolsky's network alteration theory to the large deformation range. For this extension the *non*-Gaussian network theory is used as the point of departure. A chain-rupture function and a (different) chain-growth function govern the network alteration. The general reaction laws being applied for these functions are such that creep, relaxation, time dependent flexibility changes as well as permanent set can appropriately be described. A damage function accounts for (short-term) cyclic hysteresis behaviour. A true stress controlled measuring method was developed for an appropriate model parameter characterisation. In order to be able to consider rubber products of arbitrary shape and arbitrary loading situations the constitutive model is implemented in a FEM code. An appropriate approximation method is used to avoid excessive memory and CPU-time consumption due to numerical evaluation of time-integrals involved in the constitutive model.

1 INTRODUCTION

The mechanical behaviour of filler-reinforced vulcanised rubbers depends primarily on extremely complex associations being formed during the fabrication process. A so-called primary (or chemical) bonding of molecular chains is merely attributed to vulcanisation and polymerisation. This chemical bonding process, primarily activated near the processing temperature, becomes inactivated or slowed-down after the cooling down stage. Further a secondary (or physical bonding) is thought to be due to formation of entanglements, and of rubber-filler and filler-filler attachments.

It is assumed that the relatively weak secondary bonding may rupture or slip due to the application of relatively small external loading (Eyring 1943), exerting a phenomenon known as the rubber hysteresis behaviour in short-duration cyclic loading situations (Mullins 1965-1969).

For short term loading situations the stronger primary bonding remains unaltered. The network of bonded chains thus exerts a hyper-elastic behaviour up to extremely large deformations.

However for long term situations, due to continuing attacks of oxygen molecules, especially in combination with a high temperature influence, the primary bonding may suffer from rupture and (re-) formation processes (Tobolsky 1944-1946), Murakami & Ono 1979). This so-called chemical ageing phenomenon affects the long-term mechanical behaviour, exhibiting time-dependent phenomena such as creep and/or relaxation-like processes, permanent-set after the release of loading, and ageing-induced flexibility changes.

In order to account for time-dependent behaviour Tobolsky *et al.* (1944-1946) proposed a molecular-based constitutive model, based on the Gaussian network theory. Many investigators (Lubliner 1985, Wineman et al. 1990-1994 & Drozdov 1997) have further exploited his so-called network alteration theory.

However the applicability of Tobolsky's theory remained restricted to relatively moderate strains and relatively short duration situations since it was based on the Gaussian network theory and because less accurate descriptions of the chain-formation and chain-breakage processes were used. Further time-dependent stiffness variations could not be described since the absolute rates of chain-formation and chain-breakage were chosen to be equal.

A more general approach for modelling time-dependent constitutive behaviour of rubbers, which overcomes these restrictions, was presented by Septanika *et al* (1998c). Here a combination of the network alteration theory with any "hyper-elastic" continuum model. phenomenologically based or molecularly based, was proposed. Besides, for short duration cyclic loading situations the Mullins effect could be described by adding a rubber hysteresis model based on a continuum damage approach similar to the model as proposed by Simo (1987). A combination of the general time-dependent concept and the eight chain non-Gaussian model was described in Septanika *et al* (1998d). A more strict combination of the rubber hysteresis model and the time-dependent theory finally was presented in Septanika *et al* (1998e).

Straightforward application of the time dependent constitutive models as presented in Septanika *et al* (1998c, d, e) within a finite-element theory, could be rather memory and CPU-time consuming. This is caused by the fact that time integrals involved in the model are to be evaluated by numerical integration within each time increment. However, through appropriate approximation of the kernels of these integrals, memory and CPU-time consumption can be considerably reduced. Through this approximation the theory becomes more appropriate for practical simulations of time-dependent behaviour of rubber products.

The present paper especially focuses on the actual finite element implementation and on the approximation of stress integrals involved in the theory as originally proposed in Septanika *et al* (1998c, d, e). Some characteristic example problems will show the validity of the involved approximations.

In order to make this paper self-contained, essential parts of the theory will be recapitulated. In chapter 2 we start out from a well-accepted formulation for short-duration (time-independent) loading situations. Next, in chapter 3, the extension to time-dependent rubber behaviour is discussed. Here the concept of rubber-network alteration is applied. Further evaluation on the bases of the eight chains non-Gaussian network theory is presented in chapter 4. In chapter 5 the reformation process is specified formally by introducing a chain breakage function and a chain formation function. For a selected modified SI rubber, used for experimental verifications, the established model parameters and the selected functions for chain breakage and chain formation are presented in chapter 6. Chapter 7 deals with the finite element formulation and the approximations in the numerical evaluation of time-integrals involved in the iterative FEM-equations. Finally, in chapter 8, some results from numerical simulations concerning long term time-dependent behaviour are compared to experimentally obtained results.

2 TIME IN-DEPENDENT MODELING OF RUBBER BEHAVIOUR

For short-duration (time-independent) loading situations of rubber materials, constitutive relations can be derived from an appropriate expression for the Helmholz free-energy. For the nearly incompressible case and accounting for rubber hysteresis behaviour, through a continuum damage approach (Simo 1987) the (Helmholz) free-energy can be formulated as:

$$\Psi(\mathbf{C}_d, J, \Theta^m) = g(\Theta^m)\Psi_d(\mathbf{C}_d) + \Psi_v(J) \tag{2.1}$$

where $\Psi_d(\mathbf{C}_d)$ and $\Psi_v(J)$ are the deviatoric and volumetric parts, respectively.

$\mathbf{C}$ = the right Cauchy-Green tensor.

$\mathbf{C}_d = J^{-2/3}\mathbf{C}$ = its deviatoric part.

$$J = \sqrt{det(\mathbf{C})} = \text{its Jacobian} \tag{2.2}$$

The damage function $g(x)$ and the damage parameter Θ^m is chosen as (α and β are so-called hysteresis parameters):

$$g(x) = \beta + (1-\beta)\frac{1-e^{-x/\alpha}}{x/\alpha}, \quad \beta \in [0,1], \quad \alpha \in [0,\infty]$$

$$\Theta^m = \max_{s \in (-\infty, t)} \Theta(s) \quad \text{, where} \quad \Theta(s) = \sqrt{2\Psi_d(\mathbf{C}_d(s))} \tag{2.3}$$

The 2nd Piola-Kirchhoff stress tensor $\mathbf{S}$ is defined as:

$$\mathbf{S} = 2\frac{\partial \Psi}{\partial \mathbf{C}} \equiv 2\frac{\partial\{g(\Theta^m)\Psi_d(\mathbf{C}_d) + \Psi_v(J)\}}{\partial \mathbf{C}} \equiv [\mathbf{S}_d^*] + [\mathbf{S}_v]$$

where: (2.4a)

$$\mathbf{S}_v = 2\frac{\partial \Psi_v(J)}{\partial \mathbf{C}} \equiv Jp\mathbf{C}^{-1} \tag{2.4b}$$

$$p = \frac{\partial \Psi_v(J)}{\partial J} \tag{2.4c}$$

$$\mathbf{S}_d^* = 2\frac{\partial g(\Theta^m)}{\partial \mathbf{C}}\Psi_d(\mathbf{C}_d) + g(\Theta^m)\mathbf{S}_d \tag{2.4d}$$

$$\mathbf{S}_d = 2\frac{\partial \Psi_d(\mathbf{C}_d)}{\partial \mathbf{C}} \equiv 2\frac{\partial \Psi_d(\mathbf{C}_d)}{\partial \mathbf{C}_d} : \frac{\partial \mathbf{C}_d}{\partial \mathbf{C}} \tag{2.4e}$$

$\mathbf{S}_d$ represents the deviatoric part of $\mathbf{S}$. The asterisk (*) refers to "modified for hysteresis". The symbol p is the so-called hydrostatic pressure. Through relation (2.4c), generally, a pressure function $w_p(J)$ can be defined such that:

$$p = \frac{\partial \Psi_v(J)}{\partial J} \equiv \kappa \cdot w_p(J) \tag{2.4f}$$

where the so-called bulk modulus κ is supposed to be constant within a large range of volume changes. Matrix/vector notations of tensor components will be frequently used, for example:

$$[\mathbf{S}]^T = [S_{11}\ S_{22}\ S_{33}\ S_{12}\ S_{23}\ S_{31}]$$

$$[\mathbf{C}]^T = [C_{11}\ C_{22}\ C_{33}\ C_{12}\ C_{23}\ C_{31}] \quad \text{etc.} \tag{2.5}$$

With these notations the above stress tensor can be given as follows:

$$[\mathbf{S}] = [\mathbf{S}_d^*] + [\mathbf{S}_v] \tag{2.6a}$$

$$[\mathbf{S}_v] = Jp[\mathbf{C}^{-1}] \tag{2.6b}$$

$$[\mathbf{S}_d^*] = 2\left[\frac{\partial g(\Theta^m)}{\partial \mathbf{C}}\right]\Psi_d + g(\Theta^m)[\mathbf{S}_d] \tag{2.6c}$$

$$[\mathbf{S}_d] = 2\left[\frac{\partial \Psi_d(\mathbf{C}_d)}{\partial \mathbf{C}}\right] = 2\left[\frac{\partial \mathbf{C}_d}{\partial \mathbf{C}}\right]\left[\frac{\partial \Psi_d(\mathbf{C}_d)}{\partial \mathbf{C}_d}\right] \tag{2.6d}$$

3 TIME DEPENDENT MODELING OF RUBBER BEHAVIOUR

To describe long-term mechanical behaviour, it is now assumed, that the material fractions are subject to a re-formation process, where the microstructure is replaced with a new different microstructure. In the sequence we will distinguish between the "original microstructure", state ϕ_0 and "microstructures being reformed within certain time intervals, the states $\phi_{\tau i}$. Characteristics of these microstructures are given below:

original micro structure time t_0	state ϕ_0
radius vector at time t_0	$\vec{\mathbf{X}}$
radius vector at time t	$\vec{\mathbf{x}}(t)$
(overall) deformation gradient at time t	$\mathbf{F}_{\phi_0} = \mathbf{F}(t,t_0) = \partial\vec{\mathbf{x}}(t)/\partial\vec{\mathbf{X}}$
right Gauchy Green tensor at time t	$\mathbf{C}(t,t_0) = \mathbf{F}^T(t,t_0)\cdot\mathbf{F}(t,t_0)$
the deviatoric part	$\mathbf{C}_d(t,t_0) = J(t,t_0)^{-2/3}\mathbf{C}(t,t_0)$
the Jacobian	$J(t,t_0) = \sqrt{det(\mathbf{C}(t,t_0))}$

(3.1)

new micro structure formed within time interval $\tau_i \in [t_{i-1}, t_i]$	state ϕ_{τ_i}
radius vector at time τ_i	$\vec{\mathbf{x}}(\tau_i)$
radius vector at time t	$\vec{\mathbf{x}}(t)$
(overall) deformation gradient at time τ_i	$\mathbf{F}(\tau_i,t_0) = \partial\vec{\mathbf{x}}(\tau_i)/\partial\vec{\mathbf{X}}$
relative deformation gradient at time t	$\mathbf{F}_{\phi_i} = \mathbf{F}_r(t,\tau_i) = \partial\vec{\mathbf{x}}(t)/\partial\vec{\mathbf{x}}(\tau_i)$ $\equiv \mathbf{F}(t,t_0)\cdot\mathbf{F}^{-1}(\tau_i,t_0)$
relative right Gauchy Green tensor at time t	$\mathbf{C}_r(t,\tau_i) = \mathbf{F}_r^T(t,\tau_i)\cdot\mathbf{F}_r^T(t,\tau_i)$ $\equiv \mathbf{F}^{-T}(\tau_i,t_0)\cdot\mathbf{C}(t,t_0)\cdot\mathbf{F}^{-1}(\tau_i,t_0)$
the deviatoric part	$\mathbf{C}_{rd}(t,\tau_i) = J_r(t,\tau_i)^{-2/3}\mathbf{C}_r(t,\tau_i)$
the Jacobian	$J_r(t,\tau_i) = \sqrt{det(\mathbf{C}_r(t,\tau_i))}$ $\equiv \sqrt{det(\mathbf{C}(t,t_0))}/\sqrt{det(\mathbf{C}(\tau_i,t_0))}$

(3.2)

Note that a "relative" right Gauchy-Green tensor is introduced because of the assumption that the newly formed microstructure is stress free at the moment of nucleation. (The subscript r denotes "relative".)

It is now stated that the deviatoric part of the (Helmholtz) free-energy can be made up of an unreformed fraction (from the original micro-structure) plus the summation of fractions of microstructures which are reformed at subsequent time intervals:

$$\Psi_d = (\Psi_d)_{\phi_0} + \sum_{i=1}^{N}(\Psi_d)_{\phi_{\tau_i}} =$$
$$= \Psi_d(\mathbf{C}_d(t,t_0)) + \sum_{i=1}^{N}\Psi_{rd}(\mathbf{C}_{rd}(t,\tau_i)) \quad \text{where} \quad \tau_i \in [t_{i-1}, t_i] \tag{3.3}$$

Through substitution into expression (2.4) the 2nd Piola-Kirchhoff stress tensor at time t is straightforwardly found. The time-dependent equivalents of the stress expressions (2.6) are thus found as:

$$[\mathbf{S}(t)] = [\mathbf{S}_d^*(t)] + [\mathbf{S}_v(t)] \tag{3.4a}$$

$$[\mathbf{S}_v(t)] = J(t)p(t)[\mathbf{C}^{-1}(t)] \tag{3.4b}$$

$$[\mathbf{S}_d^*(t)] = [\mathbf{S}_d^*(t)]_{\phi_0} + \sum_{i=1}^{N}[\mathbf{S}_d^*(t)]_{\phi_{\tau_i}} \tag{3.4c}$$

$$[\mathbf{S}_d^*(t)]_{\phi_0} = 2\left[\frac{\partial \bar{g}(\Theta^m)}{\partial \mathbf{C}}\right][\Psi_d(\mathbf{C}_d(t,t_0))] + 2\bar{g}(\Theta^m)\left[\frac{\partial \Psi_d(\mathbf{C}_d(t,t_0))}{\partial \mathbf{C}}\right] \tag{3.4d}$$

$$[\mathbf{S}_d^*(t)]_{\phi_{\tau_i}} = 2\left[\frac{\partial \bar{g}(\Theta^m)}{\partial \mathbf{C}}\right][\Psi_{rd}(\mathbf{C}_{rd}(t,\tau_i))] + 2\bar{g}(\Theta^m)\left[\frac{\partial \Psi_{rd}(\mathbf{C}_{rd}(t,\tau_i))}{\partial \mathbf{C}}\right] \tag{3.4e}$$

Further evaluation of the above stress matrixes and the derivation of the time-dependent moduli matrix based on these relations is given in Septanika *et al* (1998b,c).

4 APPLICATION TO THE EIGHT CHAINS NON-GAUSSIAN NETWORK THEORY

The time-dependent theory as presented in chapter 3 can be used together with any "hyper-elastic" continuum model, phenomenologically based or molecularly based. In order to allow (very) large deformations, in the sequence, the eight chains non-Gaussian model is selected for further evaluation. According to this theory, the deviatoric part of the (Helmholz) free-energy can be specified as follows (for the time in-dependent and compressible case) (Aruda *et al* 1993 & Wu *et al* 1992):

$$\Psi_d(\mathbf{C}_d) = C^R N\left(\frac{\hat{\lambda}_c}{\sqrt{N}}\hat{\beta} + \ln\frac{\hat{\beta}}{\sinh\hat{\beta}}\right) \tag{4.1a}$$

where:

$$\hat{\beta} = \mathcal{L}^{-1}\left(\frac{\hat{\lambda}_c}{\sqrt{N}}\right) \tag{4.1b}$$

$$\mathcal{L}(x) = \coth(x) - 1/x \qquad \text{(=Langevin function)} \tag{4.1c}$$

$$\hat{\lambda}_c = (I_{1(C_d)}/3)^{1/2} = \frac{J^{-1/3}}{\sqrt{3}}\left\{(\lambda_{(1)})^2 + (\lambda_{(2)})^2 + (\lambda_{(3)})^2\right\}^{1/2} \tag{4.1d}$$

n = number of chains / unit volume

N = number of rigid links in a chain

$\lambda_{(i)}$ = principle stretch

$C^R = nkT$ = rubber modulus:

k = Boltzmann's constant

T = absolute temperature

Through substitution of (4.1a) into (2.6d), following expression for the deviatoric part of the 2nd Piola-Kirchhoff stress tensor is obtained:

$$[\mathbf{S}_d] = 2\left[\frac{\partial \mathbf{C}_d}{\partial \mathbf{C}}\right]\left[\frac{\partial \Psi_d(\mathbf{C}_d)}{\partial \mathbf{C}_d}\right] =$$

$$= 2\left[\frac{\partial \mathbf{C}_d}{\partial \mathbf{C}}\right]\left[\frac{\partial \Psi_d(\mathbf{C}_d)}{\partial \hat{\lambda}_c}\frac{\partial \hat{\lambda}_c}{\partial \mathbf{C}_d}\right] \equiv \frac{1}{3}\sqrt{N}\frac{\hat{\beta}}{\hat{\lambda}_c}\left[\frac{\partial \mathbf{C}_d}{\partial \mathbf{C}}\right][I] \tag{4.2}$$

where: $[I]^T = [1 \;\; 1 \;\; 1 \;\; 0 \;\; 0 \;\; 0]$

For the time-dependent case, in chapter 3, we have assumed, that due to the alteration process, the deviatoric part of the (Helmholtz) free-energy can be made up of an unreformed fraction (from the original micro-structure) plus the summation of fractions of microstructures which are reformed at subsequent time intervals. According to (3.3) and (4.1a) we thus find:

$$\Psi_d = (\Psi_d)_{\phi_0} + \sum_{i=1}^{N}(\Psi_d)_{\phi_{\tau_i}} \qquad \tau_i \in [t_{i-1}, t_i] \tag{4.3a}$$

where

$$(\Psi_d)_{\phi_0} = \Psi_d(\mathbf{C}_d(t,t_0)) = C_0^R(t)N\left(\frac{\hat{\lambda}_c}{\sqrt{N}}\hat{\beta} + \ln\frac{\hat{\beta}}{\sinh\hat{\beta}}\right) \tag{4.3b}$$

$$(\Psi_d)_{\phi_{\tau_i}} = \Psi_{rd}(\mathbf{C}_{rd}(t,\tau_i)) =$$

$$= C_i^R(t)\sqrt{N}\left[\frac{\hat{\lambda}_{rc}(t,\tau_i)}{\sqrt{N}}\hat{\beta}(t,\tau_i) + \ln\frac{\hat{\beta}(t,\tau_i)}{\sinh\hat{\beta}(t,\tau_i)}\right] \tag{4.3c}$$

$$C_0^R(t) = n_0(t)kT \tag{4.3d}$$

$$C_i^R(t) = n_i(t - t_i)kT \tag{4.3e}$$

$n_0(t)$ = number of (original) chains per unit volume, that are left at time t.

$n_i(t - t_i)$ = number of chains per unit volume, newly formed at time interval $\tau_i \in [t_{i-1}, t_i]$, that are left at time t.

Through substitution of this expressions for the (Helmholz) free energy (4.3) into the (time-dependent) stress-expressions (3.4), the 2nd Piola-Kirchhoff stress tensor can be specified as follows:

$$[\mathbf{S}(t)] = [\mathbf{S}_d^*(t)] + [\mathbf{S}_v(t)] \tag{4.4a}$$

$$[\mathbf{S}_v(t)] = J(t)p(t)[\mathbf{C}^{-1}(t)] \tag{4.4b}$$

$$[\mathbf{S}_d^*(t)] = [\mathbf{S}_d^*(t)]_{\phi_0} + \sum_{i=1}^{N}[\mathbf{S}_d^*(t)]_{\phi_{\tau_i}} \tag{4.4c}$$

$$[\mathbf{S}_d^*(t)]_{\phi_0} = 2\left[\frac{\partial \bar{g}(\Theta^m)}{\partial \mathbf{C}}\right]C_0^R(t)N\left(\frac{\hat{\lambda}_c}{\sqrt{N}}\hat{\beta} + \ln\frac{\hat{\beta}}{\sinh\hat{\beta}}\right) +$$

$$+ \bar{g}(\Theta^m)\frac{1}{3}C_0^R(t)\sqrt{N}\frac{\hat{\beta}}{\hat{\lambda}_c}\left[\frac{\partial \mathbf{C}_d}{\partial \mathbf{C}}\right][I] \tag{4.4d}$$

$$[\mathbf{S}_d^*(t)]_{\phi_{\tau_i}} =$$

$$= 2\left[\frac{\partial \bar{g}(\Theta^m)}{\partial \mathbf{C}}\right]C_i^R(t)\sqrt{N}\left[\frac{\hat{\lambda}_{rc}(t,\tau_i)}{\sqrt{N}}\hat{\beta}_r(t,\tau_i) + \ln\frac{\hat{\beta}_r(t,\tau_i)}{\sinh\hat{\beta}(t,\tau_i)}\right] +$$

$$+ \bar{g}(\Theta^m)\frac{1}{3}C_i^R(t)\sqrt{N}\frac{\hat{\beta}_r(t,\tau_i)}{\hat{\lambda}_{rc}(t,\tau_i)}\left[\frac{\partial \mathbf{C}_{rd}(t,\tau_i)}{\partial \mathbf{C}}\right][I]$$

where (4.4e)

$$\hat{\lambda}_{rc}(t,\tau_i) = J^{-1/3}(t,\tau)\sqrt{\frac{1}{3}tr[C(t)C^{-1}(\tau_i)]} \tag{4.4f}$$

$$\hat{\beta}_r(t,\tau_i) = \mathcal{L}^{-1}\left(\frac{\hat{\lambda}_{rc}(t,\tau_i)}{\sqrt{N}}\right) \tag{4.4g}$$

It should be noted that the expressions for the deviatoric part of the 2nd Piola-Kirchhoff stress tensor, (4.4c-4.4e), are evaluated further in chapter 5.

5 CHAIN BREAKAGE AND CHAIN FORMATION RELATIONS

The reformation process can be specified by formal description of the chain breakage and chain formation phenomena. According to the network alteration theory (Tobolsky 1944-1946), the number of (original) chains per unit volume that remains at time t, can be represented by:

$$n_0(t) = n \cdot R(t) \tag{5.1}$$

$R(t)$ = the chain breakage or relaxation function

The number of chains being formed within the time interval $\tau_i \in [t_{i-1}, t_i]$ can be modelled as:

$$n_{\phi_{\tau_i}} = n\int_{t_{i-1}}^{t_i}\frac{\partial P(\tau)}{\partial \tau}d\tau \tag{5.2}$$

$P(\tau)$ = the chain formation or chain growth function

Due to continuing chain breakage, a fraction of the newly formed chains again are broken in the period between time t_i and time t. It is assumed that the reduction of newly formed chains is also governed by $R(t)$. The number of chains per unit volume, originally formed at time interval $\tau_i \in [t_{i-1}, t_i]$, that remains at time t can thus be modelled by:

$$n_i(t-t_i) = n_{\phi_{\tau_i}} \cdot R(t-t_i) \tag{5.3}$$

Substitution of (5.1) into (4.3d) and using the definition for the rubbery modulus $C^R = nkT$ yields:

$$C_0^R(t) = C^R R(t) \tag{5.4}$$

Substitution of (5.2) and (5.3) into (4.3e) and again using the definition of the rubbery modulus, $C^R = nkT$ yields:

$$C_i^R(t) = C^R \int_{ti-1}^{ti} \frac{\partial P(\tau)}{\partial \tau} d\tau \cdot R(t-t_i) \tag{5.5}$$

Substitution of (5.4) and (5.5) into expressions (4.4c-4.4e) gives the final formulation of the deviatoric part of the 2nd Piola-Kirchhoff stress tensor:

$$\begin{aligned}[\mathbf{S}_d^*(t)] &= [\mathbf{S}_d^*(t)]_{\phi_0} + \sum_{i=1}^{N} \int_{ti-1}^{ti} [\mathbf{S}_d^*(t)]_{\phi_{\tau i}} d\tau = \\ &\equiv [\mathbf{S}_d^*(t)]_{\phi_0} + [\mathbf{S}_d^*(t)]_{\mathbf{\Phi}}\end{aligned} \tag{5.6a}$$

where

$$\begin{aligned}[\mathbf{S}_d^*(t)]_{\phi_0} &= 2\left[\frac{\partial \bar{g}(\Theta^m)}{\partial \mathbf{C}}\right] C^R N\left(\frac{\hat{\lambda}_c}{\sqrt{N}}\hat{\beta} + \ln\frac{\hat{\beta}}{\sinh\hat{\beta}}\right) R(t) + \\ &+ \bar{g}(\Theta^m)\frac{1}{3}C^R\sqrt{N}\frac{\hat{\beta}}{\hat{\lambda}_c}\left[\frac{\partial \mathbf{C}_d}{\partial \mathbf{C}}\right][I] R(t)\end{aligned} \tag{5.6b}$$

$$\begin{aligned}&[\mathbf{S}_d^*(t)]_{\mathbf{\Phi}} = \\ &= 2\left[\frac{\partial \bar{g}(\Theta^m)}{\partial \mathbf{C}}\right] C^R\sqrt{N}\int_0^t \left[\frac{\hat{\lambda}_{rc}(t,\tau)}{\sqrt{N}}\hat{\beta}_r(t,\tau) + \ln\frac{\hat{\beta}_r(t,\tau)}{\sinh\hat{\beta}(t,\tau)}\right]\frac{\partial P(\tau)}{\partial \tau} R(t-\tau)d\tau + \\ &+ g(\Theta^m)\frac{1}{3}C^R\sqrt{N}\int_0^t \frac{\hat{\beta}_r(t,\tau)}{\hat{\lambda}_{rc}(t,\tau)}\left[\frac{\partial \mathbf{C}_{rd}(t,\tau)}{\partial \mathbf{C}}\right][I]\frac{\partial P(\tau)}{\partial \tau} R(t-\tau)d\tau\end{aligned} \tag{5.6c}$$

Due to the differences in time scales, for short term cyclic loading situations the integrals in the above expressions often can be dropped, while for many long term loading cases, even when combined with short term load-variations, the damage terms (the first terms in expressions (5.6) can be neglected (while assuming $g(\Theta^m)$ to be steady).

6 MATERIAL PARAMETER SPECIFICATION

Model evaluations and verifications were performed for test specimen made up from a specially selected modified SI-rubber (without anti-oxidant components) at a temperature of 80 ^{0}C. The model parameters, as shown in the following table, were obtained from special true stress controlled experiments (Septanika *et al* ,1998a, b):

Hyper-elasticity constants	C^R=0.875 ; N=72
hysteresis parameters	α=4.5 ; β=0.2
Chain-growth function	$R(t)=0.026\,e^{-1.616\,t}+0.038\,e^{-0.1325\,t}$ $0.022\,e^{-0.0743\,t}+0.84\,e^{-0.00022\,t}+0.083\,e^{-0.005\,t}$
Chain breakage function	$P(t)=0.0818+(2.1\cdot10^{-4})\,t-(2.41\cdot10^{-8})\,t^2+(1.01\cdot10^{-12})\,t^3$

7 FINITE ELEMENT IMPLEMENTATION

7.1 *Finite element framework*

For numerical simulation purposes the constitutive model has been implemented into the DIANA finite element package, using a formulation similar to the formulation for the time independent case as implemented before by Van den Bogert (1991). First in section 7.2 relations for the increments in stress, strain, pressure and time are being prepared. Next, in section 7.3, the incremental finite element procedure is being derived. In section 7.4, approximations in the numerical evaluation of time-integrals involved in the iterative FEM-equations are discussed.

7.2 *Increments in stress and strain*

Starting out from the FEM formulation as employed in DIANA, for the time-independent case, by Van den Bogert (1991), the Lagrangian strain tensor, E, will be used as the primary measure of strain. This strain measure is related to the right Gauchy-Green tensor through:

$$\boldsymbol{E} = \frac{1}{2}(\mathbf{C} - \boldsymbol{I}) = \boldsymbol{E}_L + \boldsymbol{E}_{NL} \qquad \mathbf{C} = \mathbf{F}^T\mathbf{F} \tag{7.1a}$$

$$[\boldsymbol{E}_L]_{ij} = (u_{i,j} + u_{j,i})/2 \tag{7.1b}$$

$$[\boldsymbol{E}_{NL}]_{ij} = u_{k,i}\ u_{k,j}/2 \tag{7.1c}$$

Here $\boldsymbol{E}_L$ and $\boldsymbol{E}_{NL}$ denote the linear and non-linear parts respectively. (u_k represent the displacement components with respect to the base vectors in the undeformed state) $\boldsymbol{F}$ represents the deformation gradient tensor and $\boldsymbol{I}$ is the 2nd order unit tensor.

Increments of the Lagrangian strain tensor are:

$$\Delta\boldsymbol{E} = \Delta\boldsymbol{E}_L + \Delta\boldsymbol{E}_{NL} \tag{7.2a}$$

$$[\Delta\boldsymbol{E}_L]_{ij} = (\Delta u_{i,j} + \Delta u_{j,i} + \Delta u_{k,i}\cdot u_{k,j} + u_{k,i}\ \Delta u_{k,j})/2 \tag{7.2b}$$

$$[\Delta E_{NL}]_{ij} = \Delta u_{k,i} \cdot \Delta u_{k,j} / 2 \qquad (7.2c)$$

Focussing on long term loading cases, here for simplicity, in the sequence the damage terms in expressions 4.4a and 5.6 are neglected (The FEM implementation of the damage terms is described in Septanika 1998b). The time dependent stress tensor is thus given as:

$$[\mathbf{S}(t)] = [\mathbf{S}_d(t)] + [\mathbf{S}_v(t)] \qquad (7.3a)$$

with:

$$[\mathbf{S}_v(t)] = J(t)p(t)[\mathbf{C}^{-1}(t)] \qquad (7.3b)$$

$$[\mathbf{S}_d(t)] = [\mathbf{S}_d(t)]_{\phi_0} + [\mathbf{S}_d(t)]_{\mathbf{\Phi}} \qquad (7.3c)$$

$$[\mathbf{S}_d(t)]_{\phi_0} = \frac{1}{3} C^R \sqrt{N} \frac{\hat{\beta}}{\hat{\lambda}_c} \left[\frac{\partial \mathbf{C}_d}{\partial \mathbf{C}}\right][I]R(t) \qquad (7.3d)$$

$$[\mathbf{S}_d(t)]_{\mathbf{\Phi}} = \frac{1}{3} C^R \sqrt{N} \int_0^t \frac{\hat{\beta}_r(t,\tau)}{\hat{\lambda}_{rc}(t,\tau)} \left[\frac{\partial \mathbf{C}_{rd}(t,\tau)}{\partial \mathbf{C}}\right][I] \frac{\partial P(\tau)}{\partial \tau} R(t-\tau)d\tau$$

(7.3e)

It should be noted that within the DIANA program, for nearly incompressible materials, the hydrostatic pressure p is treated as an independent variable (Van den Bogert 1991). The total stress increment is thus considered with respect to the variables $\boldsymbol{E}$, t and p:

$$[\Delta \mathbf{S}] = [\Delta \mathbf{S}_d] + [\Delta \mathbf{S}_v] \qquad (7.4a)$$

where

$$[\Delta \mathbf{S}_d] = 2\left[\frac{\partial \mathbf{S}_d}{\partial \mathbf{C}}\right][\Delta \boldsymbol{E}] + [\dot{\mathbf{S}}_d]\Delta t \qquad (7.4b)$$

$$[\Delta \mathbf{S}_v] = 2\left[\frac{\partial \mathbf{S}_v}{\partial \mathbf{C}}\right][\Delta \boldsymbol{E}] + \left[\frac{\partial \mathbf{S}_v}{\partial p}\right]\Delta p + [\dot{\mathbf{S}}_v]\Delta t \qquad (7.4c)$$

or alternatively:

$$[\Delta \mathbf{S}] = [\boldsymbol{M}][\Delta \boldsymbol{E}] + [\boldsymbol{d}]\Delta p + [\dot{\mathbf{S}}]\Delta t \qquad (7.5a)$$

where

$$[\boldsymbol{M}] = 2\left(\left[\frac{\partial \mathbf{S}_d}{\partial \mathbf{C}}\right] + \left[\frac{\partial \mathbf{S}_v}{\partial \mathbf{C}}\right]\right) \qquad (7.5b)$$

$$[\boldsymbol{d}] = \left[\frac{\partial \mathbf{S}_v}{\partial p}\right] = J[\mathbf{C}^{-1}] \qquad \left(\equiv 2\frac{\partial J}{\partial \mathbf{C}}\right) \qquad (7.5c)$$

$$[\dot{\mathbf{S}}] = \left[\frac{\partial \mathbf{S}_d}{\partial t}\right] + \left[\frac{\partial \mathbf{S}_v}{\partial t}\right] \qquad (7.5d)$$

Here $[\boldsymbol{M}]$ and $[\boldsymbol{d}]$ represent the (time-dependent) material moduli and the volumetric stiffness matrixes, respectively.

7.3 Incremental Finite Element procedure

In the non-linear finite element procedure an incremental iterative approach is employed. In each increment an equilibrium solution is being estimated by using an appropriate iteration scheme. $[\mathbf{S}^{(t)}]$ denotes the converged solution for the "previous" time. For the subsequent time, $t+\Delta t$, the following virtual work condition must be satisfied:

$$\int_{Vo} [\delta(\Delta \boldsymbol{E})]^T [\mathbf{S}^{(t+\Delta t)}] dV_o = \int_{Vo} \rho_o \delta(\Delta u)^T [f] dV_o + \int_{Ao} \rho_o \delta(\Delta u)^T [t] dA_o$$

(7.6)

Here $[f]$ and $[t]$ represent the body force and surface traction fields, while ρ_o is the density in the initially undeformed state. Assuming the additive decomposition $\mathbf{S}^{(t+\Delta t)} = \mathbf{S}^{(t)} + \Delta \mathbf{S}$, for an iteration at time $t+\Delta t$ the following scheme is used:

$$[\mathbf{S}^{(j),(t+\Delta t)}] = [\mathbf{S}^{(i),(t+\Delta t)}] + \Delta \mathbf{S} \quad \text{with} \quad j=i+1 \qquad (7.7)$$

while according to 7.5a:

$$[\Delta \mathbf{S}] = [\boldsymbol{M}^{(t+\Delta t)}][\Delta \boldsymbol{E}] + [\boldsymbol{d}^{(t+\Delta t)}]\Delta p + [\dot{\mathbf{S}}^{(t+\Delta t)}]\Delta t \qquad (7.8)$$

Substitution of expressions 7.2a and 7.8 for the strain and stress increments into the virtual work expression 7.6 and subsequent linearization yields:

$$\begin{aligned}
&\int_{Vo} [\delta(\Delta \boldsymbol{E}_L)]^T \left([\boldsymbol{M}^{(i),(t+\Delta t)}][\Delta \boldsymbol{E}_L] + [\boldsymbol{d}^{(i),(t+\Delta t)}][\Delta p]\right) dV_o + \\
&+ \int_{Vo} [\delta(\Delta \boldsymbol{E}_{NL})]^T \left([S^{(i),(t+\Delta t)}] + [\dot{S}^{(i),(t+\Delta t)} \Delta t]\right) dV_o = \\
&= \int_{Vo} \rho_o \delta(\Delta u)^T [f] dV_o + \int_{Ao} \rho_o \delta(\Delta u)^T [t] dA_o + \\
&- \int_{Vo} [\delta(\Delta \boldsymbol{E}_L)]^T \left([S^{(i),(t+\Delta t)}] + [\dot{S}^{(i),(t+\Delta t)} \Delta t]\right) dV_o
\end{aligned} \qquad (7.9)$$

Since the hydrostatic pressure p is treated as an independent variable, additional (constraint) equations are formulated to ensure a stress evolution according to the applied hydrostatic model (as earlier defined in expression 2.4f). Therefore, the hydrostatic pressure at time $t+\Delta t$, given the volumetric stress-strain relation 2.4f, is decomposed according to:

$$p^{(t+\Delta t)} = p^{(t)} + \Delta p = \kappa \cdot w_p(J^{(t)}) + 2\kappa \cdot w'_p(J^{(t)}) \cdot \left[\left(\frac{\partial J}{\partial \mathbf{C}}\right)^{(t)}\right]^T [\Delta \boldsymbol{E}]$$

or, when invoking 7.5c: (7.10a)

$$p^{(t)} + \Delta p = \kappa \cdot w_p^{(t)} + \kappa \cdot w'^{(t)}_p \cdot [\boldsymbol{d}^{(t)}]^T [\Delta \boldsymbol{E}] \qquad (7.10b)$$

where the prime denotes differentiation with respect to J. The weak form of this relation reads:

$$\int_{Vo} \delta p\left[\kappa \cdot w_p^{(t)} + \kappa \cdot w_p'^{(t)} \cdot \left[d^{(t)}\right]^T [\Delta E] - p^{(t)} - \Delta p\right] dV_o \quad (7.11)$$

Shape functions for the displacement and pressure fields are formally introduced by:

$$u = [H]\{a\} \qquad p = [N]\{p\} \quad (7.12)$$

where $[H]$ and $[N]$ represent the so-called interpolation matrixes, $\{a\}$ is the nodal displacement vector and $\{p\}$ the vector of pressures in the element-pressure nodes. Expressions 7.2 for the Lagrangian strain tensor are discretised through substitution of the shape functions:

$$\Delta E = \Delta E_L + \Delta E_{NL} \quad (7.13a)$$

with:

$$\Delta E_L = [B_L]\{\Delta a\} \quad (7.13b)$$

$$\Delta E_{NL} = \frac{1}{2}\{\Delta a\}^T [B_{NL}]^T [B_{NL}]\{\Delta a\} \quad (7.13c)$$

Here the matrixes $[B_L]$ and $[B_{NL}]$ are derived from the shape functions $[H]$ and its derivatives. Through substitution of the above discretised strains and the shape functions 7.12, the virtual work equation 7.9 and the weak form of the volumetric stress-strain relation, 7.11, and applying the standard procedure of calculus of variations the following equations are obtained:

$$\left[\int_{Vo}\left\{[B_L]^T\left[M^{(i),(t+\Delta t)}\right][B_L] + [B_{NL}]^T\left[\tilde{M}^{(i),(t+\Delta t)}\right][B_{NL}]\right\}dV_o\right]\{\Delta a^{(j)}\} +$$
$$\left[\int_{Vo}[B_L]^T\left[d^{(i),(t+\Delta t)}\right][N]dV_o\right]\{\Delta p^{(j)}\} =$$
$$= \int_{Vo}\rho_o[H]^T[f]dV_o + \int_{Ao}\rho_o[H]^T[t]dA_o - \int_{Vo}[B_L]^T\left[\tilde{S}^{(i),(t+\Delta t)}\right]dV_o$$

with: (7.14)

$$\left[\tilde{S}^{(i),(t+\Delta t)}\right] = \left[S^{(i),(t+\Delta t)}\right] + \left[\dot{S}^{(i),(t+\Delta t)}\right]\Delta t \quad (7.15)$$

and:

$$\left[\int_{Vo}[B_L]^T\left[d^{(i),(t)}\right][N]dV_o\right]^T\{\Delta a^{(j)}\} +$$
$$\kappa^{-1}\left[-\int_{Vo}\left(w_p'^{(i),(t)}\right)^{-1}[N]^T[N]\,dV_o\right]\{\Delta p^{(j)}\} = \quad (7.16)$$
$$= \int_{Vo}\left(w_p'^{(t)}\right)^{-1}[N]^T\left(w_p'^{(t)} - \kappa^{-1}p^{(t)}\right)dV_o$$

These equations can be expressed as a single matrix equation as follows:

$$\begin{bmatrix} K_{aa} & K_{ap} \\ K_{pa} & \kappa^{-1}K_{pp}\end{bmatrix}\begin{bmatrix}\Delta a^{(j)} \\ \Delta p^{(j)}\end{bmatrix} = \begin{bmatrix}F_a \\ F_p\end{bmatrix} \quad (7.17a)$$

where:

$$K_{aa} = \int_{Vo}\left\{[B_L]^T\left[M^{(i),(t+\Delta t)}\right][B_L] + [B_{NL}]^T\left[\tilde{M}^{(i),(t+\Delta t)}\right][B_{NL}]\right\}dV_o \quad (7.17b)$$

$$K_{ap} = K_{pa}^T = \int_{Vo}[B_L]^T\left[d^{(i),(t+\Delta t)}\right][N]dV_o \quad (7.17c)$$

$$K_{pp} = -\int_{Vo}\left(w_p'^{(i),(t)}\right)^{-1}[N]^T[N]\,dV_o \quad (7.17d)$$

$$F_a = \int_{Vo}\rho_o[H]^T[f]dV_o + \int_{Ao}\rho_o[H]^T[t]dA_o - \int_{Vo}[B_L]^T\left[\tilde{S}^{(i),(t+\Delta t)}\right]dV_o \quad (7.17e)$$

$$F_p = \int_{Vo}\left(w_p'^{(t)}\right)^{-1}[N]^T\left(w_p'^{(t)} - \kappa^{-1}p^{(t)}\right)dV_o \quad (7.17f)$$

In order to reduce the number of equations and thus to save memory and processing time, the pressure degrees of freedom are usually eliminated at element level. In order to avoid spurious stress patterns, as a consequence of this elimination, a force norm, $\Delta\{[F_a]^T[F_a]<\varepsilon)$, is selected within the iteration process. After the elimination process the modified incremental equations read:

$$\left[\tilde{K}_{aa}^{(i)}\right]\{\Delta a^{(j)}\} = \{\tilde{F}_a^{(i)}\} \quad (7.18a)$$

where:

$$\left[\tilde{K}_{aa}^{(i)}\right] = \left[K_{aa}^{(i)}\right] - \kappa\left[K_{ap}^{(i)}\right]\left[K_{pp}^{(i)}\right]^{-1}\left[K_{pa}^{(i)}\right] \quad (7.18b)$$

$$\{\tilde{F}_a\} = \{F_a\} - \kappa\left[K_{ap}^{(i)}\right]\left[K_{pp}^{(i)}\right]^{-1}\{F_p\} \quad (7.18c)$$

The iteration process is terminated on the bases of a selected convergence criterion. The iteration procedure has to be considered for each time step.

7.4 Numerical evaluation of time-integrals involved

A major problem in solving the iterative FEM equations 7.17 (or 7.18) is, that for the evaluation of the elements in K_{aa}, K_{ap}, K_{pp}, F_a and F_p , the actual values of the stress matrix, $[S^{(i),(t+\Delta t)}]$, and subsequently the material moduli matrix, $[M^{(i),(t+\Delta t)}]$, and the volumetric stiffness matrix, $[d^{(i),(t+\Delta t)}]$, must be calculated. For these calculations the 2nd Piola-Kirchhoff stress tensor, $[S]$, (according to expressions 7.3) has to be evaluated for each time step (and actually also for each iteration). However, for the deviatoric part of this tensor, $[S_d]$, a simple update procedure is currently not available. This is due to the integral part of this tensor, $[S_d]_{\Phi o}$, see expression 7.3e. For the times t and $t^{*}=t+\Delta t$, this integral part can subsequently be expressed as:

$$[S_d(t)]_{\Phi}=\int_0^t \Re\{C_r(t,\tau),t,\tau\}d\tau \tag{7.19}$$

and

$$[S_d(t+\Delta t)]_{\Phi}=[S_d(t^*)]_{\Phi}=\int_0^{t^*} \Re\{C_r(t^*,\tau),t^*,\tau\}d\tau \tag{7.20}$$

such that:

$$[\Delta S_d]_{\Phi}=[S_d(t+\Delta t)]_{\Phi}-[S_d(t)]_{\Phi}\neq\int_t^{t^*} \Re\{C_r(t^*,\tau),t^*,\tau\}d\tau \tag{7.21}$$

Thus, for an accurate evaluation of the integral 7.18, the history of converged strain data from all previous time steps is required. Since this integration must be performed for each time step (and each iteration), this involves a huge data storage problem. The amount of data is progressively increasing with each successive time step, which in turn objects the practical application of the model within a finite element environment. Hence, an approximate method is sought to circumvent these huge data storage and data handling problems. A more accessible manner for the integration problem is achieved by employing a (very) limited number of strain history data, by adopting a so-called multi-points approach. Here the kernel of the integral 7.19 is approximated by a polynomial through a limited number of time points. Of cause, the selection of time points for which strain data have to be stored and, for the current time step the actual selection of time points for the polynomial representation (of the kernel), requires a good engineering judgement of the simulation. The choices made may affect the results of integration and thus can influence the (converged) solutions. For the sample problems, as presented in chapter 8, results of a simple 4-points polynomial approach are referred to as "the DIANA results". These results are experimentally verified and sometimes compared to so-called "accurate results", which are obtained through direct numerical integration of the integral 7.19, involving all the intermediate data points. Although, For the selected sample problems, quite reasonable results are obtained (except for the discontinuous loading case of section 8.4), for more arbitrary loading situations a more detailed investigation of the use of the multi-point approach still remains advisable.

8 EXAMPLE PROBLEMS

8.1 *Introduction*

In order to show the capabilities of the constitutive model and its finite element implementation, some comparisons between results from numerical simulations and experiments are presented. First of all, in section 8.2, the capability of the model to describe large deformation rubber hysteresis behaviour is shown. In section 8.3 a stress relaxation test is considered. Here typically the chain breakage process prevails, while the effects of chain formation cannot yet be observed. In section 8.4 the influence of chain formation (together with chain breakage) is illustrated considering multi-step relaxation. A case where intrinsic chain breakage and chain formation are essential, is the creep problem as considered in section 8.5. A sample problem with a typically inhomogeneous state of stress concerns the perforated rubber sheet as discussed in section 8.6.

8.2 *Short term cyclic behaviour*

For short-term cyclic loading, it is assumed that chain alteration is not yet relevant. Under this assumption the integral parts in the expressions 5.6 are omitted.

FEM simulations are performed under progressive cyclic loading programs, using DIANA-plain-stress rubber elements CQ22E for a quarter of a sheet-stretch specimen and using DIANA-3D rubber elements HX64 for a rubber block of a shear test specimen (Figures 1 and 2). The update of the damage pa-

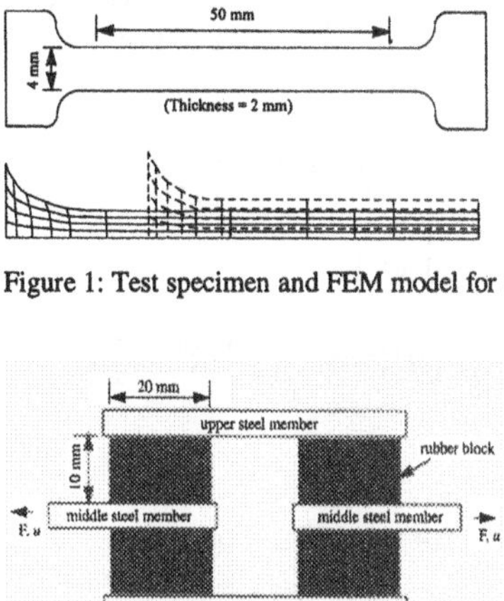

Figure 1: Test specimen and FEM model for stretch testing.

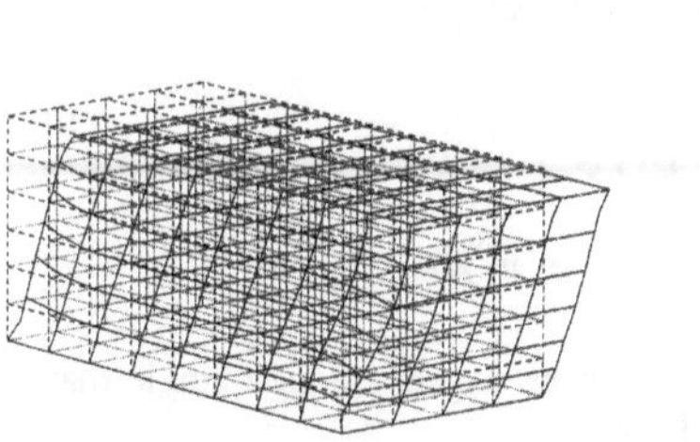

Figure 2a: Test specimen for shear testing

Figure 2b: FEM model for shear testing.

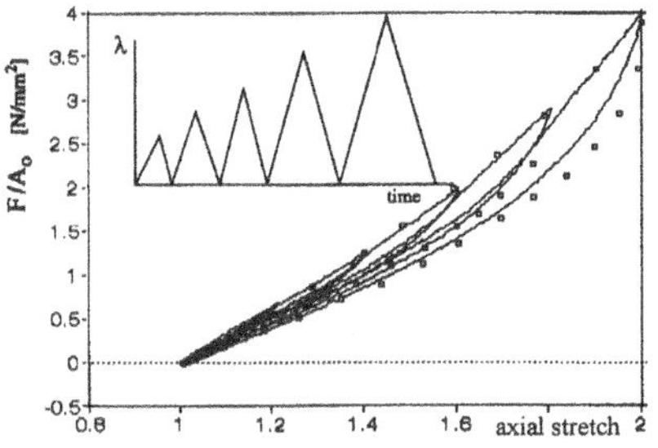

Figure 3: Hysteresis behaviour for progressive cyclic stretching

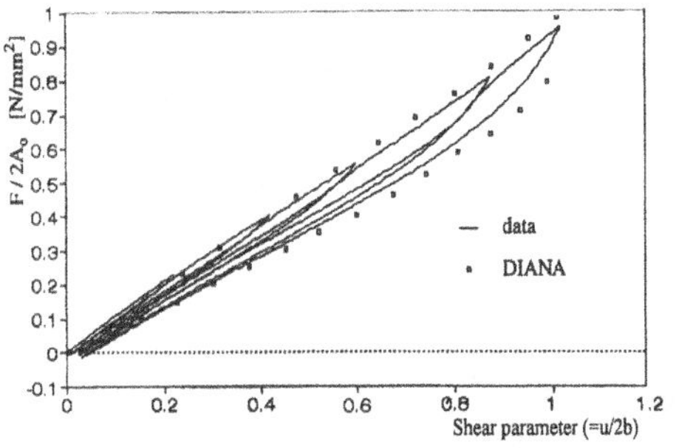

Figure 4: Hysteresis behaviour for progressive cyclic shearing.

rameter Θ^m is performed according to:

$$\Theta^m(t+\Delta t) = \max\{\Theta(t+\Delta t), \Theta^m(t)\} \qquad (8.1)$$

Experimentally and numerically obtained hysteresis diagrams for a progressive cyclic loading program are presented in Figure 3. Figure 4 represents the hysteresis data for a progressive cyclic shear test. Here the stress data represent the shear force divided by twice the undeformed horizontal area of a single shear block. From both experiments it can be observed that the hysteresis behaviour is adequately described up to relatively large deformations.

8.3 *Relaxation*

In the stress relaxation test, typically the chain breakage process prevails, since newly formed chains remain unloaded during time. Tests and simulations are performed for the specimen (+ model) of Figure 1.

Results for the relative stress (= current stress / initial stress) versus time are presented in figure 5. The solid line represents the model solution. Experimental data are given for three strain levels (20%, 50% and 70%). Here the model solution is due to the chain breakage process only. This process is assumed to be strain level independent. Since there is a good correspondence between the numerical and experimental results, here this assumption is experimentally confirmed.

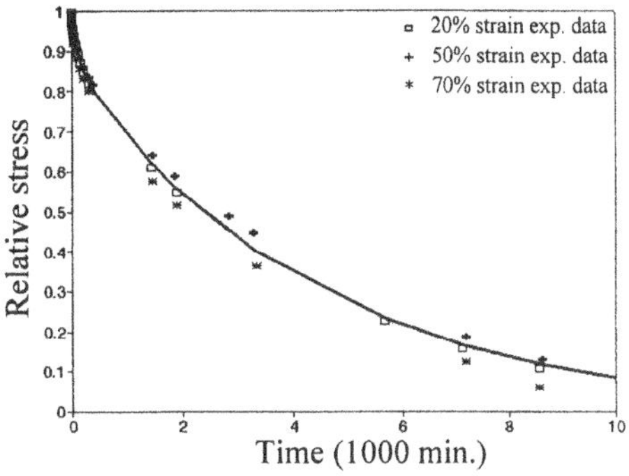

Figure 5: Relaxation behaviour for different strain levels

8.4 *Long term discontinuous loading.*

In the (previous) relaxation case, the influence of chain formation could not be observed in the response data, since there typically the newly formed chains remained unloaded. Contrarily, in a multi-step relaxation test the newly formed chains do contribute to the response, since after load variation these chains will not remain unloaded. The influence is easily shown, using the stretch histogram as presented in Figure 6. Experimentally and numerically obtained results (axial stress versus time) are presented in Figure 7. The "accurate results" and the experimental results compare quite well. However, the "Diana results" become less accurate after the load discontinuities. The simple 4-points polynomial approach then becomes insufficiently accurate and a more adequate approximation of the kernel of integral 7.18 (within DIANA) is required.

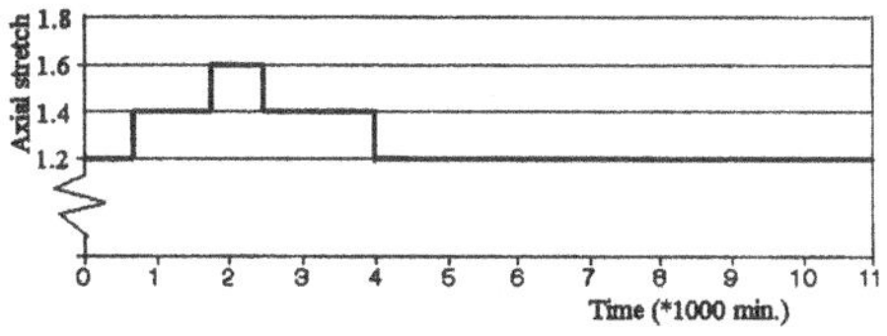

Figure 6: Axial stretch versus time (multi step relaxation).

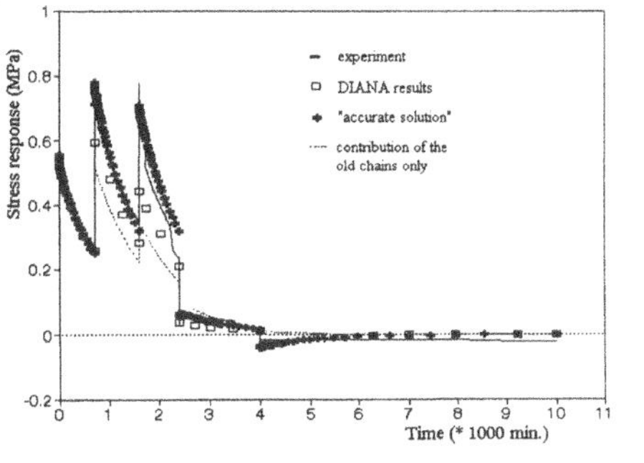

Figure 7: Gauchy stress versus time (multi step relaxation).

The dotted curves in Figure 7 represent a modified case, where the chain growth terms were omitted (These curves were also calculated according to the "accurate scheme"). From comparison with the experimental results and the other numerical results, it can be observed that the chain growth terms now are essential for a correct representation. Only by using the complete (with chain breakage as well as with chain growth) theory an acceptable correspondence between numerical and experimental results is obtained.

8.5 *Creep*

Another case where chain breakage and chain formations both are essential is the creep problem. Here the specimen (+ model) of Figure 1 is axially loaded, while the true stress is kept constant. Experimentally and numerically obtained results are presented in Figure 8. In this case of slowly varying strains, the "Diana results" (based on the simple 4-points polynomial approximation of the kernel of integral 7.18) are quite acceptable up to high strain levels.

8.6 *Perforated plate*

The capability of the model and its FEM implementation to describe a typical inhomogeneous time-dependent state of stress will be illustrated on a relaxation problem of a perforated rubber sheet. The specimen dimensions are defined in Figure 9a (the thickness is 2 mm). The FEM mesh (made up of plane-stress rubber elements CQ22E) used for the simulations is presented in Figure 9b. In this Figure the area around the hole, which in the sequence will be used for presentation of stress results, is indicated. Through vertical movement of the specimen ends a prescribed elongation of the specimen is reached and kept constant during time.

In a large part of the specimen a nearly homogeneous strain distribution is reached. However in a boundary zone around the hole an inhomogeneous stretch-stress distribution will be present. Therefore, in the experiment, stretches are measured at positions nearby the hole. For this purpose a contactless measurement is realised in order to circumvent spurious strains originating from the gauches. Here the area of interest is provided with small polystyrene globules (diameter of ~0.1 mm) which are stuck on to the specimen with a special glue. The area of interest is monitored through a servo-controlled CCD camera with microscope, coupled to an image processing system. Stretches are calculated from the mutual orientations of the centres of the globule images. More details of this measurement method are presented in Septanika *et al* (1998a, b).

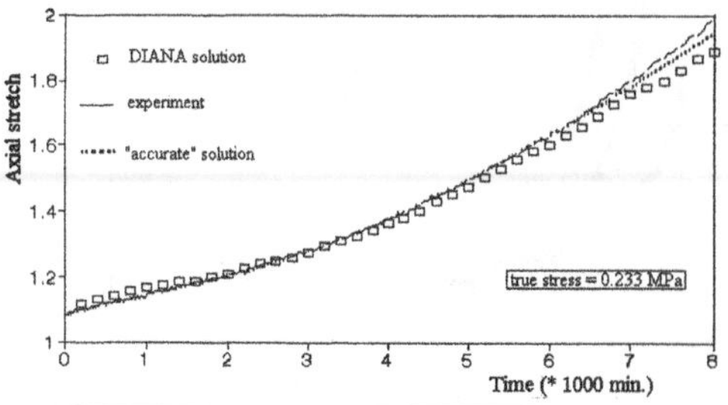

Figure 8: Creep curves

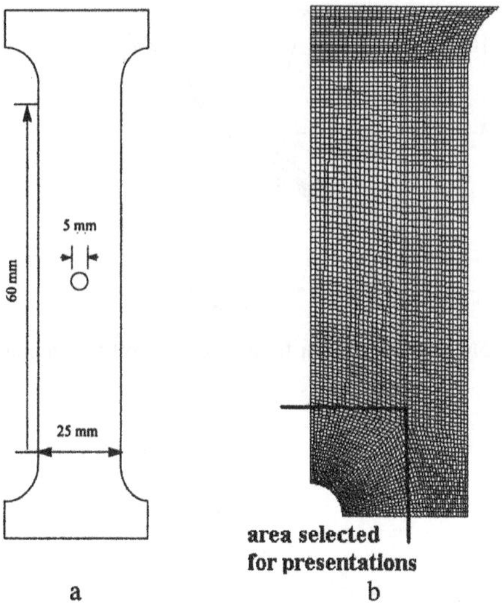

Figure 9: Specimen (a) and FEM model of a quarter (b)

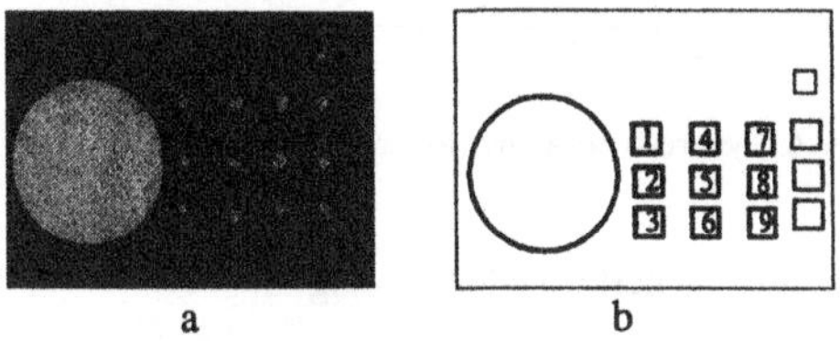

Figure 10: Measurement area with globules (a),
Globule identifications used (b).

The specimen area, selected for strain measurements, together with globule identifications, is shown in figure 10. Here the camera focus area being selected was rather large (The globule image area represents about 12 pixels. For the creep and relaxation experiments used for the model parameter identification the globule image area corresponded to about 185 pixels). As a consequence, for the present test the measurement error in the local stretches is larger than 5%.

Since the globule distances are finite, for the boundary zone around the hole, where the stretch

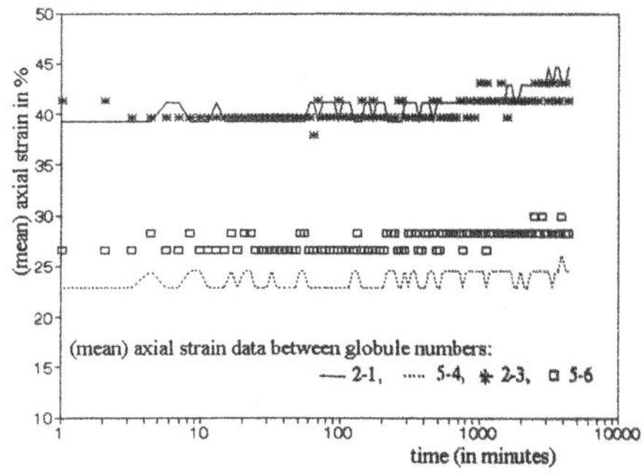

Figure 11: "Mean" axial strains (versus time) for various mutual globule positions. (relative elongation 20%).

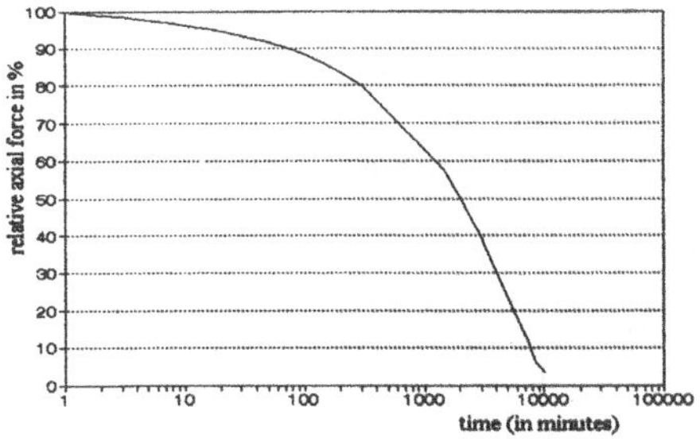

Figure 12: Measured relative force-decay.

pattern is quite inhomogeneous, just local "mean" stretch data are obtained from the measurements. For such local "mean" stretch data the mutual globule identifiers must always be specified. For a stress relaxation test, with a relative elongation of 20 %, "mean" axial strains (versus time) are presented in Figure 11. From these data it can be observed, that hardly any strain re-distribution occurs during this relaxation test.

From the measurements it appears that the axial force relaxes with time. Figure 12 shows the relative axial forces (=axial force at time t / axial force at time 0). After about 10000 minutes almost no axial force is left and after relief of the specimen it remains permanently in the final (deformed) position. This phenomenon is known as "permanent-set".

In the sequence computational results for a relative elongation case of 60% are presented. It turns out, that the deformed shape of the model, as obtained directly after elongation, does hardly change with time. This was also observed in the experiments. The deformed shape of the presentation area (which was defined in Figure 9) is shown in Figures 13a and b, for the experiment and the FEM calculation, respectively

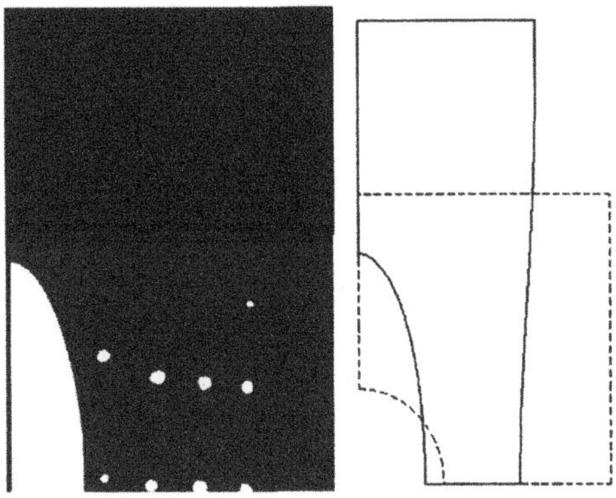

a b

Figure 13: Deformed shapes of the presentation area.

Figures 14, 15 and 16 show principle Gauchy stress distributions for times t=0, t=5000 sec and t=10000 sec. It turns out, that the shape of the distributions does not change so mutch in time. However the stress levels decay with time (while the maximum strains remain relatively steady).

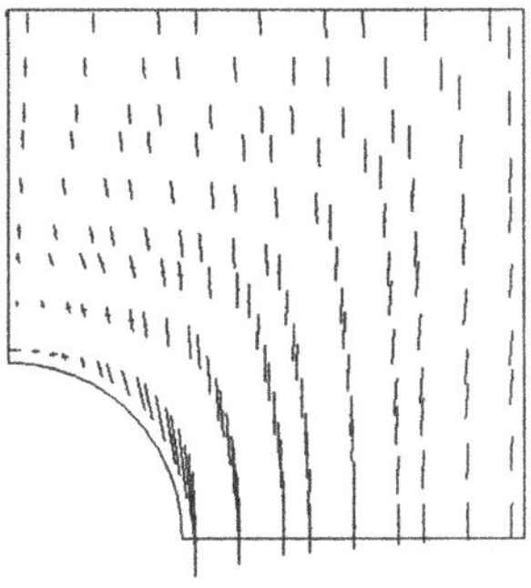

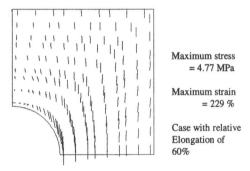

Maximum stress = 4.77 MPa

Maximum strain = 229 %

Case with relative Elongation of 60%

Figure 14: Principal Gauchy stress distribution at t=0

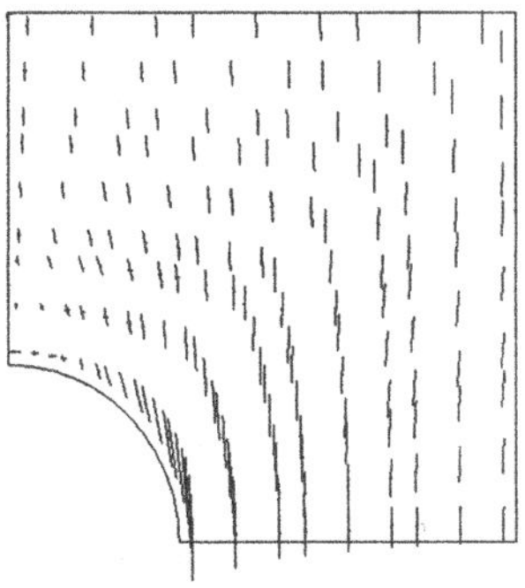

Maximum stress = 1.324 MPa

Maximum strain = 233 %

Case with relative Elongation of 60%

Figure 15: Principal Gauchy stress distribution at t=5000sec.

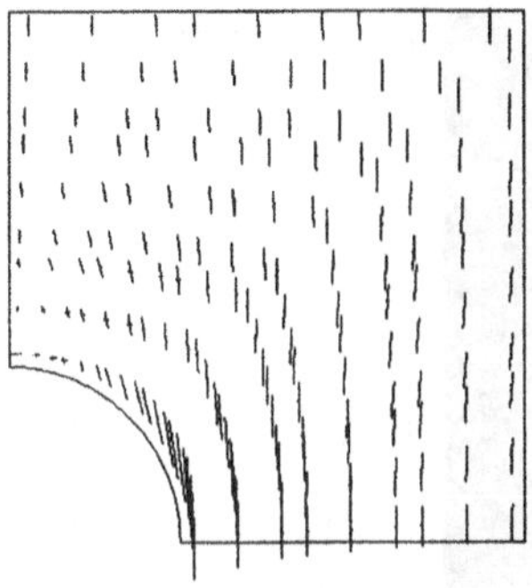

Figure 16: Principal Gauchy stress distribution at t=10000sec.

9 CONCLUSION

A new adequate model for the description of long term as well as short term constitutive behaviour of filled vulcanised rubber materials is derived and implemented in a FEM-code. An approximation method is used to avoid excessive memory and CPU-time consumption due to numerical evaluation of time-integrals involved in the constitutive model. In cases with slowly varying strains (with time) the approximations made are quite acceptable up to high strain levels.

REFERENCES

Arruda, E.M. & Boyce, M.C. 1993. A Three-Dimensional Constitutive Model for Large Stretch Behavior of Rubber Materials. In *J. Mech. Phys. Solids*, 41:389.

Bogert, P.A.J. van . 1991. *Computational modeling of rubber like materials*. Ph.D. thesis, Delft University of Technology.

Bueche, F. 1960. Molecular Basis for the Mullins Effect. In *J. Appl. Poly. Sci.*, 4(10):107.

Bueche, F. 1961. Mullins Effect and Rubber-Filler Interaction. In *J. Appl. Poly. Sci.*, 5(15):271.

Drozdov, A.D. 1997. The non-isothermal behavior of polymers. In *Europ. J. Mech. A/Solids*, 16(6): 947-991.

Eyring, H. 1943. Viscosity, plasticity and diffusion as examples of absolute reaction rates. In *J. of Chem. Phys.*, 4: 282-291.

Lubliner, J. 1985. A model of rubber viscoelasticity, In *Mech. Res. Comm.*, 12: 93

Mullins, L. & Tobin, N.R. 1965. Stress softening in rubber vulcanizates, part I. In J. Appl. Polymer Sci., 9: 2993.

Mullins, L. 1969. Softening of rubber by deformation. In *Rubber Chem. Technol.*, 42: p.339.

Murakami, K. & Ono, K. 1979. *Chemo-rheology of Polymers*, Amsterdam: Elsevier

Septanika, E.G. & Ernst, L.J. & Van den Hooff, L.A.C.M. 1998a. An Automatic and Interactive Large Deformation Measurement System Based on Image Processing. In *J. of Exp. Mech.*, 38(3): 181-188

Septanika, E.G. 1998b. *A time-dependent constitutive model for vulcanized and filled rubber materials*, Ph.D. thesis, Delft University of Technology

Septanika, E.G. & Ernst, L.J. 1998c. Application of the network alteration theory for the modeling the time-dependent constitutive behavior of rubbers, Part 1, "General Theory". In *Mech. Mat.*, 30(4): 253-263

Septanika, E.G. & Ernst, L.J. 1998d. Application of the network alteration theory for the modeling the time-dependent constitutive behavior of rubbers, Part 2, "Experimental Verification". In *Mech. Mat.*, 30(4): 265-273

Septanika, E.G. & Ernst, L.J., 1998e Hysteresis and Time-dependent Constitutive Modeling of Filled Vulcanized Rubber. In: *Proc. of the 3rd European Mechanics of Materials Conference, Oxford.*

Simo, J.C. 1987. On a Fully Three-Dimensional Finite-Strain Viscoelastic Damage Model: Formulation and Computational Aspects. In *Comp. Meth. Appl. Mech. Eng.*, 60: 153.

Treloar, L.R.G. 1975, *The Physics of Rubber Elasticity*, 3rd ed., Oxford: Clarendon

Tobolsky, A.V. & Prettyman, I.B. & Dillon, J.H. 1944. Stress Relaxation of Natural and Synthetic Rubber Stocks. In *J.Appl.Phys.*,15: 380.

Tobolsky, A.V. & Andrews, R.D. 1945, System Manifesting Superposed Elastic and Viscous Behaviour. In *J.Chem.Phys.*,13: 3.

Tobolsky, A.V. & Green, M.S. 1946. A New Approach to the Theory of Relaxing. In Polymeric Media, *J. Chem. Phys.*, 14(2):80.

Tobolsky, A.V. & Andrews, R.D. & Handson, E.E. 1946. The Theory of Permanent Set at Elevated Temperatures in Natural and Synthetic Rubber Vulcanizates. In *J.Appl.Phys.*,17: 352.

Wineman, A.S. & Rajagopal, K.R. 1990, On a Constitutive Theory for Materials undergoing Microstructural Changes. In *Arch. Mech.*, 42(1): 53.

Wineman, A.S. & Huntley, H.E. 1994. Numerical Simulation of the Effect of Damage Induced Softening on the Inflation of a Circular Rubber Membrane. In Int. J. Solids Structures, 31(23): 3295.

Wu, P.D. & Giessen, E. van der . 1992, On Improved 3-D non-Gaussian network models for Rubber Elasticity. In *Mech. Res. Comm.*, 19: 427.

Wu, P.D. 1993. On improved network models for rubber elasticity and their applications to orientation hardening in glassy polymers. In *J. Mech. Phys. of Solids*, 41: 427.

Constitutive Models for Rubber, Dorfmann & Muhr (eds) © 1999 Taylor & Francis ISBN 90 5809 113 9

Experimental determination of model for liquid silicone rubber: Hyperelasticity and Mullins' effect

A. H. Muhr & J. Gough
Tun Abdul Razak Research Centre, MRPRA, Brickendonbury, UK

I. H. Gregory
University of Hertfordshire, Hatfield, UK

ABSTRACT: The stress-strain, and in particular strain-softening, behaviour of liquid silicone rubber (LSR) is reported and compared to that of natural rubber (NR) with and without filler. The LSR exhibits strain-softening of the type described by Mullins particularly clearly. Uniaxial results can be fitted reasonably well using the Ogden-Roxburgh model, but the model cannot describe the anisotropy induced by a uniaxial pre-strain.

1 INTRODUCTION

Work on constitutive modelling of elastomers has mainly focussed on natural rubber (NR). The question arises as to how applicable the broad conclusions of this work are to other types of rubber. Rivlin & Saunders (1951) carried out a carefully planned set of experiments to determine the strain-energy function for unfilled crosslinked NR, a highly elastic and extensible material, up to maximum extension ratios of the order of 2.8. For their material and strain range the effect of strain history is barely discernible, so they could carry out a sequence of biaxial strain measurements, of varying ratios for the orthogonal strain, on the same testpiece. This has the great advantage that the problem of sample variability is eliminated, as well as the labour of making many testpieces. Of course, it also implies that modelling the material using a strain-energy function is valid.

In previous papers (Gough et al 1998, Gough et al 1999) we pointed out the problems that arise when the work is extended to include NR materials incorporating filler. The departures from elasticity are especially striking if "reinforcing" fillers, such as small particles of carbon black or silica, are used. One aspect of these departures from elasticity is the Mullins' effect (Mullins & Tobin, 1957), according to which the second strain cycle will correspond to lower stresses until the maximum strain of the first cycle is reached; the stress-strain curve then rejoins that of a virgin testpiece pulled to a higher strain. Provided the material is first "scragged" - ie. - deformed to a strain beyond that of interest for modelling purposes - it may still be reasonable to describe the subsequent stress-strain behaviour using a strain-energy function. However, the function will depend on the strains used in the "scragging" procedure and there are generally additional inelastic effects complicating the behaviour such as set, further (but less gross) softening on repeated stressing, time-dependent (if slow) recovery from the original Mullins softening, and hysteresis even for steady-state strain cycles.

It thus seems that modelling elastomers using a strain-energy function, as currently advocated in commercial finite element (FE) software packages, is fraught with difficulty. In previous work (Gough et al 1998, Gough et al 1999) we have suggested that at least the first cycle stress-strain behaviour may be modelled using the formalism of a "strain-energy" function. The purpose of this paper is to present experimental results showing the effect of subsequent strain-cycles, with a view to developing models to describe the behaviour. The materials investigated were NR, filled and unfilled, and liquid silicone rubber (LSR).

LSR is supplied as two liquid components, both having substantially lower viscosity than that of conventional uncrosslinked "dry" rubbers. On mixing the components, usually in a 1:1 ratio, and injecting them into a mould preheated to 150 to 200°C cure by poly-addition takes place very rapidly (in 15 seconds or so for thin mouldings) to yield a crosslinked polysiloxane polymer or copolymer. The processing procedures are thus quite unlike those of conventional dry rubbers. We also noticed that the stress-strain characteristics of the cured LSR elastomer are remarkable, showing a much more pronounced Mullins effect than is seen with NR.

2. EXPERIMENTAL

2.1 *Materials*

The elastomers used in the stress-strain experiments were unfilled NR with a conventional sulphur cure system, NR filled with 45 parts per hundred of rubber (pphr) of carbon black (N330 grade) and a LSR grade designated as 9280-50E obtained from Dow Corning. The NR materials were compounded and moulded as 225 x 225 x 2mm (nominal dimensions) sheets in-house, following the details given (TARRC, 1980) in EDS 19 and EDS 16 for the unfilled and filled materials respectively. The LSR material was obtained as a cured sheet (again 225 x 225 x 2mm) directly from the Dow Corning, to whom compounding details are proprietary. However, technical representatives of Dow Corning described the material as a silicone elastomer, incorporating about 30% by weight of fumed silica as a filler, with a nominal hardness of 50IRHD.

Methods

The LSR was first characterised using the standard tensile test to failure (ISO37:1994(E)). This test is performed on "dumbbell" specimens (Figure 1) in triplicate. The purpose of the widened ends (Figure 1) is to reduce the probability of rupture at the stress concentrations associated with the clamps. However, this has the disadvantage that the strain cannot be deduced from the clamp separation. It must instead be deduced from the separation of ink marks drawn on the parallel section of the dumbbell, which in the TARRC tensile test machine are monitored using optical followers.

Operation of the tensile test machine is a highly automated routine, and thus has the advantage that the tensile stress-strain characteristics may be measured right up to break very rapidly. Checks of anisotropy and strain rate may also be investigated. The disadvantage of the technique is that, being a general purpose machine, the load cell and the optical gauge mark followers must have range capacities adequate for, respectively, the stiffest or most extensible rubbers, impairing their resolution for small to moderate strains. Furthermore, the machine is not readily programmable for tests other than a single extension to break.

In addition to using the tensile test machine, strain-cycling experiments were carried out on parallel sided strips (150 x 6 x 2mm nominal dimensions) in an Instron universal test machine. The cross-head speed was set at 336mm min^{-1}, corresponding to a nominal tensile strain rate of $0.04s^{-1}$. The testpiece thickness and width were carefully measured before clamping it in the machine. The unstrained testpiece length was deduced from the clamp separation at zero force, and the strained length from the subsequent crosshead displacement. The machine signals for load and crosshead displacement were plotted on a xy recorder. Two testpieces were used for each rubber; one was strained continuously until it either broke or slipped from the clamps, thus giving the stress-strain curve for a virgin testpiece to as high a strain as possible. The second testpiece was strained to 100%, and then the crosshead was returned to the original position at 336mm min^{-1} as for the extension. After a few

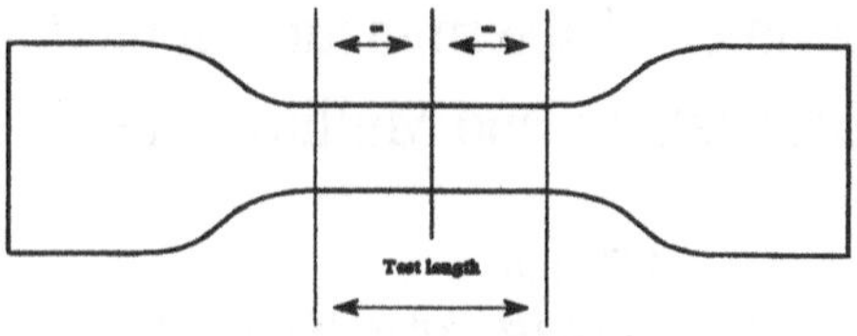

Figure 1: Shape of dumbbell testpieces

seconds the testpiece was extended to 200% and then retracted as before. The sequence was then repeated at successively higher strains, being 300%, 400% and 500% for EDS 19 and 9280-50E, and 300% and 400% for EDS 16, 500% and 400% respectively being the highest strains achievable in this sequence for these materials and this technique. Finally, the testpiece that had suffered the strain cycles was cycled six times to 300% strain.

Stress-strain measurements of LSR were also carried out in pure shear. First, a testpiece was cut from the 2mm sheet and clamped in the split pure-shear test jig previously described (Gough et al 1998) but adapted with clamps for samples cut from sheet. The dimensions of the testpiece between the clamps were 196mm wide and 20mm long. This testpiece was first strained in the Instron machine to 100% extension at a strain rate of about $0.008s^{-1}$ then the crosshead was returned to the starting position. The straining was then repeated to a higher strain. A second testpiece of the same dimensions was cut, and first strained to 100% simple extension in the length direction at a strain rate of $0.04s^{-1}$. It was then clamped in the split pure shear jig and strained as before until slippage occurred in the clamps, again at a strain rate of $0.04s^{-1}$. Finally, the testpiece was strained in tension once again, in the length direction.

3. RESULTS

Table 1 gives stress-strain results from the standard tensile tests performed on the materials. There is little anisotropy (≤5% for stress at a given strain) and there is only a very small increase in stiffness as the strain rate is increased.

The results of the strain-cycling experiments are shown in Figures 2 to 5. The stresses are reported as nominal stresses, that is load divided by the original cross-sectional area. For clarity, the retraction curves are not plotted all the way back to zero force, except for the final retraction curve, for which a dashed line is used. In contrast to the "correction" procedure for set applied by Mullins and co-workers, the results are presented as drawn directly on the plotter, so that the set resulting from extension is readily apparent from the zero-force intercepts on the extension ratio axes.

Unfilled NR (Figure 2) shows relatively little set or Mullins' effect for the strains used, the most striking departure from elasticity being the hysteresis

Table 1: Stress-strain and failure properties for LSR from standard tensile test

testpiece orientation	Rate mm/min	Stress at							
		100% MPa	200% MPa	300% MPa	400% MPa	500% MPa	600% MPa	EB %	TS MPa
short side	500	1.57	3.28	4.81	6.29	7.87	9.73	648	10.7
long side	500	1.68	3.41	5.00	6.50	8.29	9.63	587	9.85
long side	50	1.57	3.16	4.55	5.88	7.32	9.16	720	11.6
long side	5	1.58	3.07	4.38	5.67	7.13	8.87	690	10.7

Note: each entry is the mean from three replicates

which becomes very substantial for cycles to 400 or 500% strain. This phenomenon is usually ascribed to strain-crystallization and does seem to be accompanied by some stress softening. However, in contrast to the Mullins' effect, the softening applies to strains above as well as below those to which the testpiece has previously been subjected. This type of "softening" would be less apparent if the set "correction" procedure were applied.

Figure 3 shows that the introduction of N330 carbon black as a filler greatly increases the stiffness, hysteresis and set of NR. The Mullins effect is also quite evident; we see an increase in the slope of the extension curve of the softened material as the previous maximum strain is approached, and then a fall in slope to close to that of the virgin material as the maximum strain is exceeded.

However, the LSR exhibits the Mullins' effect much more clearly (Figure 4). There is relatively little set but a substantial Mullins-type softening for all strains used. As for Figure 2, the drop below the virgin stress-strain curve above 300% would be less apparent if the set "correction" procedure were applied.

Figure 5 shows that relatively little further stress softening occurs for any of the materials for subsequent cycles to a stain (300%) substantially below that of the previous maximum (500% for unfilled NR and LSR, or 400% for filled NR). However, for the filled NR there is still substantial hysteresis and increase in set.

Figure 6 shows the effect of previous cycles to 100% - either co-directional pure shear or tension in the orthogonal direction - on the main stress in the pure shear experiments on LSR. The behaviour in pure shear, with the curves for retraction from the first straining cycle and extension for the second cycle lying close together, is similar to those previously

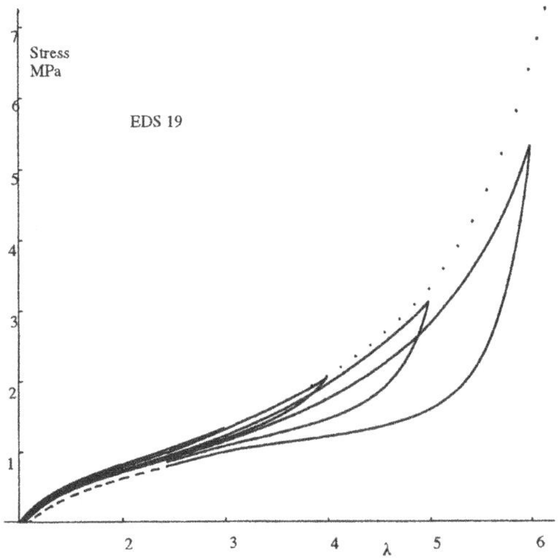

Figure 2: Tensile stress-strain curve for unfilled NR (EDS 19)
. . . virgin testpiece to high strain, ___ cycled testpiece

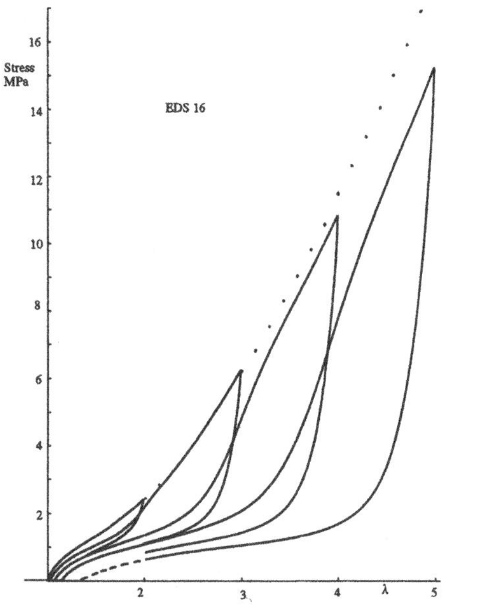

Figure 3: Tensile stress-strain curves for filled NR (EDS 16).
. . . virgin testpiece to high strain, ___ cycled testpiece

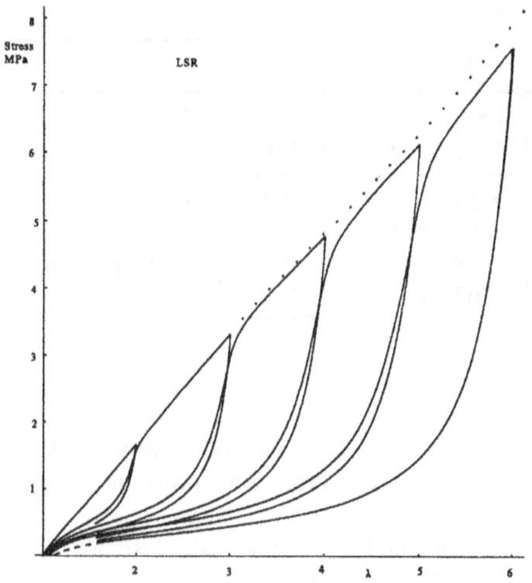

Figure 4: Tensile stress-strain curves for LSR
. . . virgin testpiece to high strain, ___ cycled testpiece

given for tension (Figure 4). Surprisingly, the effect of extension to 100% on the subsequent pure shear stress-strain curve in the orthogonal direction is just a small softening, relative to the virgin pure shear behaviour. No abrupt change in slope is seen at a strain of the order of 100%. Similarly, the effect of pure shear to 200% is only a softening of the tensile stress-strain curve in the orthogonal direction, without any obvious change in slope at around 200% strain. In fact, the strongest hint of a Mullins-type shoulder as the curve tends to rejoin the virgin stress-strain curve occurs at around 100% strain - the value of the previous maximum co-directional tensile strain.

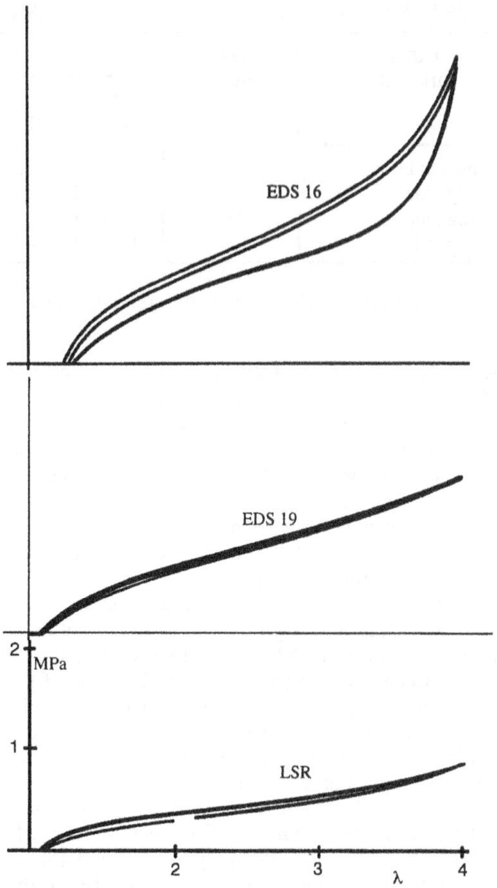

Figure 5: Tensile stress-strain curves for the first and sixth subsequent cycles to 300% carried out on the strain-cycled testpieces of Figures 2 to 4. The scales are the same on all three graphs.

4. MODELLING THE STRESS-STRAIN BEHAVIOUR

Ogden & Roxburgh (1999) have recently presented a promising way of modelling the Mullins' effect, treating the material as "pseudo-elastic". On the first loading curve its stress-strain behaviour may be described using a strain-energy function in the usual way for an incompressible hyperelastic material:

$$\sigma_1 - \sigma_3 = \lambda_1 \partial W / \partial\lambda_1$$
$$\sigma_2 - \sigma_3 = \lambda_2 \partial W / \partial\lambda_2 \qquad (1)$$

where the σ_i are principal Cauchy stresses and W is a function of the first two extension ratios λ_1, λ_2, the substitution $\lambda_3 = 1/(\lambda_1 \lambda_2)$ having been made to eliminate λ_3 as a variable. W is also taken to be a function of η, an additional variable which takes the value unity on the first loading cycle, with the form:

$$W(\lambda_1, \lambda_2, \eta) = \eta \tilde{W}(\lambda_1, \lambda_2) + \phi(\eta) \qquad (2)$$

where ϕ is a smooth function with $\phi(1) = 0$. On subsequent cycles, η is unity only if $\tilde{W}(\lambda_1, \lambda_2) \geq \tilde{W}_m = \tilde{W}(\lambda_{1m}, \lambda_{2m})$ where $\tilde{W}_m$, λ_{1m} and λ_{2m} correspond to the previous maximum energy state. The Cauchy stresses are now given by:

$$\sigma_1 - \sigma_3 = \eta \lambda_1 \partial\tilde{W}/\partial\lambda_1$$
$$\sigma_2 - \sigma_3 = \eta \lambda_2 \partial\tilde{W}/\partial\lambda_2 \qquad (3)$$

We may note that since on the first cycle $W = \tilde{W}$, comparison of (3) with (2) shows that η serves as a strain-dependent scale factor for all the stresses for strains corresponding to $W < W_m$. Ogden & Roxburgh (1999) show that η is a function of λ_1 and λ_2 complying with:

$$-\phi'(\eta) = \tilde{W}(\lambda_1, \lambda_2) \qquad (4)$$

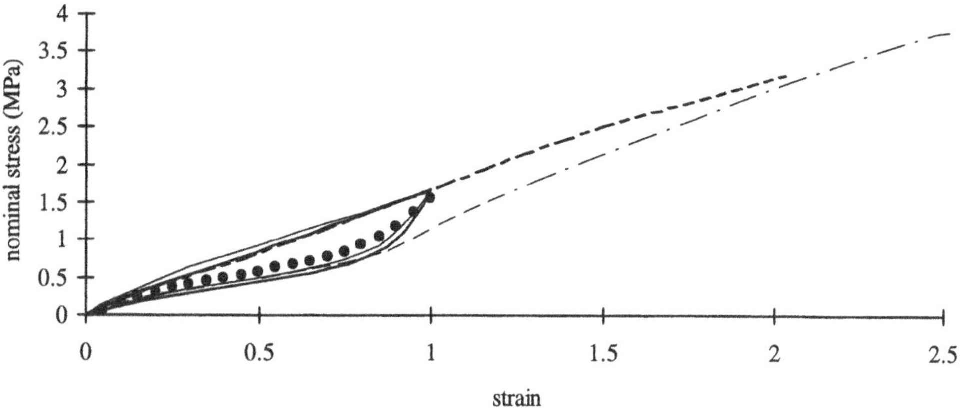

Figure 6: Effect of changes in testpiece orientation on Mullins type softening.
— previously unstrained testpiece in tension, ---- pure shear in orthogonal direction after tension to 100% , ——.—— tension after tension to 100% and orthogonal pure shear to 200%, ___ previously unstrained testpiece in pure shear, • pure shear after pure shear to 100%

so that the material behaves elastically in all subsequent strain cycles for which $W < W_m$ but with a strain energy function which differs from $W(\lambda_1, \lambda_2, 1)$.

They suggest a particular form for ϕ (η), requiring two parameters to be fitted empirically, which may enable reasonable fits with experiment and leads to the following form for η:

$$\eta(\lambda_1, \lambda_2) = 1 - \frac{1}{r}\mathrm{erf}\left[\frac{1}{m}\left(\tilde{W}(\lambda_{1m}, \lambda_{2m}) - \tilde{W}(\lambda_1, \lambda_2)\right)\right] \quad (5)$$

where $m > 0$ and $r > 1$.

We have programmed a computer such that the movement of the mouse results in a smooth change in a parameter, causing a screen plot of an equation incorporating the parameter to be adjusted in real time. This enables a rapid "feel" for the effect of parameters such as r and m in equations (5) and (3) to be developed, and quickly leads also to a subjective fit to empirical data. The fact that the procedure is subjective can be a considerable advantage, avoiding the need to "massage" the data to produce a uniform density over the strain range or to eliminate the effect of rogue data.

To try to model the data of Figures 2 to 4, the CURVEFIT program was used to plot a tensile nominal stress-strain curve based on equation (1) and the neo-Hookean form $\tilde{W} = \frac{1}{2}G\left(\lambda_1^2 + \lambda_2^2 + 1/(\lambda_1\lambda_2)^2 - 3\right)$, together with a family of retraction curves for strains of 100, 200, 300, 400 and 500%, based on equations (3) and (5). The mouse can be used to vary three parameters: the shear modulus G (MPa), r (dimensionless) and m (Mpa).

The initial extension curve depends only on the modulus and was the first to be adjusted to achieve a fit with the experimental results. It was then noted that the parameter m has dimensions of modulus, and that in expression (5) it appears, at least for a neo-Hookean W, as G/m. It would be better, therefore, to treat G/m as a single dimensionless parameter. The shape of the retraction curves would not then depend on the modulus, but would just scale uniformly in the stress direction as the modulus is varied.

Figure 7 shows the best fit achieved for filled NR (EDS 16, see Figure 3 for the experimental results).

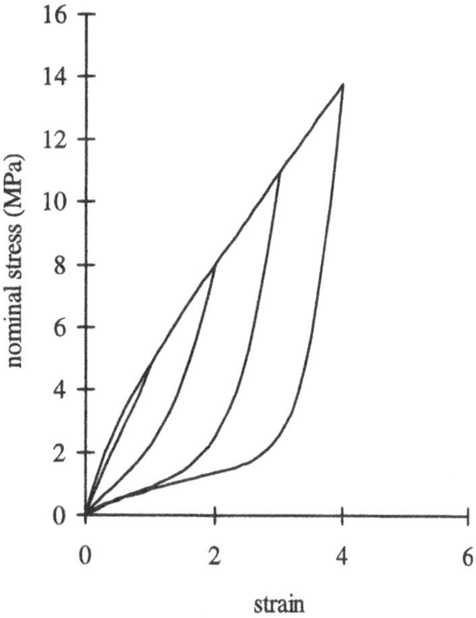

Figure 7: Best neo-Hookean - Ogden-Roxburgh fit for filled NR (G = 2.78MPa, r = 1.20, m = 9.95MPa)

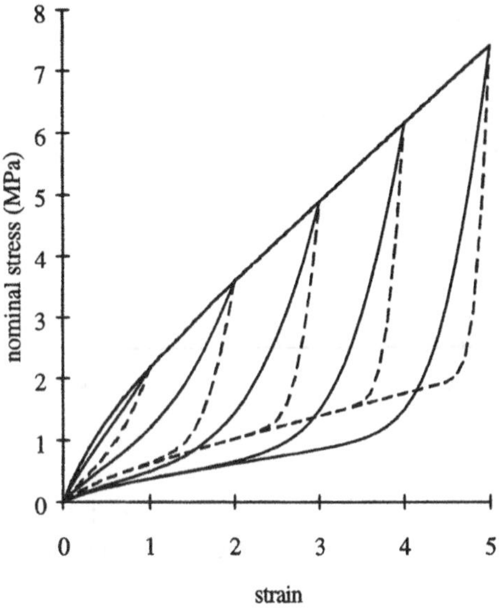

Figure 8: Best neo-Hookean - Ogden Roxburgh fit for LSR
—— G = 1.24MPa, r = 1.21, m = 5.5MPa
- - - - G = 1.24MPa, r = 1.40, m = 1.6MPa

The ability of the Ogden-Roxburgh model to fit the data has been rather impaired by the choice here of a neo-Hookean form for $\tilde{W}$. The substantial departure of the experimental retraction curves from the subsequent experimental extension curves to a larger extension shows that the shortcomings of the model are more deep-rooted; nor would a "correction" procedure for set help in this matter.

The model is considerably more successful for describing the strain cycling results for LSR (compare Figures 4 and 8). With r = 1.21 and m = 5.55MPa the fit for the 500% retraction is good, but the initial gradient at the start of retraction from 200% is too small. If the parameters are adjusted to r = 1.40 and m = 1.60MPa, the fit for the 200% retraction curve is much better but the higher strain curves then have "elbows" which are much too sharp.

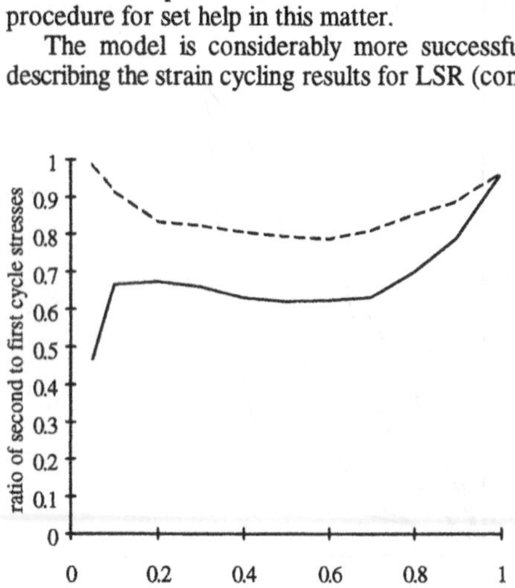

Figure 9: Ratio of first and second cycle main and transverse stresses in pure shear
—— main stress, - - - - - transverse stress

5. DISCUSSION

LSR exhibits very little set or, on strain cycles subsequent to the first, hysteresis. Thus the hallmarks of the Mullins effect - that the stress-strain curves of subsequent cycles tend to follow close to the retraction curve of the first cycle, rejoining the virgin stress-strain curve at the previous maximum strain, stand out clearly.

Dow-Corning suggest that a post-cure of four hours at 200°C may be used to improve the physical properties of the material. While we found that post-curing the material increased the stiffness by about 25%, the relative magnitude of the Mullins effect was not altered. The softening is at least partially recoverable. For LSR we found after two months that about 20% of the softening had recovered.

Mullins (1969) argued that while the Mullins effect is much more pronounced in vulcanizates containing high proportions of filler, it is not primarily due to breakdown of filler-filler or filler-rubber structures. Rather, the effect is intrinsic to the rubber network itself and is exhibited by a wide range of polymeric substances but is rendered more obvious at moderate strains in filled materials due to the strain-amplifying effect of the filler. It seems that LSR is a particularly convenient material for assessing the applicability of phenomenological models of the effect, although since we do not know the nature of the LSR in detail any mechanistic interpretation of the stress-strain results is inappropriate.

The Ogden-Roxburgh model provides quite a good fit to uniaxial stress-strain data of LSR. It is a vast improvement over the application of a scale-factor to $\tilde{W}$ to represent the softening effect of "scragging", since it allows the changed shape of the stress-strain curves to be modelled. The fit could no doubt be improved by modifying the choice of function for $\phi(\eta)$, together with choosing a more elaborate form for $\tilde{W}$ than the simple Neo-Hookean form. However, the experimental results reveal two more fundamental problems with the model.

Firstly equations (3) imply that all the stresses should scale by the same factor, η. We have found this not to be the case. Figure 9 shows the second cycle main (σ_1) and transverse (σ_2) stresses in pure shear, divided by their values in the first cycle, as measured using the split pure-shear jig. Since $\sigma_3 = 0$, the curves should both give $\eta(\lambda, 1/\lambda)$. However, the softening for the transverse stress is less than that for the main stress. Secondly, equation (5) implies that the softening

Table 2: Values of λ in pure shear (λ_s) and tension (λ_t) corresponding to given values of I_1 (and hence to $\tilde{W}$, assuming $\tilde{W}$ depends only on I_1 and not on I_2)

I_1	λ_t	λ_s
5	2	1.93
9.67	3	2.92
16.5	4	3.93
25.4	5	4.94

Note: $I_1 = \lambda_t^2 + \frac{2}{\lambda_t}$

$\lambda_s = \left(\sqrt{I_1 - 3} + \sqrt{I_1 + 1}\right)/2$

depends only on $\tilde{W}_m$ and not on the particular values λ_{1m}, λ_{2m}. In fact, when the second cycle consists of pure shear in a direction orthogonal to a first cycle in tension to 100% the shape of the stress-strain curve is quite unlike that following a first cycle in pure shear to 100% in the same direction as the second cycle (Figure 6). The difference is so striking that we need not worry about precise identification of λ-values in tension and pure shear giving the same values of $\tilde{W}$. In any case these λ-values are expected to be fairly similar if the dependence of $\tilde{W}$ on the strain invariant I_2 is small (Gregory et al., 1997), as we have generally found to be the case for NR compounds (Gough et al 1998, 1999); Table 2 gives values of λ in pure shear and tension corresponding to a range of values of $I_1 = \lambda_1^2 + \lambda_2^2 + 1/(\lambda_1 \lambda_2)^2$.

Overall, it may be concluded that the strain-softening of LSR is smaller in directions orthogonal to the strain-cycle causing the softening. Thus the "scragged" material is in general anisotropic even if the virgin material is isotropic. We have previously observed similar effects for filled NR in simple shear - after an initial "scrag" in one direction, the softening is greatest for simple shear in the same direction, least for simple shear in the opposite direction and intermediate for shear at 90°

For many practical purposes the simplification of a fully isotropic model may be justifiable. However, bearing in mind that such a model is imperfect, it is doubtful if it is worth developing a more elaborate function for $\phi(\eta)$ to improve the fit, and it is certainly not worth introducing extra parameters to this end.

6. CONCLUSIONS

1. The stress-strain behaviour of LSR is reasonably isotropic with little effect of strain rate. Post-cure results in an increase in stiffness of about 25% but the character of the stress-strain behaviour is not altered.

2. LSR exhibits a very clear Mullins effect, with little set or, on strain cycles subsequent to the first, hysteresis. The strain - softening recovers slowly - about 20% recovery after two months.

3. The Ogden-Roxburgh model for the Mullins effect can provide quite a good fit to uniaxial stress-strain cycling data of LSR. However, fitting would be easier if their parameter "m" were replaced by a dimensionless parameter "G/m" where G has the dimension of stress and serves as a scale factor for W.

4. After a first uniaxial strain-cycle, LSR becomes anisotropic. A similar effect has previously been observed for filled NR. The Ogden-Roxburgh model describes an incompressible material which remains isotropic after any strain cycle, and cannot be used to model such effects.

5. For many practical modelling purposes, it may be adequate to retain the simplification of isotropy, and the Ogden-Roxburgh model then provides a promising way of modelling the strain-softening behaviour.

REFERENCES

Gregory, I.H., Muhr, A.H. & Stephens, I.J. (1997). "Engineering applications of rubber in simple extension", Plastics, Rubber & Composites Processing and Applications, **26**, pp.118-122

Gough, J., Gregory, I.H. & Muhr, A.H. (1999), "Determination of constitutive equations for vulcanized rubber" in Finite Element Analysis of Elastomers. D. Boast & V.A. Coveney (eds) Prof. Eng. Publ., London pp.5-26

Gough, J., Muhr, A.H. & Thomas, A.G. (1998), "Material characterization for finite element analysis of rubber components", J. Rubber Research, **1**, pp.222-239

Mullins, L. (1969), "Softening of rubber by deformation", Rubber Chem. & Tech., **42**, 339-362

Mullins, L. & Tobin, N.R. (1954), "Theoretical model for the elastic behaviour of filler-reinforced vulcanized rubbers", Proc. Third Rubber Technol. Conf., London, p.397

Ogden, R.W. & Roxburgh, D.G., "A pseudo-elastic model of the Mullins' effect in filled rubber", Proc. R. Soc., in press

Rivlin, R.S. & Saunders, D.W., (1951). "Large elastic deformations of isotropic materials. IV Fundamental Concepts", Phil Trans. R. Soc., **A240**, pp.459-490

TARRC (1980), "Natural Rubber Engineering Data Sheets", published by TARRC, MRPRA, Brickendonbury, Hertford, SG13 8NL, UK

Constitutive Models for Rubber, Dorfmann & Muhr (eds) © 1999 Taylor & Francis ISBN 90 5809 113 9

Aspects of stress softening in filled rubbers incorporating residual strains

G.A. Holzapfel & M. Stadler
Institut für Baustatik, Technische Universität Graz, Austria

R.W. Ogden
Department of Mathematics, University of Glasgow, UK

ABSTRACT: The goal of this communication is to formulate a phenomenological continuum model which extends to the inclusion of residual strains the pseudo-elastic model for the (idealized) Mullins effect presented by Ogden & Roxburgh (1999a). The proposed model is based on a strain-energy function describing the loading path and an additive damage function responsible for the unloading path. Three damage variables govern the *anisotropic* damage mechanism which is controlled by the evolution of the principal stretches. The pseudo-elastic model described has been implemented in a finite element program. Finally we demonstrate two representative examples: a simple tension test and an inhomogeneous tension test of a rubber strip showing analytical and finite element results.

1 INTRODUCTION AND OVERVIEW

Nearly all practical *thermoplastic elastomers* (actually rubberlike materials) consist of polymer chains chemically connected to other chains at different locations to produce a cross-linked monolithic three-dimensional network. Engineering rubbers contain reinforcing fillers such as *carbon black* (in natural rubber) or *silica* (in silicone rubber). These finely distributed fillers, with typical dimensions of the order of $1.0 - 2.0 \cdot 10^{-12}$m, form physical and chemical bonds with the polymer chains. Carbon-black-rubber vulcanizates have important applications in the manufacture of automotive tyres and many other engineering components.

A piece of filled rubber vulcanizate under cyclic loading and unloading typically displays pronounced (strain-induced) *stress softening* associated with damage. This phenomenon, known as the *Mullins effect*, was pointed out in the early work by Mullins (1947) and is regarded as essentially being caused by the fillers (see Mullins & Tobin 1957, Mullins 1969). For a detailed description see, e.g., Johnson & Beatty (1993a,b) or Ogden & Roxburgh (1999a,b). For typical loading/unloading curves in simple tension showing the idealized Mullins effect the reader is referred to Figure 1 in the paper by Ogden & Roxburgh (1999b) which is included in this volume. For a review of the recent activities concerning the stress response of filled rubbery polymers from both the experimental and the theoretical point of view the reader is referred to the works by Lion (1996), Ogden & Roxburgh (1999a,b) and Miehe & Keck (1999).

In addition to the Mullins effect many other inelastic effects occur under cyclic loadings. In particular, one such effect that a piece of filled rubber exhibits during cyclic loading is emphasized, namely, that the shape of the rubber after unloading in general differs significantly from its virgin shape. *Residual strains* are responsible for the change of shape in a carbon-black filled rubber. Nearly all engineering rubbers exhibit some degree of Mullins effect and residual strain (in the rubber industry also called *permanent set*). For a schematic representation of the (nominal) stress/stretch curve which is based on simple tension experiments see Figure 2 in Ogden & Roxburgh (1999b).

In this short paper we are concerned with a continuum formulation of the Mullins effect in filled rubbers that accounts for residual strains. (Mechanical and thermal hysteresis effects which are essentially rate dependent are neglected in this study.) In particular, we idealize the stress response in such a way that unloading coincides with re-loading, as illustrated in Figure 1.

We apply the theory of *pseudo-elasticity* in which the material is treated as one elastic material in loading and another elastic material in unloading. Hence, finite elasticity is able to describe an inelastic material. This has the significant advantage of convenient and simple description of the stress-strain relationships in cyclic loading and their numerical (finite element) realization. The idea of pseudo-elasticity was used, for example, by Fung et al. (1979) within the context of modelling arteries and successfully adapted for a phenomenological description of the Mullins effect (considered as an idealized phenomenon) by Ogden & Roxburgh (1999a).

We propose a material model which extends to the inclusion of residual strains the pseudo-elastic model for the (idealized) Mullins effect in filled rubber presented by Ogden & Roxburgh (1999a). It is based on a (standard) strain-energy function describing the loading path from the unstressed virgin state, referred to as the *primary loading path*, and an additive *damage function* of a form able to describe residual strains. The damage function is responsible for the unloading path which is initiated from any point on the primary loading path. The stress-strain relations for isotropic and incompressible materials are derived and specialized for the case of plane stress.

The damage mechanism is described by three (discontinuous) *damage variables* and controlled by the evolution of the principal stretches (rather than the strain-energy function which was used by Ogden & Roxburgh (1999a)). We take the damage variables to be related to the values of the principal stretches reached at the end of the current primary loading process so that on the unloading path the material response is governed by a modified strain-energy function depending on these principal stretches. Note that the three damage variables make it possible to characterize anisotropic damage, but a general discussion of anisotropy is not included here. If the principal directions vary under cyclic loading and unloading the maximum stretches must be associated with the new directions by appropriate transformations. A different approach is considered by Ogden & Roxburgh (1999b) and presented in this volume. These authors develop an energy-based pseudo-elastic model for the Mullins effect which also incorporates residual strains but is based on a single damage variable for the description of isotropic damage.

The pseudo-elastic model described has been implemented in the finite element program FEAP, partly documented by Zienkiewicz & Taylor (1989). We used a simple four node element and the same shape functions for the approximation of the initial geometry and the displacements (see, e.g., Gruttmann & Taylor 1992). For each finite element the maximum stretches are stored computationally in a history field. Two representative examples dealing with the study of a thin sheet of incompressible material under tension are examined briefly in Section 3. One leads to a homogeneous deformation state and the other to an inhomogeneous deformation state. A comparison of analytical expressions with numerical results for the homogeneous deformation case shows excellent agreement.

2 RESIDUAL STRAIN PSEUDO-ELASTICITY

We consider *incompressible isotropic materials* capable of accommodating Mullins-type damage mechanisms. For the primary loading path, which describes the response of isotropic hyperelastic materials without damage, we adopt the strain-energy function $\hat{\Psi}$, although there is dissipation on the primary loading path. The dissipation is calculated retrospectively as described in Ogden & Roxbrough (1999a). Equation (1) below is used to describe the unloading and re-loading paths, on which there is no dissipation (since, in the considered idealization, damage occurs only during *primary* loading). We postulate a pseudo-energy function Ψ of the form

$$\Psi(\lambda_1, \lambda_2, \lambda_3, \xi_1, \xi_2, \xi_3) = \hat{\Psi}(\lambda_1, \lambda_2, \lambda_3)$$
$$+\Omega(\lambda_1, \lambda_2, \lambda_3, \xi_1, \xi_2, \xi_3) - p(J-1)\,, \qquad (1)$$

where the (classical) *strain-energy function* $\hat{\Psi}$ associated with the primary loading path is supplemented by the function Ω describing the damage mechanism. The *principal stretches* are denoted by $\lambda_1, \lambda_2, \lambda_3$, the (discontinuous) *damage variables* by ξ_1, ξ_2, ξ_3 and the indeterminate Lagrange multiplier (hydrostatic pressure) by p. The latter is associated with the incompressibility constraint $J = \lambda_1\lambda_2\lambda_3 = 1$ satisfied by the principal stretches, where J characterizes the volume ratio. This condition leaves just two independent stretches.

For the unloading path we regard ξ_a as a function of the deformation so that equation (1) also describes a strain-energy function but different from that on the primary loading path. This requires a prescription of ξ_a in terms of the principal stretches λ_a. It is convenient to give a prescription in the form

$$\frac{\partial \Omega}{\partial \xi_a} = 0 \tag{2}$$

(see the work by Lazopoulos & Ogden (1998) for a more detailed explanation). This results in relatively simple equations for the principal second Piola-Kirchhoff stresses, namely

$$S_a = \frac{1}{\lambda_a}\frac{\partial(\hat{\Psi}+\Omega)}{\partial \lambda_a} - \frac{p}{\lambda_a^2}, \tag{3}$$

for $a = 1, 2, 3$. For the case of plane stress ($S_3 = 0$) elimination of p enables the two non-vanishing stresses to be expressed as

$$S_\alpha = \frac{1}{\lambda_\alpha}\frac{\partial(\hat{\Psi}+\Omega)}{\partial \lambda_\alpha} - \frac{\lambda_3}{\lambda_\alpha^2}\frac{\partial(\hat{\Psi}+\Omega)}{\partial \lambda_3}, \tag{4}$$

for $\alpha = 1, 2$.

2.1 *A specific pseudo-elastic model*

Our aim now is to particularize the above model for the damage mechanism by taking the function Ω in the form

$$\Omega = \sum_{a=1}^{3}\left[\xi_a(\lambda_a - \overline{\lambda}_a) + \phi(\xi_a)\right], \tag{5}$$

where ϕ denotes the *damage function* and $\overline{\lambda}_a$, $a = 1, 2, 3$, are the values of the principal stretches located on the primary path at the end of the (current) primary loading process. On use of (5) equations (2) yield

$$(\lambda_a - \overline{\lambda}_a) = -\phi'(\xi_a), \qquad a = 1, 2, 3, \tag{6}$$

where $\phi'(\xi_a) = \mathrm{d}\phi(\xi_a)/\mathrm{d}\xi_a$.

Next, we choose $\phi'(\xi_a)$ to have the simple form

$$-\phi'(\xi_a) = c\xi_a, \qquad a = 1, 2, 3, \tag{7}$$

where c is a positive dimensionless (material) parameter which is associated with the slope of the unloading path relative to the loading path. Using (7), we conclude from (6) that

$$\xi_a = \frac{1}{c}(\lambda_a - \overline{\lambda}_a), \qquad a = 1, 2, 3. \tag{8}$$

On the primary loading path ξ_a is inactive ($\xi_a = 0$), while on the unloading path the damage variables ξ_a are switched to the active state according to equation (8).

2.2 *Plane stress specialization*

From (4) and (5) we compute the stresses S_α as

$$S_\alpha = \frac{1}{\lambda_\alpha}\frac{\partial\hat{\Psi}}{\partial\lambda_\alpha} - \frac{\lambda_3}{\lambda_\alpha^2}\frac{\partial\hat{\Psi}}{\partial\lambda_3} - \frac{1}{\lambda_\alpha}\left(\xi_\alpha - \frac{\lambda_3}{\lambda_\alpha}\xi_3\right), \tag{9}$$

for $\alpha = 1, 2$.

3 REPRESENTATIVE EXAMPLES

For the strain-energy function $\hat{\Psi}$ we may choose any standard form. The material considered in the following examples is based on the Ogden model

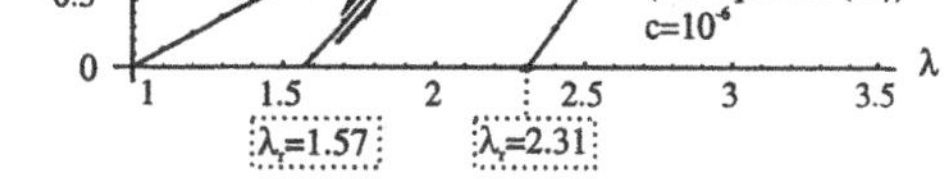

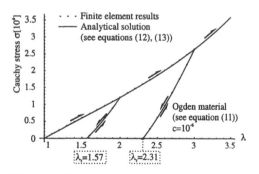

Figure 1. Simple tension loading/unloading curves for the Ogden material with Cauchy stress σ plotted against principal stretch λ. After two loading/unloading cycles a (residual) stretch $\lambda_r = 2.31$ remains.

$$\hat{\Psi} = \sum_{p=1}^{3} \frac{\mu_p}{\alpha_p}(\lambda_1^{\alpha_p} + \lambda_2^{\alpha_p} + \lambda_3^{\alpha_p} - 3) \tag{10}$$

(Ogden 1972, 1997), where μ_p and α_p are material constants such that $2\mu = \sum_{p=1}^{3} \mu_p \alpha_p$, and, for each p, $\mu_p \alpha_p > 0$, μ being the shear modulus in the reference configuration. We take the values of the constants to be

$$\left.\begin{array}{ll} \alpha_1 = 1.3 & \mu_1 = 6.3 \cdot 10^5 \mathrm{N/m^2} \\ \alpha_2 = 5.0 & \mu_2 = 0.012 \cdot 10^5 \mathrm{N/m^2} \\ \alpha_3 = -2.0 & \mu_3 = -0.1 \cdot 10^5 \mathrm{N/m^2} \end{array}\right\} \tag{11}$$

as in (Ogden 1972). Hence, $\mu = 4.225 \cdot 10^5 \mathrm{N/m^2}$. In addition, we use a damage parameter $c = 10^{-6}$.

3.1 *Simple tension test*

In this example we study a thin sheet of incompressible material under simple tension, with $\lambda_1 = \lambda$. From the incompressibility constraint $\lambda_1\lambda_2\lambda_3 = 1$, we obtain the stretch ratios in the transverse directions as $\lambda_2 = \lambda_3 = \lambda^{-1/2}$. With this kinematics and the strain energy (10) we may derive from equation (9) the non-vanishing Cauchy-stress $\sigma = \sigma_1 = \lambda_1^2 S_1$ as

$$\sigma = \sum_{p=1}^{3} \mu_p(\lambda^{\alpha_p} - \lambda^{-\alpha_p/2}) + \lambda\xi_1 - \lambda^{-1/2}\xi_3 \tag{12}$$

with, from (8),

$$\xi_1 = \frac{1}{c}(\lambda - \overline{\lambda}), \quad \xi_3 = \xi_2 = \frac{1}{c}(\lambda^{-1/2} - \overline{\lambda}^{(-1/2)}) \tag{13}$$

for the unloading/re-loading path (and $\xi_1 = \xi_2 = \xi_3 = 0$ for primary loading).

The thin sheet of material is loaded to a stretch ratio of $\lambda = 2$, then unloaded to a stress-free configuration leaving a residual stretch, λ_r say, of about 1.57. The sheet is then re-loaded up to $\lambda = 2$ by re-tracing the unloading path and then up to $\lambda = 3$ along the primary loading path, followed by a further unloading to the unstressed configuration. The resulting Cauchy stress σ is shown in Figure 1 plotted against λ for the two loading/unloading cycles. The final (residual) stretch of the sheet in the direction of the applied load is $\lambda_r = 2.31$. The finite element results (the dots shown in the figure) are in excellent agreement with the analytical solution.

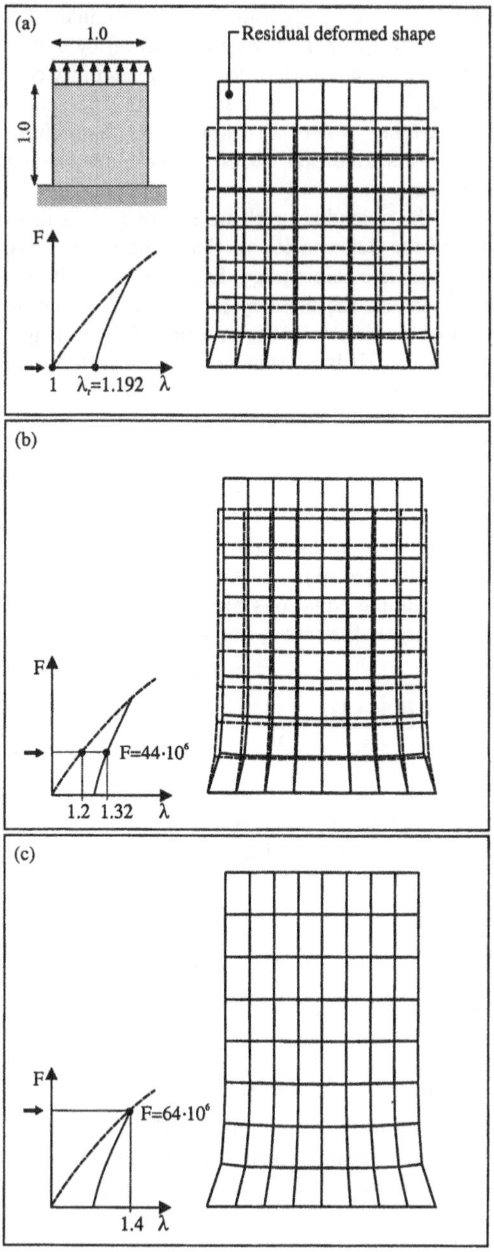

Figure 2. Finite element results from an inhomogeneous tension test for the Ogden material. Dashed curves show the geometry on the primary loading path and solid lines that on the unloading path at the values of F indicated by the arrows.

3.2 *Inhomogeneous tension test*

In order to illustrate the predictions of the model for an inhomogeneous deformation we consider a thin square sheet of incompressible material with the geometry shown in Figure 2(a). The sheet is kept fixed at the lower edge and discretized with 64 2D-finite elements, as indicated in the dashed grid shown in Figure 2(a).

A load of $F = 64 \cdot 10^6$ is applied to extend the sheet to an overall stretch of $\lambda = 1.4$. Figure 2(c) shows the resulting deformed shape. After removal of the load an overall (residual) stretch of $\lambda_r = 1.192$ remains. The dashed curves in the inset graphs shown in Figure 2(a)-(c) represent the primary loading path while the solid lines show the unloading path with F plotted against the overall stretch λ.

Figure 2(b) shows the deformed sheet at an intermediate load level of $F = 44 \cdot 10^6$ for primary loading (dashed lines and an overall stretch of $\lambda = 1.2$) and unloading (continuous lines with an overall stretch of 1.32), while Figure 2(a) presents the residual deformed shape after complete unloading ($F = 0$) compared with the original square sheet.

ACKNOWLEDGEMENTS

Support for this research was partly provided by the Austrian Fonds zur *'Förderung der wissenschaftlichen Forschung (FWF)'* under START-Award Y79. This support is gratefully acknowledged.

REFERENCES

Fung, Y.C., K. Fronek & P. Patitucci 1979. Pseudoelasticity of arteries and the choice of its mathematical expression. *Amer. Phys. Soc.* 237:H620-H631.

Gruttmann, F. & R.L. Taylor 1992. Theory and finite element formulation of rubberlike membrane shells using principal stretches. *Int. J. Num. Meth. Engrg.* 35:1111-1126.

Johnson, M.A. & M.F. Beatty 1993a. The Mullins effect in uniaxial extension and its influence on the transverse vibration of a rubber string. *Cont. Mech. Thermodyn.* 5:83-115.

Johnson, M.A. & M.F. Beatty 1993b. A constitutive equation for the Mullins effect in stress controlled uniaxial extension experiments. *Cont. Mech. Thermodyn.* 5:301-318.

Lazopoulos, K. A. & R.W. Ogden 1998. Nonlinear elasticity theory with discontinuous internal variables. *Math. Mech. Solids* 3:29-51.

Lion, A. 1996. A constitutive model for carbon black filled rubber: experimental investigations and mathematical representation. *Cont. Mech. Thermodyn.* 6:153-169.

Miehe, C. & J. Keck 1999. Superimposed finite elastic-viscoelastic-plastoelastic stress response with damage in filled rubbery polymers. Experiments, modelling and algorithmic implementation. *J. Mech. Phys. Solids*, in press.

Mullins, L. 1947. Effect of stretching on the properties of rubber. *J. Rubber Research* 16:275-289.

Mullins, L. 1969. Softening of rubber by deformation. *Rubber Chem. Technol.* 42:339-362.

Mullins L. & N.R. Tobin 1957. Theoretical model for the elastic behavior of filler-reinforced vulcanized rubbers. *Rubber Chem. Technol.* 30:551-571.

Ogden, R. W. 1972. Large deformation isotropic elasticity: on the correlation of theory and experiment for incompressible rubberlike solids. *Proc. R. Soc. Lond.* A326:565-584.

Ogden, R. W. 1997. *Non-linear elastic deformations.* New York: Dover.

Ogden, R.W. & D.G. Roxburgh 1999a. A pseudo-elastic model for the Mullins effect in filled rubber. *Proc. R. Soc. Lond.* A, in press.

Ogden, R.W. & D.G. Roxburgh 1999b. An energy-based model of the Mullins effect. Paper included in this volume.

Zienkiewicz, O.C. & R.L. Taylor 1989. *The Finite Element Method, Basic Formulation and Linear Problems.* 4th edn., Volume 1. McGraw-Hill: London.

Constitutive Models for Rubber, Dorfmann & Muhr (eds) © 1999 Taylor & Francis ISBN 90 5809 113 9

Modelling inelastic rubber behavior under large deformations based on self-organizing linkage patterns

Jörn Ihlemann
Institute of Mechanics, University of Hannover, Germany

ABSTRACT: Up to now only phenomenological constitutive models are available to simulate the inelastic behavior of rubber under large deformations. The physical reasons of the observed macroscopic effects are largely unknown. Just in the case of complicated material behavior phenomenological models are connected with marked uncertainties of their predicitions. In addition a lot of measurements are needed to fit the parameters to an individual material.
Reducing these disadvantages a new theory is written on, which explains the mechanical behavior of rubber as emergent effects of complex interactions of submicroscopic units. A screening program is developed to inspect the predictions of the theory. The yielded results contain many important effects of rubber behavior.

1 INTRODUCTION

The simulation of the load-/deformation-behavior of technical products increases in value for their developing and optimization and also for analysing mechanisms of damaging and predicting the lifetime. The Finite Element Method is used more than other numerical procedures for these purposes. To simulate structures, for all participated materials a constitutive model is needed, which allows the calculation of the complete stress tensor for any deformation. The Finite Element Method supplies by the evaluation of the constitutive model at single points an approximate solution of the distribution of stresses and strains to the whole continuum. The crucial problem often consists in the developing of a suitable constitutive model, especially in the case of complicated material characteristics.

The methods to search for a constitutive model can be divided into three groups: The phenomenological method, the derivation of the constitutive model from a physical motivated model and the direct determination of the complete constitutive model from physical given facts.

In the case of the phenomenological method a mathematical coherence is guessed from measurements considering physical restrictions. After that this coherence is generalized to any deformations with attention to physical restrictions. The constitutive model should be adjustable to several materials with similar behavior (e.g. varying rubber compounds). This is tried to achieve by a set of free parameters, but the selection of suitable parameters is very difficult in the case of phenomenological models.

If the basic physical processes at the molecular level are sufficiently known to predict the characteristical relations between stresses and deformations, this physically motivated model can be used to produce a constitutive model. Because in this case at least the functional relations come close to the real ones, only a few measurements are necessary to adjust the parameters to the individual materials.

If a complete physical model exists, which is even able to determine concrete stress- and strain-values based on physical foundations and if also the molecular structure of the material is precisely known, than the constitutive model can be directly and completely determined without adjustment of parameters.

As a rule phenomenological descriptions are most easily to find and in the case of relatively simple material properties their use is unproblematic. This applies for example to isotropic elastic materials. With regard to more complicated materials like inelastic rubber which is strongly influenced by the loading-history the phenomenological method shows unavoidable disadvantages. Even with good correspondence between measured and calculated data for several load paths the phenomenologically assumed coherence may strongly differ from the physical facts. In addition other disadvantages of the phenomenological method appear:

1. The optimum number of the parameters is unknown. Too many parameters usually cause bad re-

sults in the case of loads, which are not included in the parameter identification. Not enough parameters result in a too small variety of the model to describe several materials of the same group.

2. A physical interpretation is missing for the parameters. A simultaneous optimization of geometry and material of a structure is not possible, because it is unknown, which parameter combinations represent realizable materials.

3. Conclusions on physical parameters, which cannot be measured, are not possible.

Nevertheless, phenomenological descriptions are often the only available way to handle complicated materials, because the physical reasons of the material behavior are not sufficiently known.

Physically motivated constitutive models do not have most of the disadvantages the phenomenological constitutive models show. The only problem of this class of models consists in the necessity of carrying out an identification of the corresponding parameters for every new material. This requires a representive set of messurements. Apart from this fact the physically motivated constitutive models are an appropriate tool for the simulation of the material behavior of rubber using the Finite Element Method.

A complete physical material description represents the ideal case of a constitutive model. However, it is hardly possible to achieve. Moreover, the direct determination of the parameters needs detailed knowledge of the material structure so that the analysis of the material structure often could require more time than the identification of the parameters on the basis of measurements.

In the case of rubber till now the physical reasons of its complicated material behavior under large deformations is unknown for the most part. However, just these large deformations occur very often in technical applications of rubber. Therefore, the only possibility of modelling the behavior of rubber materials under large deformations in the past consisted in following up various phenomenological approaches to constitutive descriptions. One successful example of this kind of constitutive models in detail was presented by Besdo and Ihlemann (Besdo & Ihlemann 1996).

Apart from the purely phenomenological constitutive models for rubber other approaches exist, which include physically motivated parts. Above all some constitutive models for elasticity belong to this class of material descriptions (e.g. entropy elasticity). To these physically motivated approaches for rubber elasticity very often dissipative terms are added, which are developed from phenomenological reflections. Such kinds of material descriptions assume that filler causes exclusively energy dissipating effects. They do not take into account that filler also has energy storing properties and consequently can influence the elastic behavior.

Thus, the question, whether the introduced physical connections really determine the behavior of filled rubbers, remains still open. To find out this, the physical causes of the dissipative effects have to be clarified.

To explain the inelastic behavior of rubber a new theory is presented in the following. This theory reduces marked effects of rubber behavior to emergent effects of complex interactions of submicroscopic units. If this assumption is correct the interactions of a high number of relatively simple elements produce macroscopic properties, which are completely different from the properties of the singe elements themselfes. Thus, not the characteristics of the basic units will be examined more and more precisely (e.g. polymer-filler-contact, filler-filler-contact, etc.), but it will be tried to understand the long-range effects of the interactions between the basic units.

The prediction of emergent effects based on theoretical considerations is naturally difficult. Therefore, the new theory will be examined with a screening program composed of many simple units. Actual first results of the program show a behavior similar in many effects to the behavior of rubber. In comparison with the molecular structur of rubber an extreme abstraction is used to develope the program. Considering the simplifications the multitude of similar effects is amazing. The screening program, which can only be described here in parts, is not intended as a constitutive model for the FEM, but it is intended to be the basis for the development of physically motivated constitutive models.

2 MATERIAL BEHAVIOR

At first the typical behavior of industrial used rubber materials under large deformations shall be illustrated by the results of measurements carried out with a spe-

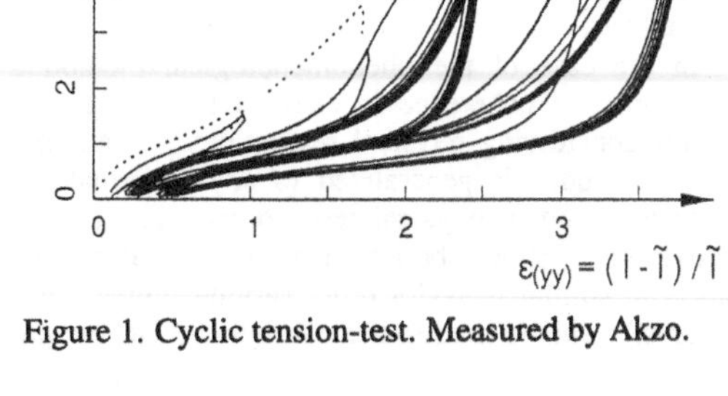

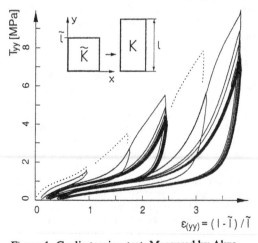

Figure 1. Cyclic tension-test. Measured by Akzo.

cial kind of rubber being used in air springs. It is a carbon black filled chloroprene rubber. The measurements were carried out by Akzo Nobel at Arnheim (Netherlands). The used specimen was initially unloaded, so that the influence of the loading-history could be found out.

The extensiv experiment was a cyclic, purely uniaxial tension-test with no restriction to the cross section. It took place under room temperature. The straining velocity was about $40\%/sec$. The upper bound of loading cycles was prescribed in terms of elongations whereas the lower bound was denoted by almost vanishing forces. At first a cyclc loading with an upper strain bound of around 100% was applied until stationarity was reached, than this was repeated for several times with increasing upper bounds. The highest strain bound was 380%. For each of these upper bounds the stresses decreased from cycle to cylce until the stationary one was reached. The size of this stress decrease grows up with the size of the upper strain bound.

The results of the tension-test are shown in Figure 1. The 1. Piola-Kirchhoff-stress T_{yy} plotted versus the strain ε is strictly proportional to the applied loading force. The highly nonlinear loading and unloading curves, expecially of the stationary cycles form a significant hysteresis loop. The enclosed area is due to the dissipation of mechanical energy. As a result of this hysteresis the deformation after unloading does not vanish. A complete recovery of this remaining deformation is possible after a long time by creap only. The remaining deformation increases with the upper strain bound.

Comparison of the five stationary cycles shows that the stress values and the shape of the hysteresis loop for equal strain values depends strongly on the prestraining, i.e. the maximum of the strain in the past. Higher prestraining softens the material.

The first cylce which reaches a new upper strain bound, called primary cycle, shows another characteristic as all the other cycles of the same strain bound, the secondary cycles. The form of the loading curve of an unloaded specimen up to the strain bound can be estimated from the primary cycles of the previous strain bounds. In contrast to all secondary cycles these loading curves would only show a slight curvature. The loading curves of all secondary cycles start with a slight degressiv curvature, followed by a long, nearly linear phase, a short phase with a strong progressiv curvature and finally again a nearly linear phase. The unloading curves of all cycles are similar to each other, also those of the primary cycles. In the case of constant upper strain bound the unloading curves nearly coincide. In the case of the presented measurements the upper strain bounds of the secondary cycles slightly differ from the one of the primary cyle and also the unloading curves are shifted horizontally. The nonlinearity of the unloading curves is even stronger than the one of the loading curves. Also in this case a change of curvature exists.

3 SELF-ORGANIZING LINKAGE PATTERNS

3.1 *Molecular structur of rubber*

The basic strength elements of rubber at the submicroscopic level are very long chain molecules which are cross-linked with each other at some points. These molecules form an intricate ball, in which the individual molecules have an irregular winding form. Between the chain molecules a filler dispersion is located. The filler particles interact with the chain molecules and with each other primarily by physical linkages. Those linkages take also effect between the chain molecules, but usually in a lower level. Strength and quantity of the physical linkages depend on the choice of ingredients, but also e.g. on the surface of the filler particles and their size.

The chemical linkages inside the chain molecules and between the chain molecules and the cross-linkage-elements have a high stability. Those linkages arise only under supply of external energy such as during the vulcanizing. Damages to those linkages are usually permanent. But normally the cross linkage is as wide-meshed, that the basic structure of cross-linked chain molecules is able to bear large external deformations without damaging the chemical linkages by straightening and dissolving loops. These deformations of the chain molecules need an external work, because the chemical linkages come into constellations with higher energy content.

Because of its wide-meshed characteristic the basic structure has several possibilities to follow an external deformation. Only in the case of very large deformations these possibilities are restricted by the high resistance against further deformation of nearly completely straightened molecular segments, caused by direct loading of the chemical linkages. In this case the deformation will be shifted to other areas with lower loading. When further shifting is not possible, single chemical linkages will be damaged. Deformations which cause this kind of damaging should not be treated in this context.

In comparison to the chemical linkages the physical linkages are far weaker and they have a very little range of influence. Therefore, physical linkages have a lower capability to store energy, but they can be present in large numbers. They arise spontaneously as soon as two suitable elements come into contact. In contrast to the chemical linkages a supply of external energy is not necessary.

Thus, rubber contains two different energy storing elements. On the one hand the stable chemical linkages with their large energy capacity are able to transfer forces across relatively large distances along the chain molecules. On the other hand exist the numerous physical linkages with lower capacity and a very little range of influence, which are able to arise spontaneously.

3.2 *Basic idea*

The presented measurements and the outlined knowledge of the submicroscopic structure of industrial used rubber materials are the basis of a new theory, which interprets important characteristics of the material behavior of rubber by an emergent effect of the interactions of basic units. In this case the global mechanical behavior would not correspond with the average value of the properties of the basic units. Instead of this, complete new properties arise from complex interactions of numerous relatively simple units (Ebeling et al. 1998). After a short description of the theory first results of a screening program which serves to test the theory will be shown.

During deforming of a rubber specimen movements of the winding chain molecules to each other start. Strength and direction of this movements strongly depend on the local conditions of the individual segment relating to e.g. contact to filler particles, near cross linkages, curvature of the segment and the direction of the segment at the moment. In this way arises a large variety of local movements, which average value agrees with the change of the external deformation.

These varying local movements are the basis of the idea presented here. On this level a change to average values is not allowed, otherwise the essential effects will be lost, because the variety of movements causes complex interactions between the stable meshwork of cross-linked chain molecules and the flexible physical linkages.

A large variety of local movements is equivalent to highly different conditions for the physical linkages. In areas with high relative velocities between neighbouring molecules the physical linkages will be destroyed very fast, their lifespan is short. In other areas with at the moment small velocities more physical linkages are able to arise and to hold. Due to this differences the distribution of the physical linkages takes effect back to the local movements. The stiffness of some areas increases, whereas in other areas the deformation increases. Thus, the differences in the local movements will grow up.

Therefore, beside the global deformation a special distribution of the physical linkages, called linkage pattern, arises. The stable network and the linkage distribution influence one another and both influence the conditions for arising and destroying of physical linkages which again intensify the irregularity of the linkage distribution. This is a complex feedback, which can be understood as a selection of the physical linkages. In this context the evolving linkage distribution is called self-organizing linkage pattern.

The forming of an individual pattern for each deformation is not a quality of the basic units, but it is a new quality caused by the numerous interactions between the basic units inside the entire system. Thus, the organizing of the linkage pattern is an emergent quality of the system.

For each course of deformations an individual linkage pattern arises. In the case of deformations with constant directions of its Lagrangean eigenvectors, e.g. tension- or compression-tests, an intensification of the linkage pattern without changing its characteristic is expected. On the other hand the pattern will gradually changes itself during deformations with changing Lagrangean eigenvectors like simple shear, because the local conditions will permanently change, too.

Which consequences would have such an evolution to the stress-strain-behavior of the system? During the first loading of a specimen the initial unspecific linkage distribution which results from the vulcanizing, would gradually evolve to the typical linkage pattern of the current deformation. Thus, many linkages permanently arise and others are being destroyed. These processes are corresponding to a large energy dissipation at the macroscopic level. The resulting course of the stress-/strain-curve is not easy to predict, apart from the point, that the stress values will be much higher than the theoretical values result from the same system without physical linkages.

In the case of unloading the feedback of the linkage distribution to the local deformations prevents the complete destruction of the existing pattern. Therefore the resistance of the system to the next loading will be less than the first time, assumed the deformations are equal. Not until approaching the maximum deformation reached in the past, the remains of the initial unspecific linkage distribution determine the system behavior again, because the pattern is not yet completely developed in this area. Thus, in the case of repeated loading stresses will be far smaller than at the first loading, except in the near of the prestraining, where the stresses will grow up near to the stresses of the first loading. Probably the stress-/strain-curve will be similar to the unloading curve. Repeated unloadings should be similar to the first one.

These speculations correspond to the observed behavior of rubber. Of course, the correctness of the theory and the speculations about its consequences to the material behavior of rubber must be checked. To make this check possible, an extensiv computer program were developed. This screening program is not designed for being used as constitutive model within Finite Element Analysis. It serves only to inspect, whether the presented theory is able to produce some effects similar to those effects, which characterize the mechanical behavior of rubber. To develop a constitutive model for Finite Element Analysis based on the theory and the results from the program is planed for the near future.

3.3 *Screening program*

It is inherent in emergent processes that a multitude of mutually influencing basic units has to act in combination. Accordingly, it is necessary to include a multitude of linkage carriers in the computer progam to make sure that the emergent effects can occur in the simulation. In consequence it has to be taken into account an extremly high need of computer memory and calculating time.

Considering these facts a nearly true to nature model of the molecular structure including filler particles, a multitude of chain molecules and the attempt to model the interactions between the single atoms in a realistic manner would not be useful. The only possibility to achieve meaningful results despite of the limited computer capacities is to realize an extrem abstraction of the processes to simulate. This abstraction here is restricted to those elements and effects, which are necessary to the characteristic emergent action. These are the following:

1. Variety of local deformations by movements of the winding chain molecules,
2. Transfer of forces across large distances along the chain molecules connecting local areas,
3. Flexibility of the winding chain molecules until they are fully stretched and after that they have a high resistance against further extension,
4. Coupling the chain molecules by cross linkages,
5. Physical linkages as energy storing elements with a very little range of influence which arise spontaneously as soon as two suitable elements come into contact.

The realized model cannot be completely presented here. But some important facts should be explained to demonstrate the necessary grade of abstraction. On the other hand should be shown, that the results actually arise from emergent effects and not from the properties of the single elements.

In the model deformations are stepwise changed, when static equilibrium is determined at each step. Released energy quantities, which would cause dynamical effects are interpreted as dissipated energy in the form of released heat. The model contains no effects like friction or damping, which are emergent themselves.

Part of the necessary abstraction is the restriction to one kind of simulated deformation. For this the uniaxial tension-test is selected. It is suitable because of its constant Lagrangean eigenvectors and the relatively small influence of threedimensional effects to the result. As an effect of the constant eigenvectors similarity of the local deformations to themselves during the external deformation progresses is expected. Thus, the evolving linkage pattern will be intensified continuously. In this property the model corresponds directly to the tension-test, otherwise similarities exist only at a highly abstract level.

Basic elements of the model are called linkage carriers. These elements represent the atoms of the chain molecules participated in physical linkages as well as the surfaces of the filler particles. Filler itself is exclusively represented by a large number of possibilities for creating physical linkages. To reduce the system of equations the linkage carriers have only one degree of freedom. Hereby the natural reasons for the differences in the local deformations cannot be simulated. Therefore, the variety of local movements have to be generated separately.

The characteristic curves of all linkages are simplified as far as possible. Cross linkages are modelled as non elastic connections. The physical linkages are represented as simple springs come into being spontaneously and being destroyed when they are overstretched. Only this property causes energy dissipation in the whole model. The resistance of the chain molecules against stretching is represented by linear springs, too. Progressing extension is limited by an upper bound. Higher extensions are prevented with a penalty method. This upper bound represents the fully stretched state of the chain molecules. After all these simplifications only the properties of the linkages as energy storing elements approximately remain.

The stiffnesses of the participated elements, the numbers of cross linkages and locations for physical linkages and the extent of the variety of movements are controllable with model parameters. All irregular distributions of elements like cross linkages are determined with the help of random numbers.

3.4 *Results of the screening program*

The computer program does not yield perfectly smooth curves, because of the permanent inserted and otherwise erased springs, which represent the physical linkages. Only the numerous degrees of freedom and in case of need the simultaneous calculation of several versions of the model with different random numbers yield curves with sufficient precision to recognize the global effects.

The model was loaded in a similar manner like the rubber specimen in the presented tension-test. In view of the high calculating time for all upper strain bounds fewer cycles were calculated than measured. Besides the first and smallest upper bound was leaved out, because the start conditions of the program are still unsuitable at the moment. But this defect have marked consequences only for a small range of deformation. Despite the extreme abstraction, needed to realize the program, first results (Fig. 2) reproduce an amazing multitude of effects, which are typical for the behavior of rubber:

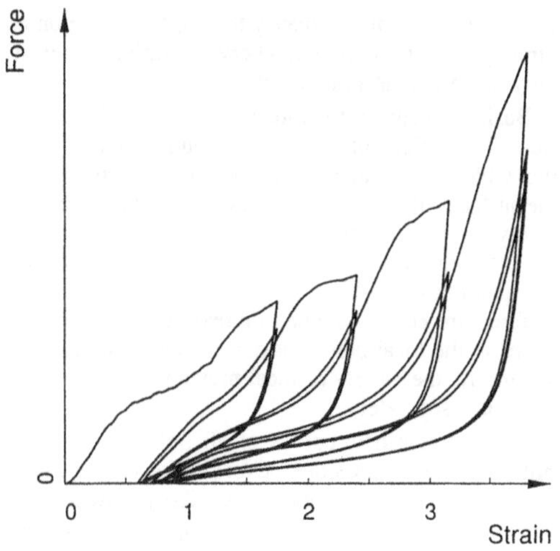

Figure 2. Results of the screening program.

1. Highly nonlinear loading and unloading curves with correct changes of curvature,
2. Increasing softening with growing up prestraining,
3. Proportions of hysteresis loop,
4. Stress decrease from cycle to cycle under constant upper bound, growing up with the size of the upper bound,
5. Remaining deformations increasing with the upper strain bound,
6. different characteristics of primary and secondary cycles,
7. similar unloading curves of all cycles.

Certainly, there are marked differences between the calculated and the measured data, but surely they can be reduced considerably by an adjustment of the model parmeters. Also the improvement of the start conditions will produce better results. Up to now an adjustment of the parameters is carried out as far as necessary to recognize the effects. Fundamental effects, which are not recognizable in the presented example do not turn up in other parameter constellations, but the intensity of the effects changes in many ways. An optimization of the parameters needs very much calculating time, but an optimization is not the primary target.

The most important fact is, that many typical effects of the material behavior of rubber are also qualities of the drastically abstracted model. In view of these facts it seems to be sure, that emergent effects, described in the presented theory, determine the mechanical behavior of rubber for the most parts and represent the only explanation for these parts. Just the extreme reduction of the model to the substantial qualities excepts a coincidental correspondence to the behavior of real rubber in so many important effects.

Thus, the computer model represents an effective tool to intensify the understanding of the processes taken place in rubber during large deformations and to develop physical motivated constitutive models. In this context the connections between model parameters and structure qualities like cross linkage density or filler incidence is especially convenient. Besides the model provides information about the development of internal parameters like stored and dissipated energy, which can not be measured or only under large difficulties.

4 CONCLUSIONS

The presented theory of self-organizing linkage patterns, which explains the mechanical behavior of rubber as an emergent effect of the complex interactions of the participated submicroscopic units, has confirmed by the results of a special developed screening program. In spite of necessary drastic simplifications of the physical mechanisms within the model, the results show in many aspects marked similarities to the behavior of rubber.

Thus, the theory together with the screening program represent an effective tool to develop physically motivated constitutive models, with its help the mechanical behavior of rubber can be predicted more precisely as before. Constitutive models of such kind are able to make the development of technical products more effective, because in principle within the simulation not only the geometry, but also approximately the material properties can be directly optimized.

The next planed step of research is to improve the initial state of the screening program, because it is not satisfactory at the moment. Afterwards the model will be used to calculate series of tests with varying parameter values to determine the influence and interaction of the model parameters. This will be the preparatory work to develope a new physical motivated constitutive model.

REFERENCES

Besdo, D. & J. Ihlemann 1996. Zur Modellierung des Stoffverhaltens von Elastomeren. *Kautschuk Gummi Kunststoffe* 49: 495-503.

Ebeling, W. & J. Freund & F. Schweitzer 1998. *Komplexe Strukturen: Entropie und Information.* Stuttgart, Leipzig: Teubner.

Constitutive Models for Rubber, Dorfmann & Muhr (eds) © 1999 Taylor & Francis ISBN 90 5809 113 9

Experimental and computational aspects of cavitation in natural rubber

S.L. Burtscher
Institute of Structural Concrete, Vienna University of Technology, Austria

A. Dorfmann
Institute of Structural Engineering, University of Applied Sciences, Vienna, Austria

ABSTRACT: General constitutive equations for hyperelastic materials are obtained from a strain energy function expressed either in terms of strain invariants or principal stretches. For most applications the strain energy functional does not need to include dilatational components. However, the pressure-volume relation for nearly incompressible materials needs to be explicitly accounted for when rubber components are highly constrained. Thus, the hyperelastic response needs to be expressed in terms of dilatational and deviatoric components. Experimental evidence is reviewed to show that rubber is subjected to a loss of stiffness attributed to cavitation damage when subjected to a hydrostatic tensile stress state. The critical pressure is identified for which microscopic material imperfections will tear open to form internal bubbles and cracks.
Cavitation damage in rubber is associated with a significant reduction in the bulk modulus. Thus, a variable bulk modulus can best be used to describe the behavior of rubber when cavitation damage occurs. The introduction of a cavitation-damage modulus does suggest a simple approach to realistically represent the mechanics of cavitation in rubber solids.

1 INTRODUCTION

The response of rubber is often represented in numerical modeling as isotropic and hyperelastic. Thus, the loading and unloading paths coincide by definition and energy dissipation is not accounted for. Furthermore, such characteristic behavior as Mullins effect, strain rate dependence or different stiffness in static or dynamic response is not taken into consideration either. In spite of these shortcomings, hyperelastic formulations are commonly used for the analysis of rubber components.

General constitutive equations for hyperelastic response are formulated in terms of a frame indifferent stored energy function expressed either in terms of invariants of the right Cauchy-Green deformation tensor or as a function of principal stretches. Historically, these mathematical formulations are based on a "phenomenological" approach, which is based not on a molecular or physical justification, but purely on mathematical reasoning. Phenomenological models use the most convenient formulation to describe the observed properties without being concerned with their physical explanation or interpretation.

The hyperelastic response is expressed in terms of dilatational and deviatoric components. This formulation follows from the multiplicative decomposition of the deformation gradient into isochoric and dilatational parts introduced originally by Flory (1961). The strain energy density function may be simplified by neglecting the dilatational term for (nearly) incompressible materials. However, when the components are subjected to large hydrostatic stresses, the dilatational term need to be accounted for.

Material damage in natural rubber is a complex process which may involve chain and multichain damage, microstructure damage and microvoid formation (cavitation). The purpose of this study is to focus on the loss of experimentally observed stiffness when rubber is subjected to hydrostatic tension or dilatant stress. In fact, a critical state may be reached when internal imperfections or cavities suddenly grow inside the rubber. Microscopic observations suggest that rubber-like solids always contain cavities with a wide range of sizes which will tear open to form a running crack when the maximum extensibility of the rubber is reached. Currently, we know little about these precursors. Are they really submicroscopic bubbles, inclusions of dust or possible weak cross-linked regions? In any case, these microscopic material imperfections are the origin or starting point for bubbles and crack formation when local hydrostatic tensile stresses larger then a material dependent critical value are generated.

In spite of the importance outlined above, the behavior of rubber subjected to hydrostatic tension did not receive the necessary attention, even though rubber components may show softening in the stress-strain response which ultimately may turn into premature material failure. Typical applications where stress levels capable to initiate cavitation may be approached are multilayer elastomeric bearings. Thin rubber layers are highly restraint by vulcanization to top and bottom steel plates, which are subjected to bending or tilting, as discussed by Gent & Meinecke (1970). Elastomeric bearings with an anisotropic layout given by the inclusion of V-shaped steel plates are another situation where cavitation damage has been noted, HARIS (1999).

2 HYDROSTATIC LOADING OF RUBBER

The shear modulus of natural rubber varies with temperature and straining between a lower value of 0.5 up to 6.0 MPa. On the other hand, the bulk modulus is in general assumed strain independent with values between 2000 and 3000 MPa depending on the vulcanisate. Because of this high ratio of bulk to shear modulus rubber can often be regarded as an incompressible material restricted by the constraint $J = \det(\mathbf{F}) = 1$. Thus, the strain energy functional depends only on the first and second strain invariant I_1 and I_2. Analysis results using fully incompressible material models show good accuracy for applications involving plane stress states like shell or membranes (e.g. inflated balloon), however it can be a problem where the material is highly constrained like in a plane strain condition. In this case the assumption of strictly incompressibility must be relaxed by introducing a dilatational term to the strain energy functional.

The condition of highly constraint rubber can be found in multilayer elastomeric bearings where the pressure volume relation of the material must be accounted for to obtain realistic results. In Figure 1 the cross-section of a single rubber layer under plane strain condition and vulcanized to top and bottom steel plates is shown. The deformed shapes and the corresponding stress-states on the free edges and in the center region are indicated schematically for compressive and tensile loading. Due to symmetry no out-of-plane deformation does exist. On the free edges stress components exist along the loading as well as in the out of plane direction, but the material is not constrained in the third direction. This situation changes toward the center where the material is confined in all three directions and a positive or negative hydrostatic stress state develops. In compression the rubber can easily withstand high pressure without damage to occur. On the other hand, in hydrostatic tension internal cracking nucleated by precursors is shown analytically to expand to an indefinitely large size under a hydrostatic stress of approximately 5E/6, where E is the elastic modulus of rubber, (Gent 1990). The initial size of the existing precursors has proven to have an important influence on the minimum cavitation stress. It was observed that small cavities are much more resistant to expansion and fracture than larger ones. For small voids in the order of about 0.5 μm large pressures up to 3E are necessary for cracks to open. Since microscopic voids with a wide range of sizes are always present, cavitation damage starts to develop at relatively low stress values. Crack initiation is not visible from the outside, it can be detected only by a sudden drop in load-extension data and in some cases by audible cracking sounds.

3 EXPERIMENTAL INVESTIGATIONS

Busse (1938) and Yerzeley (1939) first reported that rubber subjected to a hydrostatic tensile stress causes internal rupture known as cavitation. Experimental results by Gent & Lindley (1958) showed that the critical stress for rupture is related to the elastic characteristics of the material and independent of strength properties. Taking advantage of the incompressibility of rubber, Gent & Tompkins (1969) were able to design a complimentary method to confirm these findings. Soluble gas was applied under high pressure to rubber blocks filling all microscopic voids within. It was found that upon a sudden release of the external gas pressure, the cavities within the rubber did expand and ultimately tore open into gas-filled bubbles, Figure 2. In this second method, a hydrostatic pressure is applied to the interior of the cavities which has the same dilating effect than the application of an equivalent far field triaxial tension, see Figure 3.

Pond (1995) showed that with sufficient cyclic loading at a stress level below that of the minimum cavitation stress, permanent damage could be prevented or at least reduced.

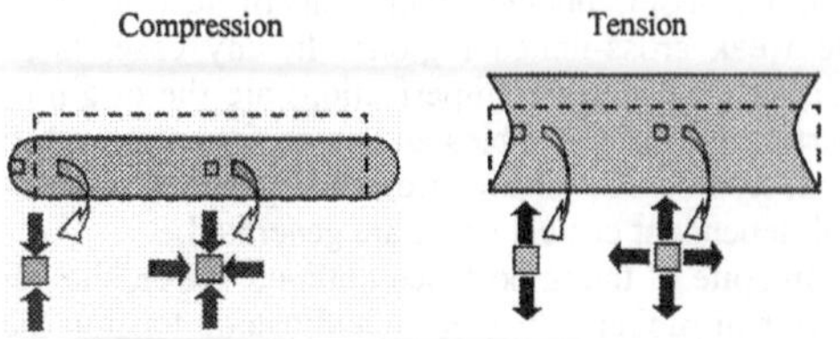

Figure 1. Deformed shapes of a single rubber layer.

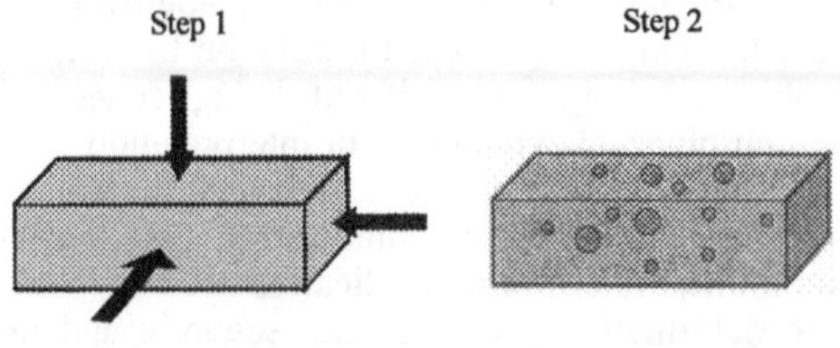

Figure 2. Dissolved gas in rubber under pressure before and after cavitation, (Gent 1990).

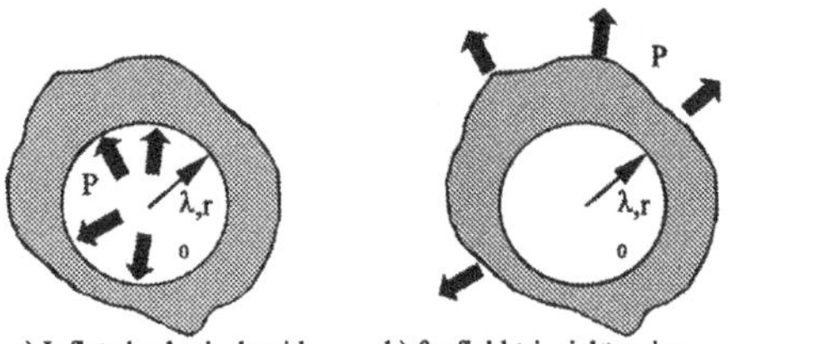

a) Inflated spherical void b) far field triaxial tension

Figure 3. (a) Sketch of an inflated spherical void. (b) Spherical void under far-field triaxial tension.

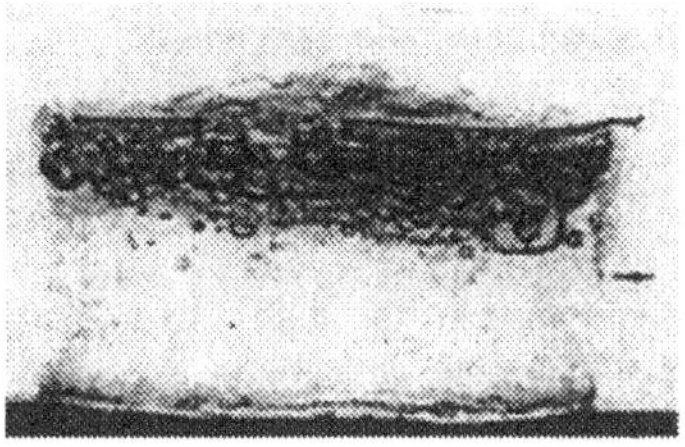

Figure 4. Bubble formation of specimen subjected to pressure gradient, (Gent & Tompkins 1969).

3.1 *Cavitation tests using dissolved gases*

In the complimentary method suggested by Gent & Tompkins (1969) a hydrostatic pressure is applied to the interior of cavities rather than a tensile stress to the rubber at infinity. Because of the incompressibility of the material, these two situations are equivalent. Gent and Tompkins were able to proof the existence of a critical pressure P_c of the internal gas that is sufficient to cause submicroscopic voids to expand, the value depending upon the initial size. It is shown that the value of P_c is extremely large for small voids, having radii in the order of 1-10 Å, but they approach the lower limit of 5G/2 for cavities of the order of 10^{-7} m initial radius. In other words, the minimum supersaturation pressure of a dissolved gas to from visible bubbles is 5G/2. These openings will be filled when gas is dissolved in the rubber. When the external pressure is then suddenly released, the internal gas pressure will inflate the bubbles. The kinetic of expansion is complicated depending among others on the surface energy of rubber, on the coefficient of diffusion, on the solubility of gas in rubber and on the material properties. As the bubbles expand in volume, the inside pressure reduces and diffusion into the cavity will occur eventually providing a means of relieving the supersaturation pressure in the specimen. In laboratory applications, the holes do not expand to infinite size for two reasons: a) the limited supplies of gas in the rubber specimen and b) diffusion of gas outward through the edges of the rubber block.

Trial end error experimental analysis provided an upper and lower bound of the supersaturation pressure that initiates bubble formation. Determination of the necessary minimum cavitation pressure was provided by an experimental arrangement that created a linear pressure gradient through a rubber specimen. On two opposite sides of the rubber sample different gas pressures were applied for a sufficient period of time to attain pressure equilibrium. After the sudden release of pressure on both sides, bubbles did form starting form one side of the specimen only. The specimen was sectioned afterwards, see Figure 4. The pressure in the transient region was determined based on a linear variation between the to applied end values. A pressure value

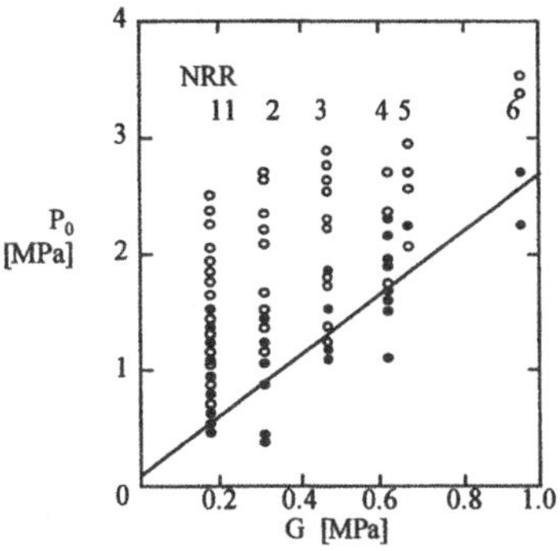

Figure 5. Cavitation results for natural rubber. Open Circles: visible bubbles formed. Filled-in Circles: No bubbles formed, (Gent & Tompkins 1969).

of approximately 5G/2 was determined as the minimum pressure value P_c.

To validate the critical conditions for bubble formation different pressures were applied to compounds characterized by different values of G. The lowest gas pressure at which bubble formation was observed was in good agreement with the 5G/2 value found before. Table 1 summarizes the shear modulus G for the rubber compounds used by Gent and Tompkins. Values for G are taken as one-third of Young's modulus determined from small extension stress-strain tests.

In many cases however, no bubbles appeared even at substantially higher pressures. The reason might be that for this particular compound no initial cavities of 10^{-7} m or larger were present. Figure 5 shows typical test results for a natural rubber compound. The solid line is the relation 5G/2.

Similar findings were reported in later studies done by Gent & Park (1984), Cho & Gent (1988) and Gent & Wang (1990).

Table 1. Shear Modulus G for the natural rubber compounds used in the experiments of Gent & Tompkins (1969).

Material	NR 1	NR2	NR 3	NR4	NR5	NR6
G (MPa)	0.19	0.31	0.47	0.62	0.68	0.96

3.2 *Cavitation tests on constraint specimens*

Pond (1995) performed a series of tension tests on different natural rubber vulcanisate, Table 2. The experiments were conducted on single rubber discs bonded between rigid plates and repeatedly pulled in tension. In the center part of the disk a hydrostatic pressure will be created to investigate cavitation damage to the rubber. Each rubber disc had a diameter of 50 mm and a thickness of 5 mm, see test setup in Figure 6.

Different loading routines were designed in order to quantify cavitation. Initially, the specimens were subjected to stress levels just above the initial cavitation stress. Accumulated internal damages due to a sudden grow of cavities and cracks could be detected by deviations in the load extension response. A subsequent microscopic analysis did in fact confirm a number of cracks and bubbles in the center of the cross section. There, the hydrostatic tensile stress is obviously larger than near the free surface. Test specimens were also subjected to repeated tensile loading just below the critical cavitation stress. Upon microscopic analysis, no cavitation damage was noticed.

Figure 6 shows stress strain data of the rubber disk using compound 1 (Table 2) repeatedly pulled in tension to a maximum constant stress value of 1.25 MPa. Initial cavitation occurs around 0.97 MPa during the first load cycle. It is interesting to note that no further damage occurs during the remaining 9 cycles. Cavitation damage is clearly visible by the fluctuating stress-strain response. Similar observation can be made for the natural rubber vulcanisate containing carbon black filler indicated as compound 2. It is subjected to 10 loading cycles each up to a maximum constant stress of 2 MPa, see Figure 7. It is noticed again that cavitation damage occurs in cycle 1 at around 1.6 MPa with no further damage during the remaining 9 cycles.

Figure 8 shows test results of a rubber disc using again compound 2. In contrast to before, the test specimen is subjected to 28 load cycles with an increase in the maximum stress per cycle equal to 0.025 MPa. The load sequence is designed for the first cycle to have a maximum applied stress of 1.4 MPa, just lower than the minimum cavitation stress of 1.6 MPa found previously for the same compound. After completion of all 28 cycles a maximum

Table 2. Physical properties initial cavitation and failure stress, when loaded continuously up to failure.

Material	Comp 1	Comp 2	Comp 3
HAF black, parts p.h.r.	-	20	80
Young's modulus, E [MPa] *	1.6	2.6	5.6
Initial Cavitation [MPa] #	0.94	1.6	2.6
Failure Stress [MPa] #	1.40	1.8	3.2

* Young's modulus found in uniaxial tension up to ±0.1 MPa.
# with an constant increase in load, no cycling.

applied stress of 2.1 MPa was reached and no cavitation damage observed. Subsequent microscopic analyses did confirm this finding.

This seems to indicate that cyclic loading below the initial cavitation stress could completely prevent

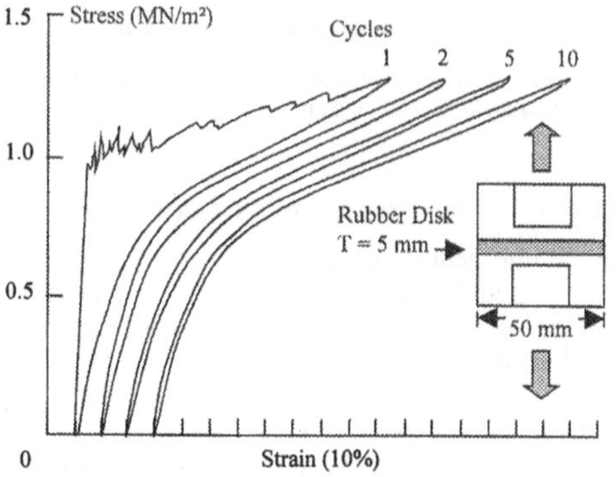

Figure 6. Unfilled rubber disk repeatedly pulled in tension to 1.25 MPa and test setup, (Pond 1995).

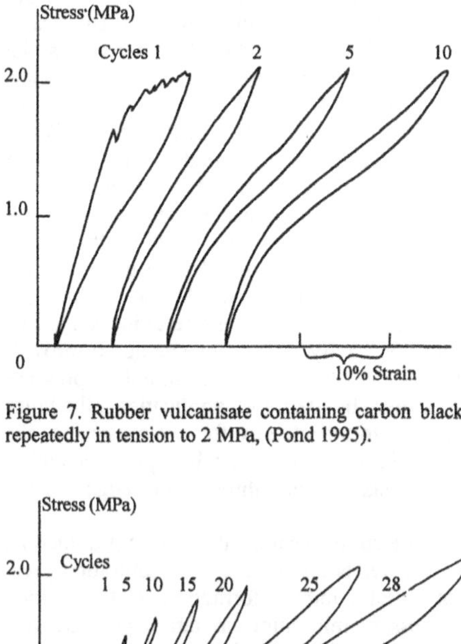

Figure 7. Rubber vulcanisate containing carbon black pulled repeatedly in tension to 2 MPa, (Pond 1995).

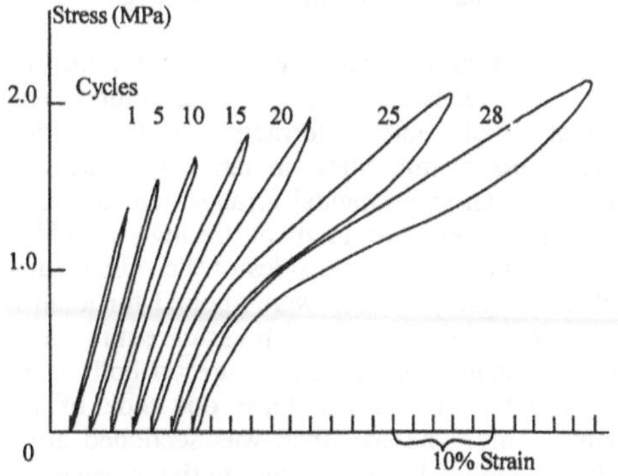

Figure 8. Reinforced rubber disk pulled in tension with an increase in maximum stress per cycle, (Pond 1995).

cavitation damage. The number of such loading cycles depends among others from the type of vulcanisate.

One additional fact is worth mentioning. In Figure 6 and Figure 7 the stress-strain cycles do not show a significant hysteresis despite cavitation damage. In fact the loading cycles do not show permanent plastic deformation due to bubble formation and can be considered a closed strain cycles with start and end point coinciding. In view of these results, cavitation damage can be associated with a significant variation of the bulk modulus K.

3.3 *Theoretical considerations on cavitation*

Experimental observations indicate that a hydrostatic stress state of approximately 5E/6 in tension suffices to form internal bubbles and cracks starting from existing precursors. The value 5E/6 can also be obtained from theoretical considerations of a spherical cavity subjected to an inflating pressure. The relation between the inflating pressure P and the expansion ratio $\lambda=r/r_0$ (Fig. 3) of a spherical cavity in a rubbery solid can be found as

$$P/E = \left(5 - 4\lambda^{-1} - \lambda^{-4}\right)/6 \qquad (1)$$

where E is the elastic modulus for small strains, r and r_0 the current and initial radius of the cavity. This equation shows that the relation between applied pressure and cavity radius assumes the value of 5E/6 when the cavitation is expected as an indefinitely expansion of the submicroscopic bubble. It has been mentioned before that the initial size of the existing precursor has an important influence on the minimum pressure required to initiate cavitation. Thus, the above relation is approximately true for voids with initial radii ranging from about 0.5 μm to 1 mm. Outside this range the value for the critical stress given by Equation 1 is increasingly inaccurate. In general however, rubber components not small in size do always contain voids or holes within the size range mentioned above. Referring to the rule of the weakest link, the critical pressure will depend on the largest holes in the rubber. Another uncertainty in the determination of the minimum cavitation pressure is the value of the elastic modulus E, which should be selected corresponding to the strain values on the surface of the largest cavity. Usually these strains are unknown and an appropriate value need to be selected based on experience.

If the rubber specimen is small in size the voids within may fall outside the above defined limits. In this case the surface energy of the rubber, neglected in Equation 1 needs to be accounted for, (Gent & Tompkins 1969). Gent & Wang (1990) applied fracture mechanics to find the cavitation stress in rubbery solids with voids smaller than the above mentioned size. Both studies found the critical pressure necessary to initiate cavitation rising to about 3E for a void radius of about 0.5 μm and still larger values for smaller voids.

4 FINITE ELEMENT MODELLING

The analysis of incompressible or nearly incompressible rubber components under highly constraint conditions can result into serious numerical difficulties. The reasons are manifold, some could be associated to material or mesh instability or to improper modeling of the pressure-volume response in the presence of cavitation. Cavitation damage is of importance for example in elastomeric bearings subjected to seismic loading. Thereby, reinforcing steel plates bend and/or rotate inducing a hydrostatic tensile stress in individual rubber layers. Similar considerations need to be addressed in elastomeric bearings where inclined reinforcing steel plates are used to generate anisotropy in the in-plane bearing response. Under simple shear deformation large positive and negative hydrostatic stress states are generated and must be properly accounted for in a numerical analysis, (HARIS 1999).

Rubber is often regarded in numerical modeling as isotropic and hyperelastic material. The hyperelastic response is formulated in terms of a strain energy density functional and is in general decoupled into a deviatoric part $\hat{W}$ and a dilatational part $\hat{U}$

$$\Psi = \hat{W}(I_1, I_2) + \hat{U}(I_3) \qquad (2)$$

This mathematical expression of hyperelasticity is expressed in terms of invariants of the right Cauchy-Green deformation tensor. Equivalent strain energy formulations can be expressed also in terms of principal stretches such as

$$\Psi = \hat{W}(\bar{\lambda}_1, \bar{\lambda}_2, \bar{\lambda}_3) + \hat{U}(J) \qquad (3)$$

Under isothermal conditions the total volume change at a point is given by $J = \lambda_1\lambda_2\lambda_3$, the deviatoric stretches by $\bar{\lambda}_i = J^{-1/3}\lambda_i$. Over the last thirty years extensive research went into the definition and formulation of the deviatoric part with more sophisticated strain energy density functions still under development. For a comprehensive review of these formulations many textbooks and FE user manuals are available. For example Crisfield (1997), ABAQUS Theory Manual (1998), Kaliske (1995), Liu (1994).

4.1 *Volumetric strain energy density functions*

Liu (1994) analyzed various formulations of the dilatational term in respect to their consistency with the classical theory of elasticity for small strains. In the following a linear and two nonlinear volumetric strain energy density functions are reviewed and their behavior discussed. For each functional the first and second derivation is given. The first derivation corresponds to the pressure, the second to the change in pressure. For *J=1* the second derivative is equal to the bulk modulus *K*. The linear function known as Ogden formulations is given by

$$\hat{U}(J) = \frac{K}{2}(J-1)^2 \tag{4}$$

$$\frac{\partial \hat{U}(J)}{\partial J} = p(J) = K(J-1) \tag{5}$$

$$\frac{\partial^2 \hat{U}(J)}{\partial J^2} = K \tag{6}$$

Simo & Taylor (1982) developed one of the non-linear models reviewed here. The original formulation used $K/2$ in the volumetric strain energy expression shown in Equation 7, however this would lead to inconsistency in Equation 9, where the second derivative of $\hat{U}$ would become $2K$ for $J=1$. Liu (1994) used $K/4$ instead. The dilatational term in the energy formulation then writes

$$\hat{U}(J) = \frac{K}{4}\left((J-1)^2 + (\ln J)^2\right) \tag{7}$$

$$\frac{\partial \hat{U}(J)}{\partial J} = p(J) = \frac{K}{2}\left(\frac{\ln J}{J} + J - 1\right) \tag{8}$$

$$\frac{\partial^2 \hat{U}(J)}{\partial J^2} = \frac{K}{2}\frac{1-\ln J + J^2}{J^2} \tag{9}$$

The third volumetric strain energy density function reviewed was proposed by Liu (1994) and has the form

$$\hat{U}(J) = K(J\ln J - J + 1) \tag{10}$$

$$\frac{\partial \hat{U}(J)}{\partial J} = p(J) = K\ln J \tag{11}$$

$$\frac{\partial^2 \hat{U}(J)}{\partial J^2} = \frac{K}{J} \tag{12}$$

The three formulations are compared to each other using a realistic bulk modulus of *K=2500 MPa* and a small volume variation between *J=0.98* and *1.01*. In fact, (nearly) incompressible materials undergo very small volumetric strains even under large applied loads. When multilayer elastomeric bearings were subjected to compressive loads, the initial thickness of each rubber layer reduces from the original 4 mm to approximately 3.975 mm for a compressive stress of 10 MPa, (HARIS 1999). Each rubber layer was characterized by a shape factor *S = 12.5*, (S = loaded area / force-free area). This reduction in thickness is mainly due to the bulging of the rubber along the free edges. If this phenomenon is neglected the thickness variation corresponds to J=0.993. In Figures 9 and 10 the strain energy density and the corresponding pressure are shown for the three functions given before.

5 CAVITATION-DAMAGE MODEL

The pressure-volume relations in Figure 10 indicate that a hydrostatic stress in excess of 20 MPa is required to obtain a 1% increase in volume. However, natural rubber is not capable to withstand this hydrostatic stress state, without cavitation damage and associated softening of the material. In general, cavitation does not occur uniformly throughout a specimen and an increase in the applied pressure causes the cavitation region to expand. This fact was confirmed experimentally by observing satellite bubbles formed in the vicinity of growing cavities, (Gent & Tompkins 1969).

Cavitation damage in rubber is associated with a significant change in the bulk modulus of the mate-

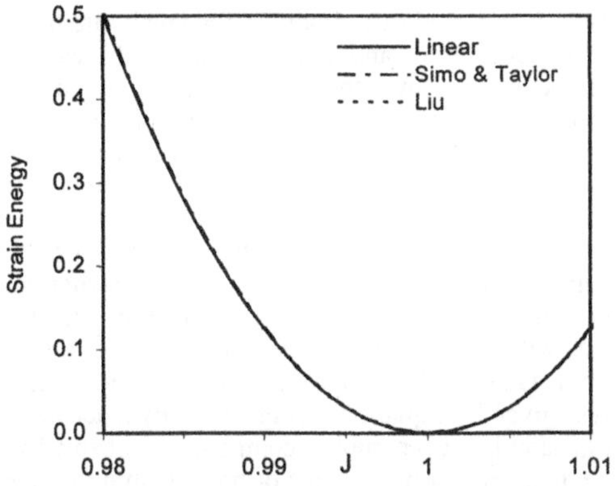

Figure 9. Strain energy for hydrostatic pressure and tension.

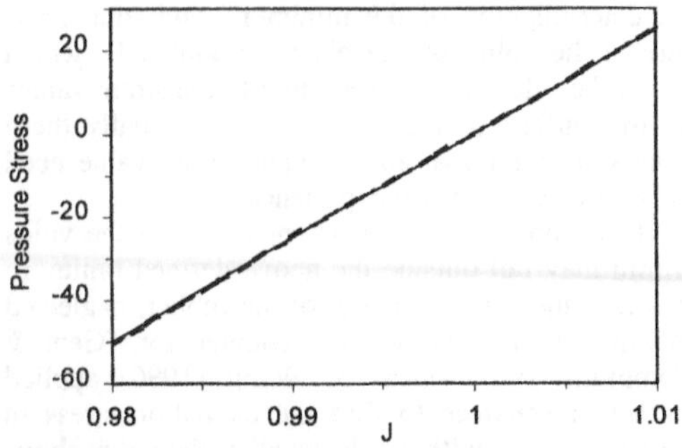

Figure 10. Hydrostatic pressure - volume relation.

rial. Thus, the introduction of a variable bulk modulus seems to best describe the behavior of rubber when cavitation damages need to be accounted for. Let the material be described by a constant bulk modulus for compressive and tensile stress states up to the cavitation threshold. This limit point is characterized by a critical dilatational strain energy $\hat{U}_{crit}$ and by a significant reduction from the original bulk modulus K_o to an experimentally determined cavitation-damage modulus K_{cav}.

For simplicity let us consider the linear Ogden function in Equation 4 which assumes the value $\hat{U}_{crit}$ at the onset of cavitation

$$\hat{U}_{crit}(J) = \frac{K_o}{2}\left(J_{crit} - 1\right)^2 \tag{13}$$

Experimental characterization tests can be used to determine the minimum cavitation pressure similar to Gent & Tompkins (1969) or Pond (1995). The corresponding critical volume change is then given by Equation 5 as

$$J_{crit} = \frac{p_{crit}}{K_o} + 1 \tag{14}$$

Inserting Equation 14 into Equation 13 gives the dilatational critical strain energy density for cavitation

$$\hat{U}_{crit}(J) = \frac{p^2_{crit}}{2K_o} \tag{15}$$

For strain energy values larger than $\hat{U}_{crit}$ the bulk modulus is given by the cavitation-damage modulus K_{cav}. The volume change accumulated during cavitation is given by $J_{cav} = J - J_{crit}$. The volumetric strain energy density describing the behavior of the rubber in compression and tension becomes

$$\begin{aligned}\hat{U}(J) &= U_o\left(J_o\right) + U_{cav}\left(J_{cav}\right)\\ &= \frac{K_o}{2}\left(J_o - 1\right)^2 + \frac{K_{cav}}{2}\left(J_{cav}\right)^2\end{aligned} \tag{16}$$

where for $\hat{U} < \hat{U}_{crit}$

$$J_o = J \qquad K_{cav} = 0 \tag{17}$$

when $\hat{U} > \hat{U}_{crit}$ after onset of cavitation

$$J_o = J_{crit} \qquad K_{cav} \neq 0 \tag{18}$$

The pressure is the first derivative of the dilatational strain energy function and is given by

$$\frac{\partial \hat{U}}{\partial J} = p(J) = K_o\left(J_o - 1\right) + K_{cav}\left(J_{cav}\right) \tag{19}$$

The second derivative of the energy formulation gives either the elastic bulk modulus K_o or the cavitation-damage modulus K_{cav}. With the strain energy density function in Equation 16 and the corresponding formulation of the hydrostatic stress-state, the pressure-volume relation is determined for all volumetric changes of interest to rubber applications.

5.1 *Appropriate parameters*

In experiments by Gent & Tompkins (1969) discussed previously, the minimum stress for rupture was found to be dependent upon the elastic properties of the material and independent of the strength characteristics. A stress limit of approximately 5G/2 can be identified capable to initiate cavitation damage. Muhr (1998) suggested using a cavitation-damage modulus K_{cav} to describe the pressure-volume relation as

$$K_{cav} = K_o/200 \tag{20}$$

The variation of the bulk modulus is shown in Figure 11. The pressure-volume relation is then fully determined for the compressive as well as the cavitation region of the rubbery material. These assumptions lead to the critical volume change and the critical energy

$$J_{crit} = \frac{5E + 6K_0}{6K_0} \tag{21}$$

$$U_{crit} = \frac{25E^2}{72K_0} \approx \frac{E^2}{3K_0} \tag{22}$$

6 APPLICATION

The above described cavitation-damage model is validated using the experimental data provided by Pond (1995) and summarized for convenience in

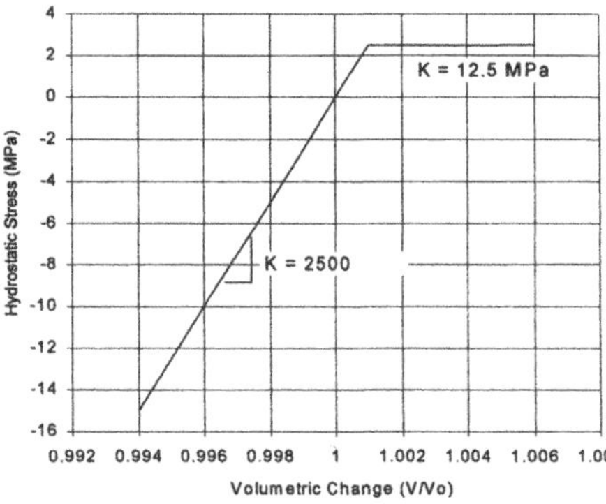

Figure 11. Hydrostatic stress before and after onset of cavitation.

Paragraph 3.2. For the numerical analysis the compound 2 (Table 2) characterized by a shear modulus $G = 0.866$ MPa and bulk modulus $K_o = 2500$ MPa is selected.

The single rubber disc in Figure 6 is vulcanisate onto two 20 mm thick steel plates and subsequently loaded in tension. The stiff steel plates do not allow bending effects to influence the test results. As mentioned before, the disc has a diameter of 50 mm and a thickness of 5 mm. Thus, the given shape factor S [=radius/(2 x thickness)] is 2.5. Symmetry reduces the model of the rubber disc to an eight, see Figure 14.

The deviatoric part of the strain energy potential is described in this analysis by the Neo-Hookean form

$$\hat{W}(I_1) = C_{10}\left(\bar{I}_1 - 3\right) \tag{23}$$

where I_1 is the first deviatoric strain invariant defined in term of deviatoric stretches as

$$\bar{I}_1 = \bar{\lambda}_1^{\,2} + \bar{\lambda}_2^{\,2} + \bar{\lambda}_3^{\,2} \tag{24}$$

and the material parameter C_{10} is dependent upon the initial shear modulus

$$G = 2C_{10} \tag{25}$$

As suggested in the previous paragraph, the cavitation damage modulus is reduced to 1/200 of the initial value K_o.

Figure 12 shows the three-dimensional state of stress of an element located at the center of the disk. The minimum cavitation stress of 5G/2 suggested by Gent is clearly indicated by the change in slope. Figure 13 shows the response of the entire disk and is compared to the test data provided by Pond. The stress information is the total applied load divided by initial undeformed cross sectional area and the strain is the change of thickness over original thickness. Finally, even though no bending effects are present during the experimental and/or numerical analysis, cavitation damage starts to develop at the

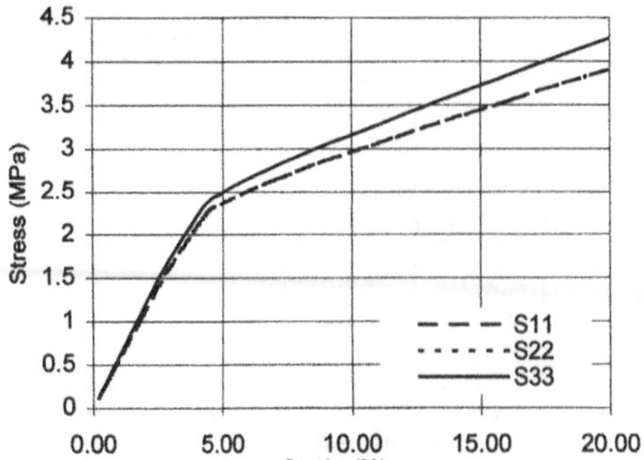

Figure 12: Three dimensional state of stress at the center.

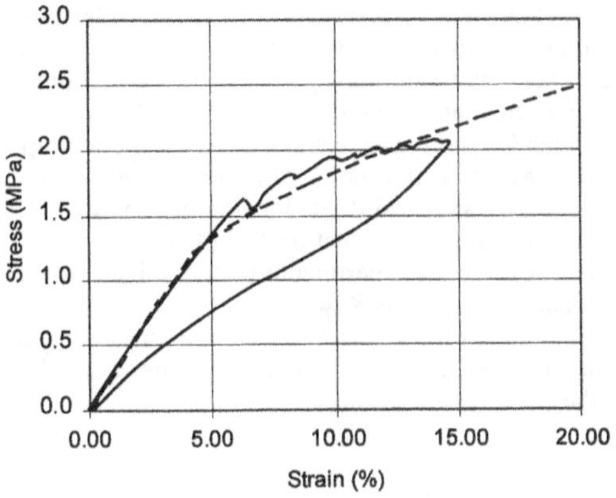

Figure 13: Experimental and numerical effect of cavitation.

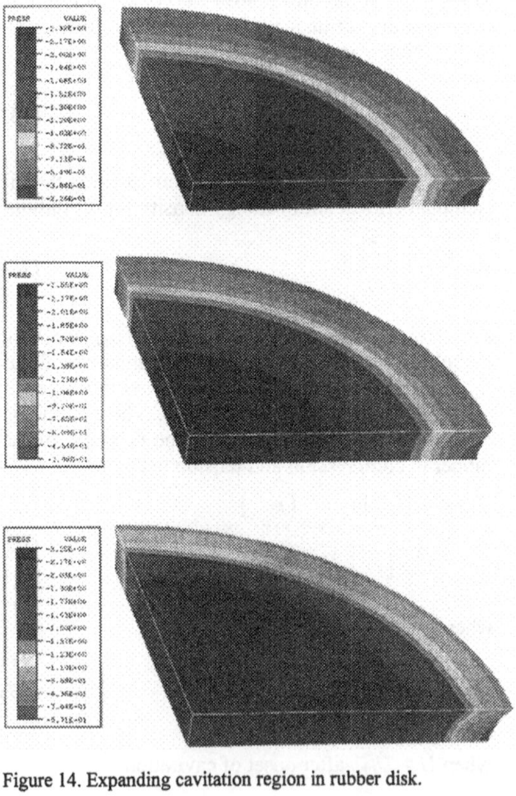

Figure 14. Expanding cavitation region in rubber disk.

center region where a purely hydrostatic stress state does exist. An increase of load causes the cavitation region to expand as shown in Figure 14.

7 CONCLUSIONS

In this study the experimental and numerical aspects of stiffness softening attributed to cavitation is reviewed. Experimental evidence is summarized first to quantify the importance of damage associated with bubble growth at the inside of rubbery materials. Cavitation tests based on supersaturated soluble gases are shown that a hydrostatic tensile stress approximately equal to 5G/2 is sufficient to tear open microscopic voids present in the vulcanisate. The experimental results by Pond (1995) show that with sufficient cyclic loading at a stress level below that of the minimum cavitation stress, bubble growth could be prevented.

Strain energy density functions for hyperelastic material models are addressed. It is shown that the dilatational term in the energy expression must be modified to account for the material softening associated with cavitation. The damage model proposed in this study introduces a variable bulk modulus to account for cavitation induced softening. In the numerical formulation the critical hydrostatic tensile stress necessary to initiate cavitation in the material is characterized by a critical dilatational strain energy $\hat{U}_{crit}$ and by a significant reduction of the original bulk modulus. Finally, validation of the damage model is done on a single rubber disk pulled in tension consistent with the experimental evidence provided by Pond (1995).

ACKNOWLEDGMENT

Partial support for this work was provided by the European Community under Contract N°: BRPR-CT95-0072. The authors would like to express their appreciation.

REFERENCES

ABAQUS Theory Manual. 1998. *Hibbitt, Karlsson & Sorensen, Inc.*

Busse, W.F. 1938. Physics of Rubber as Related to the Automobile. *Journal of Applied Physics, Vol. 9, 438*

Cho, K. & Gent, A.N. 1988. Cavitation in model elastomeric Components. *J. Mater. Sci. 23, 141.*

Crisfield, M.A. 1997. Nonlinear Finite Element Analysis of Solids and Structures. *John Wiley & Sons Ltd., UK.*

Flory, R.W. (1961). Thermodynamic Relations for High Elastic Materials. *Trans. Faraday. Soc. 57, 829-838.*

Gent, A.N. 1999. Private Communication.

Gent, A.N. 1990. Cavitation in Rubber: A Cautionary Tale. *Rubber Chem. Technol. Nr. 63, G49-G53.*

Gent, A.N. & Lindley, P.B. 1958. Internal Rupture of Bonded Rubber Cylinders in Tension. *Proc. R. Soc., A249, 195.*

Gent, A.N. & Meinecke, E.A. 1970. Compression, Bending, and Shear of Bonded Rubber Blocks. *Polymer Engineering and Science, Vol. 10, No. 1.*

Gent, A.N. & Tompkins, D.A. 1969. Nucleation and Growth of gas Bubbles in Elastomers. *Journal of Applied Physics, Vol. 40, Nr. 6.*

Gent, A.N. & Wang, C. 1990. Fracture Mechanics and Cavitation in Rubber.like Solids. *J. Mat. Sci., 26, 3392.*

HARIS, Highly Adaptable Rubber Isolating System. Final Technical Report, *Sponsoring Agency: European Community,* Contract #BRPR-CT95-0072, Project #Be-1258, 1999.

Kaliske, M. 1995. Zur Theorie und Numerik von Polymerstrukturen unter statischen und dynamischen Einwirkungen. *Mitteilungen des Instituts für Statik der Universität Hannover, Hannover.*

Liu, C. 1994. Traction of Automobile Tires on Snow, An Investigation by Means of the Finite Element Method. *PhD. Thesis at Institute of Strength of Materials, Technical University of Vienna.*

Muhr, A. 1998. Private communication.

Pond, T.J. 1995. Cavitation in Bonded Natural Rubber Cylinders Repeatedly Loaded in Tension. *Journal of Natural Rubber Research, Vol. 10, Nr. 1, 14-25.*

Yerzely, F.L. 1939. Adhesion of Neoprene to Metal. *Ind. Eng. Chem. (Industr.), 31, 950.*

Constitutive Models for Rubber, Dorfmann & Muhr (eds) © 1999 Taylor & Francis ISBN 90 5809 113 9

An advanced micro-mechanical model of hyperelasticity and stress softening of reinforced rubbers

M. Klüppel & J. Schramm
Deutsches Institut für Kautschuktechnologie e.V., Hannover, Germany

ABSTRACT: An advanced micro-mechanical concept of hyperelasticity and stress softening of reinforced rubbers is presented that combines a non-Gaussian molecular statistical approach to rubber elasticity with a damage model of stress-induced filler cluster breakdown. A generalized tube model of rubber elasticity is applied for a description of the polymer network that considers tube-like, topological constraints (packing effects) as well as finite chain extensibility relevant for real networks. The effect of the filler is taken into account via hydrodynamic reinforcement of the rubber matrix by rigid, self-similar filler clusters. It allows for a quantitative description of stress softening by means of a strain or pre-strain dependent hydrodynamic amplification factor, respectively. Thereby, the pronounced stress softening or high hysteresis of reinforced rubber is referred to an irreversible breakdown of filler clusters during the first deformation cycle. It is shown that the developed concept is in fair agreement with uniaxial stress-strain data of unfilled NR-samples of variable cross-link density and carbon black filled E-SBR- samples.

1 INTRODUCTION

The micro-mechanical modellization of quasi-static stress-strain properties of reinforced elastomers involves different influences and mechanisms that have been discussed by a varity of authors, but in most cases only on a qualitative level. Beside the action of the entropy elastic polymer network that is quite well understood on a molecular-statistical basis [1,2], the impact of filler particles on stress-strain properties is of high importance, but so far the micro-mechanical effects of the filler are not totally understood [3,4]. On the one hand side the addition of hard filler particles leads to a stiffening of the rubber matrix that can be described by a hydrodynamic strain amplification factor [5-7]. On the other hand side the constraints introduced into the system by filler-polymer bondings result in a decreased network entropy and hence, the free energy that equals the negative entropy times the temperature increases linear with the effective number of network junctions [8,9]. A further effect can be obtained from the formation of filler clusters or a filler network due to strong attractive filler-filler bondings [3,4,6-10].

A complication for the modellization of reinforced rubbers is the pronounced stress softening during quasi-static deformations that is also termed Mullins effect due to the extensive measurements carried out by Mullins [11-13]. Dependent on the history of straining, e.g. the extent of previous stretching, the rubber material undergoes an almost permanent change that alters the elastic properties and increases hysteresis, drastically. Most of the softening occurs in the first deformation and after a few deformation cycles the rubber approaches a steady state with a constant stress-strain behavior. The softening is usually only present at deformations smaller than the previous maximum. An example of stress softening is shown in Fig. 1, where the maximum strain is increased, successively, from one uniaxial stretching cycle to the next.

The softening has been attributed to breakdown or slippage [14-17] and dis-entanglements [18] of bonds between filler and rubber, while other authors assumed that a strain-induced crystallization-decrystallization [19,20] or a re-arrangement of network chain junctions in filled systems [13] is responsible for the large hysteresis. A quantitative description of stress-induced breakdown or separati-

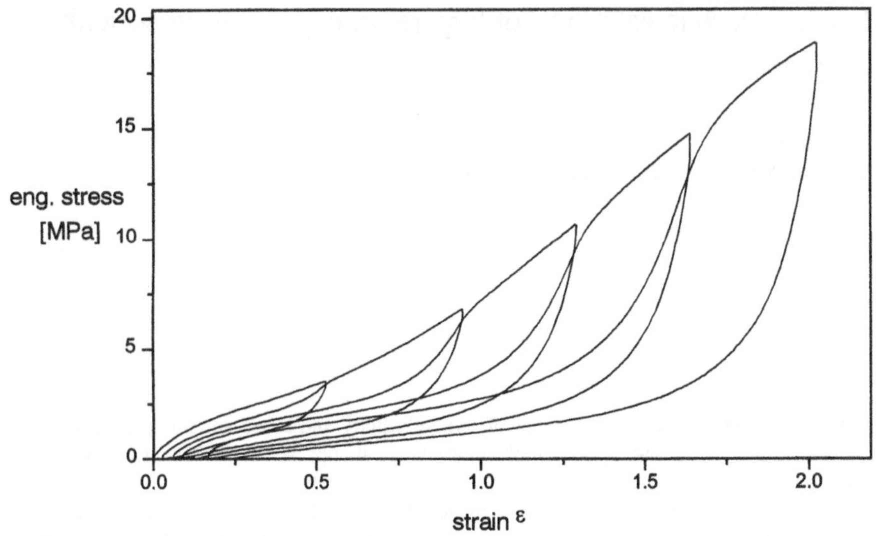

Fig. 1: Quasi-static stress-strain cycles with successively increasing maximum strain for E-SBR-samples filled with 80 phr N 339.

on of network chains from the filler surface is given in Ref. [15] where a complete macroscopic constitutive theory is derived on the basis of statistical mechanics. However, this kind of interpretation of stress softening ignores the important experimental result of Haarwood et al. [19,20], who showed by a simple mastering procedure that stress softening is related to hydrodynamic strain amplification due to the presence of the filler. A plot of stress in second extension vs. ratio between strain and pre-strain of natural rubber filled with a varity of carbon blacks yields a single master curve [13,19].

It's the aim of this paper to present a quantitative molecular-statistical model of hyperelasticity of filler reinforced rubbers that considers stress softening as a result of strain amplification by rigid filler clusters of variable size. In the first part an advanced concept of rubber elasticity is considered that combines the Edwards-Vilgis approach of finite network extensibility with a topological constraint contribution in a generalized non-Gaussian tube-model for unfilled rubbers. In the second part of the paper we will develop a micro-mechanical picture of filler cluster breakdown during quasi-static straining. It yields a damage model for the description of stress softening by means of a hydrodynamical amplification factor that depends on the applied strain during the first deformation of the virgin sample or maximum pre-strain in the following cycles, respectively.

In the third part we will present experimental results on filled and unfilled samples. In particular, we will show how the parameters of the model can be estimated by a fitting procedure for the pre-strained filled samples.

2 A GENERALIZED TUBE MODEL OF RUBBER ELASTICITY

The classical concepts of rubber elasticity consider so called phantom networks of freely fluctuating chains that are not influenced by any constraining potential appart from the cross-links [1,2]. This is a rough approximation, because in typical elastomer networks the chains cannot move freely due to the large degree of chain interpenetration. It restricts the chain fluctuations by packing effects that result from the inability of the chains to pass through its neighbours. The topological constraints on a single chain (packing effects) can be described by a tube model, i.e. a harmonic potential that forces the chain to remain in a virtuel tube around its mean position. In the case of strong topological constraints, relevant for highly molecular rubbers, the elastic free energy density can be expressed as [1]:

$$W \equiv W_c + W_e = \frac{G_c}{2}\left(\sum_{\mu=1}^{3}\lambda_\mu^2 - 3\right) + 2G_e\left(\sum_{\mu=1}^{3}\lambda_\mu^{-1} - 3\right) \quad (1)$$

where λ_μ is the strain ratio in direction of the main axis system, G_c is the elastic modulus that corresponds to the crosslink constraints and G_e corresponds to the topological tube constraints:

$$G_c = \frac{1}{2}\upsilon_c k_B T = \frac{\upsilon_s l_s^2 k_B T}{2 < R_o^2 >} \tag{2}$$

$$G_e = \frac{\upsilon_s l_s^2 k_B T}{4\sqrt{6}\, d_o^2} \tag{3}$$

Here, υ_c is the chain density, υ_s is the density of statistical segments, l_s is the length of statistical segments, $<R_o^2>$ is the average end-to-end distance of chains in the underformed state, d_0 is the tube radius (mean fluctuation radius of chain segments), k_B is the Boltzmann constant and T is temperature.

Equ. (1) with the two elastic moduli G_c and G_e is closely related to the semi-empirical Mooney-Rivlin equation with constants C_1 and C_2:

$$W = C_1\left(\sum_{\mu=1}^{3}\lambda_\mu{}^2 - 3\right) + C_2\left(\sum_{\mu=1}^{3}\lambda_\mu{}^{-2} - 3\right) \tag{4}$$

It relates the elastic energy to the first and second deformation invariants, i.e. the bracket terms of Equ. (4). Obviously, Equ. (1) does not envolve the second invariant. This is a direct consequence of a non-affine tube deformation law that leads to Equ. (1). The modified assumption of affine tube deformations reproduces the Mooney-Rivlin Equ. (4). This makes clear that the a-priori postulate of affine deformations on all length scales as applied in continuum mechanics, e.g. the Mooney-Rivlin theory, may not be fulfilled in molecular-statistical approaches of rubber elasticity.

So far, all considerations are valid in the Gaussian limit of infinite long chains. For that reason no singularity appears in the elastic free energy density Equ. (1) that could reflect the finite extensibility of real polymer networks. A singularity can be obtained if the inverse Langevin approximation is used instead of the Gaussian distribution function for the end-to-end distance of network chains [21]. Thereby, it is sufficient to consider the modifications of the cross-link term W_c in Equ. (1), because the topological constraint term W_e goes to zero at large strains ($W_e \sim \lambda_\mu^{-1}$), where the finite extensibility becomes significant [22-24]. This argument is confirmed by recent molecular-statistical investigations of tube like network models based on non-Gaussian network chains, which show that the action of tube constraints becomes weaker in the case of predominance of finite chain extensibility [25,26]. The simplest way to obtain a singularity for the free energy is the modification of Equ. (1) as proposed by Edwards and Vilgis [2]:

$$W_c = \frac{G_c}{2}\frac{\left(\sum_{\mu=1}^{3}\lambda_\mu^2 - 3\right)\left(1 - \frac{T_e}{n_e}\right)}{1 - \frac{T_e}{n_e}\left(\sum_{\mu=1}^{3}\lambda_\mu^2 - 3\right)} + \frac{G_c}{2}\ln\left(1 - \frac{T_e}{n_e}\left(\sum_{\mu=1}^{3}\lambda_\mu^2 - 3\right)\right) \tag{5}$$

Here, T_e is the Langley trapping factor [27] and n_e is the segment number of chains between successive entanglements. The singularity of W_c is found for $n_e/T_e = \sum \lambda_\mu{}^2 - 3$, i.e. if the chains between successive trapped entanglements are fully stretched out. This makes clear that the approach in Equ. (5) characterizes trapped entanglements as some kind of physical cross-links (slip-links) that dominate the extensibility of the network due to the larger number of entanglements as compared to chemical cross-links. We note, that this gives a less pronounced upturn of stress-strain curves as compared to the classical inverse Langevin approach, which can be related to the more flexible response of trapped entanglements as compared to chemical cross-links.

The final expression for the elastic free energy that considers finite extensibility together with tube constraints is found from a combination of Equ. (5) with the second term of Equ. (1):

$$W = \frac{G_c}{2}\frac{\left(\sum_{\mu=1}^{3}\lambda_\mu^2 - 3\right)\left(1 - \frac{T_e}{n_e}\right)}{1 - \frac{T_e}{n_e}\left(\sum_{\mu=1}^{3}\lambda_\mu^{-2} - 3\right)} + \frac{G_c}{2}\ln\left(1 - \frac{T_e}{n_e}\left(\sum_{\mu=1}^{3}\lambda_\mu^2 - 3\right)\right) + 2G_e\left(\sum_{\mu=1}^{3}\lambda_\mu^{-1} - 3\right) \tag{6}$$

In the limit $n_e \to \infty$ the Gaussian formulation of infinite long chains Equ. (1) is recovered. From Equ. (6) the engeneering stress $\sigma_{o,\mu}$ that relates the force f_μ in direction μ to the underformed cross section $A_{o,\mu}$

is found by differenciation $\sigma_{o,\mu} = \partial W/\partial\lambda_\mu$. In the case of uniaxial extension with $\lambda_1=\lambda$, $\lambda_2=\lambda_3=\lambda^{-1/2}$ this yields:

$$\sigma_{0,1} = G_c\left(\lambda-\lambda^{-2}\right)\left\{\frac{1-\dfrac{T_e}{n_e}}{\left(1-\dfrac{T_e}{n_e}(\lambda^2+2/\lambda-3)\right)^2} - \frac{\dfrac{T_e}{n_e}}{1-\dfrac{T_e}{n_e}(\lambda^2+2/\lambda-3)}\right\} + 2G_e(\lambda^{-1/2}-\lambda^{-2}) \qquad (7)$$

With this equations the model parameters G_c, G_e and T_e/n_e can be found from fittings to experimental stress-strain curves.

3 STRESS SOFTENING BY STRESS-INDUCED FILLER CLUSTER BREAKDOWN

For an extension of Equs. (6) and (7) to filler reinforced rubbers we have to consider hydrodynamic effects of filler particles and rigid filler clusters. The filler clusters result from an aggregation process in the rubber matrix subject to strong physical bondings between filler particles. Possible aggregation mechanisms are percolation or kinetical aggregation that both lead to a selfsimilar cluster structure [28]. We assume that due to the stabilizing bound rubber layer at the cluster surface, the strength of the filler clusters is quite high and hence, part of these clusters survive up to large deformations. With increasing strain of a virgin sample, a stress-induced successive breakdown of filler clusters takes place, during which the size of the clusters decreases. This process is almost irreversible, because for quasi-static experiments the gaps between broken filler clusters fill up with polymer that is expected to be strongly bonded to the filler surface and hence, hinders the reaggregation of the clusters when the stress relaxes during the backcycle of straining. It means that the cluster size that is reached at the maximum strain of the first cycle remaines fixed for a long time periode and almost no change of the cluster size takes place during the following cycles as long as the maximum pre-strain is not exceeded. If a larger strain is applied in a following cycle, a further breakdown of the clusters appears that is then frozen in the next cycles. This is the basic mechanism of stress softening in filler reinforced rubbers. It leads to the characteristic stress-strain behavior shown in Fig. 1, if the hydrodynamic reinforcement of the clusters is considered.

Hydrodynamic reinforcement can be described by a strain amplification factor X that relates the microscopic intrinsic strain $\lambda = 1+\varepsilon_o$ of the rubber to the macroscopic external strain $\Lambda = 1+\varepsilon$ of the sample ($X = \varepsilon_o/\varepsilon$). As mentioned above, the use of a strain amplification factor X appears appropriate for a modellization of the stress-strain behavior of pre-strained reinforced rubbers if X is coupled to the previous straining ε_{max} of the sample. Then the stress-strain curves in the second or third cycle can be described by a constant strain amplification factor $X_{max}=X(\varepsilon_{max})$ that depends on the pre-strain ε_{max} as long as the applied external strain ε is smaller than ε_{max}. It means that Equs. (6) and (7) remain valid if the intrinsic strain λ is expressed by the external strain ε as follows:

$$\lambda = 1 + X_{max}\,\varepsilon \qquad \text{for } \varepsilon < \varepsilon_{max} \qquad (8)$$

If in a following cycle the sample is strained to a higher value $\varepsilon'_{max} > \varepsilon_{max}$, then the filler clusters break further down and the hydrodynamical amplification factor $X_{max}' = X(\varepsilon'_{max})$ that is representative for the next cycles decreases ($X_{max}' < X_{max}$). In the following we will consider a quantitative model that describes the dependence $X_{max} = X(\varepsilon_{max})$ for selfsimilar filler clusters. It equals the dependence $X=X(\varepsilon)$ during the first straing of the virgin sample and hence we can restrict to this case.

Hydrodynamic reinforcement by selfsimilar, rigid filler clusters was modellized on the basis of a path integral formalism by Huber et al. [7], who found the following scaling law for the hydrodynamic amplification factor in the case of high filler concentrations ϕ with overlapping neighbouring clusters:

$$X = 1 + \text{const.}\left(\frac{\xi}{a}\right)^{d_w-d_f}\phi^{\frac{2}{3-d_f}} \qquad (9)$$

Here, ξ is the cluster size, a is the particle size, d_f is the fractal dimension and d_w the anomalous diffusion exponent of the clusters. It shows that the strain amplification factor increases with filler concentration and cluster size according to a power law with an exponent that depends on the fractal structure of the clusters.

This result Equ. (9) can be combined with a concept of stress-induced cluster breakdown. In a first approach one may assume an exponential decrease of the cluster size with increasing strain:

$$\frac{\xi(\varepsilon)}{a}=\left(\frac{\xi_o}{a}-1\right)\exp(-\alpha(\varepsilon))+1 \qquad (10)$$

where ξ_o is the initial cluster size. The second summand ensures the right infinite strain limit $\varepsilon \rightarrow \infty$, where all clusters are broken and the cluster size should equal the particle size. The exponent α in this model is purely emprical and may depend on the strength of the clusters or the elastic modulus of the rubber.

By inserting Equ. (10) into Equ. (9) we find a strain dependent amplification factor $X(\varepsilon)$ that relates the external strain $\Lambda = 1+\varepsilon$ to the internal strain λ of the polymer chains. Then, instead of Equ. (8) with a constant hydrodynamic amplification factor X_{max}, relevant for pre-strained samples, the following equation results for non-strained virgin sample:

$$\begin{aligned}\lambda &= 1+X(\varepsilon)\varepsilon \approx \\ &\approx 1+(X_\infty+(X_o-X_\infty)\exp(-z\varepsilon))\varepsilon\end{aligned} \qquad (11)$$

The exponent z is given by $z = \alpha\ (d_w - d_f)$ and abreviations for the zero and infinite strain limits for the amplification factor X_o and X_∞ are used:

$$X_o = 1+\text{const.}\left(\frac{\xi_o}{a}\right)^{d_w-d_f}\ \phi^{\frac{2}{3-d_f}} \qquad (12)$$

$$X_\infty = 1+\text{const.}\ \phi^{\frac{2}{3-d_f}} \qquad (13)$$

We note that the approximation in Equ. (11) is quite well, because in most particular cases it holds $\xi_o/a >>1$ and the exponent $d_w - d_f$ has a value close to one. In particular, for the above mentioned percolation- or cinetically aggregated clusters one finds d_w-$d_f \approx 1.3$ in both cases [28]. Hence, the stress-strain behavior of filler reinforced virgin rubber samples is described by Equs. (6) and (7) together with Equ. (11), while for pre-strained samples Equ. (8) has to be applied. Then, the hydrodynamic amplification factor X_{max} fulfills:

$$X_{max} = X_\infty + (X_o - X_\infty)\exp(-z\ \varepsilon_{max}) \qquad (14)$$

The difference between $X(\varepsilon)$ and X_{max} corresponds to the pronounced stress softening of reinforced rubbers. It results from the almost irreversibly breakdown of filler clusters during the first deformation cycle.

4 RESULTS AND DISCUSSION

Fig. 2 shows uniaxial stress-strain results of unfilled NR-samples at 100 °C that are cross-linked with a

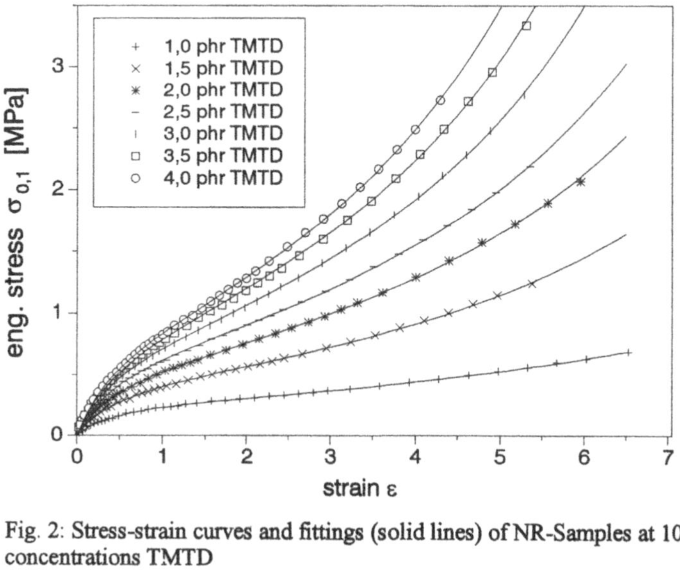

Fig. 2: Stress-strain curves and fittings (solid lines) of NR-Samples at 100°C for a varity of cross-linker concentrations TMTD

variable amount of TMTD. The solid lines correspond to fittings according to Equ. (7) that are in good agreement with the experimental data. The dependence of estimated fitting parameters on TMTD-concentration is shown in Fig. 3. As expected from Equ. (2) and (3) the cross-link modulus G_c increases with increasing TMTD-concentration, while the topological constraint modulus G_e approaches a plateau value that is characteristic for the constant entanglement density of the rubber, independent of cross-link density. The third parameter n_e/T_e decreases with increasing TMTD-concentration that can be related to an increase of the trapping factor T_e with rising cross-link density. Hence, the behavior of the model parameters for the unfilled NR-samples is well understood.

Fig. 4 shows the uniaxial stress-strain behavior of E-SBR-samples that are filled with 40 phr N339. Data for the virgin sample in first extension and for diffe-

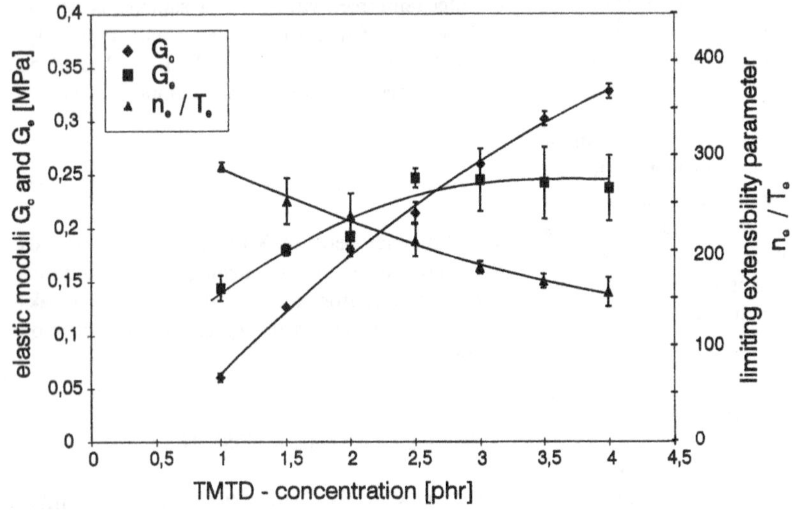

Fig. 3: Estimated fitting parameters G_c, G_e and n_e / T_e vs. TMTD-concentration from the fits in Fig. 2

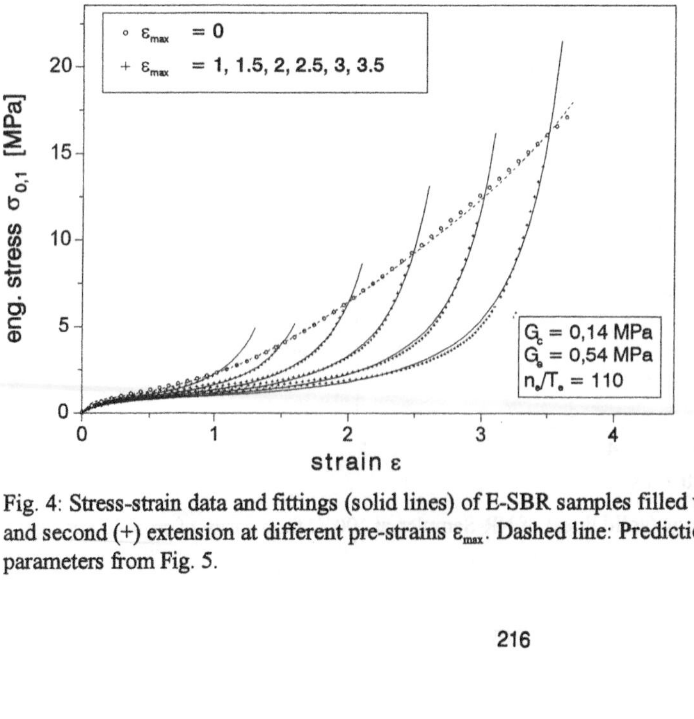

Fig. 4: Stress-strain data and fittings (solid lines) of E-SBR samples filled with 40 phr N339 in the first (o) and second (+) extension at different pre-strains ε_{max}. Dashed line: Prediction of Equs. (7) and (11) with parameters from Fig. 5.

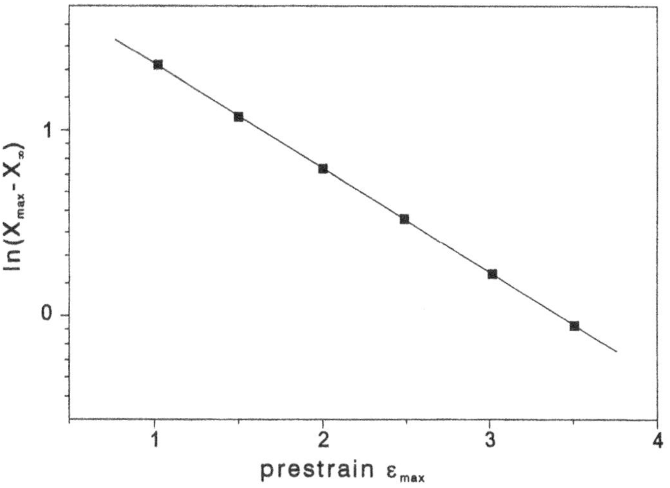

Fig. 5: Plot of log(X_{max}-X_{∞}) vs. pre-strain ε_{max} from the fits in Fig. 4. X_{∞} = 1.35 is estimated from the condition of minimum standard deviation of the regression solid line that yields X_0 = 8.15, z =0.56 (Equ.(14)).

rently pre-strained samples in second extension are shown. The cross-linking system is kept fixed for all samples (1.8 phr sulfur, 1.2 phr CBS, 0.4 phr DPG). Beside the experimental date, fitted curves are shown as solid lines for the pre-strained samples according to Equs. (7) and (8). The fitting parameters for the rubber matrix G_c, G_e and n_e/T_e are hold constant for the differently pre-strained (and virgin) samples (compare insert of Fig. 4). The impact of pre-straining is modellized by the amplification factor X_{max}. Obviously, the fittings for the pre-strained samples with only one variable parameter X_{max} that considers the hydrodynamic reinforcement of differently frozen filler cluster structures are fairly well.

The dependence of the fitting parameter X_{max} on pre-strain ε_{max} is demonstrated in Fig. 5, where according to Equ. (14) a half-logarithmic plot is chosen. The unknown parameter X_{∞} that appears on the ordinate of Fig. 5 is obtained from a least square fit to Equ. (14). We note that, due to the physical meaning, all X-parameters must be larger than one. This condition is fulfilled for the fitted value X_{∞} = 1.35 and X_0 = 8.15 that is obtained from the axis intersection of the regressin line. Furthermore, the exponent z = 0.56 is found from the slope. With these three parameters that describe the hydrodynamic reinforcement of successively broken filler clusters with increasing strain, a simulation of the first extension of the virgin sample is obtained if Equ. (7) is applied together with Equ. (11). This is shown as dashed line in Fig. 4 and confirms the developed micro-mechanical concept of stress-softening. Furthermore, we obtain for the initial cluster size $\xi_0 \approx 10a$, if the above two values for X_{∞} and X_0 are inserted into Equs. (12), (13). It shows that carbon black clusters in rubber remain relatively small.

5 CONCLUSIONS AND OUTLOOK

A micro-mechanically motivated approach to hyper-elasticity and stress softening of reinforced rubbers with well defined microscopic material parameters is given. It shows that

(i) the applied generalised tube model of rubber elasticity represents a constitutive physical basis for the description of quasi-static stress-strain properties of polymer network up to large extensions;
(ii) the introduction of stress-induced, irreversible filler cluster breakdown allows for the consideration of stress softening by means of a pre-strain dependent hydrodynamic amplification factor.

In the present approach the description of stress softening is not constituitive, yet. An extension of the model to a constituitive theory that can be applied to any quasi-static deformation mode will be a task of future works.

LITERATURE

G. Heinrich, E. Straube and G. Helmis, Adv. Polym. Sci. **85,** 33 (1988)

S. F. Edwards and T. A. Vilgis, Rep. Prog. Phys. **51,** 243 (1988); Polymer **27,** 483 (1986)

A. I. Medalia, Rubber Chem. Technol. **46,** 877 (1973); **51,** 437 (1978)

J. B. Donnet, R. C. Bansal and M.-J. Wang, Eds., "Carbon Black Science and Technology", Marcel Dekker Inc. N.Y., Basel, Hongkong (1993)

E. Guth and O. Gold, Phys. Rev. **53,** 322 (1938)

M. Klüppel and G. Heinrich, Rubber Chem. Technol. **68,** 623 (1995)

G. Huber, PhD-Thesis, University Mainz, Germany (1997)

G. Heinrich and T. A. Vilgis, Macromolecules **26,** 1109 (1993)

U. Eisele and H.-K. Müller, Kautsch. Gummi Kunstst. **43,** 9 (1990)

A. R. Payne, J. Appl. Polym. Sci. **6,** 57 (1962); **7,** 873 (1963); **8,** 2661 (1965); **9** 2273, 3245 (1965)

L. Mullins, Rubber Chem. Technol. **21,** 281 (1948)

L. Mullins and N. R. Tobin, Rubber Chem. Technol. **30,** 355 (1957)

L. Mullins, in G. Kraus, Ed., "Reinforcement of Elastomers", Intersc. Publ., N. Y., London, Sydney (1965)

F. Bueche, J. Appl. Polym. Sci, **4,** 107 (1960); **5,** 271 (1961)

S. Govindjee and J. Simo, J. Mech. Phys. Solids **39,** 87 (1991); **40,** 213 (1992)

E. M. Dannenberg, Rubber Chem. Technol. **47,** 410 (1974)

Z. Rigbi, Adv. Polym. Sci. **36,** 21 (1980)

G. R. Hamed and S. Hatfield, Rubber Chem. Technol. **62,** 143 (1989)

J. A. Haarwood, L. Mullins and A. R. Payne, J. Appl. Polym. Sci. **9,** 3011 (1965)

J. A. Haarwood and A. R. Payne, J. Appl. Polym. Sci. **10,** 315, 1203 (1966)

W. Kuhn and F. Grün, Koll. Z. Z. Polym. **101,** 248 (1946)

M. Klüppel, Prog. Colloid Polym. Sci. **90,** 137 (1992)

M. Klüppel and G. Heinrich, Macromolecules **27,** 3569 (1994)

M. Klüppel, J. Appl. Polym. Sci. **48,** 1137 (1993)

G. Heinrich and W. Beckert, Prog. Colloid Polym. Sci. **90,** 47 (1992)

J. Kovac and C. C. Crabb, Macromolecules **15,** 537 (1982)

N. R. Langley, Macromolecules **1,** 348 (1968)

M. Klüppel, G. Heinrich and R. H. Schuster, Rubber Chem. Technol. **70,** 243 (1967)

Applications

Constitutive Models for Rubber, Dorfmann & Muhr (eds) © 1999 Taylor & Francis ISBN 90 5809 113 9

Finite-element-analyses of intervertebral discs: Recent advances in constitutive modelling

R. Eberlein
Sulzer Innotec Limited, Structural Mechanics, Winterthur, Switzerland

M. Fröhlich & E. M. Hasler
Sulzer Orthopedics Limited, Biomechanical Analysis, Winterthur, Switzerland

ABSTRACT: The presentation addresses state of the art investigations of intervertebral discs. These basically consist of a fluid filled cavity (nucleus pulposus) at the disc center and surrounding tissue reinforced by collagen fibers (anulus fibrosus). Constitutive models are generally based on linear elasticity, although intervertebral discs exhibit large strains under physiological loading conditions. The major problem behind this fact is the complexity of experimental investigations. Unfortunately an overall prediction of the deformation behavior of various disc specimen is not possible. All models underlying this classical approach can therefore be applied to conforming geometries only. As a consequence the paper describes appropriate experimental investigations and measurements which finally allow an accurate constitutive modeling of the anulus fibrosus as a major mechanical component within a disc structure. Thus a basis is provided for an admissible strain energy function accounting for anisotropic effects and finite strains in the reinforced annulus tissue.

1 INTRODUCTION

Low back pain is a very common condition affecting the majority of the population in western industrialized nations at some point in life BAO *et al.* [1996]. Low back pain associated with absence from work and with the use of medical services causes enormous costs to society FRANK *et al.* [1996]. Chronic low back pain is often related to degenerative conditions in the lumbar spine such as degenerative disc disease DAWSON & BERNBECK [1998]. The most frequently used surgical treatment methods for degenerative disc disease are discectomy and fusion. Although these treatments achieve relatively good pain relief in the short term, problems in the adjacent segments due to the altered biomechanics of the treated segments account for a significant proportion of poor long-term results DEKUTOSKI *et al.* [1994], HAMBLY *et al.* [1998].

The development of new, more physiological methods such as stabilizing implants (which allow movement within a controlled range), nucleus replacement, or biological disc regeneration may offer alternatives to prevent these problems.

In the early phase of the development process of a spinal implant it may be useful to have a tool which allows analysis of the effect of the implant on the global kinetics and kinematics of the spine and which facilitates the investigation of the requirements placed on the implants in terms of stresses and strains. This is of particular importance if the implant is not intended for fusion but allows certain movement of the instrumented spinal segment. In this case, knowledge of the load sharing between the implant and the biological structure is important for the success of the implant.

A finite element (FE) model of the spine would fulfill the requirements of such a tool if the model can simulate the physiological behavior of the spine or a spinal motion segment. A spinal motion segment consists of the intervertebral disc, two adjacent vertebrae and the connecting spinal ligaments WHITE & PANJABI [1990]. In order to model the complex biomechanical behavior of a motion segment, the physical properties of its components, and in particular, of the intervertebral disc, have to be known.

The behavior of the disc is complicated because of anisotropic properties arising from the non-homogenous organization and distribution of its components URBAN & ROBERTS [1996]. Several FE models have been described in the literature in order to investigate the mechanical properties of the spine and the influence of a given implant (see GILBERTSON *et al.* [1995] for a current review). The current investigation provides state of the art FE analyses of intervertebral discs. In particular, typical disc geometry of an intact L_{2-3} lumbar motion segment is considered. An appropriate FE mesh is depicted in figure 1. Geometric non-linearities (large deformations) and in parts material non-linearities are taken into account in order to investigate the deformation behavior of natural discs under uniaxial compressive loads. Furthermore local stress and strain distributions are analyzed. This allows an evaluation of current FE models described in the literature and motivates further improvements. As a consequence the paper describes appropriate experimental investigations and measurements which finally allow an accurate constitutive modeling of the anulus fibrosus as a major mechanical component within a disc structure. Thus a basis is provided for an admissible strain energy function accounting for anisotropic effects and finite strains in the reinforced annulus tissue.

2 FEA OF INTERVERTEBRAL DISCS

The modeling of intervertebral discs discussed in this chapter relies on geometric data from literature and investigations initiated by Sulzer Orthopedics Ltd. According to these geometric data sets, finite element discretizations are chosen in such a way that the cartilaginous endplates at the superior and the inferior surfaces of the intervertebral disc as well as the anulus fibrosus and the nucleus pulposus are described accurately. That means numerical results obtained from the current finite element models verify computational results from literature EBERLEIN *et al.* [1999]. For those cases, in which experimental data differs significantly from numerical data given in literature and by the present formulation, conclusions are drawn in section 2.3. As a major consequence an experimental setup accounting for more refined constitutive modeling of the anulus fibrosus is proposed in section 2.4.

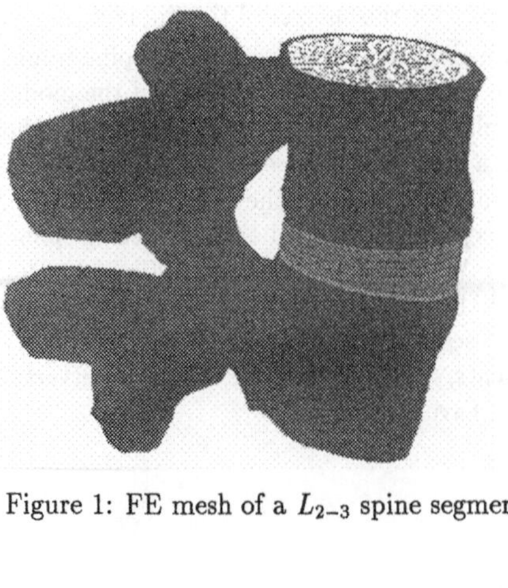

Figure 1: FE mesh of a L_{2-3} spine segment.

2.1 MECHANICAL PROPERTIES

Before the reason for the above mentioned consequence becomes obvious, it is necessary to define 'state of the art' material properties describing the mechanical behavior of a disc. In literature a wide variety of material data can be found. Nevertheless, it is important to notice that almost all constitutive descriptions used in finite element simulations are based on linear elasticity. Sometimes also transient material parameters can be found LU *et al.* [1996*b*]. Only the nucleus pulposus departs from the rule and is modeled as incompressible fluid in general. One rare case, where nonlinear elastic material behavior for the collagen fibers in the anulus fibrosus is implied, can be found in SHIRAZI-ADL *et al.* [1984].

For the current analysis some typical data from literature is used. Table 1 shows the material parameters applied in the subsequent finite element analyses and corresponding references. Due to the high water content of the nucleus pulposus (up to 80% according to BAO *et al.* [1996]), it is modeled by an incompressible fluid as mentioned above. The elastic response of the anulus fibrosus and the cartilaginous endplates is purely isotropic. Only tensile stress is sustained by the one-dimensional collagen fibers. Eventually, it must be pointed out again that the presented constitutive data is taken from literature. No assessment about the accuracy and reliability of these material properties can be given at this stage.

2.2 GEOMETRY AND LOADS

Two fresh cadaveric lumbar spines of a 56–year–old female and a 22–year–old male were investi-

Table 1: Typical material data for an intervertebral disc.

Materials	Stiffness	Poisson ratio	References
Anulus fibrosus			
• Ground substance	$E = 4.0 MPa$	$\nu = 0.4$	GALANTE [1967], WU & YAO [1976]
• Collagen fibers	$\sigma = 23\,000 \cdot \varepsilon^{1.9}$ [MPa]	—	EYRE [1979], HARKNESS [1961], HAUT & LITTLE [1972]
Nucleus pulposus	Incompressible fluid		SHIRAZI-ADL *et al.* [1984]
Cartilaginous endplates	$E = 23.8 MPa$	$\nu = 0.4$	UENO & LIU [1987]

gated. The derivation of geometric data for the first sample is according to LU *et al.* [1996*a*]. The male spine was analyzed for Sulzer Orthopedics Ltd. For the current analyses, the disc geometry of the $L2 - L3$ motion segment is chosen. The disc height (excl. endplates) is $8mm$, the disc cross–sectional area is approximately $1300mm^2$ in both samples and the ratio of nucleus pulposus area to disc cross–sectional area is some 38%.

From figure 2 the collagen fiber arrangement in the anulus fibrosus can be observed. The annulus matrix, consisting of homogenous ground substance, is reinforced by collagen fibers with an averaged $\pm 30°$ orientation towards the endplates GOEL *et al.* [1995]. An average collagen fiber content of about 16% of the disc volume is suggested by GALANTE [1967]. According to OHSHIMA *et al.* [1989] for the subsequent finite element simulations the collagen fiber content varies from 5% in the innermost layer to 23% in the outermost layer. The corresponding fiber cross-sectional areas depend on mesh refinement and are put in concrete form in section 2.3.

As loading conditions uniaxial compressive forces are considered. In ADAMS & HUTTON [1985] gradual disc prolapse was produced at average compressive loads of approximately $3500N$. A load of $3000N$ as lower limit of the range of load at fracture of the endplates is reported in ROLANDER & BLAIR [1975]. Therefore a maximum load of $3500N$ is chosen for the current analysis. Other possible physiologic loadings like flexion or torsion are not within the scope of this investigation, but will be discussed in a subsequent study, when the complete geometry of a spine segment is considered.

In order to apply the compressive load in a realistic way, it must not react at the endplates directly. Figure 3 shows a finite element geometry used in the current analysis. Not only the intervertebral disc is modeled, but also adjacent vertebrae stumps of height $10mm$ (indicated by red color). This is necessary to ensure a realistic stress distribution in the endplates which will be discussed later.

The vertebrae themselves are assumed to consist of a cancellous core and cortical bone of thickness $1.5mm$ at the periphery. Both bone parts show orthotropic material behavior. Since cortical bone has a much higher stiffness than the cancellous core, it is obvious that the normal stress distribution in axial direction at the endplates cannot be uniform. Table 2 shows the applied constitutive parameters for lumbar vertebrae and correspond-

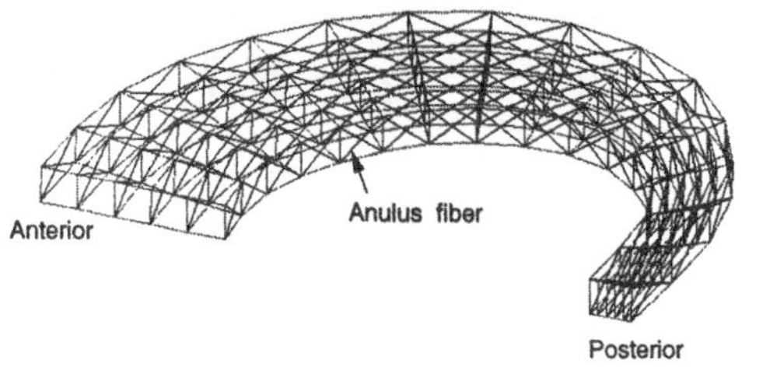

Figure 2: Collagen fiber arrangement in the anulus fibrosus.

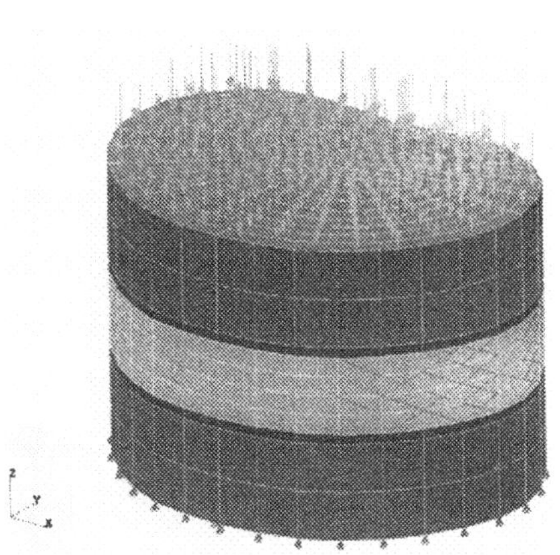

Figure 3: Finite element geometry.

Table 2: Typical material data for lumbar vertebrae.

Materials	Elastic moduli	Poisson ratios	References
Cortical bone	$E_{xx} = 11300MPa$	$\nu_{xy} = 0.484$	ASHMAN *et al.* [1984],
	$E_{yy} = 11300MPa$	$\nu_{xz} = 0.203$	KATZ & MEUNIER [1987],
	$E_{zz} = 22000MPa$	$\nu_{yz} = 0.203$	YOON & KATZ [1976]
	$G_{xy} = 3800MPa$		
	$G_{xz} = 5400MPa$		
	$G_{yz} = 5400MPa$		
Cancellous bone	$E_{xx} = 140MPa$	$\nu_{xy} = 0.450$	KEAVENY & HAYES [1993],
	$E_{yy} = 140MPa$	$\nu_{xz} = 0.315$	RAO & DUMAS [1991],
	$E_{zz} = 200MPa$	$\nu_{yz} = 0.315$	SHIRAZI-ADL *et al.* [1984],
	$G_{xy} = 48.3MPa$		UENO & LIU [1987]
	$G_{xz} = 48.3MPa$		
	$G_{yz} = 48.3MPa$		

ing references. The coordinate directions are denoted in figure 3. Otherwise the restrictions mentioned at the end of section 2.1 hold.

2.3 NUMERICAL RESULTS

This section investigates the question, whether the collagen fiber arrangement according to figure 2 can influence numerical results, if the number of collagen fiber layers is changed. Figures 4 & 5 show finite elemet meshes used for the female (1) and male (2) spine segments mentioned above, respectively. The latter one is derived from figure 1 by cutting off all facet elements.

Mesh (1) in figure 4 implies six collagen fiber layers, whereas mesh (2) in figure 5 contains five only. The corresponding cross–sectional areas are determined according to OHSHIMA *et al.* [1989] and depicted in table 3.

Before numerical results are discussed, it must be mentioned that appropriate finite element selections are discussed in EBERLEIN *et al.* [1999]. The current computations are based on this knowledge. A convergence study in EBERLEIN *et al.* [1999] also proves reliable finite element discretizations. They are applied in subsequent analyses.

Table 3: Annulus fibers' cross–sectional areas.

Layer locations	Mesh (1) $[mm^2]$	Mesh (2) $[mm^2]$
Innermost	0.030	0.066
2nd	0.055	0.078
3rd	0.082	0.109
4th	0.110	0.139
5th	0.133	–
Outermost	0.148	0.245

Figure 6 represents the nonlinear relationship of the disc in terms of axial compressive force versus posterolateral disc bulge. Obvious is a stiffening effect during loading. The current finite element analysis with mesh (1) is considerably stiffer than mesh (2). A wide variety of experimental data can

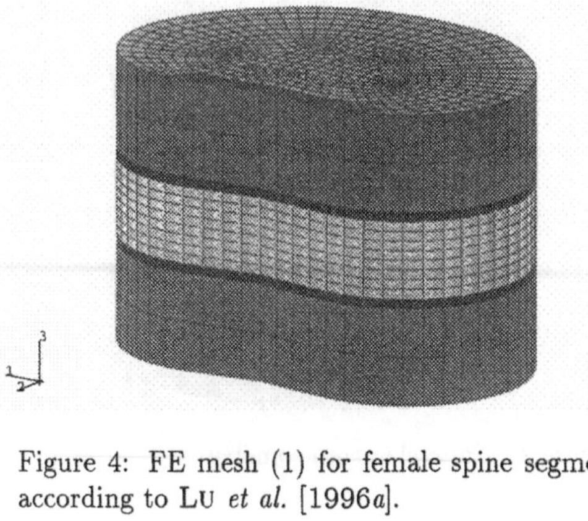

Figure 4: FE mesh (1) for female spine segment according to LU *et al.* [1996*a*].

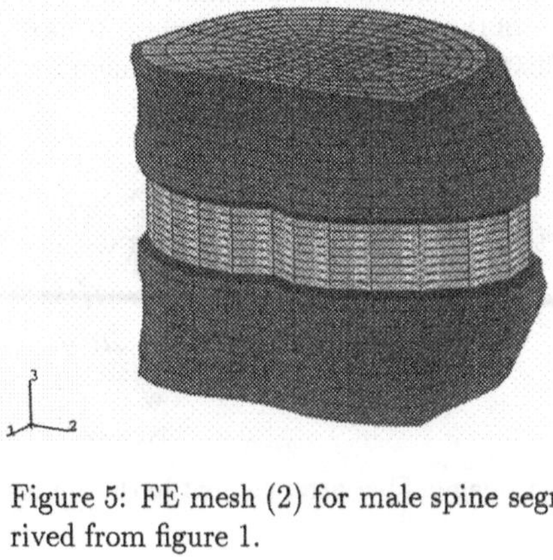

Figure 5: FE mesh (2) for male spine segment derived from figure 1.

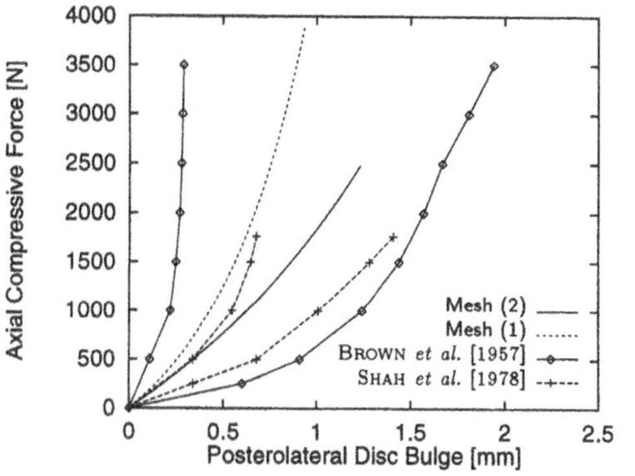

Figure 6: Axial compressive force versus posterolateral disc bulge.

be found in BROWN *et al.* [1957] and SHAH *et al.* [1978]. Therefore it can only be concluded that the computational results fall within the range of these experimental measurements. Because of this, the accuracy of disc bulge calculations remains an open question. The finite element analyses are geometry dependent, instead.

A further problem arises from the strain distribution in the anulus fibrosus. For a compressive load of $3000N$, the strains in the collagen fibers are always tensile and exhibit a continous decrease from the innermost layer to the outer ones EBERLEIN *et al.* [1999]. Large strains occur, up to 8% in the FE model with exponential fiber stiffening. The situation is more complicated for the annulus ground substance, though. From an engineering point of view strains in a range of 70% and more, as indicated in figure 7, can absolutely not be covered by a linear elastic model. On the other hand, obtaining nonlinear experimental contitutive data is rather impossible, since the so called

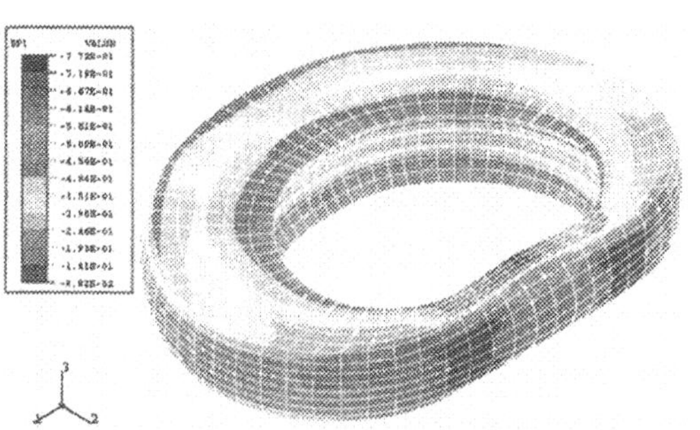

Figure 7: Maximum principal strain values in annulus matrix for $3000N$ and nonlinear collagen fibers.

ground substance can hardly be considered as an independent medium. Its physical properties primarily result from fluid particles acting on the collagen fiber bundles in radial direction. Therefore it seems promising to dissect collagen fiber layers along circumferential clefts from a disc body and determine nonlinear material properties for the resulting anisotropic composite macro material. An appropriate experimental setup is described in the following section.

2.4 EXPERIMENTAL SETUP

The experimental in vitro investigations comprise geometric boundary conditions and material behavior of human intervertebral discs. The material testing consists of quasistatic, uniaxial tension tests which are carried out on samples obtained from anulus lamellae primarily in the viscoelastic (reversible) regime. In a first step five pathologically unchanged, chemically untreated, symmetric and therefore representative discs from five different corpses are examined. Currently the first corpse is right about to be prepared.

Uniaxial extension in fiber direction yields the stress–strain relationship of annulus samples. These are taken along the circumferential direction (lateral, ventral and possibly dorsal) of the discs with three samples for each in radial direction. The influence of cortical bone at the disc edge and endplate cartilage on annulus stiffness is also considered. The final goal is to approximate the spatial distribution of mechanical properties in anulus lamellae. This finally yields a general 3D anisotropic material model for large elastic strains. It is applicable to arbitrary disc geometries and circumvents the contradictions discussed in the previous section.

The validity of the experimental concept has been proven in recent sample preparations. Corresponding publications in medical science journals will follow. From the pre–tests it became obvious that the anulus fibrosus can be analyzed in the described way. Based on the experimental data obtained, the constitutive description finally relies on a fibre model for finite elastic strains as described in GASSER & HOLZAPFEL [1998].

3 CONCLUSIONS

The current investigation deals with state of the art FE analyses of intervertebral discs. In particular typical geometry of an intact L_{2-3} lumbar disc body is considered. Geometric and partially material nonlinearities are taken into account in order to learn about the deformation behavior of natural discs under uniform compressive loads. Furthermore local strain distributions are analyzed. This allows an evaluation of state of the art FE models and motivates further improvements.

In detail the anulus fibrosus is modeled as a composite of collagen fibers embedded in a matrix of ground substance. The nucleus pulposus is represented by an inviscid incompressible fluid. Material parameters are adopted from literature. They are all based on linear elasticity except for exponential elastic stiffening of the collagen fibers in anulus fibrosus. Eventually, the significant predictions of the FE analyses can be summarized as follows:

The derivation of a general 3D, nonlinear and anisotropic material model is necessary for an accurate description of strain distributions in anulus fibrosus. This also avoids geometric solution dependencies and allows an accurate modeling of lumbar spine segments. An experimental concept for the determination of required material parameters has been discussed briefly. Thus it is possible to address the error range of purely elastic FE models currently found in literature which might not be negligible. Even though the annulus ground substance is assumed to be very soft in the current analyses, it exhibits large strains and therefore considerably contributes to the strain energy stored in the deformed system.

REFERENCES

ADAMS, M. A. & W. C. HUTTON (1985). 'Gradual Disc Prolapse'. *Spine* **10**, 524–531.

ASHMAN, R. B., S. C. COWIN, W. C. VAN BUSKIRK & J. C. RICE (1984). 'A Continuous Wave Technique for the Measurement of the Elastic Properties of Cortical Bone'. *J. Biomechanics* **17**, 349–361.

BAO, Q. B., G. M. MCCULLEN, HIGHAM P. A., J. H. DUMBLETON & A. Y. HANSEN (1996). 'The Artificial Disc: Theory, Design and Materials'. *Biomaterials* **17**, 1157–1167.

BROWN, T., R. J. HANSEN & A. J. YORRA (1957). 'Some Mechanical Tests on the Lumbosacral Spine with Particular Reference to Intervertebral Discs'. *J. Bone Joint Surg.* **39**, 1135–1164.

DAWSON, E. & J. BERNBECK (1998). 'The surgical treatment of low back pain'. *Rehabil. Clin. North. Am.* **9**, 489–495.

DEKUTOSKI, M. B., M. J. SCHENDEL, J. W. OGILVIE, J. M. OLSEWSKI, L. J. WALLACE & J. L. LEWIS (1994). 'Comparison of in vivo and in vitro adjacent segment motion after lumbar fusion'. *Spine* **19**, 1745–1751.

EBERLEIN, R., M. FRÖHLICH & E. M. HASLER (1999). Finite-element-analyses of intervertebral discs: Structural components and properties. European Conference on Computational Mechanics, Munich, Germany.

EYRE, D. R. (1979). 'Biochemistry in the Intervertebral Disc'. *Int. Rev. Con. Tissue Res.* **8**, 227–291.

FRANK, J. W., M. S. KERR, A. S. BROOKER, S. E. DEMAIO, A. MAETZEL, H. S. SHANNON, T. J. SULLIVAN, R. W. NORMAN & R. P. WELLS (1996). 'Disability resulting from occupational low back pain. Part I: What do we know about primary prevention? A review of the scientific evidence on prevention before disability begins'. *Spine* **21**, 2908–2917.

GALANTE, J. O. (1967). 'Tensile Properties of the Human Lumbar Annulus Fibrosus'. *Acta. Orthop. Scand.* **100 (Suppl.)**, 1–91.

GASSER, T. C. & G. A. HOLZAPFEL (1998). A Fibre-Matrix Model for Arteries Including Viscous Effects. 3rd World Congress of Biomechanics, Sapporo, Japan.

GILBERTSON, L. G., V. K. GOEL, W. Z. KONG & J. D. CLAUSEN (1995). 'Finite Element Methods in Spine Biomechanics Research'. *Crit. Rev. Biomed. Eng.* **23**, 411–473.

GOEL, V. K., B. T. MONROE, L. G. GILBERTSON & P. BRINCKMANN (1995). 'Interlaminar Shear Stresses and Laminae Separation in a Disc'. *Spine* **20**, 689–698.

HAMBLY, M. F., L. L. WILTSE, N. RAGHAVAN, G. SCHNEIDERMAN & C. KOENIG (1998). 'The transition zone above a lumbosacral fusion'. *Spine* **23**, 1785–1792.

HARKNESS, R. D. (1961). 'Biological functions of collagen'. *Biol. Rev.* **36**, 399–463.

HAUT, R. C. & R. W. LITTLE (1972). 'A Constitutive Equation for Collagen Fibers'. *J. Biomechanics* **5**, 423–430.

KATZ, J. L. & A. MEUNIER (1987). 'The Elastic Anisotropy of Bone'. *J. Biomechanics* **20**, 1063–1070.

KEAVENY, T. M. & W. C. HAYES (1993). 'A 20–Year Perspective on the Mechanical Properties of Trabecular Bone'. *J. Biomech. Eng.* **115**, 534–542.

LU, Y. M., W. C. HUTTON & V. M. GHARPURAY (1996*a*). 'Can Variations in Intervertebral Disc Height Affect the Mechanical Function of the Disc?'. *Spine* **21**, 2208–2217.

LU, Y. M., W. C. HUTTON & V. M. GHARPURAY (1996*b*). 'Do Bending, Twisting, and Diurnal Fluid Changes in the Disc Affect the Propensity to Prolapse? A Viscoelastic Finite Element Model'. *Spine* **21**, 2570–2579.

OHSHIMA, H., H. TSUJI, N. HIRANO, H. ISHIHARA, Y. KATOH & H. YAMADA (1989). 'Water Diffusion Pathway, Swelling Pressure and Biomechanical Properties of the Intervertebral Disc during Compression Load'. *Spine* **14**, 1234–1244.

RAO, A. A. & G. A. DUMAS (1991). 'Influence of Material Properties on the Mechanical Behaviour of the L5–S1 Intervertebral Disc in Compression: A Nonlinear Finite Element Study'. *J. Biomed. Eng.* **13**, 139–151.

ROLANDER, S. D. & W. E. BLAIR (1975). 'Deformation and Fracture of the Lumbar Vertebral end-plate'. *Orthop. Clin. North. Am.* **6**, 75–81.

SHAH, J. S., W. G. HAMPSON & M. I. JAYSON (1978). 'The Distribution of Surface Strain in the Cadaveric Lumbar Spine'. *J. Bone Joint Surg.* **60**, 246–251.

SHIRAZI-ADL, S. A., S. C. SHRIVASTAVA & A. M. AHMED (1984). 'Stress Analysis of the Lumbar Disc–Body Unit in Compression. A Three–Dimensional Nonlinear Finite Element Study'. *Spine* **9**, 120–134.

UENO, K. & Y. K. LIU (1987). 'A Three–Dimensional Nonlinear Finite Element Model of Lumbar Intervertebral Joint in Torsion'. *J. Biomech. Eng.* **109**, 200–209.

URBAN, J. P. G. & S. ROBERTS (1996). Intervertebral Disc. In 'Extracellular Matrix, Vol. 1, Tissue Function'. Comper, W. D. (Ed.). Harwood Academic Publishers GmbH. Amsterdam. 203–299.

WHITE, A. A. & M. M. PANJABI (1990). *Phasical properties and functional biomechanics of the spine*. J. B. Lippincott Company. Philadelphia. 2nd edition.

WU, H. C. & R. F. YAO (1976). 'Mechanical Behavior of the Human Annulus Fibrosus'. *J. Biomechanics* **9**, 1–7.

YOON, H. S. & J. L. KATZ (1976). 'Ultrasonic Wave Propagation in Human Cortical Bone–II. Measurements of Elastic Properties and Microhardness'. *J. Biomechanics* **9**, 459–464.

Constitutive Models for Rubber, Dorfmann & Muhr (eds) © 1999 Taylor & Francis ISBN 90 5809 113 9

High Damping Laminated Rubber Bearings (HDLRBs): A simplified non linear model with exponential constitutive law – Model description and validation through experimental activities

A. Dusi & F. Bettinali
ENEL, Milano, Italy

V. Rebecchi & G. Bonacina
ISMES, Bergamo, Italy

ABSTRACT: in this paper a parallel elasto-plastic model with exponential constitutive law for seismic isolator, able to represent the nonlinear behaviour is proposed. The model has been developed considering HDLRBs, but it can be adequately applied to other types of isolation devices having a continuously decreasing stiffness with increasing displacement. The proposed model, implemented in an existing nonlinear finite element computer code, is particularly suited to compute the time history dynamic response of base isolated structures. Its reliability is checked by comparing the numerically computed and experimentally recorder motions of both single devices and base isolated mock-ups subjected to different earthquake motions on shaking table.

1 INTRODUCTION

Since 1993 ENEL and ISMES are co-operating in the framework of a wide research program aimed at evaluating the applicability of seismic base isolation to industrial plants. Particular attention has been devoted to HDLRB devices because of their very promising features and the worldwide diffusion. These devices are manufactured by vulcanisation bonding of layers of high damping rubber to thin steel reinforcing plate, thus merging in a single mechanism both the frequency filtering and energy dissipation functions necessary to achieve an effective isolation action.

The proposed model assumes a decreasing exponential constitutive law $G(\gamma)$ for the rubber compound; this behaviour has been implemented in the finite element ABAQUS code through a parallel scheme of elastoplastic elements. The model is an improvement of a previous one developed by Serino (Serino et al. 1993) and described in Bettinali et al. 1997. A peculiar feature of developed model is the capability to reproduce the complex behaviour of HDLRB devices using only very simple information concerning the rubber compound, without any need of experimental data relevant to tests on actual devices. Static tests were carried out on single bearings to identify the force-displacement relationship, while dynamic tests were carried out using a 6 degree of freedom shaking table to reproduce seismic events on mock-ups connected to the ground by means of 4 or 6 HDLRB devices. Results obtained from the numerical model show a very good agreement with experimental displacements and accelerations, thus demonstrating the model capability to provide reliable results using only cheap and simple tests made on rubber specimens.

Recently, biaxial quasi-static tests have also been carried out at Ismes on ENEL behalf, and the comparison between numerical and experimental results pointed out that the numerical model can also reproduce the biaxial behaviour of HDLRB. A slightly more sophisticated model, taking into account the high-strain hardening for uniaxial behaviour, has also been developed and tested.

The described model for seismic isolator is able to represent the nonlinear hysteretic behaviour as well as transverse biaxial interaction and therefore is sufficiently sophisticated to correctly predict the time history dynamic response of base isolated structures. Being based on the parallel elasto-plastic concept, the proposed model is at the same time sufficiently versatile to be easily implemented in existing nonlinear finite element computer programs.

2 EXPERIMENTAL OBSERVATIONS

Experimental tests on HDLRB devices are carried out by imposing horizontal relative displacements between the base and the top faces, and by recording the value of force required. During these tests a vertical load equal to the design load is applied. Experimental tests have shown that:

i) for values of shear strain γ up to approximately 100%, the tangential modulus G_t of the bearing exponentially decreases with increasing shear strain from the initial value G_O to the asymptotic value G_∞;

ii) each time the relative motion between the two ends of the bearing changes sign, the tangential modulus takes back the maximum value;

iii) a kinematic hardening behaviour was observed, i.e. hysteretic loops remain inside an envelope which represents the isolator behaviour;

iv) under the hypothesis of small thickness of each rubber layer the rubber bearing behaviour depends on the normal area and on the total thickness of rubber layers only;

v) the energy dissipated in a cycle depends on the maximum deformation, but is almost independent from the rate of strain application, at least in the frequency range characteristic of an isolated structure (i.e. the amount of damping can't be represented by means of an equivalent viscous parameter).

3 NUMERICAL MODEL

A linear model cannot reproduce in a realistic way the behaviour of HDLRB. The proposed approach, described in Serino et al. 1993, is a non-linear hysteretic model with the following assumptions:

i) tangent modulus with decreasing exponential law, defined through 3 parameters related to the compound, i.e.:

G_∞ G value at design shear strain
G_O G value at γ=0
b exponent multiplier;

ii) overall bearing behaviour defined through 2 geometric data: normal area of bearing and total rubber thickness;

iii) no viscous effects, i.e. velocity does not affect the force-displacement relationship.

The assumed G(γ) relationship can be expressed by the following two mathematical expressions, valid for a loading and unloading curve, respectively:

$$G_t(\gamma) = \frac{d\tau}{d\gamma} = G_\infty + ae^{-b(\gamma - \gamma_{\min})}$$

$$G_t(\gamma) = \frac{d\tau}{d\gamma} = G_\infty + ae^{-b(\gamma_{\max} - \gamma)}$$

where:

$a = G_\infty - G_O$,

γ_{min} = starting strain value for increasing load phase

γ_{max} = starting strain value for decreasing load phase.

It is worth noting that the model requires the correct selection of the three rubber parameters only (G_∞, a, b) which can be simply obtained from tests on rubber specimens.

Hysteretic curves with continuously decreasing stiffness can be easily discretized using the parallel modelling concept. The above G(γ) relationship has been reproduced by means of a set of elastoplastic elements working in parallel; the parameters of each element in parallel must be calibrated to correctly reproduce the experimental bearing behaviour. The basic idea is to consider simple elastic-perfectly plastic and purely elastic elements connected in parallel, i.e. all having the same deformation but each carrying a different force. The total force acting on the assemblage is then obtained by summing the forces acting on each element. Parallel modelling is not new and has been already used in the past for representing the mechanical behaviour of inelastic materials. The idea of effective stress introduced by Terzaghi to describe the behaviour of wet soils is essentially a parallel model, as water and soil fabric both carry the total stress. The concept was first formalized by Mroz (Mroz 1963) in treating nonassociated flow plasticity and then used by Owen (Owen et al. 1974) and Pande (Pande et al. 1977) to achieve a better representation of actual material behaviour through a number of overlays of simple models. Nelson (Nelson & Dorfmann 1995) has recently used parallel mathematical models in incremental elastoplasticity to represent strain hardening in metals and to develop models of frictional materials such as soils, rock and concrete.

The concept applied to seismic isolators is very simple: the set of "n" elasto-plastic elements work all together in their elastic range at the beginning of the load history; when the relative displacement reaches a threshold value, the first elastoplastic element becomes plastic, so that the overall stiffness becomes smaller, being reduced by a quantity equal to the stiffness of the first element. When the relative displacement increases, the remaining elastoplastic elements (n-1) become plastic one at a time, and the resultant behaviour appears as a non-linear spring with decreasing stiffness. When the relative velocity changes sign, all the elastoplastic elements come back to their elastic state and the global stiffness is again at its maximum value. This scheme has been implemented through a USER MATERIAL subroutine for the ABAQUS finite element code; the HDLRB behaviour can be reproduced simply by assigning this user defined material to a truss or to a solid element having the geometric characteristic of the actual bearing. The proposed model is capable of taking into account coupling between forces and displacements lying in the horizontal plane and can therefore be used to reproduce the dynamic behaviour of an isolated system subjected to triaxial seismic events, assuming that HDLRB devices behave like a rigid element in the vertical direction.

4 STATIC MONOAXIAL TESTS

Static tests have been set up to investigate the behaviour of the proposed numerical model in case of multiple loading cycles with different amplitude. A single bearing, having a diameter equal to 125 mm and a total rubber height equal to 33 mm, composed by a "hard" compound with G = 0.8 MPa, has been tested by imposing a history of relative displacement characterised by a maximum value of 30 mm and by a series of smaller cycles with amplitude of 1, 2, 5 and 10 mm. Such a displacement history has been designed to make evident the kinematic hardening phenomena. Comparison between experimental data and numerical results is shown in Figure 1: the agreement appears quite good, highlighting a remarkable fitting of the real behaviour of the bearing even if the smallest loops of the numerical model have an area (which corresponds to the dissipated energy) smaller than the experimental one.

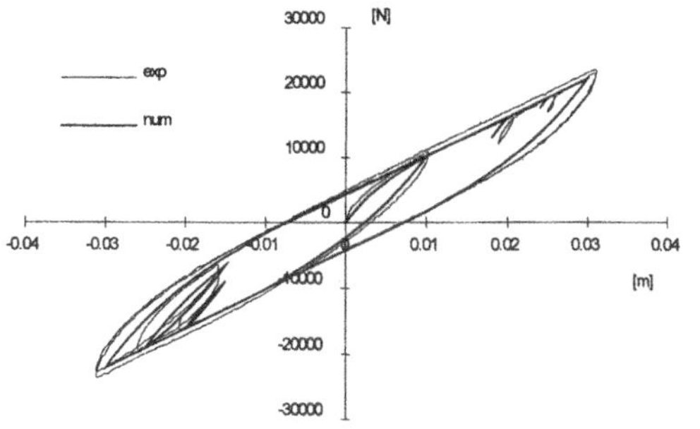

Figure 1. Experimental and numerical hysteresis loops for a rigid HDLRB (G = 0.8 MPa).

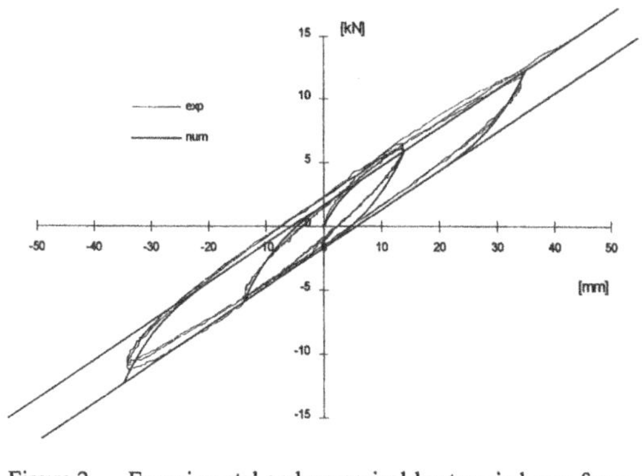

Figure 2. Experimental and numerical hysteresis loops for a soft HDLRB (G = 0.4 MPa).

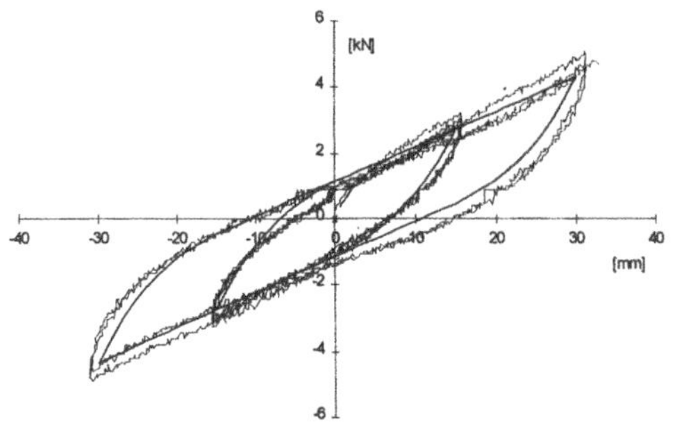

Figure 3. Experimental and numerical hysteresis loops for a soft HDLRB (G = 0.4 MPa).

Static tests have been carried out also for a "soft" compound with G = 0.4 MPa; the comparison between experimental and numerical hysteretic loops is shown in Figure 2 for a bearing having a diameter equal to 250 mm and a total rubber height of 66 mm; the agreement is also good.

The last compound tested is another "soft" compound (G = 0.4 MPa), but different from the previous one. Experimental and numerical hysteresis loops for a bearing made with such a compound, having a diameter equal to 125 mm and a total rubber height equal to 30 mm are depicted in Figure 3.

5 SEISMIC TESTS

The numerical model of elastomeric bearings has been used to reproduce the dynamic behaviour of real base isolated structures subjected to seismic excitation.

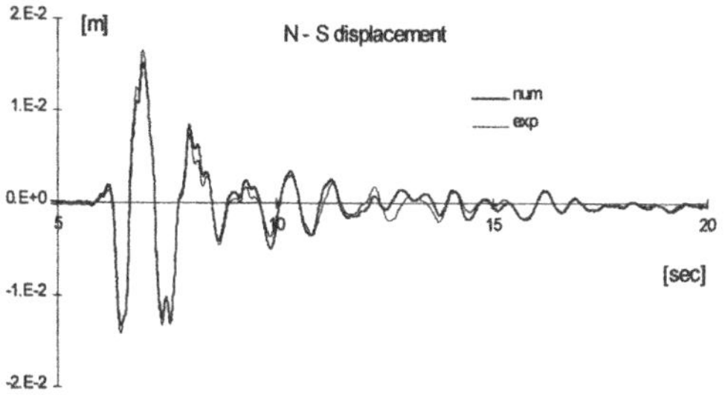

Figure 4a. Experimental and numerical relative displacements of a 400 kN isolated mock-up under the 1976 Tolmezzo NS record.

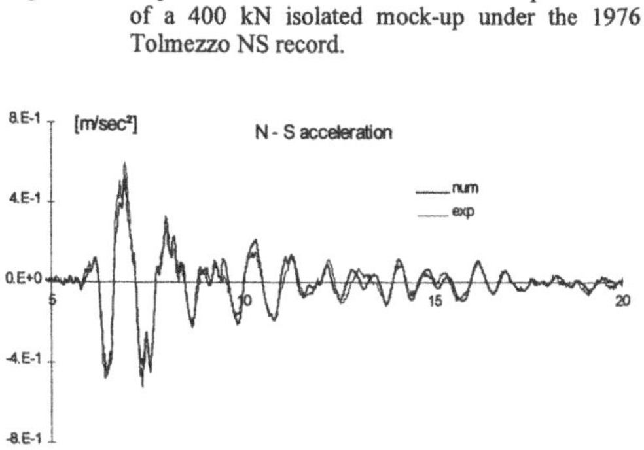

Figure 4b. Experimental and numerical acceleration of a 400 kN isolated mock-up under the 1976 Tolmezzo NS record.

Different tests have been carried out at Ismes using a shaking table.

Among other tests, shaking table experiments have been conducted on an isolated mock-up consisting in a concrete block, weighting 400 kN, supported on four bearings manufactured with rigid compound, whose experimental and analytical hysteresis loops are shown in Figure 1.

The Friuli 1976 earthquake (Tolmezzo 3D records) has been applied as triaxial excitation with a time scale of ½ to the shaking table. Comparison between numerical results and experimental data are shown in Figures 4a, 4b for the north-south record: the agreement in terms of acceleration and displacement is very satisfactory, taking also into account the presence of coupling effects.

The second seismic test considered in this paper is relevant to a four storeys steel frame, loaded with a total mass of 245 kN supported on 6 bearings, whose experimental and analytical hysteresis loops are reported in Figure 3. The scaled (-9 dB) Campano-Lucano 1981 earthquake (Calitri records) has been applied to the shaking table.

Results in terms of horizontal relative displacement of bearings are compared with the experimental data in Figures 5a and 5b.

This last case is closer to a real application, because all the three components of the isolation

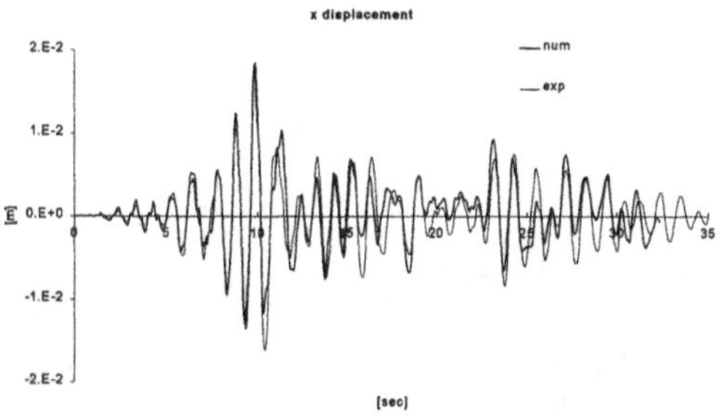

Figure 5a. Experimental and numerical relative displacement (x component) of a steel frame isolated mock-up under the 1981 Calitri record.

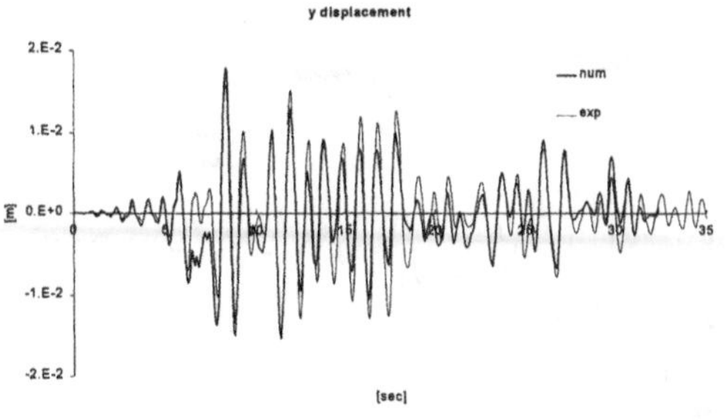

Figure 5b. Experimental and numerical relative displacement (y component) of a steel frame isolated mock-up under the 1981 Calitri record.

Table 1: biaxial tests on HRLRBs

Test Number	Phase lag [degrees]	Maximum X Shear strain [%]	Maximum Y shear strain [%]
1	15	25	25
2	15	50	50
3	15	100	100
4	30	25	25
5	30	50	50
6	30	100	100
7	90	25	25
8	90	50	50
9	90	100	100
10	90	25	25
11	90	100	100
12	30	25	25
13	30	100	100

problem are present: the excitation, the isolation system and the elastic supported structure with its own dynamic behaviour. The comparison between experimental data and numerical results can be considered satisfactory.

6 BIAXIAL TESTS

To get a deeper insight in the biaxial behaviour of HDLRB, a set of biaxial tests has been carried out at Ismes. Low frequency (0.1 Hz) time histories of sinusoidal displacements have been simultaneously imposed in the two horizontal x and y directions of a HDLRB. Three different amplitudes and three different phase lags between the horizontal components have been considered for different tests; vertical load

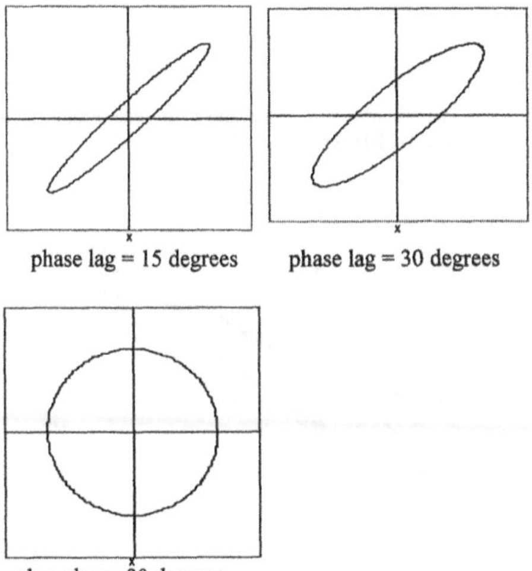

phase lag = 15 degrees   phase lag = 30 degrees

phase lag = 90 degrees

Figure 6. Typical paths for assumed phase lags.

was simultaneously applied and controlled. The description of tests is reported in table 1.

Typical paths for the three assumed phase lags are reported, for a single cycle, in Figure 6: while the path relevant to 15 degrees is similar to a monoaxial excitation, the 90 degrees path represents the most heavy biaxial working condition for the isolation device. In fact, following a circular path, the isolator is subjected to a constant shear strain amplitude, being only the direction of the shear strain variable.

The energy dissipation occurring during the tests with of 90 degrees phase lag is also qualitatively different from those relevant to 15-30 degrees phase lag because of the absence of variation in the strain amplitude. External work supplied during the different tests is reported in Figure 7: while for tests with 0-15-30 degrees phase lags, the time history of external work is periodic (i.e. there is a certain part of the external work which is recovered), for the 90 degrees test the time history of external work is an increasing linear function of time.

The amount of dissipated energy directly depends on the relative phase between force vector F and incremental displacement *ds*:

$$E_{diss} = \oint \vec{F} \cdot d\vec{s}$$

Analysis of experimental data relevant to tests with circular path showed an almost constant relative phase between force and displacement vectors equal to 11-15 degrees. The numerical simulation of these tests, carried out using the developed user subroutine, gave as result a constant phase of about 10 degrees, demonstrating that the implemented model is able to reproduce with good accuracy the observed biaxial behaviour. The comparison between experimental data and numerical results for the test n.1 (25% shear strain and phase lag equal to 15 degrees) is reported in Figure 8, together with the absolute value of imposed displacement. As it can be noted, the agreement is qualitatively and quantitatively good, with minimum relative phases around 5

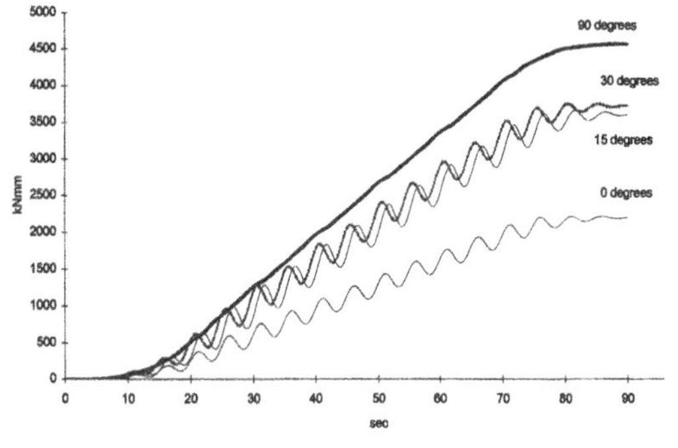

Figure 7. External work as function of time during biaxial tests.

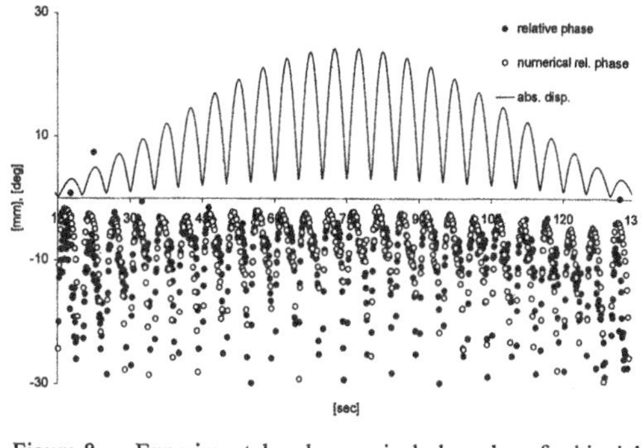

Figure 8. Experimental and numerical phase lags for biaxial test at 25% shear strain, 15 degrees phase lag.

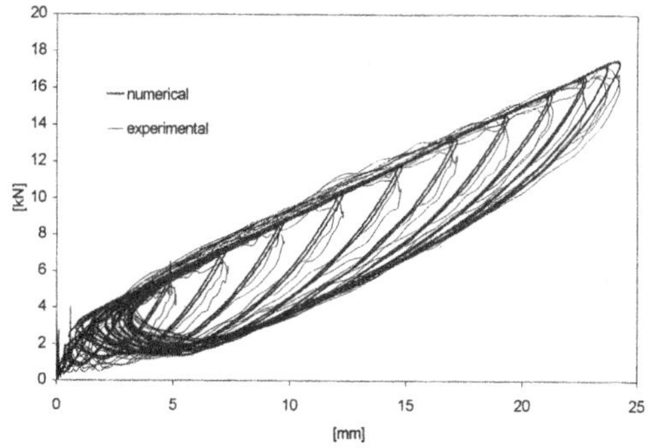

Figure 9. Experimental and numerical applied force modulus as function of displacement for biaxial test at 25% shear strain, 15 degrees phase lag.

degrees corresponding to the minimum values of displacement cycles.

As an example, for the same 15 degrees test, experimental and numerical graphs of applied force modulus versus imposed displacement modulus are reported in Figure 9. Again, the agreement can be considered satisfactory.

7 MODEL DEVELOPMENT FOR HIGH SHEAR STRAINS

The parallel model examined is able to reproduce the observed experimental behaviour of HDLRB as far as the maximum shear strain does not exceed a limit value around 100%. At larger shear strains in the rubber (above 150%) an increase of the tangent modulus is observed due to the strain crystallisation of the rubber matrix. Being the model implemented under the assumption of exponential stiffness decrement along with shear strain, the high strain hardening phenomenon cannot be taken into account.

A more refined constitutive model was therefore set up to allow for the reproduction of the high shear strain behaviour. This was obtained by putting in parallel to the original model, a second one, whose $G_h(\gamma)$ function has the following expression:

$$G_h(\gamma)=c(e^{d|\gamma|}-1)$$

where parameters c and d define slope and curvature of exponential law.

The above expression was implemented using again a set of parallel elements.

Figures 10, 11 and 12 show the comparisons between experimental hysteresis loops (obtained from quasi-static monoaxial tests) and numerical results. Figure 10 is relevant to static test with the same compound used in the example reported in Figure 1, but in this case the maximum shear strain is approximately 150%: the agreement is remarkable. It has to be stressed that the new model can reproduce the experimental behaviour for a wide range of hysteresis loops, ranging from the smallest one (20 mm) to the largest one (200 mm). It is also worth noting that both the two main effects of hardening are well reproduced: the increasing of stiffness for strains above 100%, as well as is the increasing of the "thickness" of loops for zero strain.

A test with bearing made with a "soft" compound (the same of Figure 2) is reported in Figure 11: again the agreement is satisfactory (even if the hardening effect is less evident) for a wide range of loop amplitude, ranging from 20 to 200 mm.

The last test performed is reported in Figure 12. A HDLRB made with the soft compound (the same of Figure 3), has been tested with a maximum shear strain equal to 250%. Even if the agreement is less satisfactory, compared to the previous test analysed, it can be said that the model is adequate to catch the recorded behaviour, and thus well simulate both the stiffness and the energy dissipation capacity of the isolator.

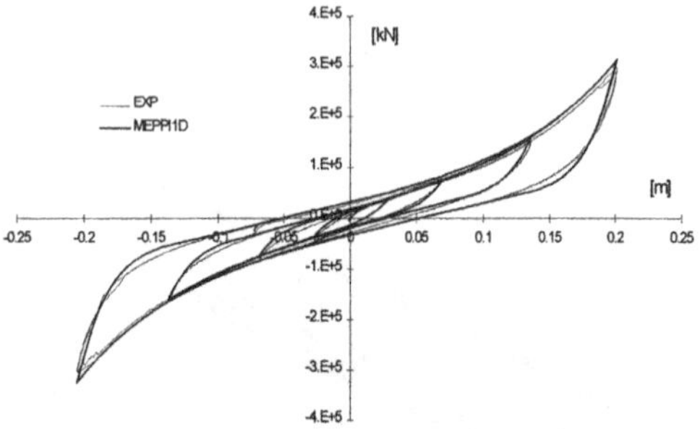

Figure 10. Experimental and numerical hysteresis loops for a rigid HDLRB (G = 0.8 MPa).

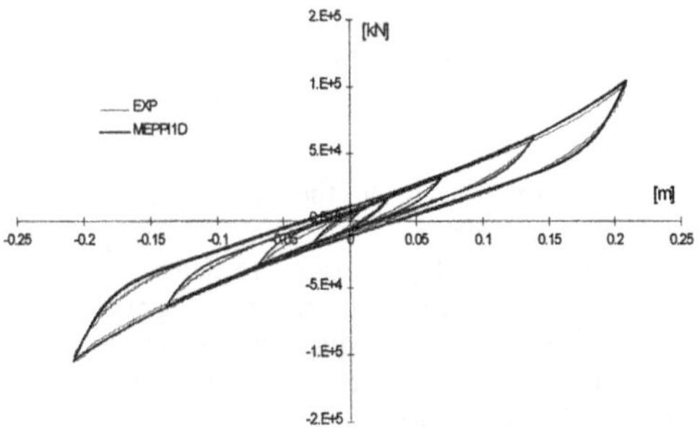

Figure 11. Experimental and numerical hysteresis loops for a soft HDLRB (G = 0.4 MPa).

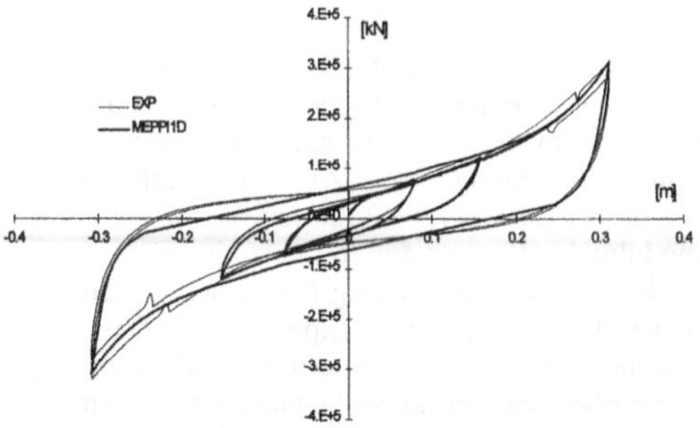

Figure 12. Experimental and numerical hysteresis loops for a soft HDLRB (G = 0.4 MPa).

8 CONCLUSIONS

Comparisons between experimental tests and numerical analyses show that the numerical model with an exponential constitutive law, implemented through the parallel elastoplastic scheme, is able to reproduce the seismic behaviour of HDLRB devices with a high degree of accuracy. Hysteresis loops, multiaxial excitations and hardening effects are accurately reproduced.

The availability of a routine implementing the HDLRB behaviour inside a commercial available finite element code (ABAQUS) allows the user to perform time histories analyses of isolated structures, simply by defining an element having the same bearing dimension and whose material characteristics (related to the compound parameters) are given in a user subroutine.

The described numerical model was also used for other applications, in order to check its applicability to other isolation devices, like wire-rope bearings and elastoplastic metallic dampers, giving very satisfactory results.

The described model for seismic isolator is able to represent the nonlinear hysteretic behaviour as well as transverse biaxial interaction and therefore is sufficiently sophisticated to correctly predict the time history dynamic response of base isolated structures. Being based on the parallel elasto-plastic concept, the proposed model is at the same time sufficiently versatile to be easily implemented in existing nonlinear finite element computer programs.

REFERENCES

Bettinali, F., Dusi, A., Bonacina, G. & Rebecchi, V. 1997. Seismic behaviour of HDLRB through a simplified non linear model with exponential constitutive law. *Proc. Intern. Conf. on New Technologies in Structural Engineering, Lisbon, Portugal, July 2-5, 1997.*

Bonacina, G., Serino, G. & Spadoni, B. 1992. Implication of shaking table tests in the analysis and design of base isolated structures. *Proc. Tenth World Conference on Earthquake Engineering, 1992,* Vol.4 pp.2405-2410.

Mroz, A. 1963. Non-Associated flow rules in plasticity. *Jour. De Mecanique,* Paris, France, Vol. 2, pp. 21-42.

Nelson, R.B. & Dorfmann, A. 1995. Parallel elastoplastic models of inelastic material behaviour. *Journal. of Engineering Mechanics,* ASCE, Vol. 121, N. 10, pp. 1089-1097.

Owen, D.R.J., Prakash, A. & Zienkiewicz, O.C. 1974. Finite Element analysis of nonlinear composite materials by use of overlay systems. *Computer & Structures,* Vol. 4, pp. 1251-1267.

Pande, G.N., Owen, D.R.J. & Zienkiewicz, O.C. 1977. Overlay models in time-dependent nonlinear material analysis. *Computer & Structures,* Vol. 7, pp. 435-443.

Serino, G., Bonacina, G. & Bettinali, F. 1993. Proposal and experimental validation of analytical models for seismic and vibration isolation devices in nuclear and non-nuclear facilities. *Proc. SMiRT-12.* K. Kussmaul Elsevier Science Publishers B.V.

Constitutive Models for Rubber, Dorfmann & Muhr (eds) © 1999 Taylor & Francis ISBN 90 5809 113 9

Implementation and validation of hyperelastic finite element models of high damping rubber bearings

M. Forni & A. Martelli
ENEA, Bologna, Italy

A. Dusi
ENEL, Milano, Italy

ABSTRACT: In this paper the reliability of finite element models for simulating the behaviour of high damping rubber bearings is presented. Since 1992, hyperelastic models of seismic isolators have been developed and implemented in the ABAQUS code by ENEA and ENEL in the framework of Italian R&D studies for seismic isolation, European Projects and international co-operations. Several rubber compounds of different mechanical characteristics and several isolators, characterised by different materials, sizes, attachment systems and geometrical features have been analysed. Seismic isolators models have been validated through comparisons of numerical results with complete bearing test data. Based on the experience achieved, ENEA and ENEL developed guidelines for the execution of tests on both specimens and bearings, computer programs for designing and qualifying seismic isolators, pre-processors for preparing three-dimensional meshes of rubber bearings. The main features of the numerical analyses are reported in the paper, together with examples of experimental validation up to very large shear strains.

1 INTRODUCTION AND SCOPE

Aim of this study is the implementation and validation of non-linear (both axisymmetric and three-dimensional) finite-element models (FEMs) of High Damping steel-laminated Rubber Bearings (HDRBs) used for the seismic isolation of civil structures and industrial plants. The work has been performed in the framework of the Italian activities on seismic isolation development and application (Forni 1997). The validation of the models is based on the results of a wide ranging experimental campaign which was carried out on single HDRBs (Martelli 1995).

2 ISOLATORS CONSIDERED

2.1 Geometry

The HDRBs are formed by alternate vulcanised rubber layers and steel plates, bonded together by use of chemical compounds. They are usually placed between the structure and its foundations. Their features provide high stiffness in the vertical direction (to support the dead load) and low stiffness in the transverse direction (which minimises amplification of ground acceleration, by leading, however, to large horizontal displacements during strong earthquakes). The isolators analysed in the above-mentioned experimental campaign were manufactured by ALGA. Three different scales (1:1, 1:2 and 1:4) and two shape factors (S = 12 and S = 24) were considered in the tests for this bearing.

2.2 Boundary conditions

Five attachment systems between the isolator and the structure were considered in the experiments: 'recess', 'bolts', 'bolts & central dowel', 'central dowel' and 'bonding'. 'Recess' and 'central dowel' attachments are very similar and are the most difficult to model because they involve monolateral contact between rigid surface and deformable body. 'Bolts', 'bolts & central dowel' and 'bonding' attachment systems are practically equivalent from the kinematic point of view and correspond to a perfect encastre of the two bases of the isolator to the structure. In this work, the 'recess' and 'bolts' attachments have been considered.

2.3 Materials

HDRBs are characterised by a highly non-linear behaviour in terms of stiffness, thus a hyperelastic model is necessary for the rubber and an elastic-plastic for steel (at large displacements for the 'recess' attachment system) to correctly describe the actual behaviour. HDRB damping is practically independent of velocity and non-linearly dependent on displacement (hysteretic damping). Its effects on the isolator stress distribution and stiffness have been neglected in the present step of the analysis. Although two different rubber compounds (soft, G = 0.4 MPa, and hard, G = 0.8 MPa), developed by TARRC, have been considered in the tests, only results relevant to the hard rubber compound are hereinafter reported. This rubber is the most difficult to model, due to the high design loads which involve compressibility and some creep and plastic effects, impossible to be calculated by the hyperelastic theory.

2.4 Loads

The full-scale isolator supports a design vertical load of 1600 kN (800 kN in the case of soft rubber). In order to have the actual stress in the rubber (which is required by a correct application of similitude laws), the values for the 1:2 and 1:4 scales turned out to be 400 kN and 200 kN respectively (the half in the case of soft rubber). The design shear strain (which is the shear strain corresponding to design earthquake) has been assumed to be 100% of the total rubber thickness (150 mm for the full scale in the tests of Martelli 1995). Qualification tests are generally carried out up to twice the design displacement (i.e. generally to at least 200% shear strain) under the design vertical load; during failure tests, 400% shear strain is often exceeded (Martelli 1995).

3 FINITE ELEMENT MODELS

The behaviour of the bearings under vertical load and shear strain has been modelled by means of FE analyses, performed using ABAQUS. The difficulties encountered dealing with the incompressibility of rubber-like materials were treated by means of a mixed FE formulation. Hybrid elements were used; in these elements the pressure stress is independently interpolated from the displacement field, making the numerical formulation of the variational problem well-behaved. An updated Lagrangian formulation was adopted. The geometry of the devices and the loading conditions make the problem symmetric. Therefore only one half of the bearing is usually modelled, by imposing appropriate boundary conditions on displacements and rotations of the nodes belonging to the plane of symmetry.

3.1 Axisymmetric models

Computation time can be saved if it can be shown that axisymmetric solution give adequate insight into bearing behaviour.

Axisymmetric models of the isolator were developed and implemented in ABAQUS. Due to the geometry considered and the loads applied, axisymmetric elements with non-axisymmetric deformation were used to model the seismic isolator.

Some detailed meshes have been analysed in order to reach stability and convergence. A minimum of three layers of CAXA4Hn elements for each rubber layer have been necessary to model the rubber parts (n=1 is generally sufficient). SAXA11 shell elements have been used for the steel plates.

3.2 Three-dimensional models

In addition to axisymmetric models, a 3D model has been implemented to verify the CAXA elements' behaviour, to model the monolateral 'recess' attachment system and to reach large non-axisymmetric shear strains. 3D models are also usefull to correctly describe the tensile stress on bolts (if present), combined compression - shear - torsion stress-strain distribution (if eccentricity is relevant), presence of defects, and effects of the construction tolerances and others asymmetries.

Analyses carried out using several kinds of FE models showed that rubber layers can be successfully modelled with eight-node element, which provide linear displacement and constant pressure interpolations, and steel plates either with eight-node elements, reduced integration, linear displacement (C3D8R), or shell elements (S4R5). At least 8 subdivisions along the radius, 32 subdivisions along the external circumference were employed in the meshing; each rubber layer had three C3D8H elements through its thickness while each steel shim had only one C3D8R (S4R5) element through the thickness. When modelling recessed HDRBs, better results were obtained using hybrid triangular elements (C3D6H), with six nodes, linear displacement and constant pressure, placed at the inner and outer bearing borders. Pre-processing of the geometry boundary conditions, materials properties and loads were undertaken using GENESIS (Dusi 1997), a pre-processor for ABAQUS, which is capable of automatically generated the rather complicated ABAQUS input file on the basis of a few input data

4. HYPERELASTIC MODEL

4.1 Generalities

Developing and defining an accurate material model of elastomers is critical for the accuracy of the FE analyses results. Bearing elastomers are typically non-linear hysteretic material. The mechanical behaviour of these materials is described in ABAQUS by means of an elastic, isotropic and approximately incompressible model; therefore the governing constitutive equations are derived from a strain energy function U.

Two hyperelastic strain energy potentials are available in ABAQUS: the Mooney-Rivlin polynomial and Ogden forms; alternatively, the user can define any hyperelatic model in a user subroutine.

The Mooney-Rivlin polynomial and Ogden theoretically yield equal results. However there is a difference between the methods in terms of their basis. The Mooney-Rivlin model assumes that the strain energy density is a polynomial function of the principal strain invariants. Whereas, the Ogden model assumes that the strain energy density is a separate function of the three principal stretches.
The form of the Mooney-Rivlin polynomial strain energy potential is:

$$= \sum_{i+j=1}^{N} C_{ij}\left(\bar{I}_1-3\right)^i\left(\bar{I}_2-3\right)^j + \sum_{i=1}^{N}\frac{1}{D_i}\left(J_{el}-1\right)^{2i} \quad (1)$$

where U is the strain energy potential, J_{el} is the elastic volume ratio, $\bar{I}_1$ and $\bar{I}_2$ are the first and the second invariants of the deviatoric strain, and C_{ij} and D_i are material constants, while N is the order of the energy function. C_{ij} describes the shear behaviour of the material, D_i introduces compressibility and is set equal to zero for fully incompressible materials. The choice of N (generally $N = 1,2,3$) provides polynomial forms of strain energy function more or less complex.

The Mooney-Rivlin type of strain energy function is the most commonly used in modelling the stress-strain behaviour of filled elastomers. The usual approach uses a function with four to five terms.

The form of the Ogden strain energy potential is:

$$U = \sum_{i=1}^{N}\frac{2\mu_i}{\alpha_i^2}\left(\bar{\lambda}_1^{\alpha_i}+\bar{\lambda}_2^{\alpha_i}+\bar{\lambda}_3^{\alpha_i}-3\right)+\sum_{i=1}^{N}\frac{1}{D_i}\left(J_{el}-1\right)^{2i} \quad (2)$$

where $\bar{\lambda}_i = J^{-1/3}\lambda_i$, λ_i are the principal stretches and J_{el} is the elastic volume ratio. The constants μ_i and α_i describe the shear behaviour of the material and D_i the compressibility.

The calculation of the invariant derivatives of the Ogden's energy function is more involved and computationally intensive than that of the polynomial form.

The calibration of the hyperelastic model consists in the definition, starting from the material test data, of the constants C_{ij} (polynomial form), μ_i, α_i (Ogden form) and D_i (if compressibility is relevant). These coefficients are defined by the data of experiments involving simple state of deformations and stress.

This can be done in particular test configurations where, due to the specimen shape and loading conditions, one or two λ_i components can be neglected or correlated between themselves, and some σ_{ij} components can be put equal to zero, which considerably simplifies the expression of U.

Experience has shown that data from more than one test is required to reproduce an accurate material model. Relying only on one test, while accurate for a specific condition, may give poor results for more complex load application.

Uniaxial, biaxial and shear experiments were used (Ogden 1984; Rebelo 1991) to determine the above mentioned constants. For all the different compounds analysed in this study, the characterisation tests on rubber specimen (Figures 1 - 4) have been carried out at ENEL laboratories (Bettinali et al. 1996). Each test was performed on 3 specimens drawn from the centre of each rubber batch used for the manufacturing of isolators, both in the 'unscragged' and 'scragged' conditions.

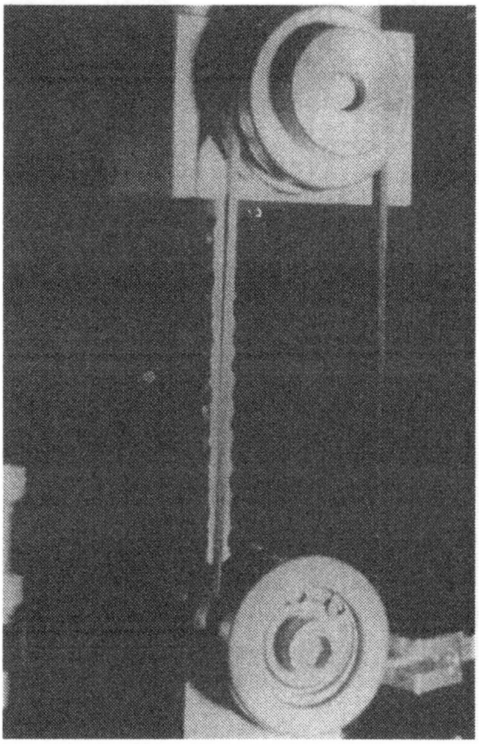

Figure 1. Tensile test (performed by ALGA according to ASTM D412 standards) on a rubber specimen ('normal' ring-shape ISO 4, mm 45 x 25 x 4).

Figure 2. Compression test (performed by ALGA according to ASTM D395 standards) on a cylindrical rubber specimen with lubricated surfaces (mm 29 x 13).

Figure 3. Planar test performed by ENEL on a rectangular rubber specimen (mm 300 x 100 x 5). Deformations were measured on the central part of the specimen by means of an optical instrumentation.

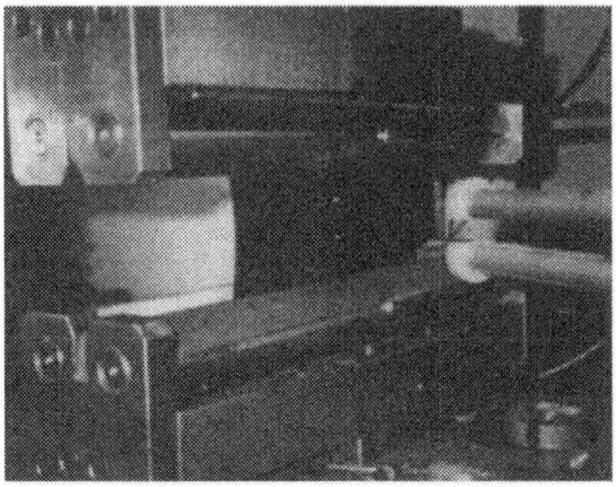

Figure 4. Equibiaxial test performed by ENEL on a square rubber specimen (mm 300 x 300 x 5). Deformations were measured on the central part of the specimen by means of an optical instrumentation.

4.2 Uniaxial tests

Uniaxial tensile and compression tests must satisfy the following conditions:

$$\sigma_{22} = \sigma_{33} = \sigma_{12} = \sigma_{13} = \sigma_{23} = 0; \quad \lambda_2 = \lambda_3 = 1/\sqrt{\lambda_1} \qquad (1)$$

The correlations (1) are strictly valid for incompressible materials only, which is the case of rubber specimens that have the possibility to strain in two principal directions at least.

The tensile tests were carried out according to ASTM D 412 on both dog-bone specimens (30mm x 6mm x 2mm in the central part) and ring specimens (Figure 1). The 'scragged' conditions consisted of four pre-cycles between 10% and 30% tensile strain. No significant scattering of test results was detected. Some simple calculations were executed on a single C3D8H element to evaluate the differences of the two hyperelastic models proposed by ABAQUS (Ogden and the above mentioned polynomial). Figure 5 shows the good agreement between both the models and with the measured curve.

The compression tests were carried out according to ASTM D 395 on 3 cylindrical specimens (diameter = 29 mm, height =13 mm) with lubricated bases (Figure 2). The 'scragged' conditions consisted of four pre-cycles between 5% and 15% compression strain. In this case also, no significant scattering of test results was present and perfect agreement between the calculated and measured results was found.

4.3 Biaxial (no shear) tests

As an alternative option to the uniaxial compression tests, equibiaxial tests may be performed. We

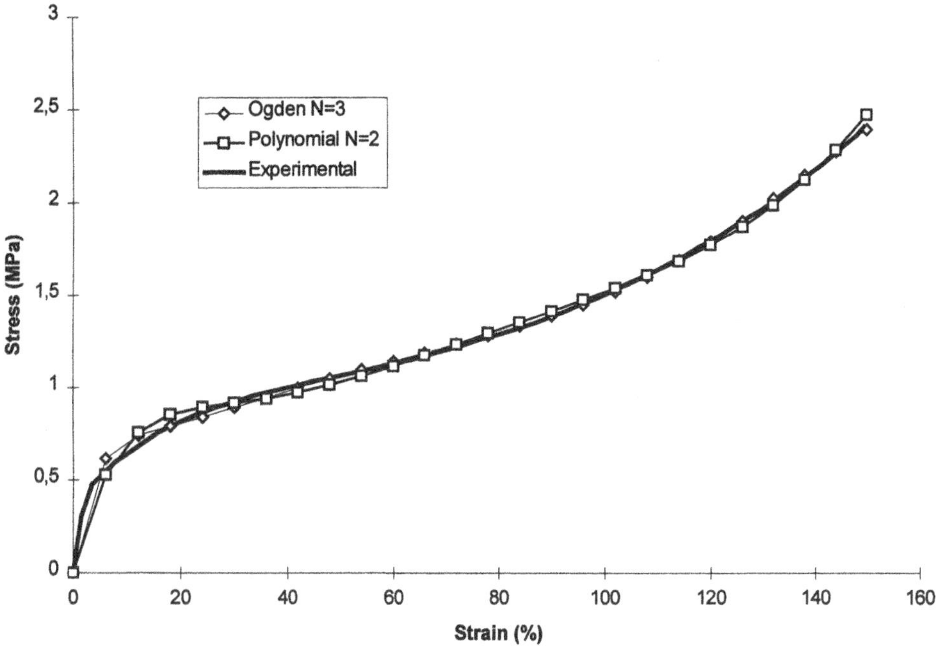

Figure 5. Measured and calculated stress-strain values (tensile test on unscragged rubber).

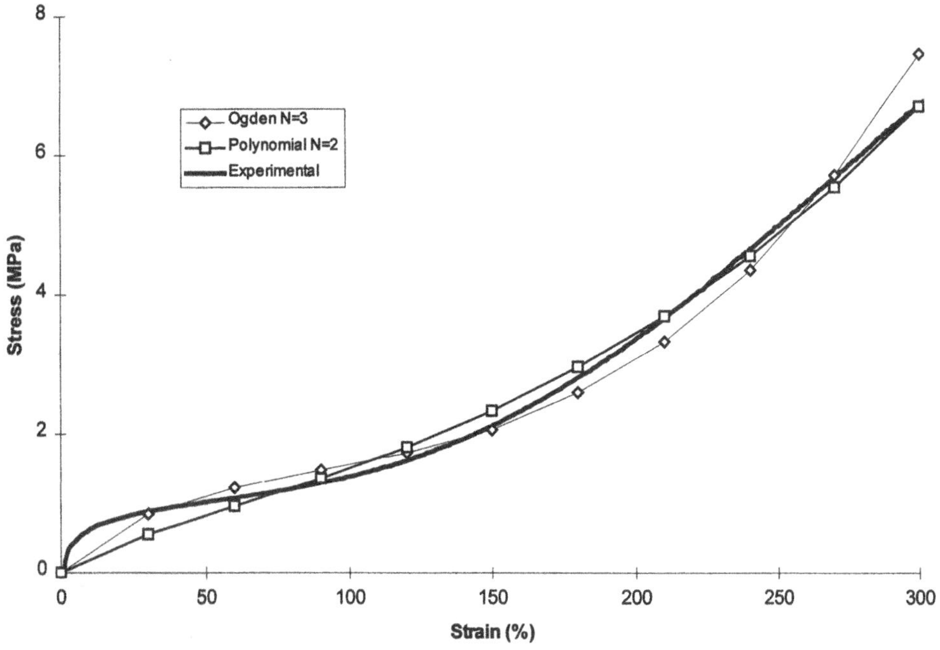

Figure 6. Measured and calculated stress-strain values (planar test on unscragged rubber).

executed such tests on thin square specimens (300mm x 300mm x 5mm), strained in the principal directions 1 and 2 (with a constant velocity of 100 mm/min), thus to satisfy the following conditions:

$$\sigma_{11}=\sigma_{22};\ \sigma_{33}=\sigma_{12}=\sigma_{13}=\sigma_{23}=0;\ \lambda_1=\lambda_2=1/\sqrt{\lambda_3} \quad (2)$$

The 'scragged' conditions consisted of four pre-cycles between 5% and 10% biaxial strain.

These tests are quite expensive and rather difficult to be executed, but allow larger deformations to be achieved with respect to the compression tests, where friction between the specimen and test device does not permit strains larger than 35% to be reached in real uniaxial conditions (see equations 1 and 2). Figure 4 shows the equipment used for the equibiaxial tests and the optical instrumentation adopted to measure the strain in the central part of the specimen, far away from boundary perturbations.

4.4 Planar (pure shear) tests

Finally, planar tests were carried out on rectangular specimens (300mm x 120mm x 5mm) strained along their longer edge (principal direction=1), using a constant velocity of 100 mm/min, in order to satisfy the following condions:

$$\sigma_{33}=0;\ \lambda_2=1;\ \lambda_1=1/\lambda_3 \quad (3)$$

The 'scragged' conditions consisted of four pre-cycles between 5% and 10% planar strain.

The test equipment shown in Figure 3 allowed for large displacements; in fact, the optical instrument measured a strain larger than 300% in the central part of the specimens. In this case also, both Ogden model and the second order polynomial model provided good agreement with the experimental results (Figure 6).

5 VERIFICATION OF THE ISOLATOR FEMs

Appropriate boundary constraints were applied to the models to simulate the actual service conditions: each bearing was first compressed with the relevant compressive load and then sheared by keeping constant the vertical force until the target value of shear strain was reached.

In order to reproduce the experimental conditions of the bolted device, the FE models assume that the top and the bottom faces of the bearings are constrained to remain parallel. While the base plate nodes are fully constrained, every node of the top plate is tied, by means of *EQUATIONS, to a pilot node located at the center of the device; either the vertical and the horizontal loads are then applied to this pilot node.

The second order polynomial strain energy function and the Ogden one have been adopted for the hyperelastic model in the numerical analyses of the experimental tests on complete seismic isolators. Parts of the curves shown in Figures 5 and 6 were used as input data. The length of the various parts used depended on the analysed range of deformation: data points were deleted at very low strains if large strains are anticipated, while some data point were crossed out at the highest strains if small strain are expected.

5.1 Vertical stiffness

The experimental test for measuring the isolator vertical stiffness (K_V) consists in the application of two conditioning cycles between 0.3 and 1.5 times the vertical design load (V), followed by a ramp up to V, used to evaluate K_V. Theoretically, the vertical stiffness is proportional to the square of S, according to the following correlation which is often used for a preliminary evaluation of K_V:

$$K_V=(1+2\,k\,S^2)\,EA/T \quad (4)$$

In (4) E is the Young modulus, A the transverse section of the isolator, T the total rubber thickness and k a coefficient depending on the rubber compound (k=0.57 in the case of the hard rubber considered). However, for the isolator with S=24, the experimental results showed a vertical stiffness of 729 kN/mm, which is only 2.3 times larger than that related to the case of S=12 (311 kN/mm), against the theoretical factor 4. This was initially attributed to rubber creep phenomena, due to the fact that the velocity of application of the load in the measuring of K_V was quite low. Thus, some tests were repeated using a loading velocity up to 20 times higher, but no change in the K_V measurement was found.

This problem was numerically analysed using both axisymmetric and 3D models with three or five elements for each rubber layer and the scragged rubber hyperelastic model with input data at 150% for uniaxial test, 120% for biaxial test and 300% for planar test. No relevant differences were found between 3D and axisymmetric models, while only a little decrease of the calculated vertical stiffness was detected using five element layers in spite of three. The first calculations provided results in agreement with the correlation (4) and well higher with respect the experimental results (Figure 7). The differences between the theoretical and experimental behaviour have been attributed to the rubber compressibility, which has the effects of decreasing the vertical stiffness, mostly in the case of high shape factors. Since no volumetric experiments on rubber specimens were carried out , the available data of a

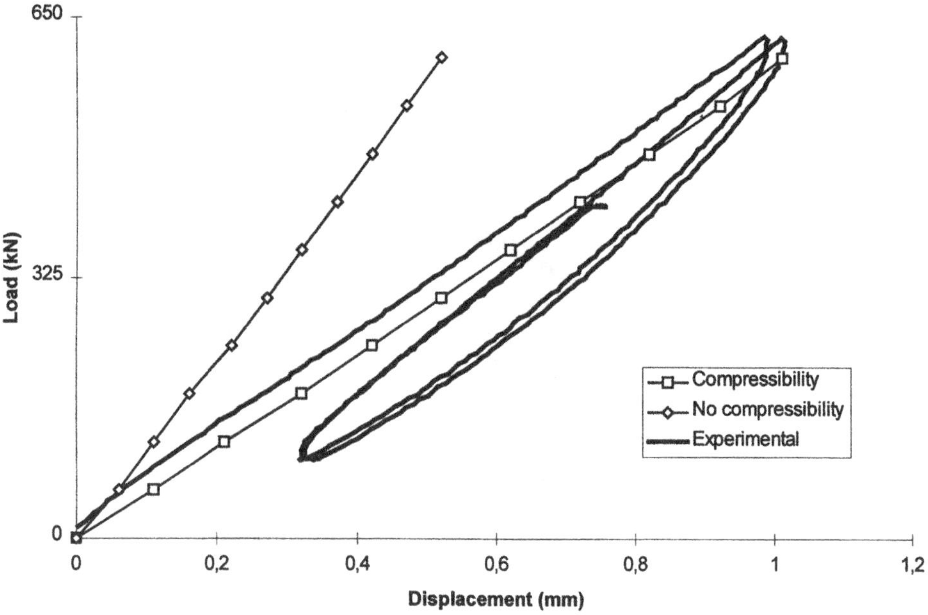

Figure 7. Measured and calculated force-displacement values for a compression test at 150% design vertical load of a complete bearing (S=24).

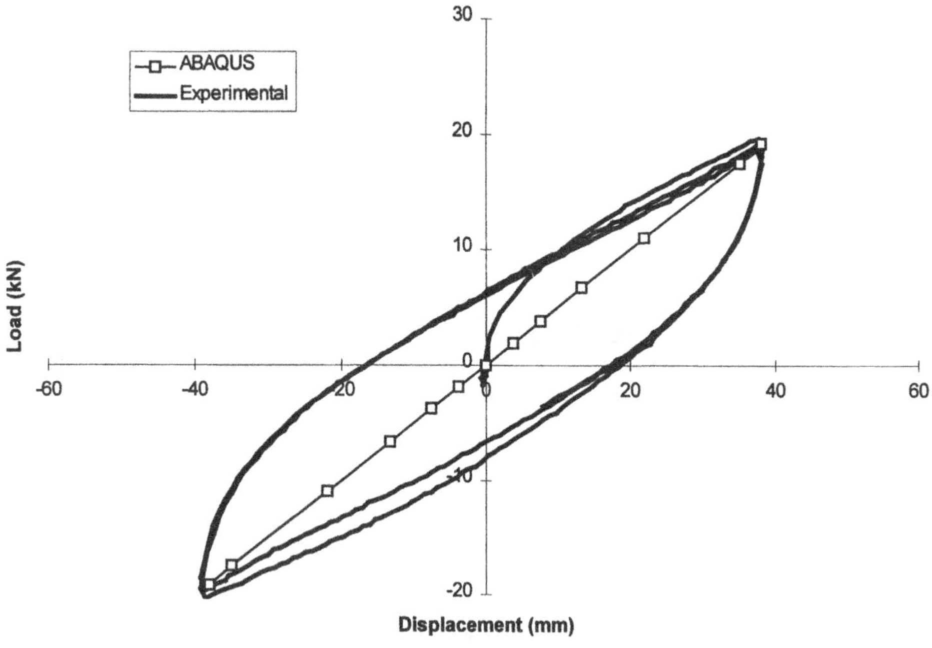

Figure 8. Measured and calculated force-displacement values for a combined compression & 50% shear strain test ('recess' attachment system, S=12).

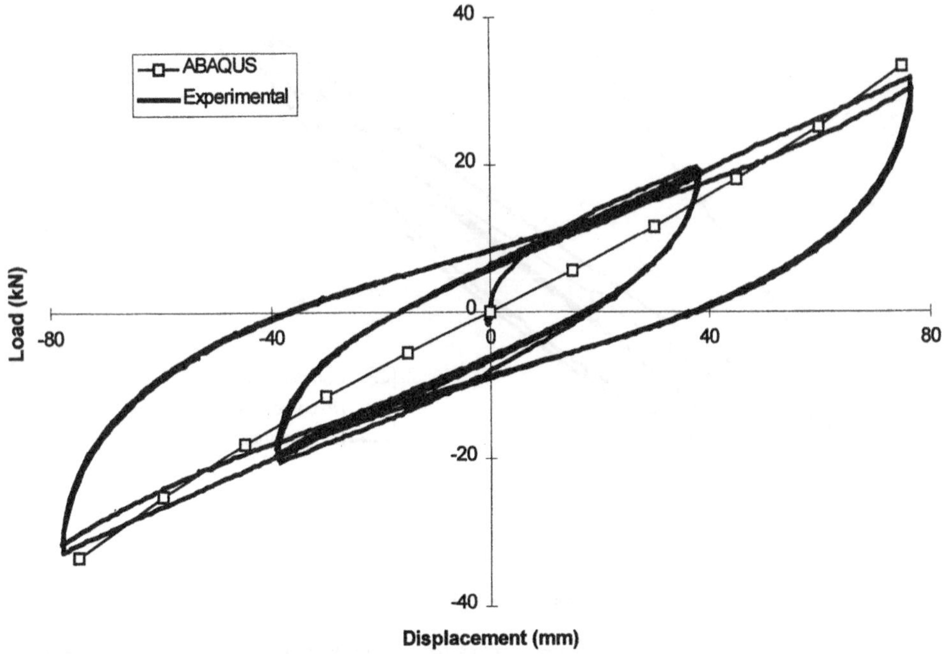

Figure 9. Measured and calculated force-displacement values for a combined compression & 100% shear strain test ('recess' attachment system, S=12).

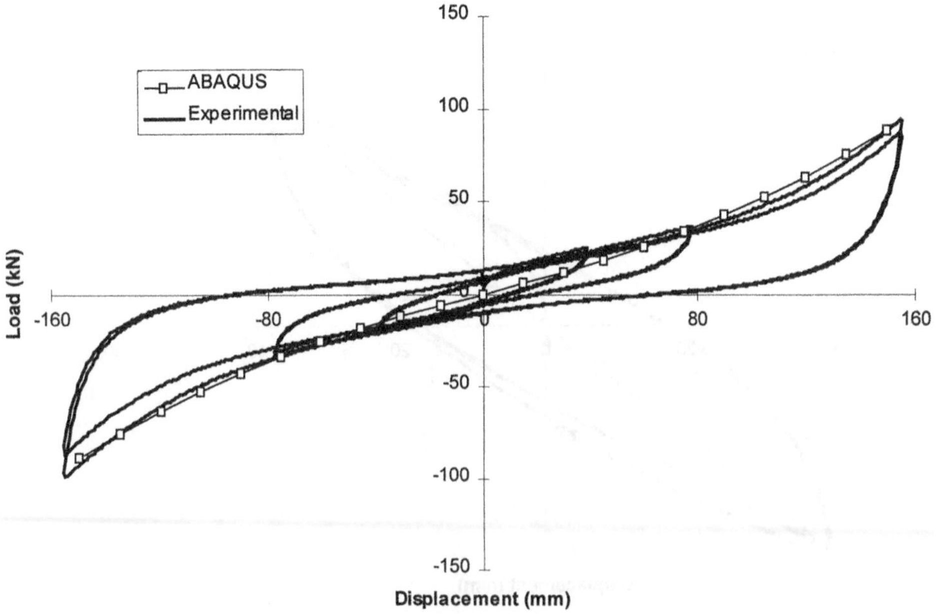

Figure 10. Measured and calculated force-displacement values for a combined compression & 200% shear strain test ('bolts' attachment system, S=24).

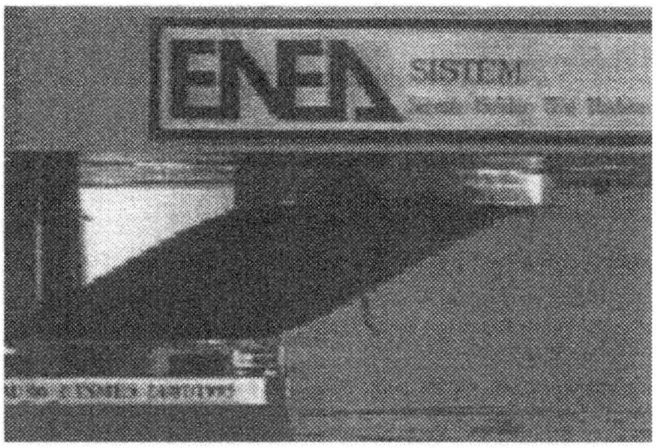

Figure 11. Compression and shear test at 300% on a bolted 'optimized' HDRB, 1:2 scale, D=250 mm, H=75 mm, S=24, G=0.8 MPa.

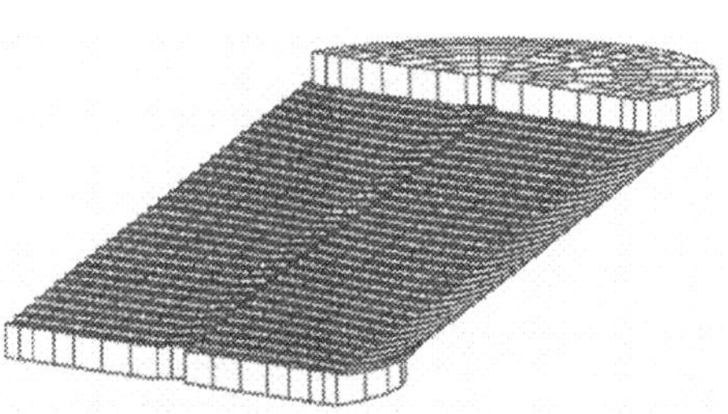

Figure 12. FEM of a bolted 'optimized' HDRB, 1:2 scale, D=250 mm, H=75 mm, S=24, G=0.8 MPa, at 300% shear strain.

similar rubber compound were used to complete the rubber hyperelastic model. In this case the agreement between the calculated and measured force-displacement values became very good for both the shape factors considered (Figure 7).

Whereas rubber can be considered to be nearly incompressible, compressibility must be taken into account to correctly evaluate the vertical stiffness of bearings when bulk stiffness plays an important role.

5.2 Horizontal stiffness

Both axisymmetric (for low shear strains) and 3D models (for high shear strains) were used for calculating the horizontal stiffness (K_h) at 50%, 100%, 200% and 300% shear strain, under the design vertical compression load (Martelli 1995). Due to the quite large range of deformation, it was not possible to use a unique hyperelastic model, but appropriate models were adopted for each deformation of interest. The effects of compressibility on K_h are negligible, due to the very large horizontal deformations, thus no volumetric data were provided as input to ABAQUS.

An axisymmetric FEM (with three layers of elements for each rubber layer) using the unscragged rubber input data not exceeding 100% for uniaxial and biaxial tests and 150% for planar test, was used to evaluate the horizontal stiffness at 50% shear strain of the isolator with 'recess' attachment system and low shape factor. Figure 8 shows that the agreement between the calculated and measured

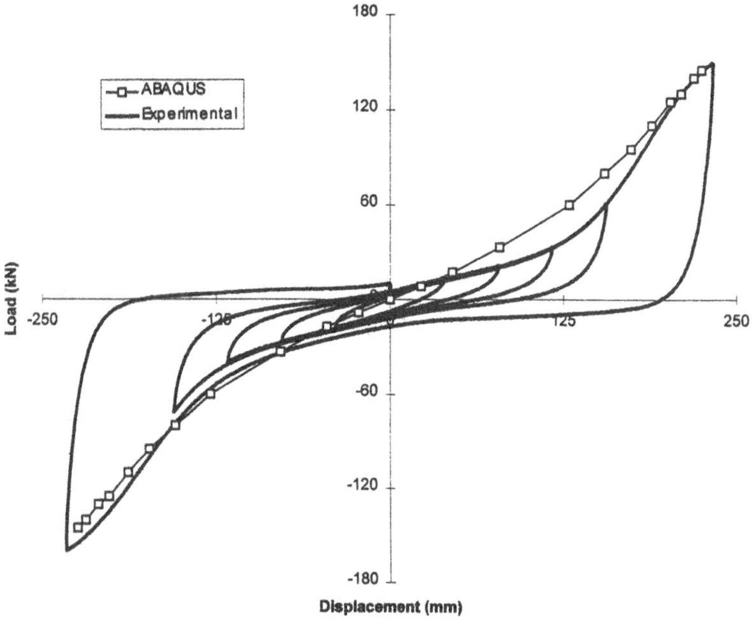

Figure 13. Measured and calculated force-displacement values for a combined compression & 300% shear strain test ('bolts' attachment system, S=24).

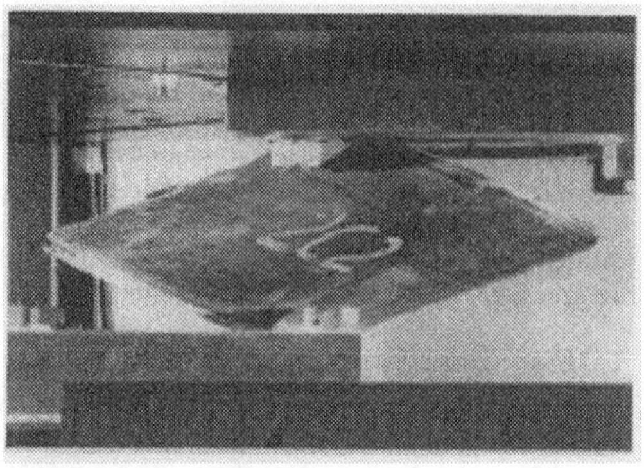

Figure 14. Compression and shear test at 300% on a bolted 'optimized' HDRB, 1:2 scale, D=250 mm, H=75 mm, S=24, G=0.8 MPa.

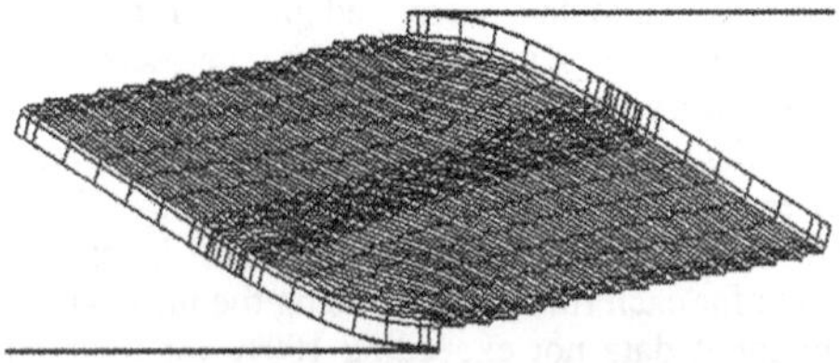

Figure 15. FEM of a bolted 'optimized' HDRB, 1:2 scale, D=250 mm, H=75 mm, S=24, G=0.8 MPa, at 300% shear strain

horizontal stiffnesses is very good, taking into account that the model has no damping (thus the hysteresis loop cannot be reproduced). For the evaluation of the horizontal stiffness at 100% shear strain of the above isolator, the same model for the unscragged rubber, including uniaxial input data up to 150%, biaxial input data up to 120% and planar input data up to 300%, was used. In this case also, the agreement between the calculated and measured values was good (see Figure 9), but the unscragged model became too rigid at high deformations, with respect to the isolator, which was subjected to the two cycles at 50% shear strain. For higher shear strains, a 3D FEM and the scragged rubber model adopted in the previous section for evaluating K_V, had to be used. Figure 10, which refers to a bolted isolator having S=24, shows that the model perfectly evaluated the horizontal stiffness at 200% shear strain under the vertical design load. Finally, the above model was used to analyse a horizontal failure test up to 300% shear strain (Figure 11) on another bolted bearing having S=24. In this case, it was necessary to adopt an elastic-plastic model for the steel plates, due to the very severe strain conditions reached during the test. In this model the Young modulus is assumed equal to 200,000 MPa; the first yeld occurs at 230 MPa, then the material hardens to 345 MPa at two percent strain, after which it is perfectly plastic. Figure 12 shows the FEM used for the analysis while figure 13 reports the comparison

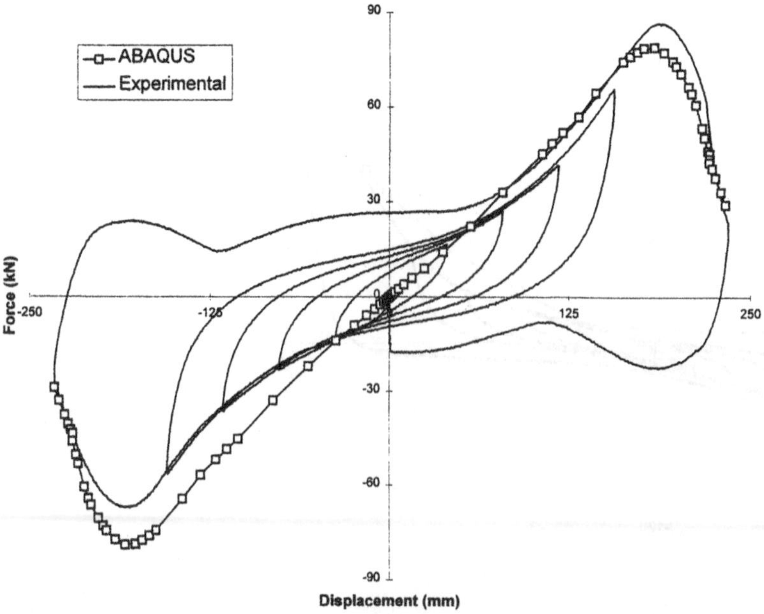

Figure 16. Experimental and numerical force-displacement values for a combined compression & 300% shear strain test on optimized HDRB (1:2 scale, diameter=250 mm, H=75 mm, S=12, G=0.8 MPa, recess attachment system).

between measured and calculated hysteresis loops.

The recess attachment system (Figure 14) is more difficult to be modelled (Figure 15) because it involves sliding of a deformable body (the bearing) against a rigid body (the recess plate). In this work, the contact problem was solved by meshing the recess plate' surface using rigid surface elements (IRS4). Figure 16 shows the results obtained for the 1:2 scale 'optimised' HDRB with recess attachment system, S=12 and G=0.8 MPa under 400 kN vertical load and 300% shear strain. The numerical-experimental agreement is really very good and demonstrates the reliability of the hyperelastic model.

5.3 Parametric analysis

Due to the encouraging results of the numerical analyses carried out in the previous section, lots of calculations at large shear strains have been performed on different HDRBs in order to evaluate the effects of some parameters on the stress distribution in both steel and rubber. This is quite important for the designer, who must correctly adopt those parameters, like shape factor, attachment system, steel plate thickness, central hole diameter etc., which have a negligible influence on the horizontal stiffness (at least in the operational range) but that have a significant effect on the internal stresses and then, on the isolator failure limit.

The results of these studies are reported in the works of Forni (1995), Martelli (1996) and Forni (1996).

6. CONCLUSIONS

In this paper the finite element modelling of elastomeric bearings subjected to vertical load and horizontal displacement has been analysed and verified against experimental evidence.

The extensive numerical simulation aimed at investigating the effects of the several variables of the problem has put into evidence which type of material model, discretisation and elements have to be adopted in order to obtain a good correlation with the experimental results at very large strains.

Particular attention should be devoted to the experimental tests on rubber specimens (expecially the planar deformation ones) in order to avoid slip or boundary effects which can affect the solution of the problem.

The availability of reliable isolators' models will allow the number of quite costly experimental tests on complete bearings to be considerably limited, not only as regards the analysis of the effects of some important parameters (e.g. temperature, ageing, vertical load on horizontal stiffness, etc.), but also for complicated experiments like for instance, failure tests and analysis of the effects of defects.

REFERENCES

ABAQUS User's Manual - Version 5.7 1998. Pawtucket: Hibbitt, Karlsson & Sorensen, Inc.

Bettinali F., Cazzuffi D., Dusi A., Fede L. 1996, Valutazione delle Caratteristiche di Mescole in Gomma Naturale Utilizzate per la Costruzione di Isolatori Sismici, *ENEL internal report n° 5187*, Milano (in italian)

Dusi, A. & Cadei, R. 1997. Genesis: a Pre-processor for the Implementation of FE Models of Seismic Isolators. *ENEL Internal report n. 5835*, Milano.

Forni, M., Martelli, A., Dusi, A. & Castellano, G., 1995. Hyperelastic Models of Steel-Laminated Rubber Bearings for Seismic Isolation of Civil Buildings and Industrial Plants. Proc. of the Int. ABAQUS USERS' CONGERENCE, Paris, France, May 31 - June 2, 1995, pp. 273-287.

Forni, M., Calabrese, R., Martelli, A., Bettinali, F., Bonacina, G., Pucci, G., La Grotteria, M. & Sobrero, E., 1996. Development and Validation of Finite Element Models of Rubber Bearings. Proc. of the Firts European Conference on Structural Control, 29 - 31 May 1996, Barcelona, Spain

Forni, M., Martelli, A., Spadoni, B. Vemturi, G., Dusi, A., Bettinali, F., Bonacina, G., Pucci, G., Marioni, A., Mazzieri, C. & Cesari, F. 1997. Development and validation of FEMs for the Design and qualification of HDRBs. *Proc. 14th SMiRT Conf. on Structural Mechanics in Reactor Tecnology, Lyon, 17-22 August 1997.* Rotterdam: Balkema.

Martelli, A., Forni, M., Spadoni, B., Marioni, A., Bonacina, G. & G. Pucci 1995. Progress of Italian experimental activities on seismic isolation. *Proc. 13th Int. SMiRT Conf., Porto Alegre, 13-18 August 1995.* Rotterdam: Balkema.

Martelli, A., Forni, M., Indirli, M., Spadoni, B., Bettinali, B., Dusi, A., Marioni, A., Bonacina, G., Pucci, G., Mazzieri, C., 1996. Italian Studies for the Optimisation of Seismic Isolation Systems for Civil and Industrial Structures. 11th World Conference on Earthquake Engineering, Acapulco, Mexico, 23-28 June, 1996.

Ogden R.W. 1984. *Non-Linear Elastic Deformations*, Ellis Horwood Limited, Chichester.Rebelo, N. 1991. Analysis of rubber components with ABAQUS. *Proc. of 2° National Congress of ABAQUS Users' Group Italia, Bologna, Italy, March 21-22.*

Constitutive Models for Rubber, Dorfmann & Muhr (eds) © 1999 Taylor & Francis ISBN 90 5809 113 9

Application of fracture mechanics for the fatigue life prediction of carbon black filled elastomers

J.J.C. Busfield & A.G.Thomas
Materials Department, Queen Mary and Westfield College, London, UK

M.F. Ngah
Centre for Composite Materials, Imperial College, London, UK

ABSTRACT: An attempt to predict the fatigue life of carbon black filled elastomer test pieces using a fracture mechanics approach is made. The fatigue failure of a tensile test specimen is predicted, using cut growth behaviour determined from a pure shear crack-growth test. This is done initially for specimens that contain centre edge cracks cut with a razor to a specified known size and then for normal materials without these initial flaws already introduced. To do this, it is necessary to make an estimate of the size of the initial flaws present in the elastomer. Using sensible estimates for the initial flaw size, permits a far better life prediction analysis to be made than is possible using the traditional S-N technique. It is also noted that to be generally applicable a wide range of flaw sizes and geometries need to be evaluated.

1 INTRODUCTION

The traditional S-N approach is used in general engineering (Callister 1994) for determining the fatigue life of components. This is the standard approach adopted by the elastomer component industry to predict component life spans. The stress term is typically a measure of the maximum principal stress, predicted somewhere in a loaded component. With elastomer components, the results obtained from this approach are specific to the particular specimen geometry and loading conditions which may not be representative of the typical service conditions for an engineering component. In addition, long testing times are required to obtain the S-N data and the scatter in the measured results can be large. Also it has been anecdotally observed that the S-N prediction and actual fatigue lives can be out by a factor in excess of 1000.

Fracture mechanics provide a more reliable approach to predicting this crack-growth behaviour. The technique is based a tearing energy concept. This approach assumes that a characteristic crack-growth rate relationship exists for an elastomer that is dependent on the tearing energy. Rivlin and Thomas (1953) showed that this relationship is a material property, which is independent of loading mode and specimen geometry. Ranges of crack-growth rate expressions are available to describe the various tearing phenomena. This work is aimed at studying the relationship in the engineering fatigue regime or for crack-growth rates that would lead to catastrophic component failure in the range of 10 thousand to 10 million cycles.

This approach has been used in the past by Gent et al. (1964) to reliably predict the fatigue failure from small razor cuts in unfilled natural rubber tensile test specimens. Here the work was repeated with typical filled engineering compounds, which exhibit different crack-growth behaviour. This investigation extended earlier work by Thomas (1966) that had investigated rapid crack-growth behaviour in high tearing energy regimes and work that had concentrated on slow crack-growth behaviour by Lake & Lindley (1966) who explored oxidative effects and ozone cracking.

In practice however components fail without the prior introduction of razor cuts. Fatigue failure in elastomers has been regarded as an accumulated crack-growth process resulting from the growth of small inherent initial flaws under a cyclic loading or deformation (Gent et al. 1964). Intriguingly, these flaws have the same order of magnitudes for different types of elastomers. Typical flaw sizes reported in the literature by Lake & Lindley (1965), Lake (1972) and Choi & Roland (1996) for unfilled elastomers are shown in Table 1. The initial flaw sizes, for carbon black filled materials, are notoriously difficult to measure. They are seen to vary dramatically depending on the dispersion of the aggregates during the mixing process. It would appear that even for well-dispersed materials that the initial flaw size is dramatically increased from that found in unfilled materials. Lindley & Thomas (1962) proposed that it

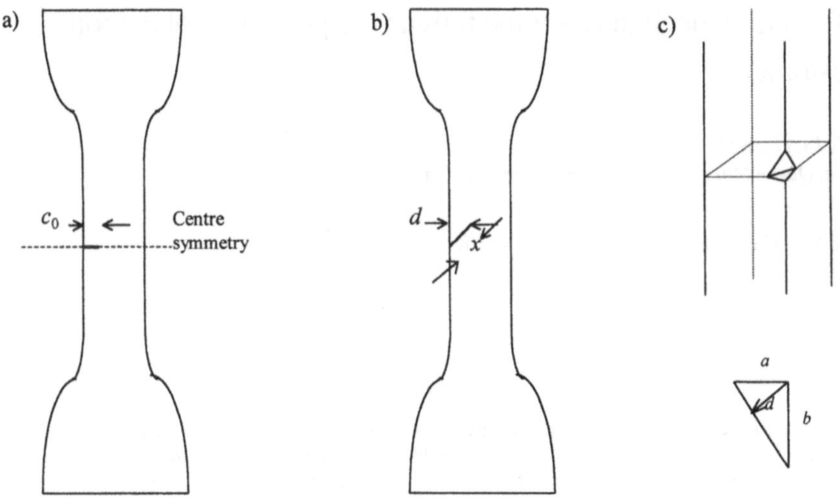

Figure 1. A tensile test piece with: (a) a razor edge-crack cut in it, (b) a corner edge-crack and (c) a slanted edge-crack. With the approximation for the equivalent crack length indicated.

Table 1. Reported c_0 values that have been taken from the literature for different types of elastomers.

Elastomer type	c_0 / μm
Unfilled butyl rubber (IIR)	50
Unfilled poly-butadiene (BR)	25
Unfilled styrene butadiene rubber (SBR)	55
Unfilled acrylonitrile-butadiene rubber (NBR)	40
Unfilled natural rubber (NR)	25
Unfilled low purity natural rubber (SMR-10)	29
Unfilled medium purity natural rubber (SMR-L)	26
Unfilled deproteinised high purity natural rubber (DPNR)	16
Unfilled guayule rubber (GR)	29
HAF black filled natural rubber (NR)	50

is not unreasonable to estimate that the initial flaw size could be as large as 50-100μm. Here two materials NR50 and NR32 were fatigue tested to failure. The first material was well dispersed and was assumed to have typically sized initial flaws. NR32 however, was prepared with a poor dispersion resulting from a reduced mixing time. It was anticipated that because of this it would have a reduced fatigue life. In addition, the flaw sizes in the NR32 mix should be large enough to be measured using a microscope following a freeze fracture procedure.

Finally, razor cuts were introduced into the tensile test piece with a different geometry to the centre edge cut. The fatigue lives resulting from these cuts were examined using a similar approach, as the centre edge cuts, to see if the relationships derived were generally applicable to any flaw geometry that may be present in the component.

2 THEORETICAL BACKGROUND

The tearing energy (Lake & Thomas 1988), T is defined as the energy released by the change in the total stored elastic energy, dU per unit area of crack extension, dA held at a fixed external deformation.

$$T = -\left(\frac{dU}{dA}\right)_l \quad (1)$$

Various tearing energy expressions are available for different test piece geometries. From dimensional considerations the expression for a centre edge-crack test piece can be derived as,

$$T=2kWc \quad (2)$$

In this equation, W is the stored elastic energy density at the region remote from the crack tip, k is a tearing energy parameter that varies with the applied strains and c is the initial flaw size.

Greensmith (1963) measured the tearing energy relationship for centre edge crack in tensile test specimens. He used his measurements to derive the strain dependent tearing energy parameter, k for a range of strains. Using a different approach Lake (1970) proposed an approximate relationship for $k=\pi/\sqrt{\lambda}$. However to be more accurate it would be valuable to derive the actual functional relationship between k and the applied strain. To do this, a finite element model of the test piece geometry was considered. Several workers including Busfield et al. (1996) had in the past used finite element models to calculate the tearing energy relationships in elas-

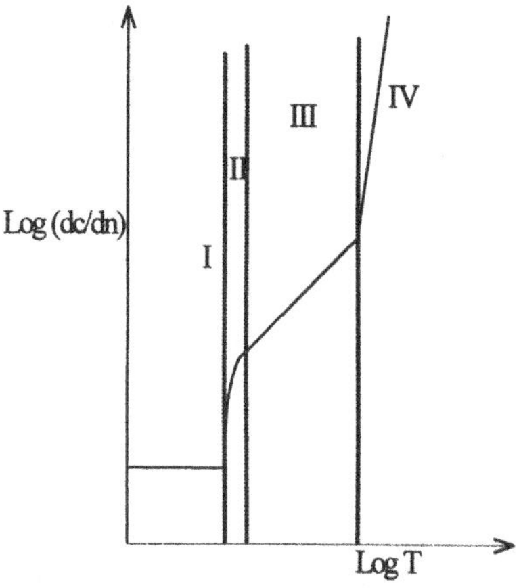

Figure 2. A double logarithmic plot of tearing energy against crack growth rate per cycle for a conventional fille natural rubber test specimen (Lake 1983). The figure is di vided into 4 regions, which exhibit different tear behaviour.

tomer components. The geometry considered here was that of the centre edge-crack test piece shown in figure 1(a). Lindley (1972) had analysed this geometry up to 150% strain. He calculated the energy by considering the change in total energy as the crack is extended in the model. Here a small cut was again modelled, this time using the J-Integral approach to deduce the tearing energy. The two methods appeared to yield similar results. This approach allowed the derivation of the definitive relationship between k and the applied strain.

A typical relationship between the tearing energy versus the crack-growth rate for an elastomer (Lake 1983) is shown in Figure 2 on a double logarithmic plot. The crack-growth rate per cycle, dc/dn in each of the four various regimes can be fitted by a specific mathematical expression. For this work region III was utilised, as this is the region that corresponds most closely with crack-growth rates found in the engineering fatigue range. In this region, a power law relationship describes the behaviour well.

$$\frac{dc}{dn} = BT^{\beta} \tag{3}$$

Here B and β are elastomer crack-growth parameters that are determined from crack-growth studies typically using a pure shear test piece. By eliminating the tearing energy, T, from Equations 2 and 3 the following expression is derived,

$$dn = \frac{dc}{B(2kWc)^{\beta}} \tag{4}$$

This relationship can be integrated to produce the following fracture mechanics expression that can be used to predict the fatigue life within the power law tearing energy regime for a centre edge-crack test piece, thus

$$N_f = \frac{1}{B(\beta-1)(2kW)^{\beta}\left(c_0^{\beta-1}\right)} \tag{5}$$

where c_0 is the initial centre, edge-crack length and N_f is the number of cycles required to fatigue fail the test specimen.

3 EXPERIMENTAL

Two typical engineering natural rubber compounds were selected for the tests, with either 32 or 50 parts per hundred (phr) HAF carbon black filler. The formulations of these materials, designated NR32 and NR50, are shown in Table 2. The mixing time for the NR32 material was reduced in order to increase the inherent initial flaw size. It is worth noting that 50pphr carbon black is equivalent to 31% by mass or 20% by volume. The elastomers were moulded into flat square sheets cured at 155°C for 10 minutes to a T100 cure state. From these sheets ASTM D-638M type II dumbbell test pieces were die-stamp and pure shear test pieces were cut.

Table 2. Elastomer formulation used throughout the study.

Composition	NR 32	NR 50
Natural rubber	100	100
Carbon black (phr)	32	50
Antioxidant (phr)	2	2
Antiozonant (phr)	3	3
Accelerator (phr)	2.4	2.4
Sulphur (phr)	1	2.7
Density (g/cm^3)	1.04	1.14
Crosslink density (10^5 mol/cm^3)	10.86	11.76

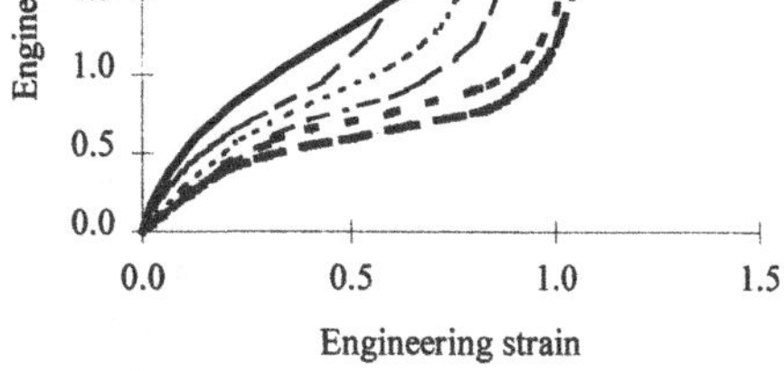

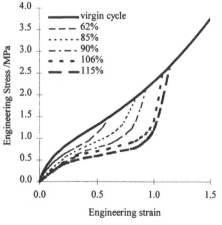

Figure 3. The first cycle stress versus strain relationship plotted for NR50. In addition, the ten thousandth stress versus strain relationship is plotted for a range of different cyclic strain values.

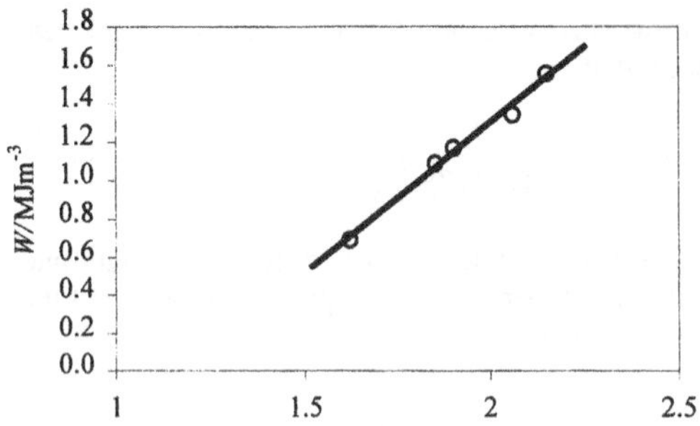

Figure 4. A plot of the ten thousandth cycle stored energy density versus the applied cyclic strain.

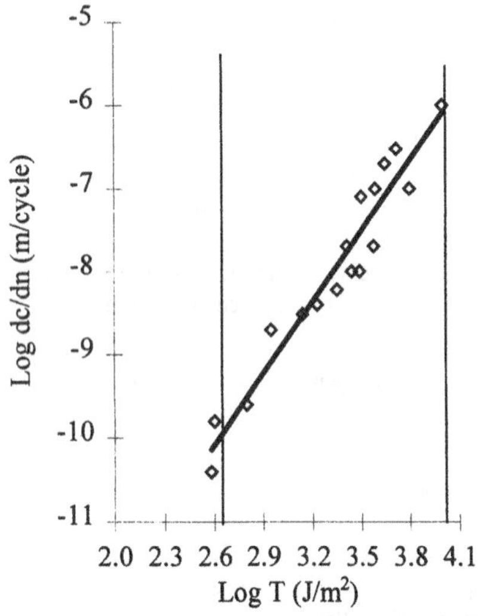

Figure 5. Experimentally derived crack growth behaviour for NR50. The thick line represents the best-fit relationship given in Equation 3. The vertical lines represent the maximum and minimum values of tearing energy that are encountered in the fatigue test work.

The stress strain behaviour of an uncut tensile test piece has to be measured in order to obtain an accurate value of the maximum stored energy density present in any given fatigue test. This approach was however complicated by the effects of cyclic stress softening, which is known to be quite large in carbon black filled elastomers. Therefore, additional tests were undertaken to determine the magnitude of the cyclic stress softening. From these tests it was observed that the stiffness behaviour appears stabilised after the first 10,000 test cycles. Therefore a series of virgin test pieces were each cycled 10,000 times at one of a number of maximum specified strains to determine the drop in the stabilised stress strain behaviour. Each test piece was therefore used to identify the behaviour at a specific applied strain. These stress strain curves are shown in Figure 3. The stored energy density required for Equation 5 was calculated as the integral of the 10,000th strain cycle. These measured results for the strain energy density at a given applied strain are plotted as a function of the maximum strain in a given cycle in Figure 4. From this investigation it was observed that the cyclic stress relaxation processes are more pronounced at higher strain amplitudes.

The parameters B and β used in Equation 5 were determined from the crack-growth measurements made by Ratsimba (in prep.) performed using the pure shear test piece described by Busfield et al. (1997) on these materials. The measured data is shown for NR50 in Figure 5. The graph shows the crack-growth rate versus tearing energy relationship plotted on logarithmic scales. It was seen that for the crack-growth rates of interest that the data can be fitted using the power law relationship in Equation 3. The best-fit line is drawn for NR50 with fitting parameters of B=2.34 x 10^{-18} and β=2.90. Similarly a best fit line for the NR32 material was calculated with the parameters of B=5.15 x 10^{-15} and β=2.00.

As was discussed earlier, Equation 5 is only applicable to the power law zone of the crack-growth rate versus tearing energy relationship. It was therefore important to ensure that this zone was sufficiently wide to account for the fatigue crack-growth behaviour that we encountered in this investigation. The minimum flaw size that was cut into the test piece was about 0.1mm, and the minimum cyclically induced strain 60%. For our material this results in a tearing energy of 350Jm^{-2}. The maximum tearing energy of 11,000Jm^{-2} was calculated as the energy that relates to a crack that has grown to 2mm in length at a cyclic strain of 115%. These extreme values were also plotted on Figure 5 as the solid horizontal lines. From this it was noted that the data in Figure 5 was measured over the appropriate tearing energy range to predict the fatigue crack-growth behaviours with which this paper is concerned.

The dumbbell specimens with the razor cuts were all cycled to failure using a servo-hydraulic displacement controlled fatigue test machine. The specimens were cycled between zero displacement and either 62% or 85% maximum strain. The number of cycles to failure, N_f for each of these specimens was recorded.

Specimens of the NR50, without razor cuts introduced, were similarly fatigued over a range of strains from 62-115% extension. The test pieces were again fully relaxed during each loading cycle. The number of cycles to failure, N_f for each of these specimens was recorded as a function of the applied strain. In order to estimate the fatigue life it was necessary to make an estimate of the initial flaw size. Attempts to measure this were made by freeze fracturing rubber specimens cooled in liquid nitrogen, which could then be examined using a microscope. However, for the NR50 specimen it was not possible to identify either flaws or agglomerates. Therefore, a second material NR32 was prepared, with a shortened mixing time and with an observed poorer dispersion. This material was characterised for both its 10,000-cycle stress-strain behaviour and its fatigue crack-growth behaviour. Tensile test strips were then fatigue tested without the introduction of razor cuts over a range of applied strains. In addition, moulded specimens were freeze fractured to identify and measure any observable voids or agglomerates.

Finally, two additional cut initial flaw geometries were investigated in the well-dispersed material NR50. These are illustrated in Figure 1(b & c). The first was a corner cut and the second was again a through thickness cut that was inclined at an angle to the horizontal. The purpose of this exercise was to investigate if a crude approximation for the tear relationship would yield an accurate prediction for the fatigue life.

4 FINITE ELEMENT ANALYSIS

A finite element model of the centre, edge-crack, test piece geometry was considered. The small cut in the test piece was modelled to deduce the tearing energy for a range of strains using a J-Integral approach. To do this a half symmetry plane stress dumbbell specimen model was made. A simple Neo-Hookean material model was implemented. The finite element model adopted is shown is in Figure 6. From the tearing energy, it was possible to use Equation 2 to calculate how k varies with the applied strain. The results of this are plotted in Figure 7 as a graph of k versus strain. On this graph is also shown the experimental data of Greensmith (1963) and the theoretical relationship of Lake. Abaqus FEA software was used to perform the calculations.

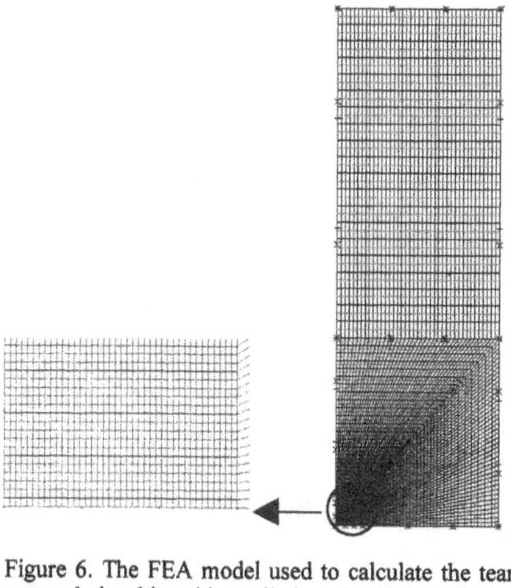

Figure 6. The FEA model used to calculate the tearing en ergy relationship with applied strain. The crack tip section (circled) is enlarged.

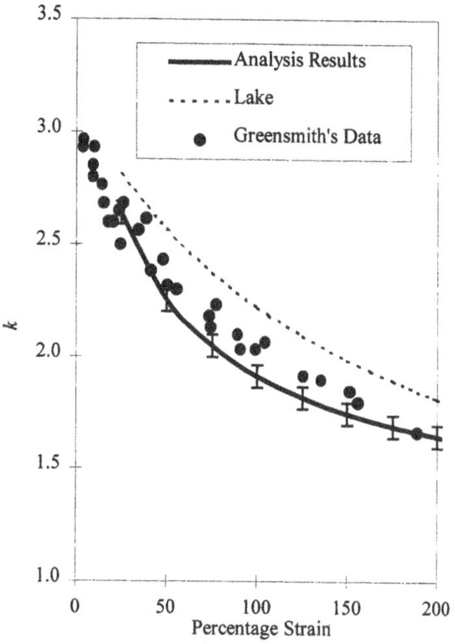

Figure 7. A plot of how k predicted by finite element analysis varies with strain. This data is compared with that measured or calculated by Greensmith (1963) and Lake (1970) in other published work.

5 RESULTS AND DISCUSSION

5.1 *Fatigue life investigation of razor cut tensile test specimens.*

Actual fatigue life measurements were obtained using a displacement controlled fatigue test machine. The test geometry adopted was a dumbbell specimen with a crack of a known initial length. These initial flaw sizes were measured using an optical microscope. These test results are shown in Table 3, with the fatigue life values predicted from Equation 5. The ratio of the predicted result divided by the measured value is also shown. It was observed that the measured results were very consistent and that the predicted fatigue lives were similar to the measured results. The predicted fatigue lives were accurate to within a factor of 1.81. As the initial measurements in crack length were only accurate to about 10%, the predictions in the fatigue lives were surprisingly consistent.

It was observed that Equation 5 consistently overestimates the actual fatigue life by a factor of about 1.65±10%. The effects of an initially sharp

Table 3. Measured fatigue life data for a range of initial razor cut flaw sizes ranging from 0.3mm to 0.8mm for two different applied strains using NR50.

Initial Cut length (± .05mm)	Applied Strain	Measured fatigue life, N_m	Predicted fatigue life, N_f	N_f/N_m
0.4	0.62	49800	87988	1.77
0.5	0.62	31800	57592	1.81
0.5	0.62	33630	57592	1.71
0.6	0.62	22680	40736	1.80
0.6	0.62	23400	40736	1.74
0.6	0.62	24390	40736	1.67
0.6	0.62	27390	40736	1.49
0.7	0.62	18000	30396	1.69
0.8	0.62	15960	23587	1.48
0.3	0.85	30510	45193	1.48
0.4	0.85	16170	26168	1.62

crack tip may have caused this discrepancy. Lake (1983) showed that the sharp crack tip caused by a razor insertion leads to a rapid initial fatigue growth of the crack. This would serve to reduce the measured fatigue life for these specimens. It was only after a number of cycles or after the crack has increased in length a certain distance that the crack-growth rate becomes typical of steady state cut growth behaviour (Thomas 1958).

Lake (1983) measured that the initial crack-growth rate resulting from the sharp tip was around 24 times higher than the final rough crack-growth rate. Using this observation it was possible to propose that a new theoretical B, 24 times larger than the measured steady state behaviour value, could be estimated for the initial sharp-tip crack-growth behaviour. It was also assumed that the β term will remain unchanged during this initial crack-growth phase.

If Equation 4 was integrated, this time not to catastrophic failure, but to calculate the number of cycles N_i that are required to make the crack grow to a length of c_i. Then it was possible to derive the following relationship.

$$\left(\frac{1}{B(2kW)^{\beta}}\right)\left(\frac{1}{\beta-1}\right)\left(\frac{1}{c_0^{\beta-1}}-\frac{1}{c_i^{\beta-1}}\right)=N_i \qquad (6)$$

This relationship could be used with the new value of B to predict the crack-growth behaviour during a period of rapid initial sharp crack-growth. At some stage after an unknown number of cycles the crack tip behaviour could then be assumed to have roughened and become more typical of the steady state fatigue crack-growth. At this time, the crack was assumed to fatigue fail from a flaw of initial size c_i using Equation 5. The new total fatigue life would now be a summation of $N_i + N_f$. Initial estimates to the scale and magnitude of N_i and c_i were made to make the predicted fatigue lives fit the measured fatigue results.

Table 4. The number N_i of initial rapid cut growth cycles and the increase in crack length (c_i -c_0) that would be required in order to make the experimental and measured fatigue lives agree. The data is presented for NR50 with a range of different razor cut lengths ranging from 0.4mm to 0.8mm cycled at 62% strain.

c_0 /mm	N_i	(c_i -c_0) /mm
0.4	337	0.143
0.5	190	0.177
0.6	111	0.188
0.7	76	0.225
0.8	44	0.186

The results of this investigation are summarised in Table 4 for each of the crack lengths cyclically fatigue tested at 62% strain. These results indicated that the sharp tip effect was evident. It was also apparent that the duration of this sharp tip behaviour was not determined by a fixed number of cycles but more probably a transition zone from sharp to conventional tear behaviour. This size of this transition zone was indicated as being about 0.15-0.2mm. Hence, it was concluded that the presence of sharp tip effects could be used to explain the apparently consistent discrepancy between predicted and measured fatigue life data.

5.2 *Fatigue life investigation of an uncut tensile test specimen.*

The fatigue results for the uncut test pieces made from NR50 are shown in Table 5. Also presented are the calculated fatigue lives that assume an initial flaw size of 0.1mm. This value was taken as the upper estimate from a review of the literature. The first observation was that the fatigue test results were again very reproducible. The standard deviations for each set of tests (with at least 10 samples) were modest for fatigue results, being in the range of 4 to 35%. It was also seen that an initial estimate of 0.1mm for the initial flaw size yields very accurate predictions for the fatigue life. The prediction was surprisingly accurate to well within an order of magnitude and was actually more reliable than anticipated. The attempt to evaluate the initial flaw size using a freeze fracture technique did not reveal any measurable flaws within the specimen.

The fatigue results of the tensile test specimens made out of NR32 are presented in Table 6. It was apparent that this material was substantially less fatigue resistant than the more highly dispersed NR50, which also had a higher reinforcing filler content. Again, the measured fatigue lives were very consistent with very low values for the percentage standard deviation. A SEM picture of the freeze-fractured surfaces is shown in Figure 8. It shows what was presumed to be the site of an internal flaw. The largest of which was observed as being 0.4mm in size. Using this as the measurement for c_0, together with

Table 5. Average fatigue life for a minimum of ten test pieces cycled to the specified maximum strain, with a measure of the percentage standard deviation for NR50.

Applied strain %	Average fatigue life, N_m	Percent standard deviation	Predicted fatigue life assuming a c_0 of 0.1mm, N_f	N_f/N_m
62	948590	4%	1552498	1.64
85	337230	10%	566104	1.68
90	447196	7%	519264	1.16
106	339321	16%	401786	1.19
115	332440	35%	272485	0.82

Table 6. Average fatigue life for a minimum of ten test pieces cycled to the specified maximum strain, with a measure of the percentage standard deviation for NR32.

Applied strain %	Average fatigue life, N_m	Percent standard deviation	Predicted fatigue life assuming a c_0 of 0.4mm, N_f	N_f/N_m
65	230014	24%	156996	0.68
82	96654	19%	104561	1.07
91	48638	10%	86741	1.78
115	55258	10%	47588	0.86

the appropriate values for the other terms in Equation 5 measured on NR32, allowed a accurate prediction for the fatigue life. The ratio between the measured and the predicted values was observer to be consistently close to unity.

It is likely that the apparently larger flaws present in filled, compared to un-filled materials results from agglomerates of carbon black particles. These agglomerates have a low strength and effectively make quite large flaws. For the case of the poorly dispersed material, these agglomerates will be considerably larger.

It was interesting that these initial flaws can be considered using a relationship, derived for a centre edge crack through the entire test-piece thickness.

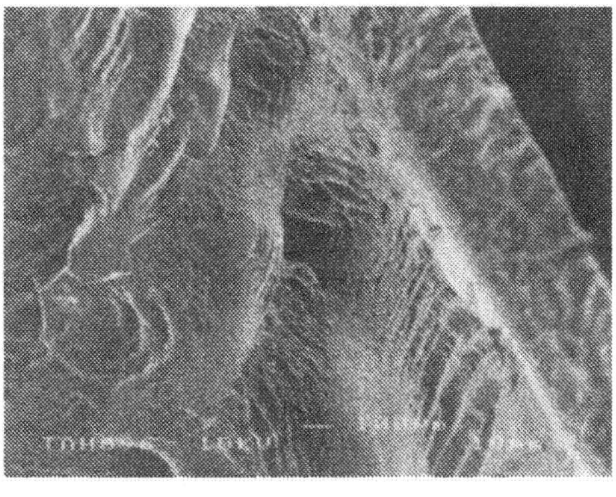

Figure 8. The SEM plot for the freeze fractured NR32 sample. It is believed that a flaw of 0.25mm is visible.

However, it is unlikely that the original flaw would either be in that location or have that geometry. From dimensional considerations, the dependence of any critical linear dimension of the specific crack geometry was always considered proportional to the length of the crack. Otherwise the strain dependence of the tearing energy and hence the crack-growth rate would not work with all geometries. Hidden in the k term is the specific dependence for a given geometry. Therefore, the c_o value used could be considered as an equivalent flaw size to the through section edge crack.

5.3 *Fatigue life investigation of razor corner and angled cut tensile test specimens.*

To critically examine this theory fatigue tensile test pieces were fatigue cycled after the insertion of two different razor-cut geometries. These cut profiles are shown in Figure 1 (b & c). The first incorporates a corner edge crack (Fig 1(b)) and the second a through section thickness crack inclined at an angle of 45° to the normal horizontal direction (Fig 1(c)). For simplicity an equivalent crack size was suggested for both of these cracks. The equivalent crack length, shown in Figure 1(b) for the corner crack, was defined as the perpendicular distance d from the crack front to the specimen corner. From trigonometry, it was seen that this distance was

$$d = \frac{ab}{\sqrt{a^2 + b^2}} \qquad (7)$$

The equivalent crack length d for the slanted edge crack is also shown in Figure 1. From trigonometry, it was seen that for a cut of length, x inclined at an angle of 45° the equivalent length $d=x/\sqrt{2}$.

The corner cut crack results are shown in Table 7. There was consistent underestimate in the prediction of the fatigue life by factors ranging from 2.5 to 7.4. The fatigue test results for the inclined crack cut lengths are shown in Table 8. Here the fatigue life was also dramatically underestimated by a factor of about 12.

Here the predicted fatigue lives were calculated using k values obtained from the plane stress model of a centre edge crack. It was considered that a more relevant k value specific to each of these cracks could be obtained in future by using appropriate 3D and plane strain, finite element models.

It was possible to deduce from dimensional considerations that the tearing energy was given by a relationship of the form, $T = k'(\lambda)Wd$. Of course, k' will not be identical to the functional relationship for k derived earlier. It seems likely that in both cases that k' would be substantially smaller.

These discrepancies between measured and predicted fatigue lives obtained using the fracture mechanics method however, were still far smaller than

Table 7. Discrepancies between the predicted and actual fatigue lives for corner cracks in NR 50 tested at 62% strain.

Actual d measured (±0.05mm)	Predicted fatigue life using the actual d and Equation 3, N_f	Experimental fatigue life, N_m	N_m/N_f
0.38	99002	253080	2.56
0.38	95030	257760	2.71
0.71	29819	179280	6.02
0.72	28813	180900	6.29
0.84	21513	158520	7.35
0.98	16032	59640	3.72
0.98	15969	81480	5.10
0.98	15969	83160	5.21
1.02	14773	82320	5.59
1.17	11434	30120	2.63

Table 8. Discrepancies between the predicted and actual fatigue lives for slanted edge cracks in NR 50 tested at 62% strain.

Actual d measured (±0.05mm)	Predicted fatigue life using the actual d and Equation 3, N_f	Experimental fatigue life, N_m	N_m/N_f
0.25	214837	2538600	11.82
0.38	98463	1617000	11.80

the discrepancies involved in the S-N approach that may be out by a factor of 100 or 1000. It is now possible to predict fatigue lives that are much more accurate than was assumed possible before.

6 CONCLUSIONS

The fracture mechanics approach for the fatigue life prediction for a carbon black filled elastomer within the 'power law' tearing energy region was studied. It was noted that it was possible to predict the fatigue failure of a cut specimen quite accurately. The small but consistent discrepancy could be accounted for by a taking into consideration the initial faster crack growth due to a sharp crack tip. The approach adopted also produces good fatigue life predictions for uncut specimens provided that suitable initial estimates are made for the internal flaws. The predictions for fatigue life were still more accurate than was possible using a traditional S-N approach.

Further improvements in the non-centre edge-crack results can be obtained using a k value derived using a finite element approach. It was thought likely that this technique could be further developed to predict the fatigue failure of real components subjected to a complex fatigue loading.

REFERENCES

Busfield, J.J.C., Davies, C.K.L. & Thomas, A.G. 1996. Aspects of Fracture in Rubber Components. *Progress in Rubber and Plastics Technology.* 12(3): 191-207.

Busfield J.J.C., Ratsimba C.H.H. & Thomas A.G. 1997. Crack-growth and Strain Induced Anisotropy in Carbon Black Filled Natural Rubber. *Journal of Natural Rubber Research.* 12: 131-141.

Callister, W.D. 1994. *An Introduction to Materials Science and Engineering - 3rd Edition.* New York: John Wiley and Sons: 203-206.

Choi, I.S. & Roland, C.M. 1996. Intrinsic Defects and the Failure Properties of Cis-1,4-Polyisoprenes. *Rubber Chemistry and Technology.* 69: 591-599.

Gent, A.N., Lindley, P.B. & Thomas, A.G. 1964. Cut Growth and Fatigue of Rubbers. I. The Relationship between cut Growth and Fatigue. *Journal of Applied Polymer Science.* 8: 455-466.

Greensmith, H.W. 1963. Rupture of Rubber X. The change in Stored Energy on Making a Small Cut in a Test Piece Held in Simple Extension. *Journal of Applied Polymer Science.* 7: 993-1002

Lake, G.J. 1970. Application of Fracture Mechanics to Failure in Rubber Articles, with Particular Reference to Groove Cracking in Tyres. *Yield, Deformation and Fracture of Polymers Conference.*

Lake, G.J. & Lindley, P.B. 1965. The Mechanical Fatigue Limit for Rubber and the Role of Ozone in Dynamic Cut Growth of Rubber. *Journal of Applied Polymer Science.* 9: 1233.

Lake, G.J. & Lindley, P.B. 1966. Fatigue of Rubber at Low Strains. *Journal of Applied Polymer Science.* 10: 341-351.

Lake, G.J. 1972. Mechanical Fatigue of Rubber. *Rubber Chemistry and Technology.* 45: 309-328.

Lake, G.J. 1983. Aspects of Fatigue and Fracture of Rubber. *Progress of Rubber Technology.* 45: 89-143.

Lake, G.J. & Thomas, A.G. 1988. Strength Properties of Rubber. In Roberts, A.D. (ed) *Natural Rubber Science and Technology*: 731. Oxford: Oxford University Press. 731

Lindley, P.B. 1972. Energy for Crack-growth in Model Rubber Components. *Journal of Strain Analysis.* 7: 132.

Lindley, P.B. & Thomas, A.G. 1962. Fundamental Study of the Fatigue of Rubbers. *The Fourth Rubber Technology Conference.* London: 1-14.

Ratsimba, C.H.H. (in prep.) PhD Thesis. Queen Mary and Westfield College, London.

Rivlin, R.S. & Thomas, A.G. 1953. Rupture of Rubber. Part 1. Characteristic Energy for Tearing. *Journal of Polymer Science.* 10: 291-318.

Thomas, A.G. 1966. Fracture of rubber. Physical basis of yield and fracture. Conference proceedings. Oxford: 134-143.

Thomas, A.G. 1958. Rupture of Rubber. V. Cut Growth in Natural Rubber Vulcanizates. *Journal of Polymer Science.* 31: 467-480

Constitutive Models for Rubber, Dorfmann & Muhr (eds) © 1999 Taylor & Francis ISBN 90 5809 113 9

Development of artificial elastomers and application to vibration attenuating measures for modern railway superstructures

D. Pichler
VCE, Vienna Consulting Engineers, Austria

R. Zindler
Getzner Werkstoffe GmbH, Grünwald, Germany

ABSTRACT: Noise and in particular vibrations induced by machines or railways, for example, can be reduced by using elastic layers to separate the source from the surrounding. The principle of the vibration reduction is based on the dynamic behaviour of a single-degree-of-freedom-system. In practise cellular polyurethane (PUR) materials are very well suited to act as the elastic layer. Depending on the intended use different properties of the material can be specified. Beside the static material properties the behaviour under dynamic loading is most important for the reduction of vibrations. Furthermore a couple of characteristics are necessary to ensure practical applicability of elastic materials for railway superstructures. Three examples are presented which give an impression about the possibilities to attenuate noise and vibration propagations caused by railways by using cellular PUR materials.

1 INTRODUCTION

Due to their high transport capacity and their effective use of energy with lowest damage to the environment railways are one of the most important means of transportation for the future.

In spite of the advantages of railways in comparison with other transport systems as for example motor cars the acceptance of new railway lines is very low especially by potential neighbours. One of the most important reasons for that is the fear of irritations from noise and vibrations induced by modern high speed trains. These problems especially occur in densely populated areas as in towns, where railway routes are in tunnels with low overburden and very close to residential buildings.

To reduce noise and in particular vibration propagation different kinds of vibration attenuating systems have been developed in the last years based on well known systems used for machine foundations.

2 THEORETIC BACKGROUND

The principle of all these systems is based on the response amplification factor of a dynamic system and can be explained in a very simple way by a linear single-degree-of-freedom system (Fig. 1, 2).

Figure 2 shows, that vibration attenuating effects occur for frequencies higher than $f_l\sqrt{2}$, where f_l stands for the natural frequency of the system. The inner damping factor has a dominant effect on the amplification of excitations with frequencies nearby the natural frequency of the system.

These basic principles of the physics of a dynamic system can be transformed into possible variation parameters for the realisation of a vibration mitigating construction. Reduced to a single-degree-of-freedom system consisting of a vibrating mass *m*, a spring *k* and a damper *c*, the insertion loss performance can be controlled by varying these three elements.

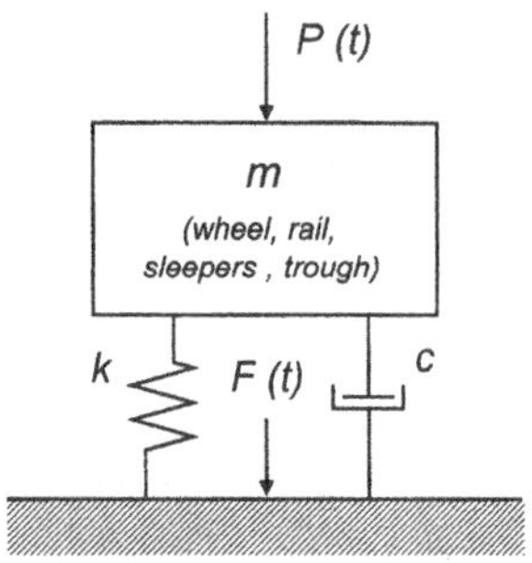

Figure 1. Mass-spring-system

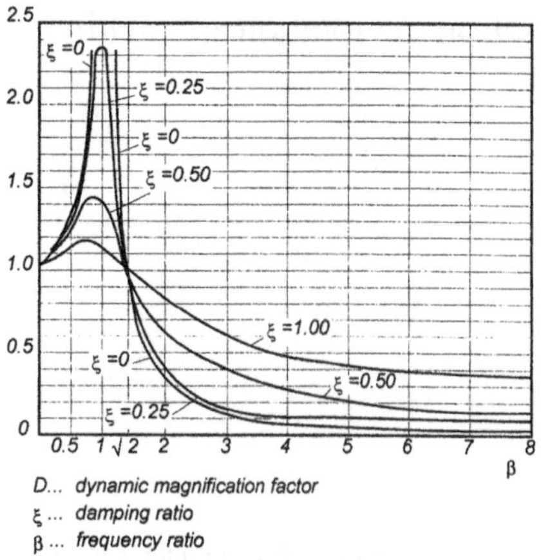

Figure 2. Dynamic magnification factor of a SDOF system

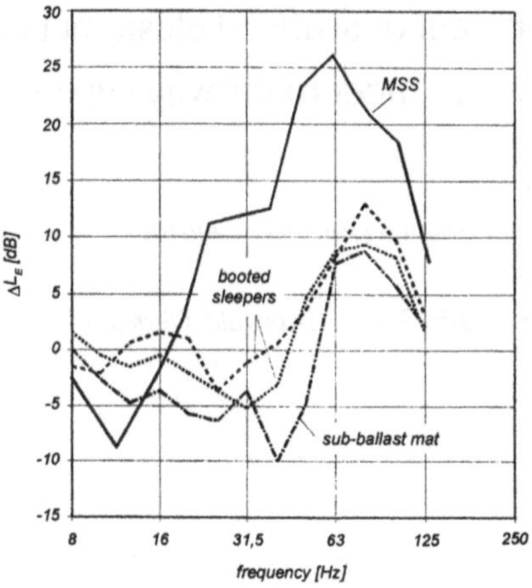

Figure 3. Insertion losses (measurement results)

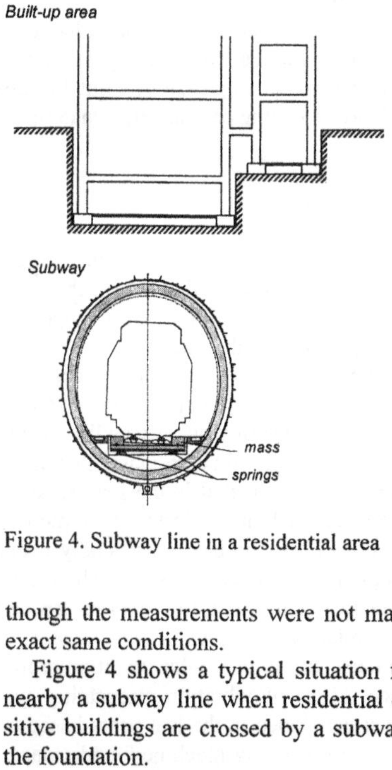

Figure 4. Subway line in a residential area

The insertion loss ΔL_E is defined as the difference of the noise and vibration level induced by trains on conventional ballast bed track systems and on a modified superstructure.

These theoretical thoughts can be summarised by the statement "elasticity in modern railway superstructures reduces noise and vibrations".

To realise this elasticity in practise different kinds of modification of a conventional superstructure (either ballast bed or solid roadway) are possible:

- Elastic rail fasteners
- Booted sleepers
- Sub-ballast mats
- Classic mass-spring-systems

All four systems mentioned are basically mass-spring-systems (MSS) – in this context "classic" means, that the whole superstructure is separated by an elastic layer from the underground.

According to the effectiveness of mass-spring-systems (MSS) the following division into three groups is commonly used:

- lightweight MSS with $m \leq 4$ t/m, $f_l \geq 15$ Hz
- mediumweight MSS with $m \leq 8$ t/m, $f_l \geq 10$ Hz
- heavyweight MSS with $m > 8$ t/m, $f_l < 10$ Hz

The damping capability of MSS increases with the increase of the mass and the decrease of the natural frequency. The first three groups of superstructure modifications – elastic rail fasteners, booted sleepers and sub-ballast masts – are all lightweight MSS.

Figure 3 gives an impression about the damping capability of booted sleepers and sub-ballast mats in comparison to a classic mass-spring-system. The insertion loss curves gave a good impression about what can be reached with the different systems although the measurements were not made under the exact same conditions.

Figure 4 shows a typical situation for buildings nearby a subway line when residential or other sensitive buildings are crossed by a subway just below the foundation.

Using a classic MSS for the subway superstructure the noise and vibration level in the buildings is reduced by a significant value. Figure 5 shows the measurement results in buildings nearby a city tunnel.

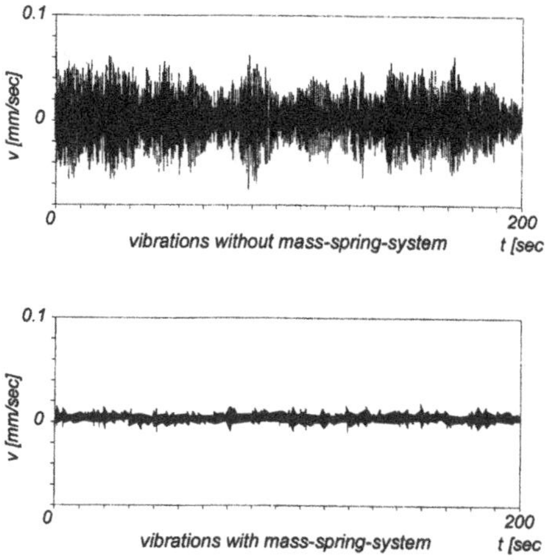

Figure 5. Vibrations in a building nearby a subway line

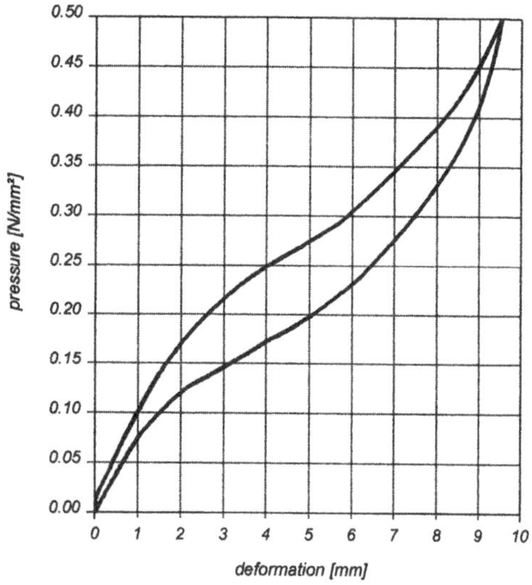

Figure 6. Stress-deformation diagram of Sylomer® P25

3 ELASTIC ELEMENTS - BEARINGS

3.1 *General*

To apply these principles in practice, there are different requirements for the materials in the elastic layer.

The first group can be summarised under the headline "vibration requirements" – the dynamic characteristics of the bearing material are needed for the design. The most important quantity is the stiffness (which depends on the load), the frequency and amplitude of excitation, and the inner damping. Furthermore, changes of the dynamic behaviour of the materials over time should be considered.

Additionally, "mechanical requirements" can be defined. Long-time stability of the elastic material has to be ensured for the relevant applied load combinations. Important for the serviceability of the bearings, are the load carrying capability, the fatigue behaviour, the deformations due to pressure and the long time settlements.

A third group of requirements must be defined to ensure the "integrity of the elastomer" under site conditions. The material has to be stable not only against water but also against chemicals like diluted alkalis and acids, and against commonly used oils and fats. To reduce the risk of installation mistakes, the bearings should be easy to manage on site.

3.2 *Mechanical properties of cellular PUR*

Cellular polyurethane (PUR) products are particularly useful for continuous elastic layers. Due to their structure these artificial materials have the necessary volume for deformations. Therefore, though lying continuously between rigid surfaces, these materials do not loose their elasticity.

Sylomer® and Sylodyn® – both products of Getzner Werkstoffe, Austria – are mixed-cellular PUR materials. The necessary cells are created during the cross-linking process by adding a propellant. The cells have either an open or a closed structure. The elasticity of the material is only a function of the stiffness of the plastic matrix and does not depend on the gas in the cells. Due to the very fine cellular structure subsequent closing of the cells by dirt or alien elements is not possible.

In general the stiffness of a bearing is described by the stress-deformation diagram. Bearings made of natural or artificial rubber materials generally show an extensive non-linear stress-deformation diagram. The tangent on the curve corresponds to the stiffness at this load level. All tangent stiffness can be plotted in a graph. Figure 6 shows a non-linear stress-deformation diagram of Sylomer®, Figure 7 shows the corresponding tangent stiffness (quasi-static deformations).

Furthermore, the stiffness depends on the frequency of excitation – all kinds of elastomers show an increase of stiffness with dynamic loading. For higher excitation frequencies these materials react with an increasing stiffness. This effect is caused by the inner material damping. The bigger the damping the more the stiffness is influenced by the frequency. In Figure 7 the two upper curves show the dynamic stiffness-increase of Sylomer® between 5 Hz and 40 Hz excitation frequency. For comparison, Figure 8 shows the corresponding curves for a profiled rubber-mat.

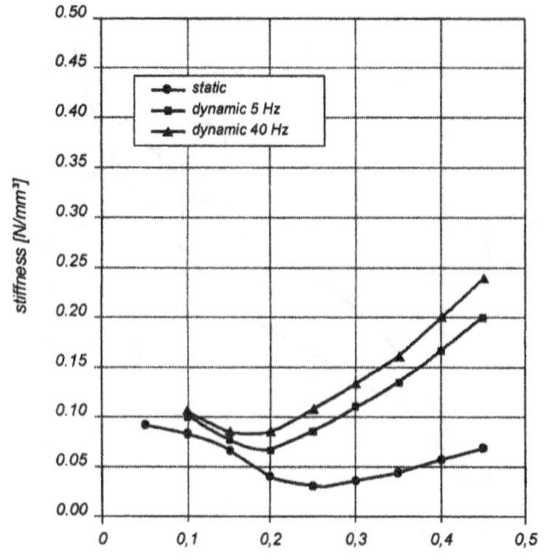

Figure 7. Tangent-stiffness diagram of Sylomer® P25

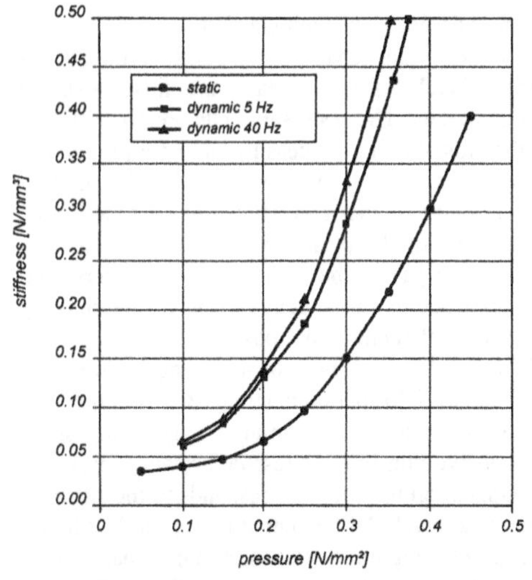

Figure 8. Tangent-stiffness diagram of a profiled rubber-mat

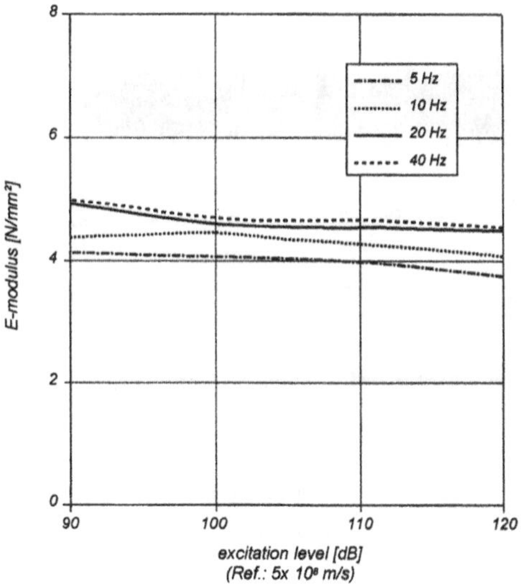

Figure 9. Stiffness dependency of excitation amplitude for Sylomer® S600

Beside the excitation frequency, the excitation amplitude influences the bearing-stiffness. In regular cases with increase of amplitude, the stiffness decreases. This dependency can be neglected for Sylomer® materials (Fig. 9). However, with very low amplitudes (e.g. vibrations of buildings) these materials react very softly.

The practical way of determining the influence of frequency on the stiffness is to consider ratio between the static k_{stat} and the dynamic stiffness k_{dyn}. These is the most important parameter which describes the quality of a bearing. The static stiffness is responsible for the deflection of the mass-spring-system under dead and live loads, whereas the dynamic stiffness is the key-parameter for the insertion loss. In summary Sylomer® shows a very reasonable k_{stat} / k_{dyn} ratio.

PUR-bearings (Sylomer® or Sylodyn®) make it possible to limit the increase of stiffness due to dynamic loading by a factor of 1.30. Using such advanced materials highly efficient mass-spring-systems with natural frequencies about 7 Hz are possible.

The inner damping of plastics is described by the loss factor η. For Sylomer® η is between 0.23 and 0.12. Close to the resonance frequency the loss factor correspondance to the damping ratio ξ can be approximated by the equation $\eta = 2\,\xi$. Sylodyn® has a two to three times lower inner damping than Sylomer.

The load carrying capacity of cellular PUR depends on the density of the material. For Sylomer® and Sylodyn® the allowable permanent stresses vary between 10 and 1,500 kN/m^2.

Mainly the long-time effects, especially the settlements, depend on the loading. Figure 10 shows the creep effects as a function of static permanent loading. Figure 11 shows the influence on the dynamic stiffness.

An optimum solution between dynamic stiffness increase, creep and the necessary natural frequency of a mass-spring-system, can be found close to the allowable static permanent load. In the stress-

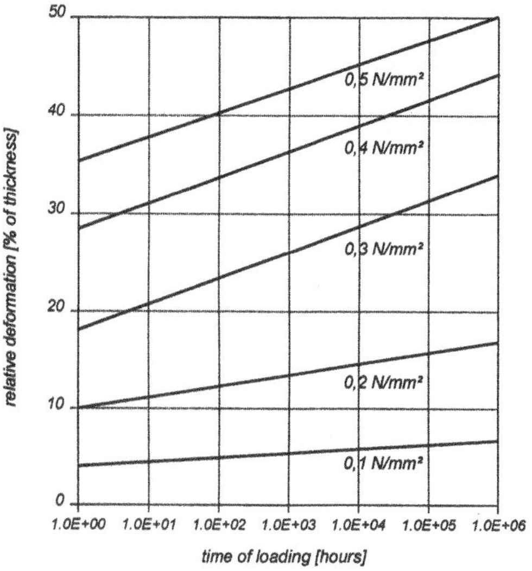

Figure 10. Creep effects for Sylomer® P

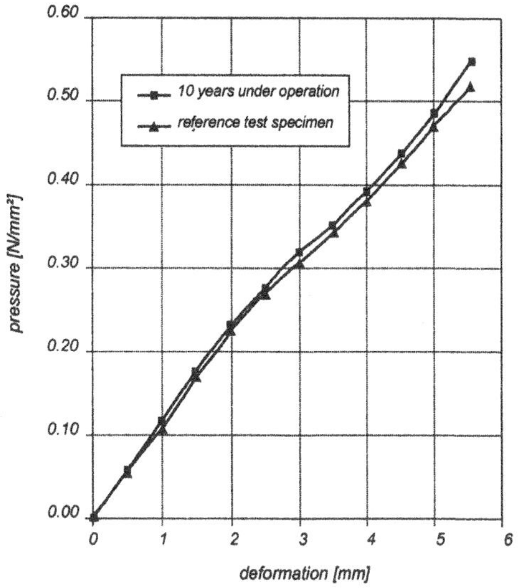

Figure 12. Comparison for Sylomer® - sub-ballast mat

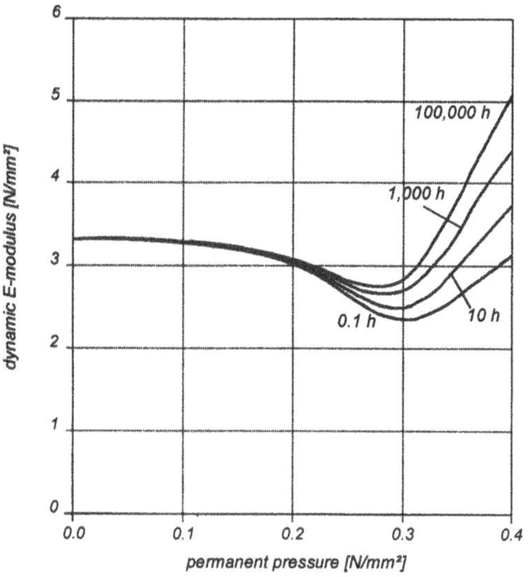

Figure 11. Dynamic E-modulus of Sylomer® P

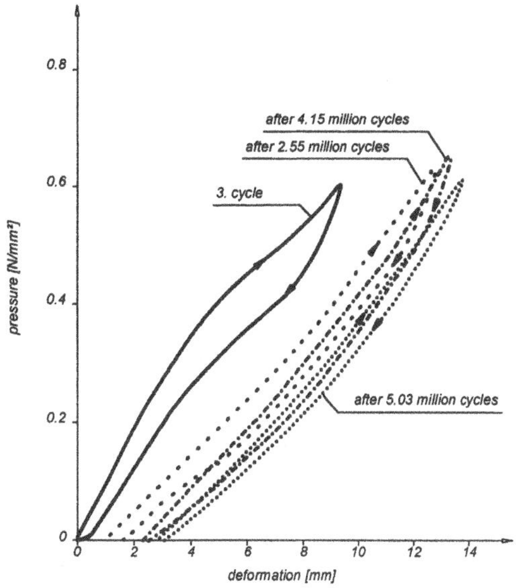

Figure 13. Dynamic fatigue test for Sylomer® S50

deformation diagram this value can be found at the beginning of the digressive sector.

Figure 12 shows measurement results for a sub-ballast mat which was under operation for ten years, in comparison with a stored reference test specimen.

Beside the static permanent loadings high dynamic loadings over a long period of time must not influence the bearing stiffness. Due to the low static masses in comparison with the mass of the train for many railway applications the excitation amplitudes are much larger than for buildings. Although the bearing stiffness should not change during the life expectancy of the whole superstructure (mass-spring-system). Laboratory tests help to find out the long-time behaviour of bearings. Figure 13 shows the documentation of a dynamic fatigue test for Sylomer® S50. 5 x 10^6 load cycles were tested with an excitation amplitude 2.5 times larger than under real conditions. Figure 13 shows the stiffness change during the test.

3.3 Chemical properties of cellular PUR

Sylomer® as well as Sylodyn® is very stable against water influence. Ice formation in open cells of wet test specimens do not damage the material.

Chemical stability against other mediums was tested, the results are summarised in tables.

3.4 Handling of Sylomer® and Sylodyn®

Both products are produced in a continuous process. They are deliverable either in form of mats, of strips or in form of confected single bearings. Modifications on site are easily possible with conventional tools. Sticking together of pieces is possible with any usual glue. Recommended are 2-component PUR-glues or glues on bituminous basis.

Very important for successful use of any kind of bearing materials in practice are simple construction principles which are appropriate to site conditions. Highly adaptable materials may help to reach this target.

4 PRACTICAL EXPERIENCES

Due to its typical stress-deformation behaviour – first linear and then degressive tendency – Sylomer® makes very effective mass-spring-systems with low static deformations possible. Short-time load maximums up to many times of the permanent loading do not damage Sylomer®-bearings. Because of the progressive ascending stress-deformation curve, additional deformations are relatively small.

As mentioned before, all kinds of bearing geometry are possible – full surface bearings, strip bearings or single bearings. Which kind of bearing is

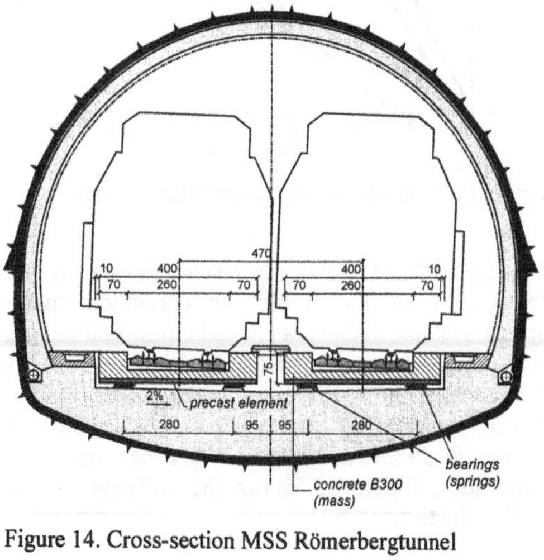

Figure 14. Cross-section MSS Römerbergtunnel

used depends on the constructive boundary conditions, on the loads and on the necessary tuning frequency of the mass-spring-system. Figure 16 shows possible constructions.

The following three examples show different applications of Sylomer® and Sylodyn® to vibration attenuating railway superstructures.

4.1 Mass-spring-system Römerbergtunnel

For the first time the new concept of a jointless mass-spring-system was put into practice, was in the Römerbergtunnel near Schwanenstadt in Upper Austria. The Römerbergtunnel with a total length of 710 m represents a section of the development programme of the Austrian Westbahn railway. It partly runs situated under existing buildings of a residential area with low overburden.

After finishing of the inner tunnel lining Vibro-Scan® tests were made by Prof. Steinhauser which showed that preventive measures for noise and vibration mitigation were necessary. To reach the limits for good noise and vibration protection in residential areas according to ON S 9012 a medium-weight mass-spring-system had to be used.

The result of the prediction model based on linearity for both main construction elements, bearings and reinforced concrete trough, was, that a natural frequency of about 13 Hz and a mass of 6 tons per metre would be sufficient. Knowing these parameters a jointless system was developed and put into practice (Fig. 14). The system has a total length of 348 metres, 192 metres supported on single bearings and the remaining parts of the system lying on a continuous bearing layer. For the horizontal stabilisation of the floating reinforced concrete trough shear keys with vertical elastic bearings were developed. At both ends of the MSS instead of rail expansion joint constructions a continuous connection between the MSS and the conventional ballast bed superstructure was designed.

Together with the exchangeable bearings and the continuous transition to the ballast bed the jointless MSS has two big advantages compared with other systems: the construction costs are very low and the expenditure for maintenance works is limited too. The single bearings are made of Sylodyn® whereas the full surface bearing layer are Sylomer® materials.

The static design concept of the MSS Römerbergtunnel is based on the new "semi-probabilistic" Austrian codes ON B 4003, B 4700 and B 4703. So partial safety factors have been used for the different loads and for the resistant values of the materials.

The internal forces and deformations due to the different load cases were calculated by using a simple finite element model consisting of 3-dimensional beam elements. For computation of these static values the static stiffness of the bearings has to be used, and according to the codes the beam stiffness can be estimated with the non-cracked cross-section of the concrete trough. Dynamic amplification effects which lead to an increase of the internal forces are taken into consideration by multiplication of the live loads (simulating the trains) with a dynamic factor. According to ON B 4003 this factor is a function of the length l_φ of the deflection line between the points of inflection. First tentative computations led to a l_φ about 30 metres which implied a dynamic amplification factor about 1.10. As a result of several discussions and because of the special wish of the client it was decided to use a factor of 1.30 for the ultimate limit state.

The serviceability limit state (deflections, crack width) was calculated using the same finite element model as before for the ultimate limit state but without any amplification factor for the loads. The deformations were limited to three criteria: maximum deflection due to live loads below 10 mm, length of deflection line versus maximum deflection over 2,500, and maximum inclination of the deflection line below 0.03%.

For computation of the deflections the real stiffness of the concrete beam has to be taken into account. This was done by using the approximation method of ON B 4703.

Additional special investigations were necessary for long-time effects due to temperature, creep and shrinkage. Because of the tunnel situation it was decided to reduce the range of temperature values according to the Austrian codes down to ±10 Kelvin. The horizontal deformations of the single bearings and of the continuous bearing layers due to temperature and shrinkage are limited to $\tan \gamma \leq 0.70$.

The dynamic characteristic of the MSS Römerbergtunnel – the natural frequencies and the mode shapes – was determined by modal analysis of the MDOF-system. Therefore the dynamic stiffness of the bearings was taken into account as well as different stiffness conditions of the mass.

Figure 15 shows the first vertical mode shapes calculated by using the 40 Hz stiffness of the bearings and the non-cracked cross-section of the concrete trough.

During construction and after finishing of the MSS in the Römerbergtunnel numerous different measurement programmes were carried out. The targets of these investigations were to get knowledge of the static and dynamic behaviour of the MSS under real conditions. The measurements can be split into four groups:

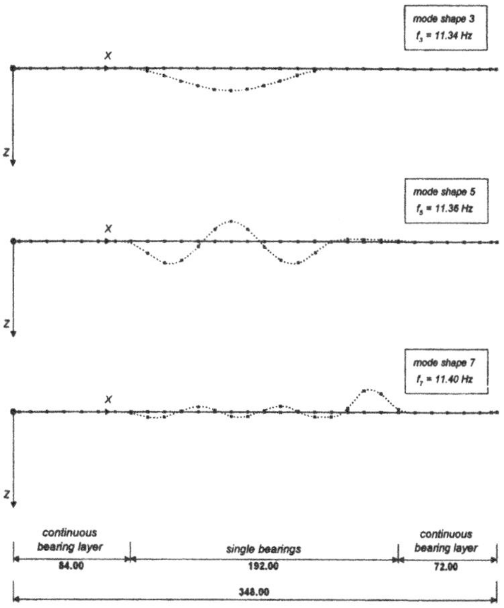

Figure 15. Vertical mode shapes MSS Römerbergtunnel

- dynamic characteristic and insertion loss
- temperature and long-time displacements
- deflections of the MSS due to live loads
- rail stresses and rail deformations

Because of the importance of the damping capability of the MSS, for the dynamic characteristic and the real insertion loss numerous different tests and measurements were done. After finishing the trough construction VibroScan® tests were made to get sure that all components of the system work correctly. These tests showed, that the prediction model for the vibration propagation through the soil and for the insertion loss of the MSS was accurate. Ambient vibration measurements and evaluations with the dynamic measurement and testing system BRIMOS, developed by VCE, showed that the natural frequencies and mode shapes fit very well to the calculated ones for the non-cracked MSS and the 40 Hz bearing stiffness.

Figure 16 shows a comparison of two measurement results of the VibroScan® tests and the predicted insertion loss.

The comparison shows, that up to 32 Hz the measured results are identical with the calculated ones, for higher frequencies the calculated insertion loss is higher than the real one. The reason for that is not an insufficient prediction model, but the fact that the VibroScan® tests were made on the pure concrete trough without the mass of the rails and the rail carrying elements. Nevertheless the spectral analysis (FFT) of the VibroScan® sweeps and the BRIMOS

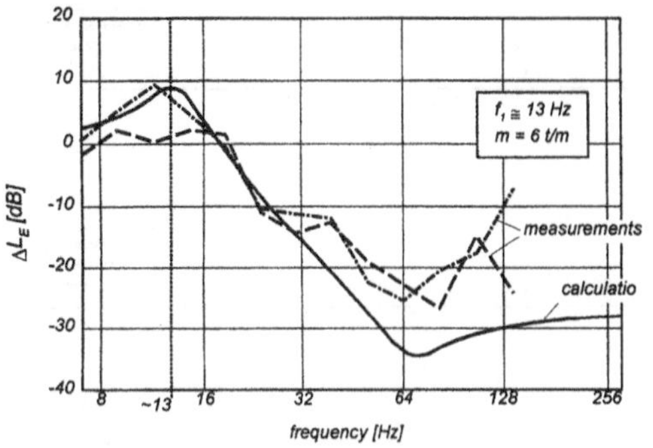

Figure 16. Insertion loss of MSS Römerbergtunnel

results fit very well together. Both investigations led to a maximum effective vertical natural frequency of about 11.8 Hz for the single bearing section.

For the effects to neighbours of railway lines, the absolute values of the noise and vibration level are responsible. After finishing the whole system, including the rails and additional equipment, many noise and vibration measurements were done in the buildings nearby the Römerbergtunnel. All these investigations showed that the limits for good noise and vibration protection according to the Austrian codes were reached.

Beside the dynamic parameters, the static behaviour of the MSS due to live loads is of special interest. Especially the real deflections and displacements of the concrete trough influence the safe operation of the whole system. Therefore several load tests with electric locomotives were done before starting with the regular traffic on the MSS.

Figure 17 shows the measured result of such a test and for comparison the calculated deflection line for the same loading.

It is obvious that the computation model used describes reality in a sufficient way.

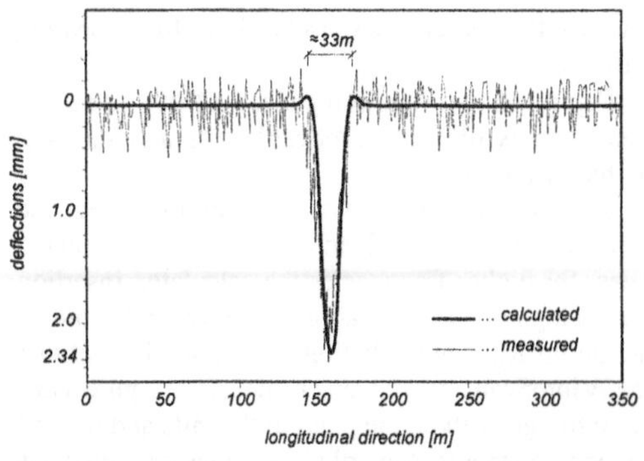

Figure 17. Measured and calculated deflections

The long-time measurement programmes are well under way and, until now, show a very good correlation with the assumptions for the calculation model. After the first two years under regular operation the temperature range of the MSS seems to be between about –5 and +15°C; the horizontal displacements are also within the calculated limit values.

Furthermore the measurement results of the stresses and deformations of the rails are within the allowed limits – there are no indications of problems with the long-time behaviour of the rails and the rail carrying system.

4.2 Mass-spring-system Zammer Tunnel

Because of the successful application of the new jointless mass-spring-system in combination with PUR-bearings in the Römerbergtunnel the Austrian

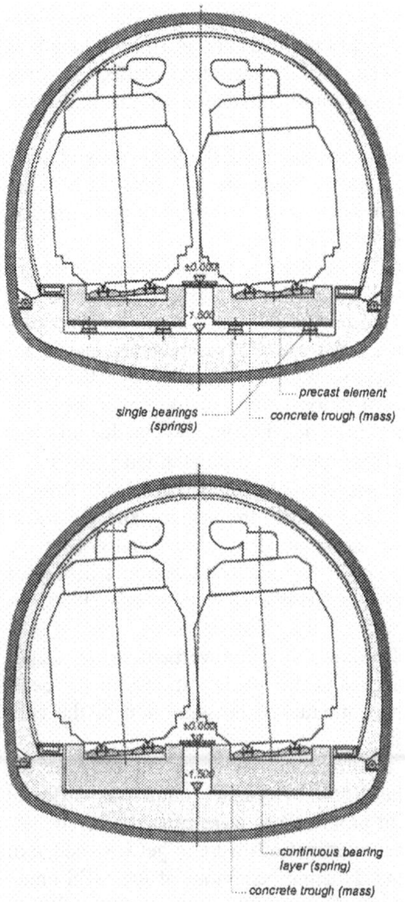

Figure 18. Cross-sections MSS Zammer Tunnel

railway company ÖBB decided to use a similar vibration attenuating system in the Zammer Tunnel in Tyrol. The Zammer Tunnel with a total length of 2.3 km is situated in a very sensitive area in the river Inn valley near Landeck.

Once more VibroScan® tests by Prof. Steinhauser lead to the input values for design. This time in the most sensitive part of the tunnel a heavy-weight mass-spring-system with a natural frequency of 7.5 Hz and a sprung mass of 10 t/m was necessary.

In other parts of the tunnel systems with natural frequencies from 10 to 24 Hz and masses from 10 t/m to 4 t/m had to be designed – in some insensitive parts conventional solid roadways were possible.

Based on the design parameters developed for the Römerbergtunnel, VCE created a modular jointless mass-spring-system. The lengths of the conventional solid roadway is incorporated with the lengths of mass-spring-systems. Thus, the final system is one continuous system without any joints. Even between the different systems no joints were arranged. This concept lead to a continuous superstructure with a lot of advantages for the construction process and also for the regular operation of the railway line. A major advantage is that maintenance costs are reduced to a minimum due to the fact that no moveable elements are part of the system, and because of the continuous transition to the ballast bed at both portals of the tunnel. For the springs of the MSS Sylomer® and Sylodyn® bearings were used once more. The newest developments made it possible to reach a natural frequency of the MSS of 10 Hz by use of full surface bearings.

The system in the Zammer Tunnel was erected in 1998 and regular operation started in spring 1999. Extensive measurements similar to those in the Römerbergtunnel proved the structural integrity of the superstructure, the correct dynamic tuning of the different MSS and the predictions of the insertion losses.

4.3 Test track Riedau

In Upper Austria the ÖBB installed a test track for sub-ballast mats and booted sleepers. The test track is part of the regular railway line between Wels and Passau.

The ÖBB and Getzner installed different kinds of Sylomer® and Sylodyn® booted sleepers and a Sylomer® sub-ballast mat in different parts of the test track. The design criteria for all these elements was to get the same deformations for the different elastomeric elements of superstructure. During a special test procedure, VCE measured the insertion loss of

Figure 19. Test track Riedau

the systems. The results of these investigations can be seen in Figure 20.

The results show, that all three modifications of the superstructure resulted in similar insertion losses. This is in accordance with theory, due to the fact that the natural frequency of all three systems is the same.

Furthermore, it is obvious that Sylomer® and Sylodyn® are effective materials for booted sleepers as well as for sub-ballast mats.

5 CONCLUSION

The importance of vibration and noise protection measures for railway is increasing. Nowadays mass-spring-systems are the most effective way to reduce these effects.

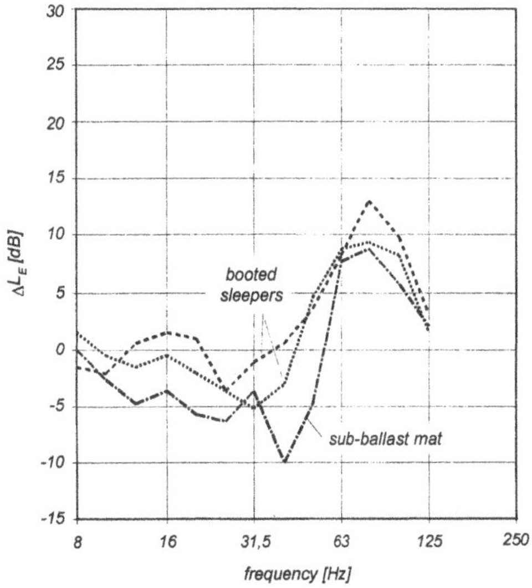

Figure 20. Insertion losses test track Riedau

To design and realise systems with large vibration attenuating capacity, spring elements with specially defined static and dynamic performance are necessary. PUR based materials give the possibility to control the static and dynamic properties during the production. Sylodyn® and Sylomer® cellular PUR-bearings have a wide range of applications for elastomeric isolation of railway superstructures. With both materials all kinds of MSS – from lightweight to very heavyweight systems – are possible. Together with a modern construction concept – jointless systems – and a sophisticated prediction and measurement programme, successful vibration reduction can be realised.

REFERENCES

Buda, R. 1996. *Elasticity in Modern Superstructures – system design aspects and typical components.* Getzner Werkstoffe Ges.m.b.H., Austria.

Clough, R.W. & Penzien, J. 1993. *Dynamics of Structures.* Second International Edition, McGraw-Hill.

Getzner Werkstoffe GmbH. Chemische Beständigkeit von Sylomer®-Standardprodukten Datenblatt W3.

Kohler, K. A. 1984. *Körperschalldämmung im Hoch- und Tiefbau.* Kunststoffe im Bau Heft 4.

Pichler, D. et al. 1997. *Entwicklung eines neuartigen Masse-Feder-Systems zur Vibrationsverminderung bei Eisenbahntunnels.* Bauingenieur 72 p. 515-521.

Pichler, D. et al. 1997. *Reduction Measures for Tunnel Lines.* Report for RENVIB II Phase 1 to ERRI, Vienna Consulting Engineers and Rutishauser Ingenieurbüro.

Pichler, D. 1998. *Concrete based floating track slab systems- Modelling and reality.* In R. de Borst, N. Bicanic, H. Mang & G. Meschke (eds), *Computational Modelling of Concrete Structures:* 665-671. Rotterdam: Balkema.

Prüfamt für Bau von Landverkehrswegen 1998. *Forschungsbericht über Eignungsversuche an unbewehrten Troglagern Typ Sylomer S50 auf PU-Basis für Masse-Feder-Systeme bei U- und S-Bahnen.* TU Munich: Report No. 1231.

Steinhauser, P. 1996. *Römerbergtunnel – Ergebnisse der VibroScan® Untersuchung zur immissionsmäßigen Abstimmung des Oberbaus.* HL-AG Vienna.

Steinhauser, P. 1997. *Römerbergtunnel – Ergebnisse der VibroScan® Untersuchung auf dem Masse-Feder-System.* HL-AG Vienna.

Steinhauser, P. 1997. *Römerbergtunnel – Ergebnisse der Erschütterungsimmissionsmessungen des Bahnverkehrs auf dem Masse-Feder-System.* HL-AG Vienna.

Wenzel, H. et al. 1997. *Reduktion von Lärm und Vibrationen durch Masse-Feder-Systeme für Hochleistungseisenbahnen.* Oral presentation at the D-A-CH-Meeting in Zürich, SIA-Dokumentation D 0145.

Wettschureck, R. et al. 1985. *Insertion loss of ballast mats.* Acustica Vol. 58.

Wettschureck, R. 1995. *Vibration and structure-borne noise insulation by means of cellular polyurethane (PUR) elastomers in railway track applications.* Rail Engineering International, Edition 1995, No. 2: 7-14.

Constitutive Models for Rubber, Dorfmann & Muhr (eds) © 1999 Taylor & Francis ISBN 90 5809 113 9

Different numerical models for the hysteretic behaviour of HDRB's on the dynamic response of base-isolated structures with lumped-mass models under seismic loading

J. Böhler & Th. Baumann
DYWIDAG Design Department, Munich, Germany

ABSTRACT: The effect of seismic base isolation by HDRB´s (**H**igh **D**amping **R**ubber **B**earings) is based on two main effects. The one is the capability of large elastic deformations which increases the fundamental period of the structure and reduces the seismic response. The second effect is the non-elastic absorption of the energy introduced by the earthquake by means of the hysteretic damping.

An adequate description of the nonlinear characteristic of the HDRB´s and a correct modelling of the overall structure are presuppositions for a realistic estimation of the mechanism and effectivity of such a base isolation.

Three different numerical approaches for the description of HDRB´s have been developed, in order to show the influence on the results. The effect of these approaches on the dynamic response are studied with lumped-mass models representing a base-isolated storage tank for **l**iquefied **n**atural **g**as (LNG). The maximum accelerations of the isolated structure are compared, in order to verify the accuracy of the numerical approaches and the boundaries of their applicability.

1 INTRODUCTION

For above-ground LNG-storage-tanks the state of the art is defined by the Full Containment Type (Figure. 1). The liquid is stored by a steel inner tank. A prestressed concrete outer tank protects the sensitive inner tank against external hazards and serves as back-up-pool in case of failure of the inner tank.

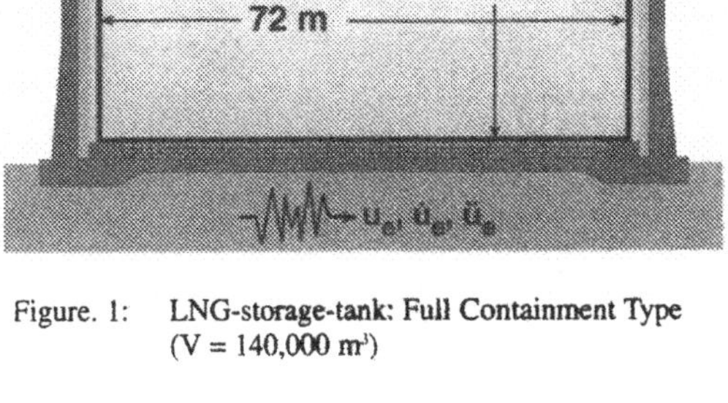

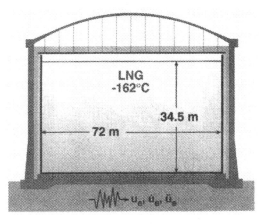

Figure. 1: LNG-storage-tank: Full Containment Type ($V = 140{,}000\ m^3$)

The steps which are required for the earthquake design of such a tank system are typical also for other structures. First we need an idea of the failure modes and the corresponding design criteria. For a containment acc. to Figure. 1, one important criterion is uplifting of the inner steel tank and consecutive buckling or elephant-footing of the tank wall due to the horizontal excitation (Figure. 2). The most important problem in this respect is whether a base isolation is required in order to reduce this risk of elephant-footing or not. For answering this question, the idealization by a so-called tuning fork-model is suited best. The overturning moments of the inner tank, which may cause uplifting and buckling, have to be sustained by a rotational spring, which may represent also the nonlinear uplifting characteristic of the inner tank. With this model, the effect of a base isolation on the uplifting of the inner tank can be analyzed in a reliable way.

For other design criteria, e.g. buckling of the cupola of the outer tank under vertical accelerations, other idealizations are required. Thus, the computational model has to be chosen always with regard to the individual design question.

The tuning fork model of figure. 2 can be simplified further by an oscillator with three degrees of freedom (Figure. 3). The relevant influences –

Uplifting of Inner Tank

Tuning Fork Model

$M_T = F_2 \cdot h_2$

$F_2 = m_2 \ddot{u}_2$

isolators

m_1

m_2

sloshing mass m_3

uplifting characteristic of inner tank

pile foundation

a) $u_e, \dot{u}_e, \ddot{u}_e$

b) $u_e, \dot{u}_e, \ddot{u}_e$

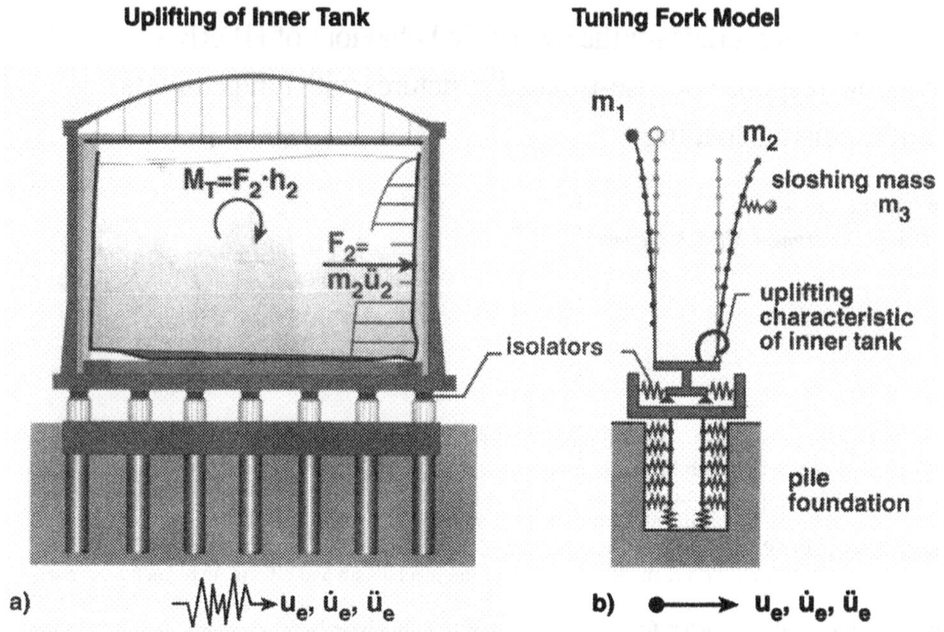

Figure. 2: Modelling of base-isolated tank as tuning fork model including uplifting effects of inner tank

characteristic of base isolation, mass of outer tank, stiffness of inner tank, impulsive and sloshing mass of liquid – can be depicted also by this model without loss of accuracy. The absolute displacement of the subsoil and the masses m_1, m_2, m_3 are denoted by u_e, u_1, u_2, u_3. The elongation of the springs K_1, K_2, K_3 are given by $u_e - u_1$, $u_1 - u_2$, $u_2 - u_3$.

It is a crucial question, in how far the reliability and accuracy of an analysis can be improved by a more discrete idealization (Figure. 4). For the practical design it seems more efficient in most cases to apply simple models which are related to the indivi-

dual design task rather than increasing the number of elements.

In the following we consider the effect of different modelling of isolators on the 3DOF-oscillator with the numeric characteristics shown in Figure. 3 (corresponding to an LNG-tank with V = 140,000 m³). The periods of the three eigenmodes are determined by

- the base isolation,
- the inner steel tank and
- the sloshing of the liquid.

$T_3 \cong 2\pi \cdot \sqrt{\dfrac{m_3}{K_3}} \approx 7s \qquad T_2 \cong 2\pi \cdot \sqrt{\dfrac{m_2}{K_2}} \approx 0.3s \qquad T_1 \cong 2\pi \cdot \sqrt{\dfrac{m_1 + m_2}{K_1}} \approx 2.5s$

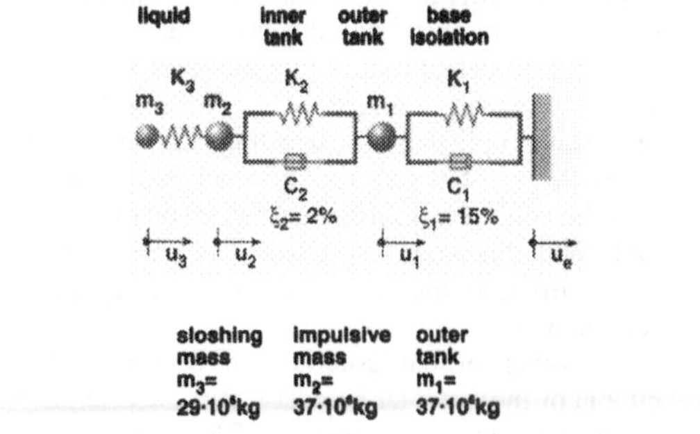

Figure. 3: 3DOF-oscillator and numeric characteristic for an LNG-tank with V = 140,000 m³

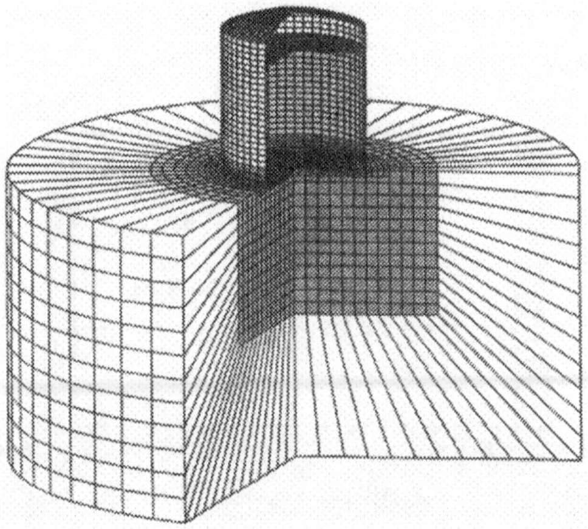

Figure. 4: Discrete idealization of containment, liquid and subsoil

The basic stiffness K_1 of the isolators has been designed in order to increase the period T_1 to 2.5 s. This can be reached by the arrangement of about 460 isolators as described below.

For the intended reduction of the seismic response, a high damping ratio of the isolators is decisive. The computational modelling of the structure and of the damping is the main subject of this report. The consideration of the different types of energy which occur during an earthquake allows a better understanding of the basic actions.

2 COMPUTATIONAL MODELLING OF ISOLATORS

We consider High Damping Rubber Bearings, briefly HDRB´s, which are suited well as isolators. Figure. 5 shows a typical characteristic, which has been found by tests. For the computational description of such a characteristic, we have studied three different models.

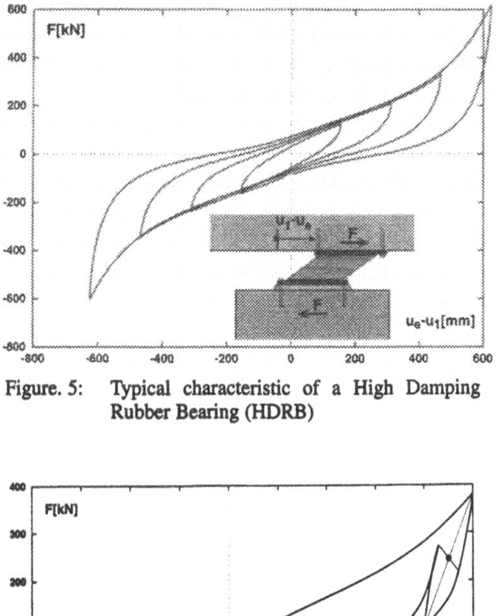

Figure. 5: Typical characteristic of a High Damping Rubber Bearing (HDRB)

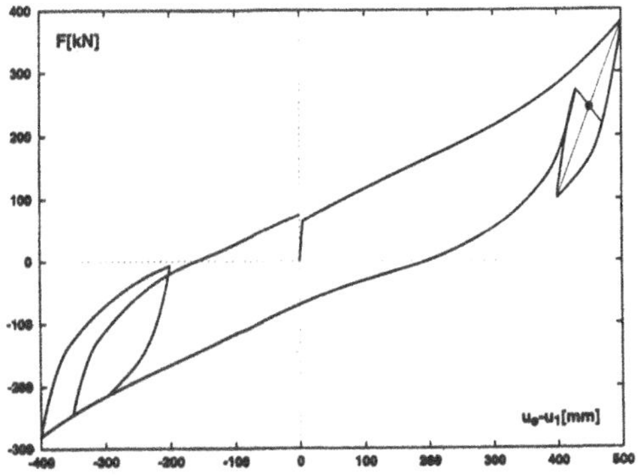

Figure. 6: Model 1: Direct numerical description

Model 1 applies a direct numerical description (Figure. 6). The envelope of the hysteresis is described by about 50 points. For a partly unloading and reloading, special procedures have been developed for shifting and reflecting the relevant parts of the envelope.

In the second model, two nonlinear springs are superimposed (Figure. 7). The elasto-plastic spring (1a) is intended to cover the hysteresis, the multilinear elastic spring (1b) simulates the stiffening of the isolators at higher loads. In this second model only seven parameters are required, which is less expensive than the detailed numerical description of model 1.

In most cases the model 3 is applied (Figure. 8). It combines an elastic spring (K_1) with a viscous damper. First a damping ratio ξ_1 is derived from the hysteresis in a semi-empiric way by comparison of the elastic energy $E_{elastic}$ and the area of the hysteresis loop. In a second step the coefficient C_1 for a velo-

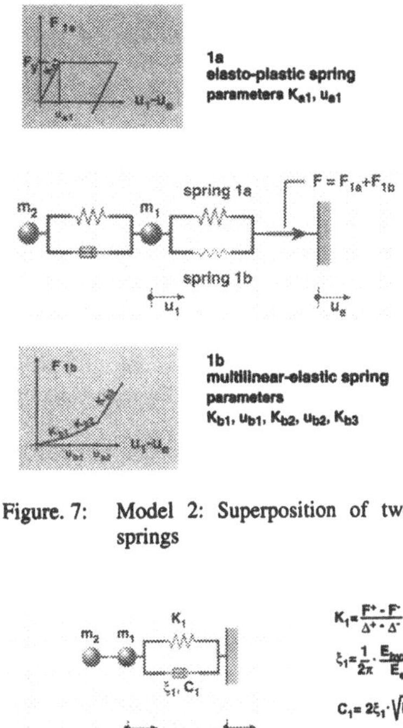

Figure. 7: Model 2: Superposition of two nonlinear springs

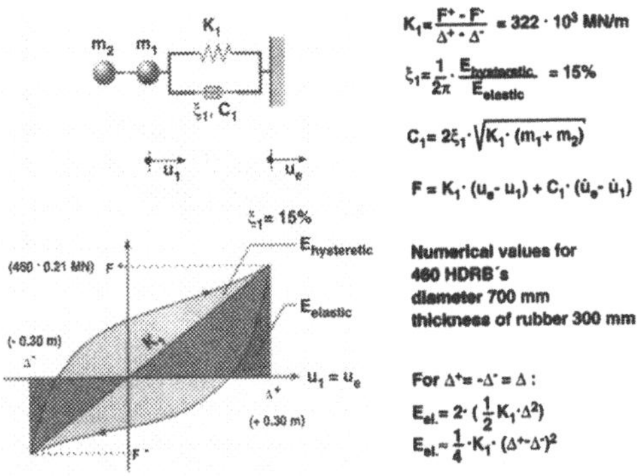

Figure. 8: Model 3: Elastic spring and velocity-dependent damper

city-dependent damping force is defined by $C_1 = 2\xi_1 \cdot [K_1 \cdot (m_1 + m_2)]^{0.5}$, the inner steel tank being considered rigid compared to the isolators for this assessment.

The overall force F which is transferred from the subsoil into the structure is given by

$$F = K_1 \cdot (u_e - u_1) + C_1 \cdot (\dot{u}_e - \dot{u}_1) \qquad (1)$$

At least for a load-history, which is not symmetric to the origin, there is no clear theoretical foundation for this procedure. Nevertheless it is practicable and delivers reasonable results, as we will see later.

In our parameter studies we assumed a base isolation consisting of 460 HDRB´s with the characteristic shown in Figure. 5. For the application of model 3 acc. to Figure. 8 we had first to estimate the maximum deformations of the bearings (± 0.30 m) and the corresponding force F (460 · 0.21 MN). From this we got the stiffness K_1 = 322 MN/m and by evaluation of the hysteretic area the damping ratio ξ_1 = 15%.

How do these three models fit to reality, that means to the test results of the bearings?

For a load history which is symmetric to the origin the direct description of model 1 delivers of course a good agreement with the test result (Figure. 9). However, the real load history of isolators during an earthquake is not symmetric to the origin at all. Figure. 10 shows the loading und unloading paths obtained as result of a time history analysis with HDRB´s acc. to model 1. The paths seem quite reasonable but we have to realize that they cannot be verified by test results as such results are not available acc. to our knowledge.

Figure. 11 shows the comparison for model 2. It can be seen that the elasto-plastic spring (1b) does not cover adequately the increase of the hysteresis for higher loads.

For model 3 (elastic spring plus damper), a direct comparison of the computational model with the test results is not possible. For the assessment of the quality of model 2 and 3, we consider the results of parameter studies on the 3DOF-oscillator of Figure. 3. The standard for the quality of these models are the results obtained with model 1, which are postulated to be correct, because model 1 describes the test results of the bearings most exactly.

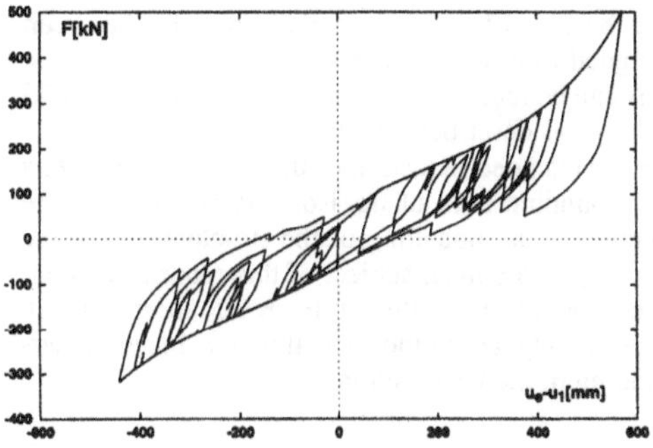

Figure 10: Loading and unloading paths of isolators during an earthquake as result of a time history analysis

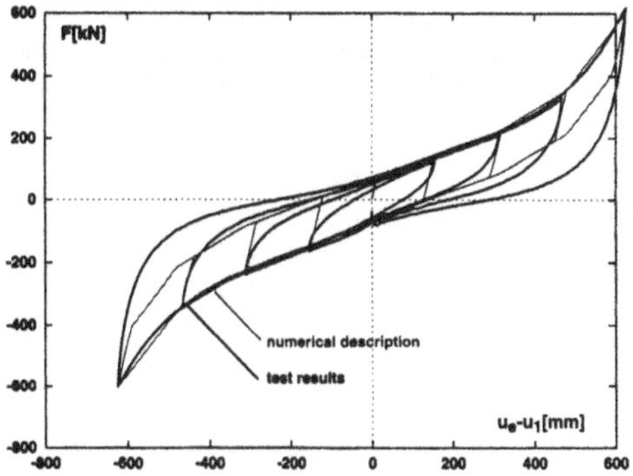

Figure. 11: Comparison of test results and numerical description of HDRB´s by model 2

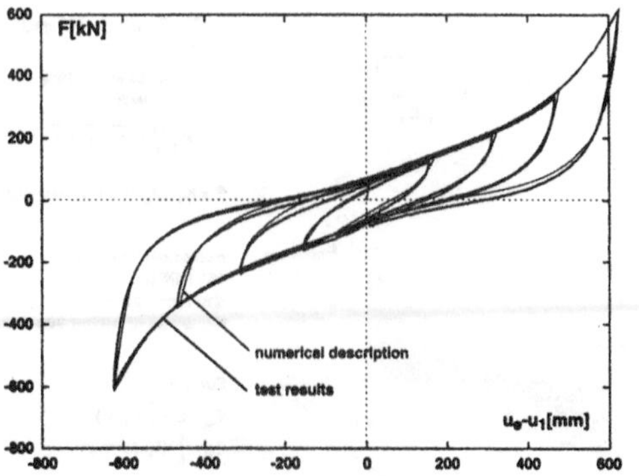

Figure. 9: Comparison of test results and numerical description of HDRB´s by model 1

3 RESULTS OF TIME HISTORY ANALYSES (PGA = 0.75 G)

Our parameter studies are based on an artificial time history fitting the response spectrum given in Eurocode 8 for soil type B (Figure. 12). The response value for a non-isolated tank is given by the period of the inner steel tank of 0.3 s to about 2.5 and for an isolated structure with a period of 2.5 s to about 0.4, both for a damping ratio of 5%. In reality the steel inner tank has a lower damping ratio and therefore a higher response, whilst for the isolated system we aim at damping ratios much higher than 5%.

Additionally we consider the natural acceleration time history of Santa Cruz / Corralitos earthquake. For the period of 2.5 s, which corresponds to an iso-

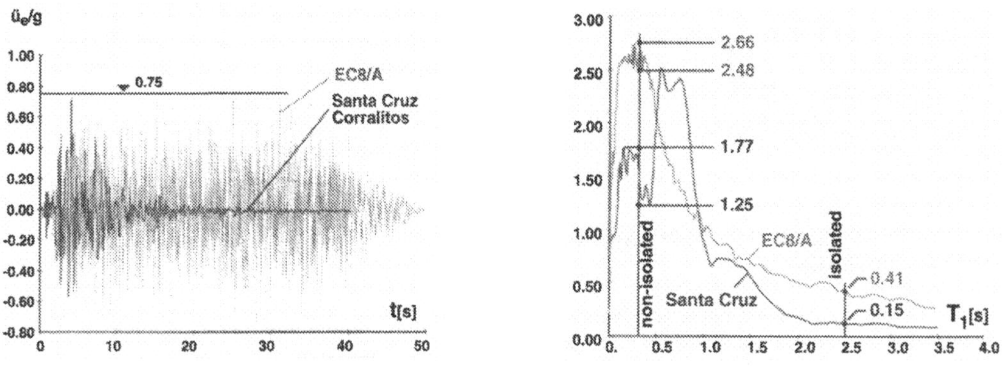

acceleration time histories (PGA = 0.75g) **response spectra for ξ_1= 5%**

Figure. 12: Acceleration time histories and response spectra for typical earthquakes

lated system, the response is only 0.15 for this earthquake. Both time histories have been scaled to a PGA of 0.75 g for this parameter study.

Figure. 13 show the displacement history of the isolators for the first 20 s of the earthquake as result of the time history analysis of the 3DOF-oscillator acc. to figure. 5. The plotted values $u_e - u_1$ describe the shear deformation of the isolators.

The results are quite similar for the three models. We recognize clearly the period $T_1 = 2.5$ s, which was intended in the design of the base isolation. The results of model 1 are postulated to be the correct scale. There is practically no difference of model 1 and model 3, the maximum being 0.30 m, which had been assumed for the evaluation of stiffness and damping ratio of model 3. Contrarily the model 2 yields in general smaller deformations.

With regard to overturning of the inner tank as main design criterion, the acceleration $\ddot{u}_2$ of the inner tank and the liquid mass m_2 is the most important result. It is shown in Figure. 14 for model 1 and model 3. We recognize again the period of 2.5 s given by the isolators. The maximum acceleration is 1.3 m/s² for model 1 and 1.6 m/s² for model 3 with a damping ratio of 15%.

The higher acceleration of model 3 is due to an undesired transfer of forces from the subsoil into the structure by the damper. This can be seen also from the more pronounced peaks, which occur with the period of the inner tank. After the end of excitation, that is after 50 s, the oscillation comes to an end much earlier for model 3 than for model 1. The reason is that in model 3 the damping ratio of 15%, which has been defined for the maximum displacement, is kept constant also for small displacements. In reality the damping of the bearings is less for small displacements. This effect is included only in model 1 and 2.

From a scientific point of view, the model 3 has some shortcomings compared with the postulated

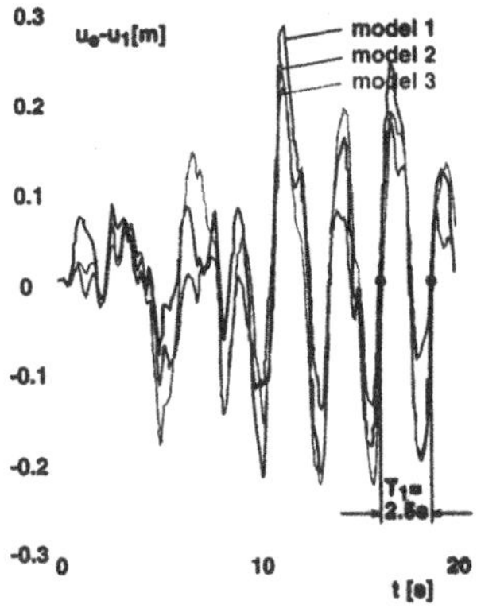

Figure. 13: Displacement history of isolators for an artificial acceleration time history (EC 8 / soil type A / PGA = 0.75 g)

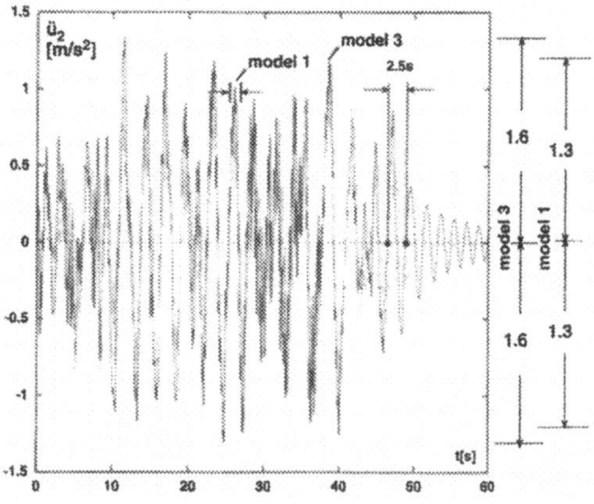

Figure. 14: Acceleration history of inner tank for model 1 and 3

reality of model 1. However, the deviation of the acceleration of the inner tank – 1.6 m/s² instead of 1.3 m/s² – remains within the accuracy, which can be reached within such a dynamic analysis. Therefore the model 3 is also acceptable for the practical design.

4 DISCUSSION OF RESULTS WITH REGARD TO LNG-STORAGE-TANKS

In Figure. 15 we see the most important dynamic responses of the tank system acc. to figures. 1 and 2, that is the maximum acceleration $\ddot{u}_2$ of the inner tank related to the PGA, in dependence of the damping ratio ξ_1 for model 3. The influence of ξ_1 is much more pronounced for the artificial time history acc. to EC 8 than for the natural time history of the Santa Cruz earthquake.

In Figure. 16, the dynamic response for a time history acc. to EC 8 / A is shown in dependence of the modelling of isolators. For a PGA of 0.75 g, model 3 yields a higher response than model 1 (1.6 m/s² instead of 1.3 m/s²). However, for a PGA of 0.4 g, the opposite tendency was found. Nevertheless, this figureure confirms the earlier statement that the three models of isolators are equivalent for the use in the practical design. Although the model 3 has some shortcomings from a theoretical point of view, it allows the most simple description of the HDRB's in the dynamic analysis.

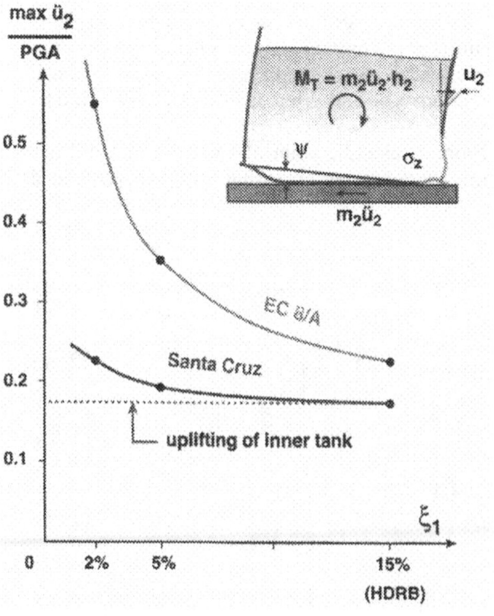

Figure. 15: Accelerations and overturning moments of inner tank depending on the damping ratio $\xi 1$ (for model 3)

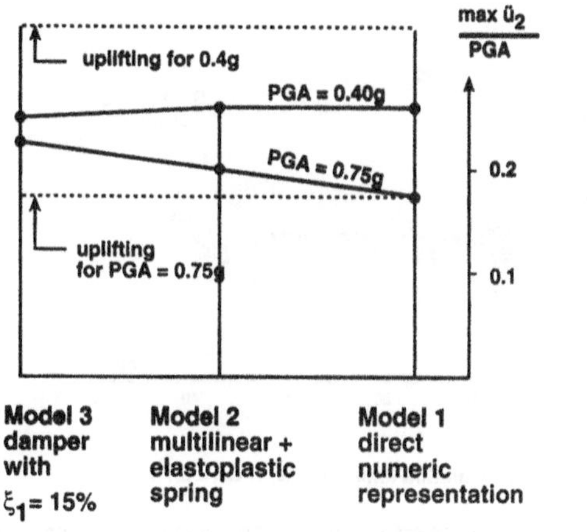

Figure. 16: Accelerations and overturning moments of inner tank depending on modelling of isolators (for $\xi 1$ = 15%, time history acc. to EC 8 / A)

The meaning of these results for the design of the inner tank can be seen from the dotted horizontal lines in Figure. 15 and Figure. 16. They mark the level, where uplifting and elephant footing occur. This is the case for an acceleration $\ddot{u}_2$ = 1.35 m/s².

That means that the inner tank cannot withstand the higher PGA of 0.75 g with a sufficient safety margin, at least, if the artificial time history acc. to EC 8

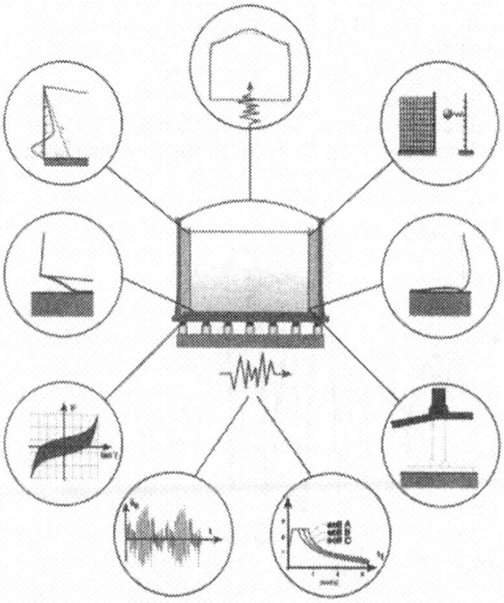

Figure. 17: Features to be delt with in the seismic design of an LNG storage tank

is assumed. In this case the base isolation should be improved by increasing the thickness of rubber.

However, an earthquake with PGA = 0.4 g could be sustained well with the base isolation assumed within this case study.

5 CONCLUSIONS

There are several possibilities for the numerical description of the damping characteristics of the seismic base isolation. The most simple model is an elastic spring combined with a velocity-dependent damper. Despite its theoretical shortcomings it is sufficient with regard to the accuracy which can be reached in any numerical analysis.

Figure. 17 shall address the complexity of the earthquake analysis of an LNG-storage tank. The appropriate modelling of the hysteresis of the isolators is only one of many tasks. Equally important are

- the modelling of the total system of the structure with regard to the individual design question
- the definition of the seismic input from the subsoil into the structure and
- the detailed design of the critical parts of the structure, for example the bottom corner of the inner tank.

These items cannot be treated separately. A reliable design requires the consistent consideration of all these features.

REFERENCES

Seismic Design of Storage Tanks: Recommendations of a Study Group of the New Zealand National Society for Earthquake Engineering, December, 1986.

Kelly, J.M.: *Earthquake-Resistant Design with Rubber*, Springer, London, 1993.

Bomhard, H.; Stempniewski, L.: LNG Storage Tanks for Seismically Highly Affected Sites, *POST SMiRT Conference*, Capri, 1993.

NEHRP Recommended Provisons For Seismic Regulations For New Buildings, 1994.

Malhotra, P.K.; Veletsos, A.S.: Uplifting Response of Unanchored Liquid-Storage Tanks. In: *Journal of Structural Engineering*, Vol. 120, No. 12, p. 3525-3547, December, 1994.

Eurocode 8: Design Provisions for Earthquake Resistance of Structures, Part 1, General Rules, Seismic Actions and General Requirements for Structures, ENV 1998-1-1, May 1994.

DYWIDAG: Performance and Design Criteria for Base Isolated Buildings and Structures on the Example of a Liquefied Natural Gas-Storage-Tank. Brite EuRam research project, 1996.

Baumann, Th.; Böhler, J.: Engineering Aspects towards Seismic Base Isolation. *POST SMiRT Conference*, Taormina, 1997.

Baumann, Th.; Böhler, J.: Performance and Design Criteria for Base-Isolated Buildings and Structures on the Example of Liquefied Natural Gas-Storage-Tanks, *Structural Engineers World Congress*, San Francisco, 1998.

Baumann, Th.; Lieb, M.; Böhler, J.: Sensivity of Seismic Response with Regard to the Computational Modelling of the Nonlinear Characteristics of Isolators. Seismic, Shock and Vibration Isolation (PVP Vol. 379), San Diego, 1998.

Douglas, H.R.; Maurer, H.: Outer Prestressed Concrete Tanks. A Vital Component in the Safety of LNG-Terminals. *LNG-Journal*, June 1998, Nelton Publications, England, 1998.

Constitutive Models for Rubber, Dorfmann & Muhr (eds) © 1999 Taylor & Francis ISBN 90 5809 113 9

Finite element analysis on bolster springs for metro railway vehicles

R.K.Luo, W.X.Wu & W.J.Mortel
Metalastik, Leicester, UK

ABSTRACT: The bolster springs investigated here were used in pairs, fitted at an angle to a bolster beam system for metro railway vehicles. The product has been used for over thirty years but recently some springs have shown splits and degradations. A finite element analysis is conducted to identify the causes. Three normal load cases (tare, mean and laden load) and three combined load cases were investigated. A stiffness curve from the FE model was compared with test results, which has shown good agreement. Natural frequencies and mode shapes were analysed. Two potential degradation areas from the combined load cases were located. The critical areas found from the stress analysis were the same as those exhibited on the actual spring components in service. It is shown that FEA method can play an important role at design stage and for failure analysis.

1 INTRODUCTION

Rubber springs are important components in railway vehicles. They can be used for both primary or secondary suspension systems. The bolster springs investigated here were used in pairs, fitted at an angle to a bolster beam system for metro railway vehicles. The product has been used in the same application over thirty years, however a refurbishment for the vehicles was coincident with a sudden increase in visual degradation on the product not previous experienced. Therefore a finite element analysis was carried out to identify the causes. A component of bolster springs is shown in Figure 1.

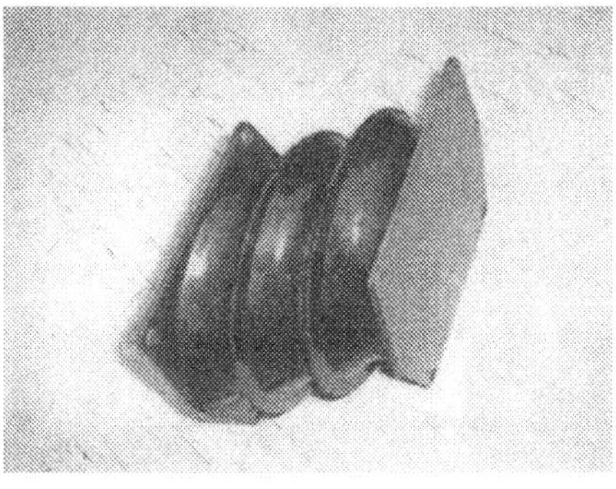

Figure 1. A bolster spring component.

In principle, there are four steps in this systematic analysis procedures:

1. Specification of material properties and measurement of component stiffness.
2. Calculation of component stiffness and verification.
3. Stress evaluation at critical areas.
4. Dynamic behaviour investigation.

The following sections will describe the above steps in detail.

2 MATERIAL PROPERTIES AND STIFFNESS MEASUREMENT

There are several hyperelastic material models which are commonly used to describe rubber and other elastomeric materials based on strain energy potential. Strain energy potential can be expressed as the following polynomial series:

$$U = \sum_{i+j=1}^{N} C_{ij}(\bar{I}_1 - 3)^i(\bar{I}_2 - 3)^j + \sum_{i=1}^{N}\frac{1}{D_i}(J_{el} - 1)^{2i} \quad (1)$$

where C_{ij} and D_i are temperature dependent material parameters, J_{el} is the elastic volume

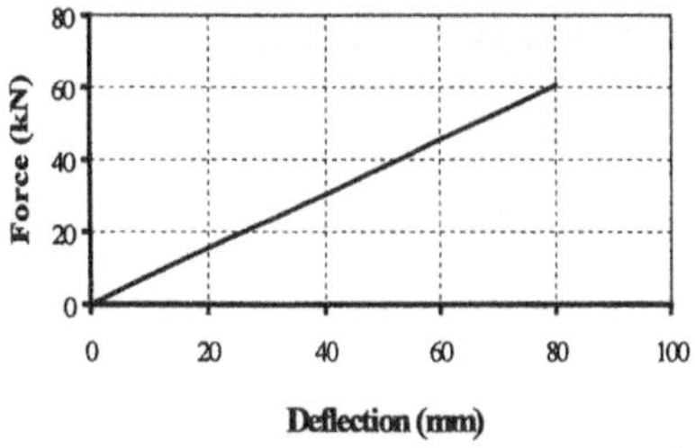

Figure 2. A typical stiffness curve measured.

strain, $\bar{I}_1$ and $\bar{I}_2$ are strain invariants. If N=1, the polynomial formulation represents the Mooney-Rivlin hyperelasticity model. The energy potential is as follows:

$$U = C_{10}(\bar{I}_1 - 3) + C_{01}(\bar{I}_2 - 3) + \frac{1}{D_1}(J_{el} - 1)^2 \quad (2)$$

The Mooney-Rivlin form was used to model the bolster spring components. The material constants in equation (2) were determined from experiments.

The stiffness of the bolster spring component was also measured in our test laboratory. During the test the components were arranged in pairs, fitted at an angle to the vertical axis. The force was applied through the loading head. The test data were recorded using a computer. A typical stiffness curve is shown in Figure 2.

3 FINITE ELEMENT ANALYSIS

3.1 *Finite element model*

It is only necessary to model one component because the layout of the components and the load are symmetrical. Three rubber layers and two layers of steel interleaves were modelled using 20-noded solid hexahedral elements. For convenience the co-ordinate system used was rotated at an angle equal to the fitted angle against vertical axis. The number of total elements is 832 and the total degrees of freedom is 64,149. The finite element model is shown in Figure 3.

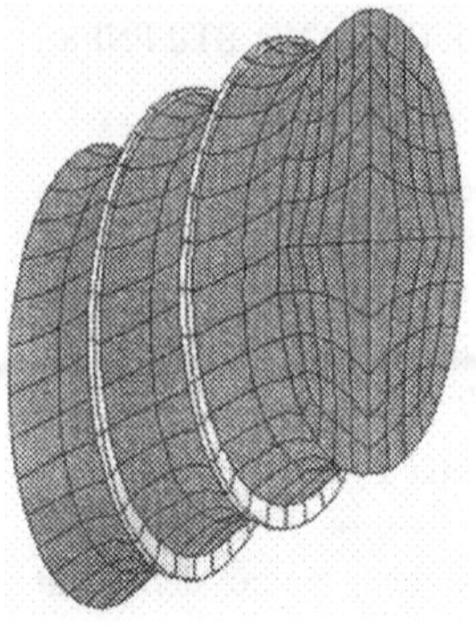

Figure 3. Finite element model of a bolster spring component.

The lower bonded surface in the bottom layer of the component was fixed in x, y and z three directions. The load was applied to the top bonded surface in the top layer of the component. Three normal load cases and three-combined load cases were investigated using ABAQUS software. The six load cases are listed below:

Table 1. Three normal load cases.

	U1	U2	U3
Case 1	-17.55	-57.38	0
Case 2	-20.47	-66.94	0
Case 3	-23.39	-76.5	0

Table 2. Three combined load cases.

	U1	U2	U3
Case 1	-17.55	-57.38	-10
Case 2	-20.47	-66.94	-10
Case 3	-23.39	-76.5	-10

The above load cases can be classified into two types: the first type is a design case (tare = 60 mm working displacement, mean = 70 mm and laden = 80 mm); the second type is the design case plus 10 mm lateral displacement.

3.2 *Stiffness calculation and verification*

A stiffness curve comparison was made between the simulation and the test in order to verify the FE model, as shown in Figure 4. The deflection range compared is up to 80 mm which is equivalent to laden load. There is a good agreement between the calculation and the test. A typical deformed shape of the component is shown in Figure 5, which is the same as that observed from the test. Therefore the stress profile derived from the model can be used to analyse the stress distribution of the component.

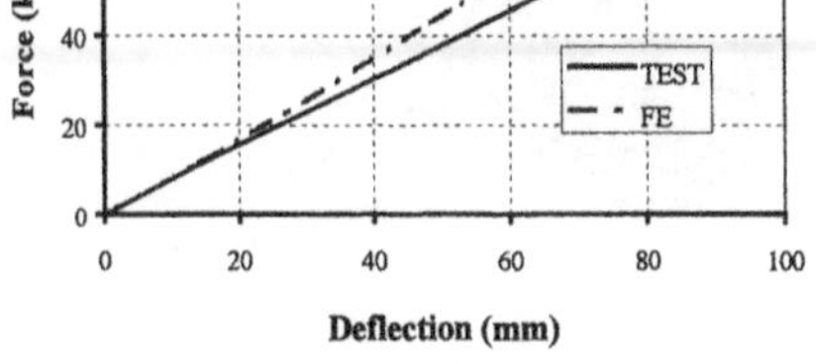

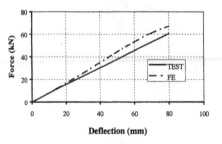

Figure 4. Stiffness curve comparison.

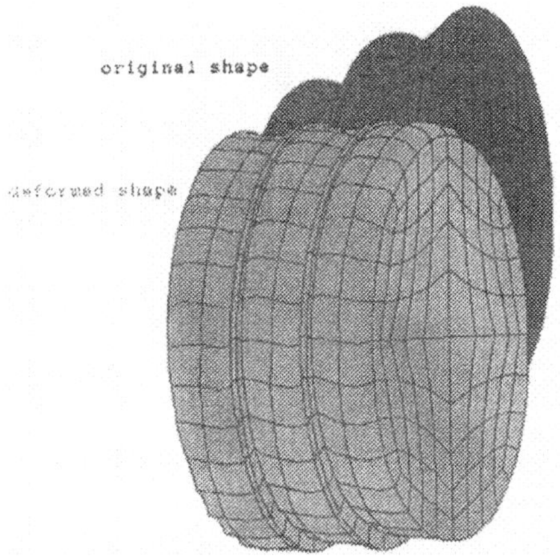

Figure 5. A typical deformed shape.

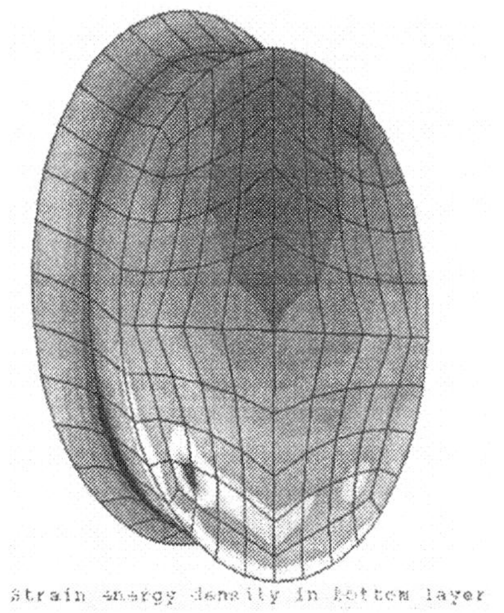

Figure 6. A typical NSED profile in bottom layer.

3.3 *Stress analysis*

Table 3 shows distribution of Nominal Strain Energy Density (NSED) at different load cases. The middle layer is taken as the calculation base. It is shown that at any load case the critical layer is the bottom layer, followed by the top layer and the middle layer. Actually the degradation areas were found mainly in the bottom and top layer of the failed components in service. This phenomenon has verified the calculation. The maximum value in NSED happened in the bottom layer at combined displacement case 2 in which NSED value from 1.19 increases to 1.51, compared with 1.3 in the top layer and 1.15 in the middle layer. Two potential failure areas caused by a typical combined displacement case are found, as shown in Figure 6. These areas are located along the rubber section edge near the metal plate and inside the robber section about 8 mm away from the edge respectively. In order to find out which stress component causes the failure different stress investigations were conducted. Figure 7 shows the fringe plot of maximum principal stress in the bottom layer, taken from combined displacement case 2. The figure indicates one of the failure areas, inside the rubber section about 8 mm away from the edge. Figure 8 shows a picture taken from a slice sample of a degraded spring component in service. A crack shown in Figure 8, having about 10-mm length, is also about 8 mm away from the edge. It is clear that the critical area based on calculation is matched well with the cracked area observed. The crack may have been caused by the excessive principal tensile stress under combination of the compression and shear. Since two potential failure areas were identified from the

Table 3 Nominal Strain Energy Density Comparison.

Tare load			**Combined displacement case 1**		
Bottom layer	Middle layer	Top layer	Bottom layer	Middle layer	Top layer
1.13	1.00	1.09	1.38	1.13	1.34

Mean load			**Combined displacement case 2**		
Bottom layer	Middle layer	Top layer	Bottom layer	Middle layer	Top layer
1.19	1.00	1.06	1.51	1.15	1.30

Laden load			**Combined displacement case 3**		
Bottom layer	Middle layer	Top layer	Bottom layer	Middle layer	Top layer
1.22	1.00	1.11	1.45	1.14	1.34

– Max Shear stress in bottom layer

simulation and one of them is verified, investigation of the other one is needed to complete the validation. Figure 9 shows the distribution of the maximum shear stress. The critical area is located along the edge near the metal plate. Also another high stress area is found at the same place as the maximum principal stress appears. A photo taken from a degraded component is displayed in Figure 10. Apparently there is a slip along the edge of the rubber section near the metal plate. The pattern of the slip is identical to that of maximum shear stress distribution in Figure 10. It is indicated that the slip along the edge of the rubber section near the metal plate may have been caused by the excessive maximum shear stress. Based on the analysis further investigation may be needed to either modify the design to accommodate the combined loads or improve the dynamic environment of the spring components.

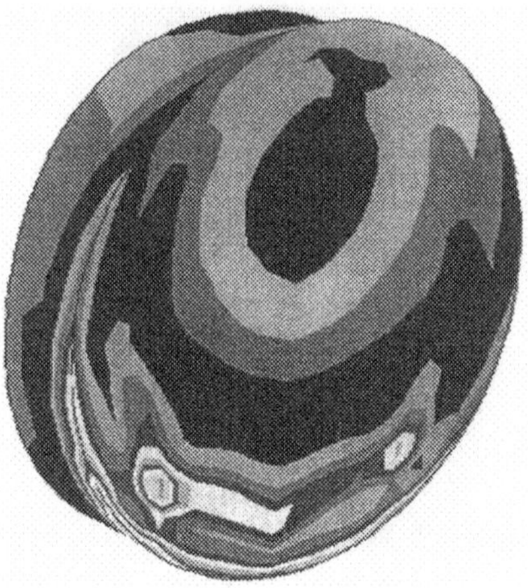
Figure 9. A shear stress profile in bottom layer.

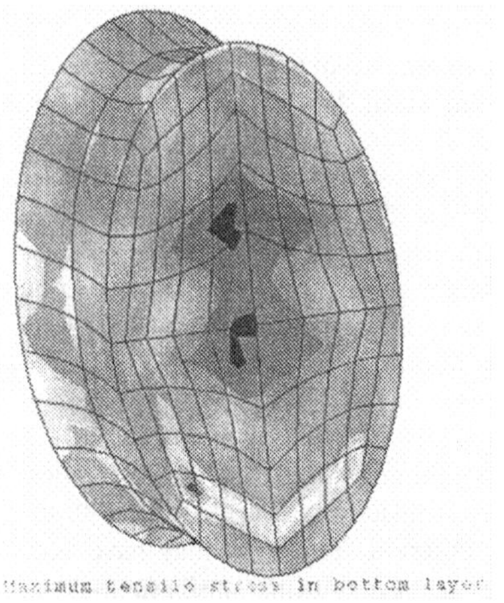

Figure 7. A maximum tensile stress profile in bottom layer.

Figure 8. A slice sample of a component.

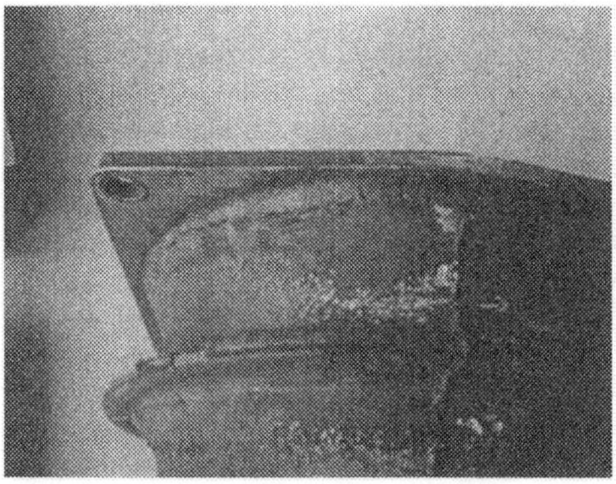
Figure 10. A slip along the edge.

3.4 *Free vibration analysis*

The most straightforward type of dynamic analysis is the determination of natural frequencies and mode shapes. This type of calculation can give considerable insight into the dynamic behaviour of the structure itself and provide information for validating the whole system. The equations of motion for the free vibration of an undamped system are:

$$[M]\{\ddot{\Delta}\} + [K]\{\Delta\} = \{0\} \qquad (3)$$

where [M] is the system mass matrix, [K] is the system stiffness matrix, $\{\Delta\}$ is the system displacement vector, $\{\ddot{\Delta}\}$ is the system acceleration

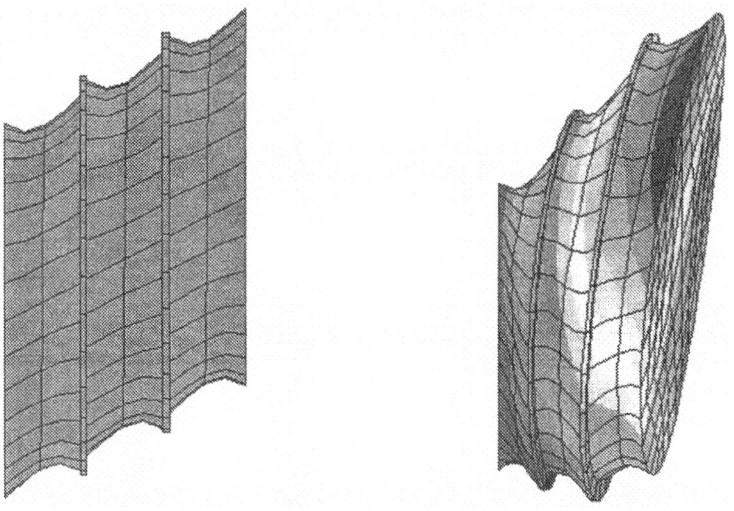

(a) Undeformed and deformed component in vertical plan.

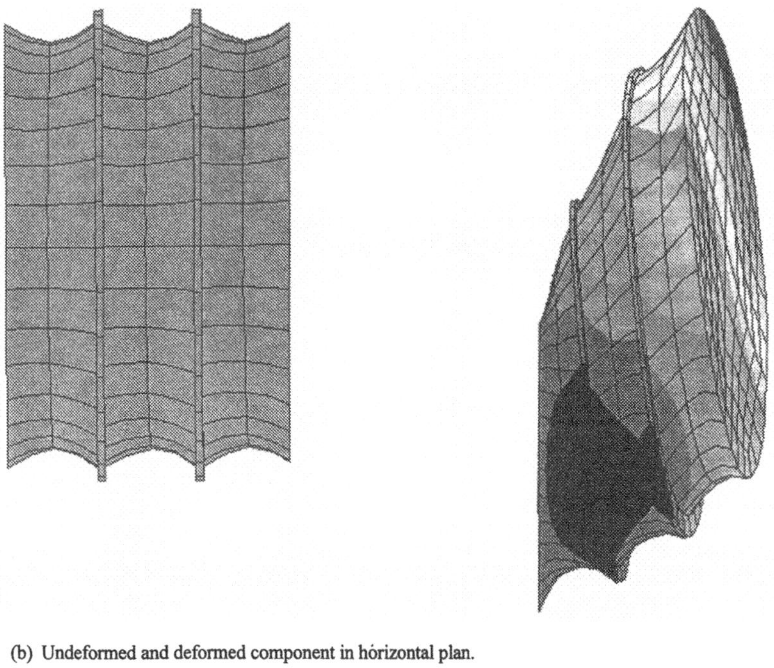

(b) Undeformed and deformed component in horizontal plan.

Figure 11. Undeformed and deformed component.

vector. The problem of vibration analysis consists of determining the conditions under which equation (3) will permit motions to occur.

A solution exists if a determinant is zero. That is

$$\| [K] - \omega^2 [M] \| = 0 \quad (4)$$

Equation (4) is called the frequency equation of the system. Expanding the determinant will give an algebraic equation of the *nth* degree in the frequency parameter ω^2 for a system having n degrees of freedom. The n roots of this equation $(\omega_1^2, \omega_2^2, \ldots\ldots, \omega_n^2)$ represent the frequencies of the

Table 4. Natural frequencies of the component

Frequency	1	2	3	4	5
Hz	34.8	40.1	49.3	89.8	101

n modes of vibration, which are possible in the system. Table 4 lists the first five natural frequencies ranged from 34.8 Hz to 101 Hz. The fundamental mode shape is dominated by the lateral movement, as shown in Figure 11. It is indicated that this mode is easily excited by the external dynamic movements, which could be added on the top of the normal design loads. In fact from 3.3 the critical load case is the combination between the vertical and lateral movements. The contribution to NSED from lateral movement is very significant.

4. CONCLUSIONS

Two potential degradation areas from the combined displacement cases are found. One is along the edge of rubber layer near the metal plate. The other is inside the rubber section about 8 mm away from the edge near the metal plate. The critical areas located are the same as those exhibited on components running in service. The degradation modes identified are the tensile and the shear, under the conditions modeled. To achieve the displacement parameters which match this high stress pattern, vehicle displacements in a lateral direction, must have occurred. Further investigation is needed from both design and application side to improve the performance of the bolster spring components.

REFERENCES

Ogden, R.W. 1984. *Non-linear elastic deformations*. Chichester: Ellis Horwood Limited

Turner, D.M. 1995. *Problems with equations for rubber elastic behaviour*. Rubber Engineering Using FEA, IMechE London, 4 April 1995

Clough, R.W. & Penzien J. 1993. *Dynamics of structures*. Singapore:McGraw-Hill

Luo, R.K., Gabbitas B. L. & Brickle B.V. 1994. *Fatigue life evaluation of a railway vehicle bogie using an integrated dynamic simulation*. Proc Instn Mech Engrs Vol 208:123-132

Hibbitt, Karlsson & Sorensen, Inc. 1997. *ABAQUS theory manual*

Constitutive Models for Rubber, Dorfmann & Muhr (eds) © 1999 Taylor & Francis ISBN 90 5809 113 9

Computational simulation of the vulcanization process in rubber profile production

M.André & P.Wriggers
Institute for Structural and Computational Mechanics, University of Hannover, Germany

ABSTRACT: The aim of this work is to develop a computational tool to simulate the thermal and mechanical behavior of rubber material during vulcanization. A viscoelastic-viscoplastic constitutive model is used in which the material parameters are dependent on temperature and degree of crosslinking. Additional pseudoplastic strain rates are introduced in order to account for new crosslinks. The temperature field, the vulcanization process and the mechanical deformations of a rubber sample are calculated using the finite element method. A staggered scheme is developed to solve the coupled thermomechanical system of equations.

1 INTRODUCTION

In rubber profile extrusion there is an increasing interest in predicting the deformation of profile shapes during the vulcanization process. Regarding the design of extrusion dies and the adjustment of heating lines, a knowledge of temperature, state of curing and deformation, as well as thermal and mechanical stresses, is very important and enables faster product development. Finite Element computations can play a complementary role to laboratory tests and help to avoid expensive testing.

The process of vulcanization usually starts after the rubber profile has left the extrusion die and has been formed to the desired shape. One then tries to heat up the rubber material as fast as possible in order to start the curing process and to avoid further residual deformations. Nevertheless, the rubber profile will usually be influenced by thermal and gravity loads and internal stresses. Therefore further deformations often can not be avoided. The aim of this work is to predict these deformations and develop a tool to optimize profile shapes and process parameters.

During the vulcanization process the material can either be heated by convective heat transfer in a hot air stream or by microwave radiation. Both possibilities are considered in this work.

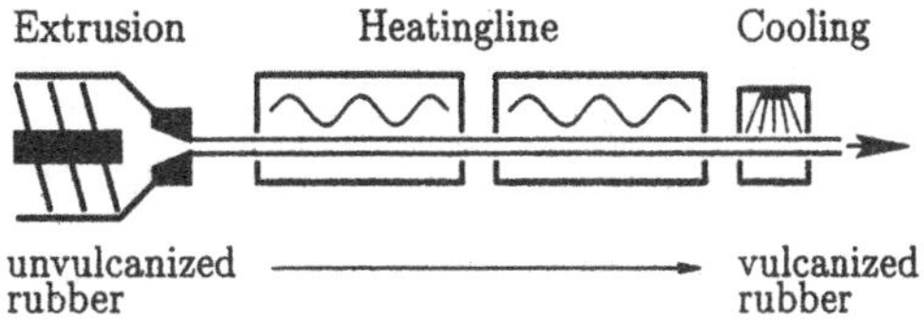

Figure 1: Extrusion and heating in rubber profile production

We use the finite element method with an implicit euler time integration scheme to calculate the temperature field as well as the mechanical deformations.

Once the time-dependent temperature field is known one can calculate the state of vulcanization by integrating the vulcanization rate. Mechanical properties such as elastic stiffness, viscosity and yield stress depend directly on temperature and state of vulcanization. Using a viscoelastic-viscoplastic constitutive model, described shortly below, it is possible to determine the mechanical stresses and deformations during the process of vulcanization.

2 CONSTITUTIVE MODEL

Unvulcanized rubber consists of a large amount of long polymer chains stochastically distributed and orientated. When applying a strain to such a rubber sample the chains will be elongated, which results in a reaction force. This is considered an elastic response. After some time the chains reorientate within the material due to the elongation of some polymer chains and thus the stress will decrease. This is known as relaxation phenomena (e.g. Ferry 1980). As long as the chains are not forced to change their position but only their orientation, we obtain a reversible deformation. This viscoelastic strain disappears for sufficiently long time scales when the body is unloaded.

In addition to this, unvulcanized or only lightly crosslinked rubber allows the movement of polymer chains. This happens when a certain yield stress is exceeded and results in a plastic deformation.

Plastic deformation is intentionally used when forming the rubber material into its desired shape, but usually is undesired during further processing. Nevertheless, we have to take into account that further plastic deformations during the first seconds of the vulcanization process are possible, since the heated material is extremely soft at this stage.

We are interested in mechanical deformations of the rubber profile only after having left the extrusion die. We assume that all deformations are small and use linearized strain measures. All the preceding effects caused by the extruder, and its extrusion die, are neglected in this work. Also, the viscoelastic effects are approximated by a linear model, since the strain rates are comparatively small.

To simulate the described process we have to consider viscoelastic, viscoplastic and elastic effects (Kaliske & Rothert 1997). Also thermal expansion has to be included since temperature changes are quite large. Under the assumption of small deformations, we obtain a material model in which deformations can be described by an additive split corresponding to the rheological model in Figure 2. This yields

$$\boldsymbol{\epsilon} = \boldsymbol{\epsilon}^e + \boldsymbol{\epsilon}^{th} + \boldsymbol{\epsilon}^{vp} + \boldsymbol{\epsilon}^{ve} + \boldsymbol{\epsilon}^{pl} \tag{1}$$

and

$$\boldsymbol{\epsilon}^{pl} = \boldsymbol{\epsilon}^{pl,e} + \boldsymbol{\epsilon}^{pl,ve}\,. \tag{2}$$

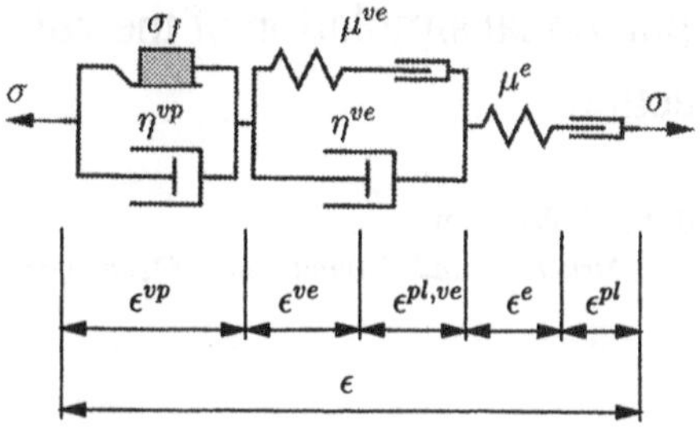

Figure 2: A five parameter rheological model with additional pseudoplastic strain-components. For clarity $\boldsymbol{\epsilon}^{th}$ is not shown.

Here $\boldsymbol{\epsilon}^e$ denotes elastic strains, $\boldsymbol{\epsilon}^{th}$ thermal strains, $\boldsymbol{\epsilon}^{vp}$ viscoplastic strains, $\boldsymbol{\epsilon}^{ve}$ viscoelastic strains and $\boldsymbol{\epsilon}^{pl,e}$ and $\boldsymbol{\epsilon}^{pl,ve}$ pseudoplastic strains from the vulcanization. The latter are defined and explained in the next section.

The main components of this model are described by the following equations:

Elastic part: The elastic behavior is represented by a *Hookean* spring element. Since deformations are small, *Hooke's* law is sufficient to calculate the stresses from the elastic strains.

$$\boldsymbol{\sigma} = \boldsymbol{C}^e : \boldsymbol{\epsilon}^e\,. \tag{3}$$

Here $\boldsymbol{C}^e$ is the fourth order elastic stiffness tensor and $\boldsymbol{\sigma}$ the Cauchy stress tensor. For known changes in temperature $\Delta\Theta$ the thermal strains can easily determined by

$$\boldsymbol{\epsilon}^{th} = \mathbf{1}\alpha^{th}\Delta\Theta\,, \tag{4}$$

where $\mathbf{1}$ is the identity tensor.

Viscoelastic part: Viscoelastic strain rates are described by

$$\dot{\boldsymbol{\epsilon}}^{ve} = B_1\boldsymbol{\epsilon}^e - B_2\boldsymbol{\epsilon}^{ve} - \dot{\boldsymbol{\epsilon}}^{pl,ve} \tag{5}$$

with

$$B_1 = \frac{\mu^e}{\eta^{ve}} \quad \text{and} \quad B_2 = \frac{\mu^{ve}}{\eta^{ve}}\,. \tag{6}$$

This viscoelastic model is known as *Voigt* element. The strain rate $\dot{\boldsymbol{\epsilon}}^{pl,ve}$ will be defined in the following section.

Viscoplastic part: When the applied load exceeds a certain limit, plastic strain rates occur (Simo & Hughes 1988). Here we use a von Mises yield criterion with a viscous flow rule. The yield function F is

$$F = \sqrt{J_2'} - k \begin{cases} < 0 & \text{no plastic flow} \\ \geq 0 & \text{plastic flow} \end{cases} \quad (7)$$

where

$$J_2' = \frac{1}{2} \boldsymbol{s} : \boldsymbol{s} . \quad (8)$$

The tensor $\boldsymbol{s}$ is the deviatoric part of the stress tensor $\boldsymbol{\sigma}$ and J_2' denotes the second invariant of $\boldsymbol{s}$. Using an associated viscous flow rule we determine the viscoplastic strain rates to

$$\dot{\boldsymbol{\epsilon}}^{vp} = \gamma^{vp} \frac{\partial F}{\partial \boldsymbol{\sigma}} = \lambda^{vp} \boldsymbol{s} . \quad (9)$$

Putting these three components together leads to a five parameter model.
We remark that all material parameters needed for this rheological model are dependent on temperature and on the crosslink density and thus on the state of vulcanization. It is reasonable to assume that the elastic and viscoelastic stiffnesses are increasing with progressive crosslink density. Also viscosity will increase due to the polymer chains having less possibilities to move with increasing crosslink density. The same reason leads to an increasing yield stress. Regarding a fully cured rubber material we assume that no plastic deformations can occur. This is equivalent to an infinite yield stress.
These considerations and assumptions on temperature influences (Aklonis & MacKnight 1983) together with experimental results lead to functions for the material parameters μ^e, $\mu^{ve}, \eta^{ve}, \eta^{vp}$ and σ_f. Materialdependent constants within these functions can be determined by a set of experiments with different temperatures and different crosslink-densities.

3 VULCANIZATION PROCESS

As mentioned, the crosslink density has a significant influence on the material stiffness. Therefore, it is possible to determine the development of the vulcanization process by measuring the elastic stiffness of a rubber sample. This is usually done by using a rheovulcameter, by which a rubber sample is heated to a certain temperature Θ and at the same time a periodic torsion $\varphi(\omega t)$ with frequency ω is applied. The measured reaction moment $M(t)$ is dependent on the stiffness of the material, which will increase with progressing vulcanization. It is common to plot the amplitude $\hat{M}(t)$ of the moment versus time t like shown in Figure 3 (Mark 1994). The result is

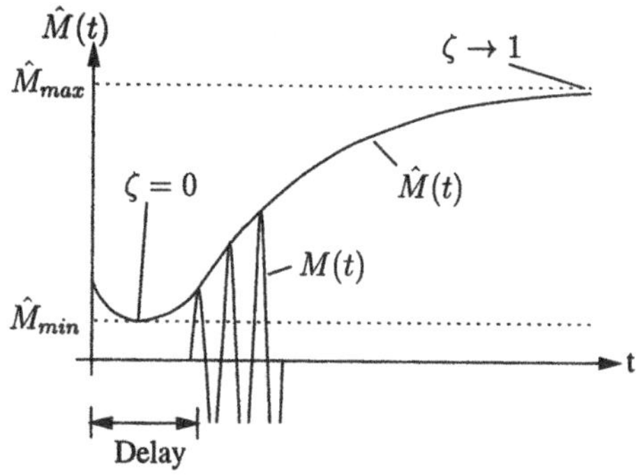

Figure 3: Rheovulcameter curve with delay

$$\zeta(t) = \frac{\hat{M}(t) - \hat{M}_{min}}{\hat{M}_{max} - \hat{M}_{min}} , \quad (10)$$

which serves as a measure for the state of the vulcanization process and the crosslink density. $\zeta = 0$ corresponds to a zero crosslink-density, i.e. the material is uncured, while $\zeta = 1$ is a completely cured material.
Note that the delay in vulcanization at the beginning of the vulcameter curve is the time the material needs to reach the desired temperature.
By comparing such vulcameter curves $\zeta(t)$, measured at different temperatures Θ, it can be shown that equation 11 is a good approximation for the vulcanization process. The evolution law is

$$\dot{\zeta} = \frac{\partial \zeta}{\partial t} = (1 - \zeta)\, \alpha^v , \quad (11)$$

with

$$\alpha^v = K_1 \exp\left(\frac{-K_2}{\Theta}\right) . \quad (12)$$

The constants K_1 and K_2 can be determined by vulcameter experiments and depend on the mixture of the rubber material.

4 STRAIN RATES DUE TO VULCANIZATION

During the vulcanization process, rubber changes its mechanical behavior. When the rubber is cured in a deformed and loaded state one realizes that the calculation of only (visco-)elastic and plastic strains, as described above, leads to incorrect results. An additional strain rate, call it pseudoplastic strain rate, has to be introduced.

Micro-mechanical motivation: Let us assume an unloaded, unvulcanized rubber sample consisting of several molecular chains in a stress free reference configuration (see Figure 4a). Applying a load to this body leads to elastic, and with increasing time, also to viscoelastic strains. The molecular chains are deformed and take a different orientation relative to each other in this deformed configuration. This causes a stress state inside the sample due to the elongation of the molecular chains. To make the considerations clear we exclude any viscoplastic strain rates at this point.
Now we start the curing of this rubber material. This means that crosslinks are built between the molecular chains. It is important to realize that these new crosslinks are unloaded in this deformed configuration (see Figure 4b).
After the material has been cured we unload it. It is obvious that the body will not move back into its reference configuration, otherwise it would not be stress free. Instead it will recover its elastic strains only up to the point where all the chains connected by the crosslinks are in equilibrium (Figure 4c). This leads to a new equilibrium configuration that is different from the reference configuration in Figure 4a.

Phenomenological motivation: Regarding the same experiment as mentioned above (curing in a deformed state) one observes the following effect: A rubber material will increase its elastic stiffness during the curing process. To avoid increasing stresses under constant strains one has to reduce the elastic strains in such a way that all stress changes due to crosslinking effects disappear. This can be done by additional pseudoplastic strains as already introduced in equations 1 and 2
Based on these considerations the evolution laws for $\epsilon^{pl,ve}$ and $\epsilon^{pl,e}$ are

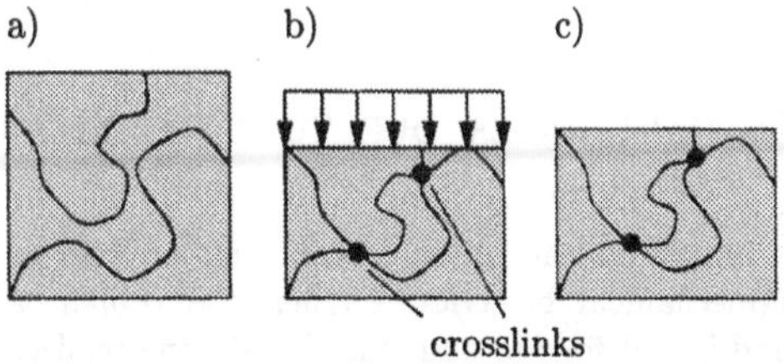

Figure 4: Deformed molecular chains during crosslinking

$$\dot{\epsilon}^{pl,ve} = B_3 \epsilon^{ve} \tag{13}$$

$$\dot{\epsilon}^{pl,e} = B_4 \epsilon^{e} \tag{14}$$

with

$$B_3 = \frac{\partial \mu^{ve}}{\partial \zeta} \frac{\dot{\zeta}}{\mu^{ve}} \tag{15}$$

$$B_4 = \frac{\partial \mu^{e}}{\partial \zeta} \frac{\dot{\zeta}}{\mu^{e}} . \tag{16}$$

Although the macroscopic effect of these strain rates is reminicent of plasticity, the physical reason is completely different and has nothing to do with such processes.

5 NUMERICAL IMPLEMENTATION

To simulate the vulcanization of usually complicated shaped rubber profiles we have implemented this constitutive model in the finite element program FEAP courtesy of Prof. R.L. Taylor.

5.1 *Temperature and Vulcanization Calculation*

As already mentioned, the temperature is the driving force for the vulcanization process. Therefore we need to calculate the time dependent temperature field $\Theta(\vec{x}, t)$ within the rubber profile.
If we assume that heat flow in direction of the profile axis is much smaller than in directions perpendicular to this axis we can reduce the problem to the two-dimensional case. Using the finite element method the temperature field can be calculated by introducing temperature Θ as degree of freedom in every node. Energy conservation with *Fourier* heat conduction and *Green's* Theorem yields to the weak form (Zienkiewicz & Taylor 1994)

$$\int_{\Omega} \left(-k\, \eta_{,i}\, \Theta_{,i} + \eta\, q - c\rho \dot{\Theta} \right) dV$$

$$+ \oint_{\bar{\Omega}} \left(\eta\, \alpha^c \left(\Theta - \Theta_U \right) \right) dA = 0 . \tag{17}$$

Here η denotes a test function, $c\rho$ the heat capacity and k the thermal conductivity of the material. q is a heat production term, e.g. through UHF-heating. Also inelastic strain rates causing dissipation thermal energy have to be included in this term but can be neglected in the following because of small strain rates.

The second integral in equation 17 describes the convective heat transfer to the surface by an environmental temperature Θ_U with heat transfer coefficient α^c.
With the iso-parametric finite element concept the test function η can be connected to a virtual temperature $\delta\Theta$ and one gets a system of equations

$$\mathbf{K}^\Theta \mathbf{v}^\Theta = \mathbf{f}^\Theta \tag{18}$$

with $\mathbf{v}^\Theta$ being the vector of unknown nodal temperatures, $\mathbf{K}^\Theta$ the tangent stiffness matrix and the load terms $\mathbf{f}^\Theta$.
When solving the transient temperature problem with an implicit time integration scheme, e.g. Euler backwards, the tangent stiffness matrix $\mathbf{K}^\Theta$ includes the damping matrix $\mathbf{D}^\Theta$ resulting from the time derivative $\dot{\Theta}$ in equation 17.
The knowledge of the temperature field enables the calculation of the local vulcanization rate according to equation 11 in every integration point. Using Euler backwards with time step size Δt this is

$$\zeta^{t+\Delta t} = \frac{\zeta^t + \alpha^v \Delta t}{1 + \alpha^v \Delta t}. \tag{19}$$

5.2 *Mechanical Calculation*

Strains $\boldsymbol{\epsilon}$, stresses $\boldsymbol{\sigma}$ and displacements $\boldsymbol{u}$ are determined by the constitutive equations, the linearized kinematics

$$\boldsymbol{\epsilon} = \frac{1}{2}\left(\operatorname{grad}\boldsymbol{u}^T + \operatorname{grad}\boldsymbol{u}\right) \tag{20}$$

and the equilibrium condition in the weak form

$$\delta\Pi = \int_\Omega (\boldsymbol{\sigma} : \delta\boldsymbol{\epsilon} + \boldsymbol{f}\cdot\delta\boldsymbol{u}) + \delta\Pi^a = 0 \tag{21}$$

in conjunction with boundary conditions and initial values by a standard iso-parametric finite element formulation. In equation 21 $\delta\Pi^a$ denotes the terms from external loads.
Using an implicit time integration scheme equilibrium has to be determined in every time step. This leads to a system of equations

$$\mathbf{K}^u \mathbf{v}^u = \mathbf{f}^u \tag{22}$$

Since the rate dependent material behavior results in a nonlinear system of equations this equilibrium state must be computed by an iteration, e.g. through a *Newton-Raphson* scheme. This usually requires the computation of the tangential stiffness $\mathbf{K}^u$ in every iteration step.

5.3 *Coupled Problem*

Regarding equations 18 and 22 it is possible to solve the whole system of equations in one iteration procedure according to

$$\underbrace{\begin{bmatrix} \mathbf{K}^u & \mathbf{K}^{u\Theta} \\ 0 & \mathbf{K}^\Theta \end{bmatrix}}_{\tilde{\mathbf{K}}} \underbrace{\begin{pmatrix} \mathbf{v}^u \\ \mathbf{v}^\Theta \end{pmatrix}}_{\tilde{\mathbf{v}}} = \underbrace{\begin{pmatrix} \mathbf{f}^u \\ \mathbf{f}^\Theta \end{pmatrix}}_{\tilde{\mathbf{f}}}. \tag{23}$$

Although the matrices $\mathbf{K}^\Theta$ and $\mathbf{K}^u$ are each symmetric the global tangent matrix $\tilde{\mathbf{K}}$ is obviously not symmetric because of the coupling terms $\mathbf{K}^{u\Theta}$. Using an unsymmetric solver usually leads to long solution times, especially for large FE-models.
Since dissipation is neglected because of the small strain rates, the thermal solution is decoupled from the mechanical equations. Thus we can solve the complete set of equations in two steps with symmetric stiffness using a staggered algorithm based on operator splitting.
First equation 18, corresponding to the lower part of equation 23 is solved in only one step since the thermal system is linear. Equation 22 (upper part of equation 23) is then solved by a Newton iteration. Both problems can be solved with symmetric stiffness. We remark that using this operator splitting avoids the need of calculating the coupling matrix $\mathbf{K}^{u\Theta}$. This algorithm leads to a significant improvement in speed compared with solving the whole set of coupled equations with unsymmetric Jacobian.

6 EXAMPLES

6.1 *Rheovulcameter Experiment*

First the finite element implementation is verified by a comparison with experimental data. Here we use rheovulcameter curves for a comparison. A 8-node brick element with the constitutive model described above is used for this calculation.
In this vulcameter experiment the unvulcanized rubber sample with initial temperature of $20°C$ is heated through the upper and lower surface with

Table 1: *Material data*

material density	ρ	$1.2\,g/cm^3$
thermal conductivity	k	$0.3\,W/mK$
specific heat capacity	c	$1.6\,J/(gK)$
thermal expansion coeff.	α^{th}	$1.0\cdot 10^{-5}\,1/K$

a temperature of $\Theta_U = 200°C$. Material data is chosen as noted in Table 1. A periodic torsion displacement $\varphi(t)$ is applied on this sample. The frequency of the torsion is $\omega = 314.16\,\frac{rad}{s}$ with an amplitude of $\hat{\varphi} = 0.5°$. The rheovulcameter testing chamber and the finite element discretization is shown in Figure 5. For reasons of efficiency only one fourth of the symmetric circular sample is calculated. Figure 6 shows the oscillating reaction moment calculated by the finite element model during the first 25 seconds. One can recognize clearly the initial decrease of material stiffness due to the rising temperature within the sample and then the rapid increase due to vulcanization.

A comparison between the amplitude measured in the experiment and calculated is shown in Figure 7.

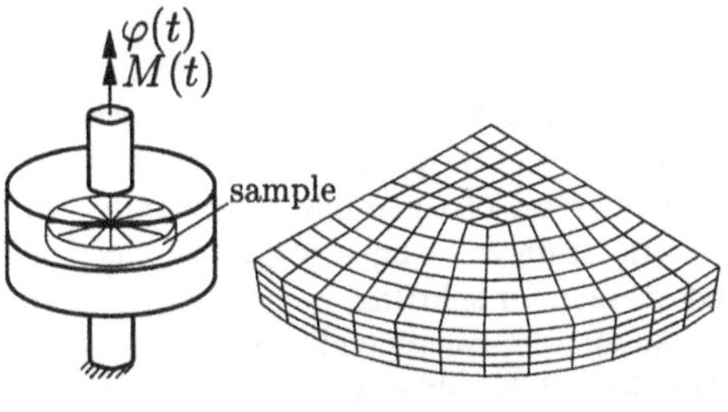

Figure 5: Rheovulcameter and finite element mesh of sample

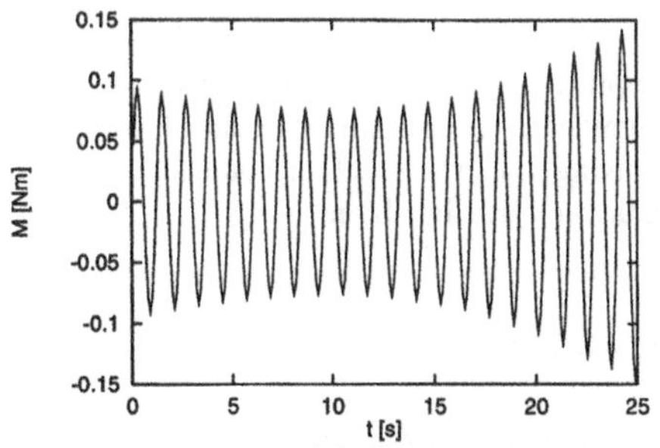

Figure 6: Reaction moment during vulcameter experiments

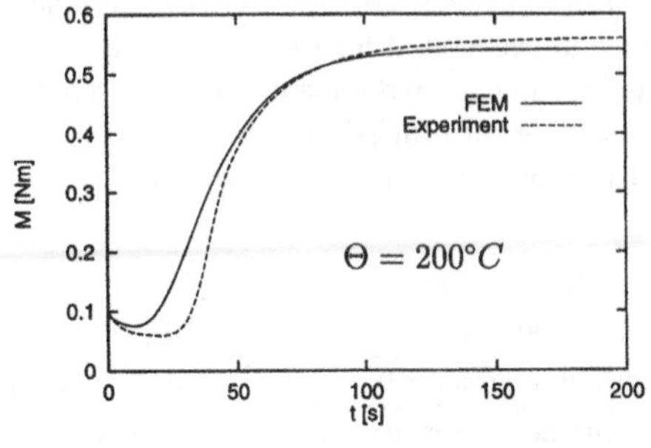

Figure 7: Comparison of rheovulcameter data and simulation

6.2 *Process Simulation*

We have used a two-dimensional, plain strain, four node finite element formulation with the same constitutive model and same material data as before to simulate the vulcanization process of a simple rubber profile. Figure 8 shows the initial shape of the profile. Here radius $r = 10mm$ and thickness $d = 2mm$. Because of symmetry only one half of the shape with symmetry boundary conditions is calculated.

The initial temperature is supposed to be homogenous at $100°C$. Convective heat transfer is assumed at the outer surface with a temperature of $\Theta_U = 250°C$ and a heat transfer coefficient of $\alpha^c = 40\,W/(m^2K)$. The sample is loaded only by its own weight.

Figure 9 shows the temperature field, the state of vulcanization and the deformation after a processing time of $5s$ and $20s$. The outline of the initial shape is plotted in these pictures for comparision. The displacement u of Point A is plotted versus time in Figure 10. The purely elastic response leads to a displacement at the beginning of the process. Then one can clearly see the increasing displacement due to viscous effects and the material becoming softer.

After a curing time of $100\,s$ the sample is cooled down to $20°C$ and all loads are set to zero. This enables the evaluation of residual strains and the resulting deformations, shown in Figure 11.

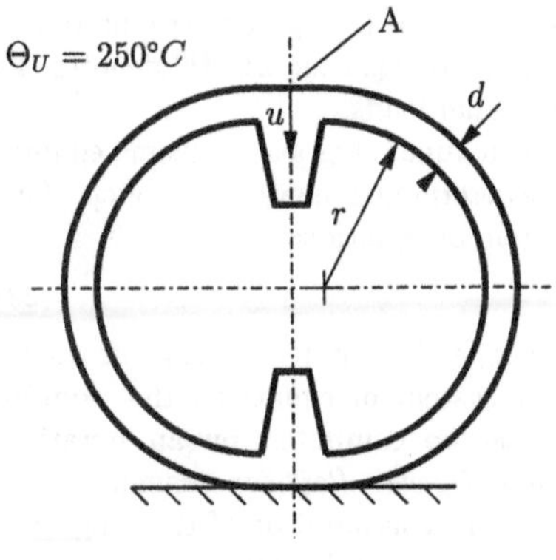

Figure 8: Shape of sample profile

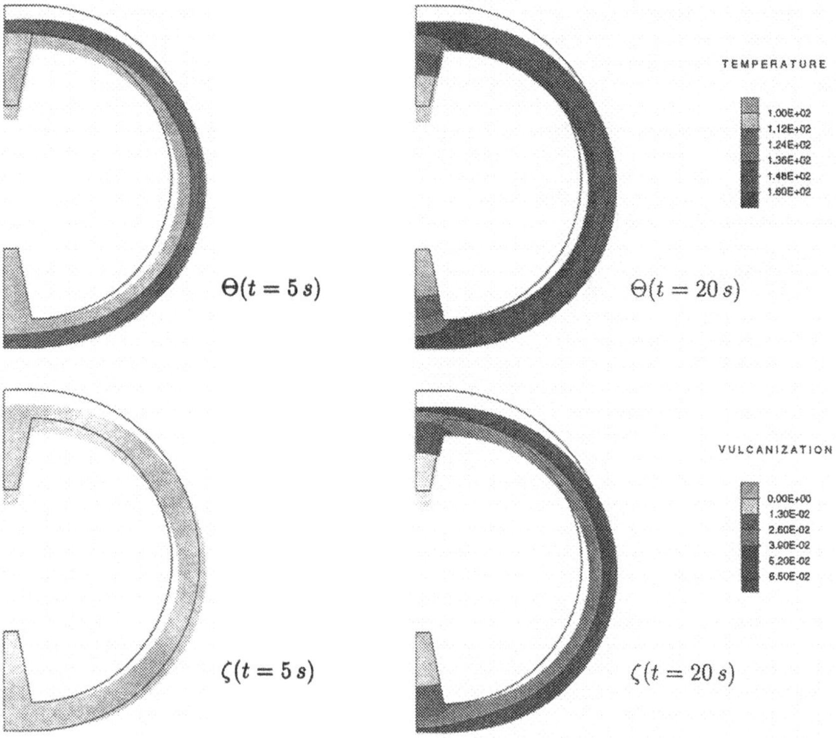

Figure 9: Temperature Θ and vulcanization ζ after $t = 5\,s$ and $t = 20\,s$

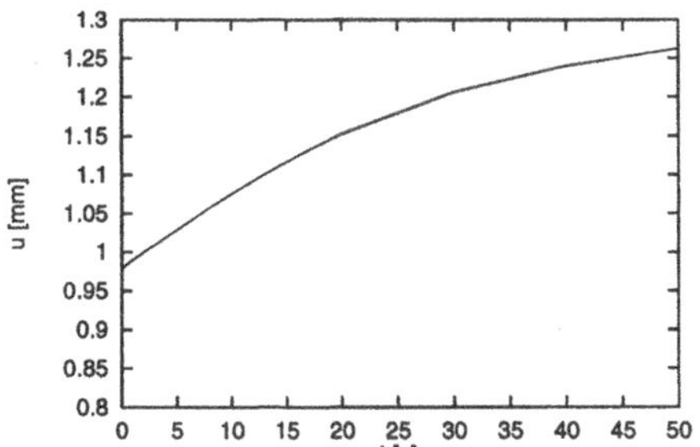

Figure 10: Displacement u of Point A

Figure 11: Residual deformation after curing and cooling

ACKNOWLEDGEMENT

This work was kindly supported by BTR Automotive Sealing Systems Group, Lindau, Germany. Special thanks is given to Dr. Ahmad, Mr. Eggler and Mr. Kotzur for supplying experimental data.

REFERENCES

Aklonis, J.J. & MacKnight, W.J. 1983. *Introduction to polymer viscoelasticity*. New York: Wiley.

Ferry, J.D. 1980. *Viscoelastic properties of polymers*. New York: Wiley.

Kaliske, M. & Rothert, H. 1998. Constitutive approach to rate-independent properties of filled elastomers. *Int. J. Solids Structures* Vol. 35, No. 17:2057 - 2071.

Mark, J.E. 1994. *Science and technology of rubber.* San Diego: Academic Press

Simo, J.C. & Hughes, T.J.R. 1988. *Elastoplasticity and viscoplasticity.* Stanford: Stanford Univ.

Zienkiewicz, O.C. & Taylor, R.L. 1994. *The finite element method* Vol. 1,2. London: McGraw-Hill.

Constitutive Models for Rubber, Dorfmann & Muhr (eds) © 1999 Taylor & Francis ISBN 90 5809 113 9

Indentation of rubber sheets with spherical indentors

J.J.C. Busfield & A.G.Thomas
Materials Department, Queen Mary and Westfield College, London, UK

ABSTRACT: The problem of indentation hardness for elastomer blocks has been examined at two levels. Initially an examination of the geometric non-linearity was undertaken. It was observed that the empirical equations adopted by the various standards organisations to predict the stiffness relationships were not always applicable. It appears that the classical Hertz solution to the problem gives a better representation of the general behaviour. A finite element approach was also adopted here to tackle the large displacement problem and the limitations of this approach have been discussed. This geometric problem is further complicated in practice by the effect of the finite thickness of the elastomer sheet. This problem has also been analysed and a suitable general relationship proposed to account for the finite thickness effects. The second problem examined was how the effects of the non-linear elasticity of the material could be tackled. It is shown that the form of the elastic stored energy function at small strains influences the indentation behaviour.

1 INTRODUCTION

Contact problems are frequently encountered in elastomer engineering applications. Typically, these problems involve rigid metal or plastic parts contacting elastomer sections to provide a progressive stiffness. In addition, elastomer against elastomer self contact is also frequently encountered. This paper considers the first case of a rigid body contacting a much softer elastomer surface. The test piece geometry considered specifically in this paper is that of a spherical indentor contacting with an elastomer sheet. However, the techniques adopted will be more generally applicable to a range of other engineering geometries.

1.1 *The indentation hardness of a semi-infinite elastomer block.*

A spherical indentor is currently used in the standard indentation hardness test. This technique has been used in the rubber industry for many years as a standard test method (ASTM D1415-88, ISO 48 & BS903) for characterising the elastic behaviour of elastomer compounds. The standard test method measures the difference in penetration of a specified dimension ball under two loads, an initial small one and a final much larger one. The difference in penetration is taken at a specified time and converted to an international rubber hardness degree (IRHD) value. Hertz first derived the small strain analytical solution to this problem in 1881 (Timoshenko 1934). From his general elastic solution for the contact of two spheres it is possible to derive the solution to the problem of the indentation of a rigid ball of radius R into a semi-infinite elastic medium of shear modulus G. To make this simplification it is assumed that the indentor is infinitely stiff and that the radius of the elastomer sheet, provided that it is of sufficient thickness, can be considered as a flat surface, which is equivalent to a sphere with an infinite radius. The resulting relationship between the depth of penetration x and the compressive force P is

$$x = \left(\frac{9}{256}\right)^{1/3} \frac{P^{2/3}}{R^{1/3} G^{2/3}} \tag{1}$$

This assumes that the elastomer is incompressible, that is, no volume change occurs under the deformation and thus for small strains Poisson's ratio can be taken as 0.5. This is a good approximation for elastomers under the present conditions. Hertz's solution also predicts that the small strain radius of contact a between the indentor and the sheet is,

$$a = \sqrt{Rx} \tag{2}$$

For incompressible materials (Gere & Timoshenko 1985) the shear modulus G and the Young's modulus E are related by

$$E = 3G \tag{3}$$

The engineer uses either the Young's or the shear modulus in order to calculate the stiffness when designing a component. When the product is transferred from design into production the modulus is however rarely specified. More typically the international rubber hardness degree (IRHD) is used by the rubber technologist to classify the material and therefore the engineer frequently has to use the hardness as a proxy parameter for the modulus.

Instead of using the analytical expression shown in Equation 1 to calculate the relationship between the change in depth of penetration of the ball to the modulus of the elastomer material, an empirical relationship is used in all three of the standards, namely

$$\frac{P}{E} = kx^{1.35}R^{0.65} \qquad (4)$$

where k is 1.9, if the force is measured in N, the modulus in MPa and the radius and the deflection in millimetres.

It is worth noting that the latest version of the British standard (BS903) contains a mistake as it stipulates that the equation has a k value of 0.0038. This would only be correct if the Young's modulus value was quoted in bar rather than MPa. However, this mistake is not reflected in the tabulated calculations relating the depth of penetration to the modulus where back calculations show that a value of 1.9 must have been used. The relationship in Equation 4 was originally derived by Scott (1935 & 1948) from the study of the indentation behaviour of various vulcanizates using two indentors and in the original work k was quoted as being 1.91. Quite clearly the relationship used in the standards is only an approximation, although no guide as to the level of the accuracy is made. It is quite possible that an engineer may specify an IRHD for the elastomer and a stiffness for a component and because of the inaccuracies in the relation between hardness scale and the modulus it may be impossible to meet both of the requirements simultaneously.

Gent (1958) investigated the relationship between the measure of hardness and the Young's modulus for elastomers. He showed that provided the indentation was sufficiently small and the elastomer sufficiently elastic and isotropic then the relationship in Equation 1 was more appropriate than Equation 4 to predict the indentation behaviour. This relationship had the double advantage that it used no arbitrary fitting constants. Expressed in a similar form to Equation 4, Equation 1 can be re-written as,

$$\frac{P}{E} = 1.78x^{1.5}R^{0.5} \qquad (5)$$

Yeoh (1984) suggested that the discrepancy between the Gent and the Scott relationships occur because of the difficulty in measuring the modulus experimentally. This is not helped by the non-linear behaviour of the elastomer. In addition, it should be noted that Gent determined his Young's modulus in a compression mode and that Scott measured his in tension.

The relationship between the IRHD value H and Young's modulus can be represented by a sigmoidal curve (Tangorra 1966) that can be described mathematically by

$$H = \left(\tfrac{100}{\sqrt{\pi}}\right)\int_{-\infty}^{\log_{10}(E/E_0)} \exp\left(-x^2\right)dx \qquad (6)$$

E_0 represents the Young's modulus corresponding to 50 IRHD. From the standards this occurs at 2.31 MPa. The function is chosen so that 0 represents a material that has a Young's modulus of zero and 100 represents a material that has an infinite Young's modulus. This function is complicated and as a result Tangorra (1966) suggested a simpler function that introduced an error of less than 1.5 in the value of the IRHD when compared with that generated using Equation 6 in the hardness range from 3 to 97. The relationship Tangorra proposed was

$$H = 100(E/E_0)/[1+(E/E_0)] \qquad (7)$$

Equation 7 is reliable to less than a quarter of an IRHD point if its application is restricted to the range between 20 and 80 IRHD. Values from Equation 6 have been tabulated for the entire hardness range in all the standards.

It is experimentally difficult to cover the entire range of hardness from 0 - 100 IRHD with a single ball size. The standards recommend that specific indentor sphere sizes be employed in specific overlapping ranges. In the normal range of 30 - 95 IRHD an indentor with a 2.5mm diameter sphere is prescribed. For the higher modulus materials of 85 - 100 IRHD a smaller diameter sphere of 1mm should be used; for lower modulus materials of 10 - 35 IRHD a larger sphere of 5mm diameter is specified. The same forces are specified throughout all the tests. The validity of Equation 4 with a k factor of 1.9 was assessed experimentally by Stiehler et al (1979). They commented that Equation 4 was only able to predict the behaviour of indentors of about 2.5 mm in diameter. It is the initial aim of this paper to solve this indentation problem using a large deformation Neo-Hookean material model in a finite element analysis programme. This complements the work of Gent (1958) and Stiehler et al. (1979) and evaluates the applicability of Equation 4 in the standards in relating the modulus of an elastomer to the depth of penetration.

1.2 *Indentations of sheets of a finite thickness.*

In practice the thickness of the sheet used in indentation hardness measurements may not be large enough for the theory for semi-infinite sheets to be sufficiently accurate. Tangorra (1966) tackled this problem using equations to represent the effect of the elastomer sample thickness t on the force-indentation relationship. He proposed the force P_{Hz} predicted by the classical Hertz relationship should be multiplied by a suitable factor thus:

$$P = P_{Hz} \exp\left[m / (x - t) \right] \quad (8)$$

where m is an empirical constant dependent on the ball diameter. Waters (1965) suggested a different approach. He noted that the linear dimensions of the stress field scales with the radius of the contact area and therefore he suggested that the thickness effect should be a function of t/a_t, where a_t is the contact area radius when the indentation involves a sheet of thickness t. From experimental observations, Waters proposed that this function could be represented by,

$$X = 1 - e^{\left(-\frac{At}{a_t}\right)} \quad (9)$$

A is a constant deduced. The two cases he explored were a lubricated lower boundary where A= 0.67, and a non-lubricated lower boundary where A= 0.41. It is worth noting that in his paper Waters (1965) has labelled his figure captions with the values of A for each of the two cases transposed. This relationship is quantitatively similar to that of Tangorra as it had asymptotes of zero and one at zero and infinite thickness respectively. However, it was more general and did not require the derivation of a term m that was related to the diameter of the ball size. In this work the Waters relationships' are explored and compared with finite element predictions.

1.3 *Material non-linearity*

The above approaches are based upon classical, small strain, elastic theory. Although in most indentation experiments using spherical indentors the strains are quite small, however filled elastomers can exhibit substantial non-linearity at strains of 10% or less. It is therefore uncertain how useful a single modulus value may be when deduced from a hardness test using relations based on small strain theories. As standard hardness tests are performed under prescribed loads, the stresses induced in a low modulus material are greater than those in one of a high modulus. This means that materials of different modulus have their hardness evaluated over different applied strain ranges. Thus, any small strain material non-linearity will complicate the interpretation of an indentation test in terms of modulus. Muhr and Thomas (1989) developed a method of deriving the representative average shear strain level corresponding to the hardness measurement. They derived from strain energy considerations an equation that relates the average shear strain γ under an indentation. This approach gave the average shear strain for an area localised under the indentor as

$$\bar{\gamma} = c^{-\frac{3}{2}} \left(\frac{64}{15} \frac{x}{R} \right)^{\frac{1}{2}} \quad (10)$$

c being the ratio of the radius of the region of deformed elastomer to the radius of the circle of contact. The value of c was found empirically to be 10. They observed that the hardness measurements corresponded to a low average shear strain even though the local deformations may be quite large. It is well known that for carbon black filled elastomers the chord shear modulus is strain dependent, decreasing as the strain increases. Therefore, when a small strain measurement made in a hardness test is used to predict the general behaviour of components in the engineering strain range, typically it will result in an overestimation of the component stiffness. This observation is backed up by data presented by Fuller and Muhr (1992) indicating that a reasonable correlation between the modulus measured at 2% shear strain and the modulus deduced from IRHD hardness measured for about 50 different engineering elastomers. This was compared with a poor correlation between the value of the modulus measured at 50% strain and that deduced from the IRHD hardness value.

An indentation study using a finite element analysis approach and a non-linear material model was made by Chang and Sun (1991). They failed to recognise that the average strain under the indentation was small. Instead they concentrated on how the Ogden stored energy function could be used to predict the indentation behaviour. This function is not suitable for describing stress-strain data in the small strain regions of interest. Actually all that they investigated was the indentation stiffness of apparently identical material models at small strains (with identical initial modulus values) but with widely differing large strain behaviour against each other. Not surprisingly, the results look remarkably consistent, as the strains observed in the indentation test piece remain small even when the deformations become quite large. A more suitable approach to understand this material non-linearity effect would be to use a stored energy function that had been fitted to the behaviour of the material at lower strains. A number of suitable stored energy functions exist. In this work, the material model proposed by Davies et al (1994) was utilised. Their stored energy function, which is suitable for carbon black filled elastomers used over the engineering strain range, is given by

$$W = \frac{A}{2(1 - n/2)} \left(I_1 - 3 + C^2 \right)^{(1-n/2)} \quad (11)$$

where W is the strain (or stored) energy density, I_1 is the first strain invariant, A is a measure of the 100%

shear strain modulus and n is a measure of the small strain non-linearity. The term C is the shear strain below which the modulus approaches a finite, small strain value. It describes a real effect (Davies et al. 1994) but the strains at which it is important are less than 1%. It is necessary to give it a non-zero value in the finite element implementation of the function so that the material does not lock by becoming infinitely stiff at zero strain. The technique for generating the data for this function was described in the paper by Davies at al (1994). They assumed that W is a function of I_1 only. The validity of this approach for filled materials has been well established (Gough et al. 1998) and it makes the determination of W much easier. Their basic method adopts the approach derived by Rivlin (1956). The engineering stress, σ, versus extension, λ, data measured in a tensile test was plotted as the reduced stress, $\partial W/\partial I_1$ versus the first strain invariant (I_1-3). Assuming that W is a function of I_1 only, the relationships for $\partial W/\partial I_1$ and I_1 are:

$$\partial W / \partial I_1 = \frac{\sigma}{2\left[\lambda - \lambda^{-2}\right]} \quad (12)$$

$$\left(I_1 - 3\right) = \lambda^2 + \frac{2}{\lambda} - 3 \quad (13)$$

The function in Equation 11 can be implemented in ABAQUS using the UHYPER subroutine. This technique requires both the definitions of W as well as its first and second derivatives with respect to I_1 thus:

$$UI1(1) = \frac{\partial W}{\partial I_1} = \frac{A}{2}\left(I_1 - 3 + C^2\right)^{-n/2} \quad (14)$$

$$UI2(1) = \frac{\partial^2 W}{\partial I_1^2} = -\frac{An}{4}\left(I_1 - 3 + C^2\right)^{-1-n/2} \quad (15)$$

2 EXPERIMENTAL

2.1 *Finite Element Analysis*

The simplest problem analysed was that of a semi-infinite rubber block subjected to an indentation test. This is equivalent to the Hertz problem of the indentation of a deformable semi-infinite block with a rigid spherical indentor as defined by Equations 1 and 2. The problem is symmetric about the axis of the indentor and lends itself to a two dimensional axi-symmetric model. Conventional four node elements were used, because linear interpolation elements handle the contact algorithms more accurately. The elastomer was defined as a large strain Hyperelastic material, whose elastic behaviour would be determined by the use of a stored energy function. Initially and for simplicity the Neo-

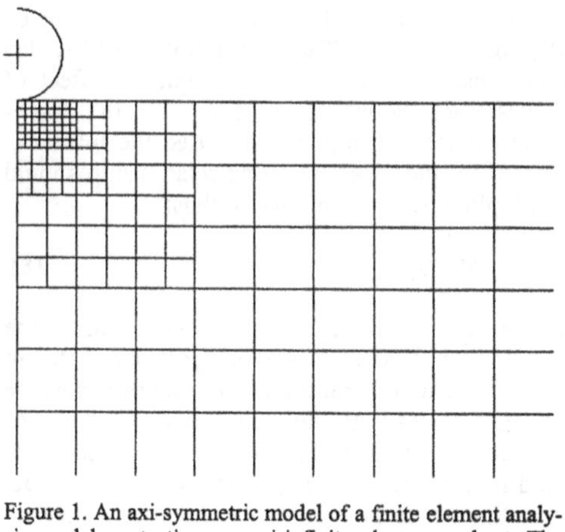

Figure 1. An axi-symmetric model of a finite element analysis model contacting a semi-infinite elastomer sheet. The semi-infinite elements are shown as the unbounded elements in the plot.

Hookean stored energy function with a shear modulus of 1MPa was used. This required the use of hybrid (in ABAQUS CAX4H) elements.

The indentor was modelled as a rigid surface whose centre was at the axis origin and which was initially in contact with the block. The indentor ball diameter modelled was comparable with the standards and therefore had a radius of 1.25mm.

A semi-infinite block cannot be easily represented using traditional finite element models. The boundary condition the borders should be stress free would normally require the use of a large model. However, semi-infinite elements are available in ABAQUS. In two-dimensional ABAQUS problems these are CINxx4 elements, where xx represents PS, PE, or AX, which are plane stress, plane strain, or axi-symmetry respectively. These elements provide a linear response, as the behaviour at infinity is linear, and are designed to provide a damped boundary in a dynamic problem. They also have an application in that they can surround a conventional model that can then represent a semi-infinite slab. A typical model used is shown in Figure 1. The elements around the outside are semi-infinite. This is indicated because they are unbounded in the direction in which they are infinite.

Figure 1 also shows the different densities of conventional elements in the elastomer model. The mid-side nodes where the element densities alter are controlled by multi-point constraints. These control the mid-side node movement to be forced by the average linear response of their two neighbouring nodes along that boundary. This is an established mesh refinement technique. The actual model adopted for most of the work was of a similar design to that shown but had a higher mesh density in the region of contact. The reason for this increased mesh density is discussed later.

Following this, investigations were made into how the force, P, versus deformation, x, behaviour changes with elastomer test piece geometry by modelling sheets of a finite thickness. The effect of friction at the indentor and elastomer interface was also investigated. The standards suggest that the friction should be minimised at this interface by dusting the elastomer surface with a 'dusting powder'. Using finite element analysis, Coulomb friction was the most easily implemented friction model. This required the specification of the coefficient of friction between the two materials. The two limiting cases, when the surface had no frictional effects, and when no slippage occurs, were compared at the contact interface. Chang and Sun (1991) reported that the effects of friction were small and played no important role performed this work. However in their investigations they did not examine the two extreme cases. They assumed that the maximum value of the Coulomb friction was only one, whereas for elastomer materials it may be possible for the value to be higher than this. In addition, they did not explore the differences caused by frictional effects at the larger deformations that may exist with the hardness test when performed on materials that are at the lower end of the IRHD range.

The effects of friction were thought likely to be more significant with sheets of a finite thickness. In this work, the two limiting cases of infinite and zero friction were considered and were modelled using a ball of 1.25mm radius. Four different sheet thicknesses (1, 2, 4 and 8mm) were considered.

2.2 *Experimental Investigation of Indentation Behaviour*

The two materials used are given in Table 1. Elastomer blocks of 13mm depth, sufficient to allow a reasonable accuracy of indentation test, and length and breadth of 150 mm, were compression moulded using a 10-minute (153°C) cure. The compound was masticated on a two-roll mill before being added to the compression mould. This reduced weld lines at the interface where the elastomer blank folds onto itself. These folds could potentially lead to laminations throughout the sample.

Indentation tests were made on the specimens using a specially adapted dial gauge calibrated in 0.01mm increments. The main return spring inside the gauge was removed. A platform was attached to the top of the dial gauge operating column on which 100g weights could be placed. The standard anvil was replaced with a receptor that could locate a steel ball bearing, used as the indentor. A 2.5mm diameter ball was used for this work, as it is the size most widely adopted in the standard. The internal friction over the operating range of the spindle movement was measured for each test, and was allowed for in the analysis of the results as was the pre-load due to the platform weight. A schematic of the set up is shown in Figure 2.

The greatest difficulty in performing this test was in determining the position of the zero load dial gauge reading. This was done by using a thin glass cover slip of a known thickness pressed onto the upper surface of elastomer under the smallest (4g) load. Repeat testing showed this value accurate to within ±0.02mm. This meant that in the measured data at the smallest values of load (45g) and with the softer material it was possible to measure the value of the deflection to ±20%. However, as the indentation increases there was a proportional improvement in the accuracy of the technique.

The surface of the elastomer test samples was carefully cleaned using a solution of detergent and water in order to ensure the removal of all traces of dust and grease. The blocks were then thoroughly dried. The base was supported on a rigid flat bed. Using finite element techniques on thin sheets it will later be shown that the frictional effects at the base of the elastomer block were significant. For consistency, the actual indentation tests were all performed on sheets having a lower surface interface lubricated by a mixture of detergent and water. The comparative finite element models assumed zero friction conditions at this interface. In our case, all the measurements were more than 40mm from the edge of the test piece. In order to minimise the complications caused by the effects of stress relaxation the rate of

Table 1. The two natural rubber elastomers used to evaluate the non-linear indentation behaviour.

Compound Name	Amount of N330 reinforcing carbon black.	Vulcanising system
NR 29	29 pph	semi EV
NR 49	49 pph	semi EV

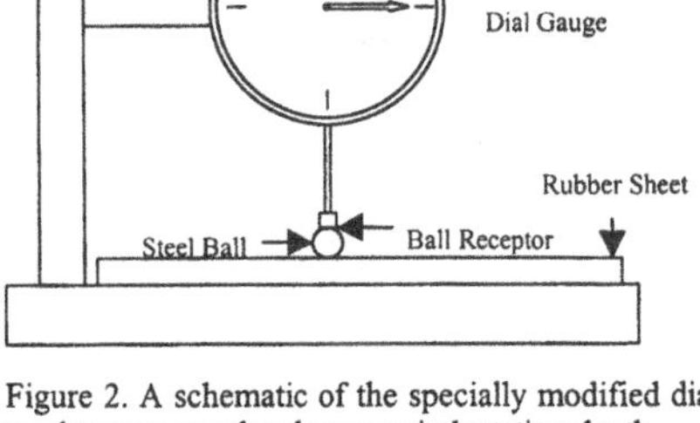

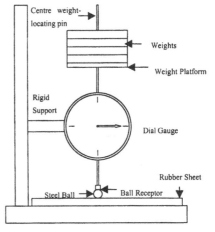

Figure 2. A schematic of the specially modified dial gauge used to measure the elastomer indentation depth.

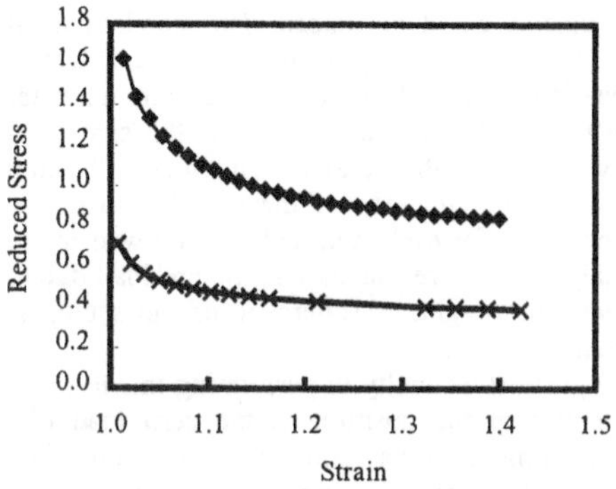

Figure 3. The engineering stress versus the engineering strain plot measured using a tensile test for the two materials investigated in this work. The x shows the measured behav iour of NR29 and the ♦ shows the behaviour of NR49.

addition of the weights and the timing of deflection measurements was carefully controlled so that an approximate average strain rate could be calculated. This also ensured that there was repeatability between successive tests.

Tensile test specimens were also cut from the blocks. These strips were then tested in a tensometer to determine the stress strain behaviour. The results are shown in Figure 3. Using Equations 12 and 13 the elastic behaviour is shown in Figure 4 as the dependence of $\partial W/\partial I_1$ on I_1. The linearity shown in this double logarithmic plot enables the ready calculation of parameters A and n Equation 14. These values are shown in Table 2. The value of c is taken as 0.003, as given (Davies et al. 1994) for materials similar to those used here. Its precise value is of little importance for the strain range involved in the present study.

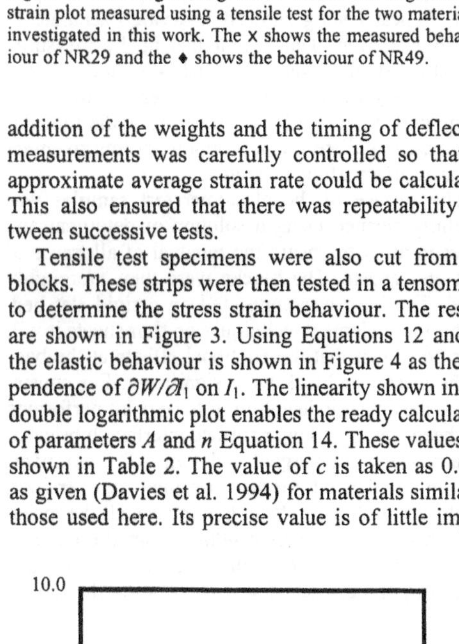

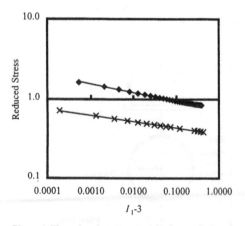

Figure 4. The reduced stress versus the first strain invariant plotted on log scales for the two materials investigated in this work. The x shows the measured behaviour of NR29 and the ♦ shows the behaviour of NR49. The power law relationship between these terms is shown by the linear na ture of the two plots.

Table 2. The curve fitting parameters measured using a tensile test for each of the two-elastomer compounds.

Parameter	NR29	NR49
A /MPa	0.725	1.520
N	0.155	0.210
C	0.003	0.003
Shear Modulus at 8% extension /MPa	1.00	2.35

The strain rate for the tensile tests was selected to be similar to that used for the indentation hardness test. To calculate the strain rate for the indentation test an average strain approach was attempted. This was at best only an approximation as the hardness test was made by adding weights at even time increments, and because of the non-linear geometric effects the displacement response was highly non-linear with load. Despite these difficulties an approximation for the strain rate of the hardness test can be calculated from the work of Muhr and Thomas (1989). Using a finite element approach it was possible to estimate that a ball of radius 2.5mm with a deflection of 0.5mm had a stress field (1% of the maximum principle stress value) that extended to a depth of 11.4mm. It was assumed that the radius of deformation ρ extended at least this far. From Equation 2 and again using an indentor of radius 2.5mm, it was possible to calculate a minimum value of c,

$$c=\frac{\rho}{a_t}=\frac{\rho}{\sqrt{Rx}}=\frac{11.4}{\sqrt{1.25x0.5}}=14.4 \tag{16}$$

To simplify further calculations the value of c was rounded to 15.

The indentation tests were loaded applied at a rate of 200 grams (1.96 N) per minute. Equation 10 was used to give an indication of the average strain rate for the test. It was observed that the average strain rate was below 2.5% strain per minute for all the tests undertaken. In light of this, the tensile tests were performed at a comparable slow strain rate of approximately 1% strain per minute. It was estimated that there was a small error in the stress versus strain measurements used in the work caused by variations in test piece geometry and errors in the measurement of strain. This resulted in a possible error of about ±3% in the calculation for the modulus of the material.

3 RESULTS & DISCUSSION

3.1 *The indentation hardness test of a semi-infinite block.*

Figure 5 shows the results of the finite element analysis calculation for the force deformation re-

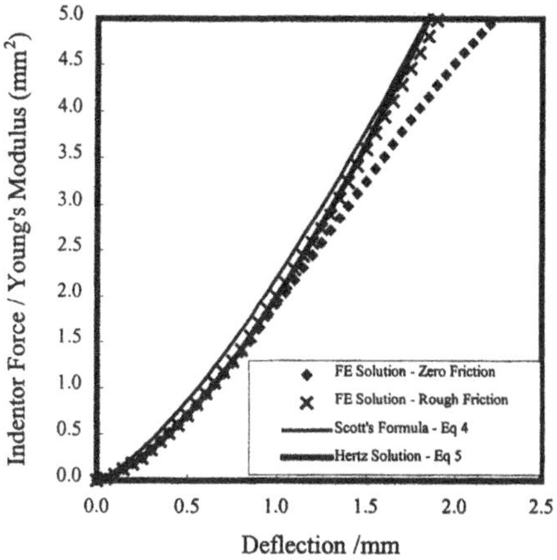

Figure 5. The indentor force predicted by the FEA (with and without friction at the elastomer / ball interface), Scott's empirical solution and Hertz analytical solution are all scaled by Young's modulus, this is then plotted against the deflection of an indentor of 2.5mm diameter into a semi-infinite slab. All these models assume that the material has a Neo-Hookean material response.

sponse of the 2.5mm diameter sphere indenting the semi-infinite block with a neo-Hookean stored energy function. The results are plotted in the form required by Equations 4 and 5 with the predicted force divided by the Young's modulus (P/E) against the prediction for the depth of penetration. The two sets of finite element data presented represent the extreme cases of zero friction at the interface between the indentor and the elastomer sheet and the alternative case of no slippage at the interface. The plot also shows Equation 4, the empirical relationship derived by Scott used in the standards with k=1.9, as the dotted line and Equation 5, the analytical solution derived by Hertz, as the solid line.

At small deflections, both sets of finite element results were very similar, and the two curves followed each other closely. At larger deformations when the depth of penetration became comparable to the radius of the indentor, the model with the frictional effects introduced became stiffer. This was because at the very large deformations the zero friction model allowed the rubber to slide around the indentor surface. This effectively stress relieved the geometry and relaxed the stresses and hence the stiffness. This effect was surprisingly small and even at a relatively large displacement of 1 mm the discrepancy between the two cases of limiting friction was only about 5%. This paralleled the observations made by Chang and Sun (1991). The reason that friction was so insignificant results from the modest shear stresses at this interface for all but the largest displacements. The normal test with an elastomer of 30 IRHD would result in a total deflection of 1.8mm; at this deflection, Figure 5 showed that the frictional effects were significant, producing a 15% difference between the predictions for the two forces. Therefore, in the standard hardness test for soft materials the surfaces should be prepared to minimise the variation due to frictional effects.

For a standard hardness test the case of zero friction is probably more appropriate if the standard recommended practice of dusting the surface in 'dusting powder' first is adopted. In contrast, it could be anticipated that the friction value at this interface would be large for normal clean surfaces. When a comparison is made later with the measured data the finite element model assumes that the friction at this interface was high.

The Hertz relation assumes that the strains are small, and it is perhaps surprising that it follows the finite element solution so closely for quite large deflections. This reflects the point made earlier, that the strains in the indentation problem are not very large. This close correspondence between the solution is consistent with observations by Gent (1958). The correlation with the Scott relation is less good. When the results were examined in more detail the difference between the calculation of the depth of penetration for both the Hertz solution and the Scott equation could be compared against the prediction of the finite element model. Using the table from the ISO standard, the IRHD value that relates to a specific value of Young's modulus can be obtained. In addition, the anticipated displacement between the maximum (5.4N) and the minimum force (0.3N) can be derived for a specified IRHD. The graph in Figure 5 was used to calculate the change in the depth of penetration between the maximum and the minimum forces at specific IRHD values. This was done for both of the limiting frictional cases modelled using the finite element technique, and using the equations employed in the standards. The results are shown in Table 3.

Table 3. The three different values for the change in depth of penetration between the minimum and maximum applied indentation forces, for a 2.5mm diameter rigid indentor. Both the zero and high friction FEA results were calculated for materials with the specified Young's Modulus. The ISO standard deformation was related to the specified IRHD. The relationship between the Modulus and the IRHD were also derived from the standard.

Young's Modulus /MPa	IRHD	D_{ISO} /mm	D_{FEA} (High Friction) /mm	D_{FEA} (Zero Friction) /mm
1.00	29.8	1.81	1.74	2.12
1.59	40.6	1.28	1.24	1.35
2.51	51.7	0.92	0.90	0.94
3.98	63.4	0.64	0.67	0.68
10.00	81.2	0.33	0.36	0.36
25.12	92.6	0.17	0.20	0.20

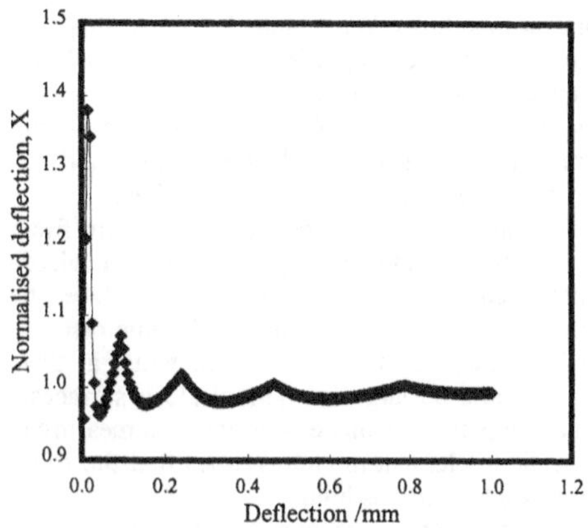

Figure 6. The normalised deflection X, calculated using a fine finite element analysis mesh plotted against deflection. This graph demonstrates the discontinuous nature of the contact problem tackled using a finite element technique.

It was apparent from the table that the three calculations for the depth of penetration produced different results. These differences are:

An under prediction of the depth of the penetration for the higher modulus materials, using the equation presented in the standards. This was equivalent to an overestimation in the IRHD. As the IRHD is the value actually recorded, then this would results in an error for the calculation of modulus of over 10%.

For very soft materials, the effects of friction became more significant. The uncertainty was now of the order of about 3 points on the IRHD scale for a material around 40°IRHD and was even greater for a material of 30°IRHD. Friction and hence surface preparation were now quite important.

The Scott solution actually lies between the two limiting friction cases as calculated by the finite element model for softer materials. Therefore, the calculation was up to 1.5 points out on the IRHD scale depending on how the surface was prepared.

The finite element data were examined more closely at the smaller deflections in Figure 6 with the concept of a normalised deflection X introduced. This was the deflection at a given load predicted by the finite element method divided by the deflection at the same load predicted using the Hertz solution. Figure 6 shows how for this model X varied against the penetration depth predicted by the Hertz solution. The graph was interesting in that it clearly demonstrated small strain fluctuations in the stiffness prediction caused by the introduction of additional nodes to the contact algorithm as the analysis proceeds. The mesh density determines the size and scale of these deviations. Provided that at least five nodes were used in the contact zone the deviations were less than 1% from the value calculated by Equation 5. For the remaining analyses used here, an even more accurate mesh density was employed that had over 15 elements in the contact zone.

This work has so far established the good agreement to within 1% between the finite element solution on the semi-infinite sheet, and the Hertz solution up to a penetration of about 80% of the ball radius. This was provided the mesh density was sufficiently refined.

3.2 *Indentation of sheets of a finite thickness.*

The indentation of sheets of finite thickness was examined for the limiting cases of zero friction and full friction at both interfaces. A standard spherical indentor of 2.5mm diameter with a range of thickness values (1mm, 2mm, 4mm, 8mm) was used. The deformation results were scaled by the prediction given by the Hertz solution for a semi-infinite sheet to calculate X_w as defined in Equation 9. These results are plotted on a Waters type diagram in Figures 7, 8 and 9 against t/a_t, where a_t was calculated from x and R using a small strain solution proposed in Equation 2. Note that in these plots a large t/a_t value corresponds to a small indentation. In Figure 7 all four thickness values of the base material are plotted for the case where both the elastomer base interface and the elastomer indentor interface friction values were taken as zero. In Figures 8 and 9 the finite element models for the thin sheets are compared with the Waters thin sheet solutions for the whole range of friction contact conditions. In Figure 8, the base friction is set to zero and in Figure 9, the base friction is high.

These diagrams give an indication of the form of the function X_w. In each of the figures is plotted the Waters solution proposed in Equation 9, for the case of the lubricated (zero friction) base rubber interface, with A=0.67, and for the case of the non-lubricated (full friction) base rubber interface with

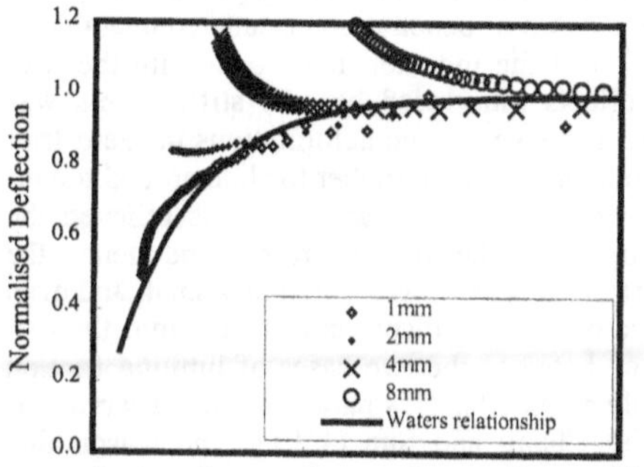

Figure 7. The finite thickness plot with zero friction at both the base and the rubber ball interfaces for a range of different rubber block thickness values (1mm, 2mm, 4mm, 8mm).

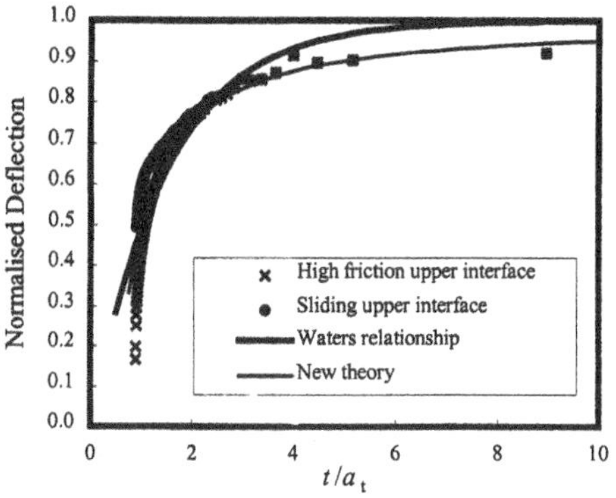

Figure 8. The finite thickness plot for a 1mm thick rubber block with zero friction at the base interface.

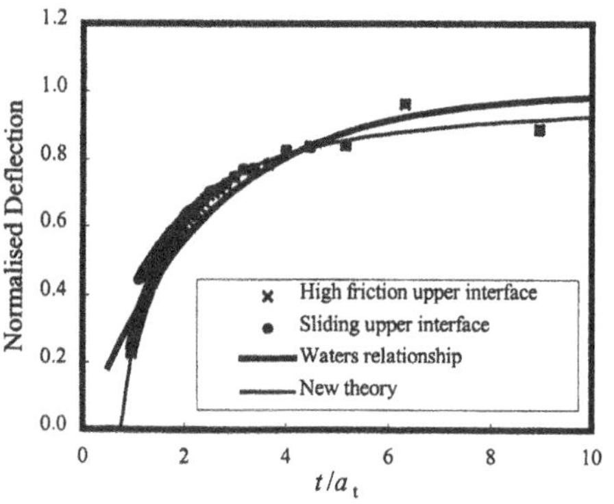

Figure 9. The finite thickness plot for a 1mm thick rubber block with high friction at the base interface.

A=0.41. As indicated earlier, in the original Waters (1965) paper the figure captions identified the coefficients the wrong way around.

A number of general observations could be made. In all of the figures a small scatter at low deformations, at the large t/a_t values, was seen. This was again due to additional nodes contacting the rigid surface that were then introduced into the contact algorithm when using the finite element method. This happened discontinuously as the depth of penetration increases. In Figure 7, the deformations for the cases of a 'small' indentor against a 'thick' rubber sheet were not predicted well. This was because the normalised deflection plot of X_W became inappropriate, as the depth of penetration was comparable to the radius of the indentor. This results from neither Equation 1 nor Equation 2 being reliable in predicting either the force deflection behaviour or the radius of contact at large deformations. From geometric considerations it was possible to recast Equation 2 to be more generally applicable over the whole range of deflections. This approach is shown in Equation 17.

$$\begin{aligned} a_t &= \sqrt{xR} \quad x < R \\ a_t &= R \quad x \geq R \end{aligned} \tag{17}$$

This correction would affect the scaling of the X-axis in the three Waters plots. It was also clear that with the larger penetrations into the 'thicker' section geometries the prediction for the stiffness by the Hertz equation was too high. This effect was shown earlier in Figure 5. This was because the predictions for the Hertz contact solution were seen to be only appropriate to penetrations about equal or less than the radius of the ball, in the case of high friction at the rubber indentor interface, and to only about half of the ball radius for the case of no friction at the ball indentor interface. Therefore, for the 8mm thick sheet with the low friction case, t/a_t would have to be above 12 for this to apply. Below this value the deviations due to the inaccuracy of the Hertz equation would dominate and mask the effects of a finite thickness. This was effectively beyond the X-axis as presented in Figure 7. This limitation indicated that the Waters theory could only work when the depth of penetration was less than the radius of the ball for high friction at the upper interface and less than half the radius for low friction. Figure 7 indicated that the sheets should be thin enough to allow finite thickness effects to dominate the behaviour and for the Waters' relationship to be applied. This was the reason why Figures 8 and 9 only give the prediction for the indentation behaviour into the 1mm sheet.

Some further general observations could also be made. The approach suggested by Waters was good for predicting the general trends for thin sheets. It appeared in general to slightly overestimate the correction required at small deflections and to underestimate the correction required at larger deflections.

The behaviour at the smallest deflections was somewhat masked by the uncertainty in the contact algorithms used in the finite element work, but it appeared to be of the right size and scale. The models with friction at the ball indentor interface were closer to the effects measured by Waters. This is due to two effects. The first being that the upper surface preparation was ignored by Waters and was therefore probably a high friction case. The second effect was the difference in the contact radius that was lower at a specified load for the case of high friction. Also the small strain Hertz solution calculated all the values by assuming that the geometry did not change, which is nearer the case for the high friction interface model.

The Waters solution for high friction at the base overestimated the correction factor. This was probably because, when Waters made his measurements, some small sliding occurred in his experiments. This slippage would have reduced the stiffness in the measured test when compared to the analytical results. The behaviour for zero friction at the base appeared more complex. At moderate deflection values, the stiffness was slightly overestimated by the Waters approach and that at the higher deflections the stiffness was underestimated. This suggests that an additional non-geometric phenomenon was occurring. It is possible that, in addition, Waters in performing his experiments observed the inherent effect of a reduced modulus as the strain increased. This material non-linearity effect is examined in more detail in the next section. The Waters equation was derived from experiments conducted on peroxide cured NR materials that probably had a low but specific material non-linearity in the stress versus strain relationship. In his work, the level of cross-linking was varied to modify the material modulus.

It was seen that the geometric effects when considered in isolation away from the material non-linearity were not modelled accurately by Equation 9. An alternative function could be,

$$f(t/a_t) = 1 - \frac{B}{(t/a_t)} \tag{18}$$

This equation is also shown in Figure 8 and 9. The best fit for the case of a lubricated base is produced with the factor with B=0.50 and for the non-lubricated case B=0.74.

It can be concluded that the Waters approach was partially valid, in that it identified the cause and the nature of the deviation from the Hertz solution for thin sheets. However, it did not clearly identify the difference between the geometric effects and the material non-linearity. This would limit the application of the numerical coefficients derived experimentally by Waters, and used in Equation 9, to the prediction of the behaviour of similar unfilled rubbers. It would not necessarily be appropriate to use this approach to model the behaviour of more non-linear materials, such as carbon black filled elastomers. In addition, the Waters approach would only work with thin sheets or with large indentors, so that the depth of penetration was maintained below a certain value. It was apparent from dimensional arguments that the analysis of the indentation problem was scalable. For the standard indentor of diameter, 2.5 mm and an indention of depth, 0.5 mm, and a maximum deviation of 10% from the Hertz solution, would require a minimum sheet thickness of about 6mm for the lubricated base case and about 9mm for the non lubricated case.

3.3 *The effects of material non-linearity on the indentation hardness behaviour.*

Figure 10 and Figure 11 show the force deflection response measured using our indentation test for both of the black filled materials. Each graph shows four test data sets. The good repeatability between the tests was achieved by careful elastomer surface preparation as outlined earlier. Because talc was not used at the elastomer indentor interface, it was assumed for the purposes of finite element modelling that this interface had high friction. The lower surface was assumed a low friction case as it was lubricated experimentally by using a mixture of detergent and water.

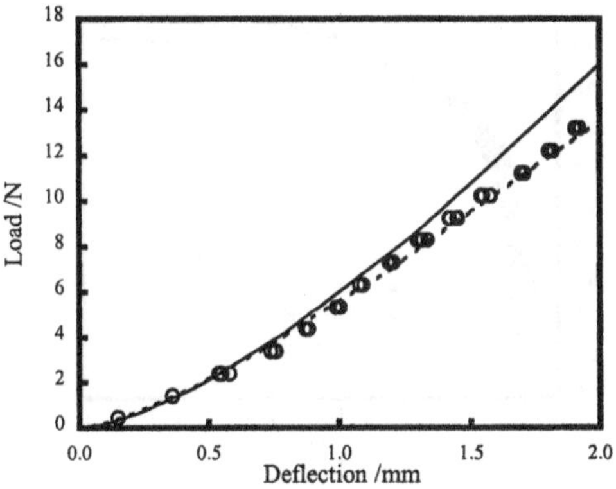

Figure 10. The measured hardness test for NR29. O represents the experimental data points and the dotted line represents the finite element prediction using the stored energy function in Equation 11 and the solid line represents the data for the finite element prediction using the Neo-Hookean material model.

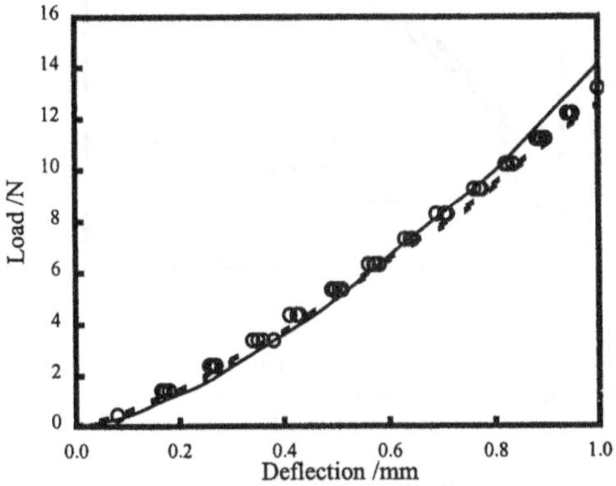

Figure 11. The measured hardness test for NR59. O is the experimentally measured data and the dotted line represents the finite element prediction using the Davies et al. stored energy function shown in Equation 11 and the solid line rep resents the data for the finite element prediction using the Neo-Hookean material model.

In both plots, a Neo-Hookean finite element model is included. The shear modulus for the Neo-Hookean confirmation was taken as the equivalent chord shear modulus at 8% strain. The equivalent strains in different deformation modes were compared using the I_1 values as suggested by Equation 11. The correlation was very good at an indentation of 0.5mm, however below this value the stiffness was underestimated and above this value it was overestimated. This indicated that the average strain under the standard indentation test, at a deformation of 0.5mm was approximately equivalent on an equivalent I_1 basis to the strain that the material would have experienced in a tensile test at about 8% tensile strain. This showed that the indentation test was actually measuring both the geometric non-linearity, as the contact area increased, and the reduction in the shear modulus as the depth of penetration (and hence the level of strain) increased.

The level of correlation between the finite element analysis for the non-linear stored energy function in Equation 11 was good. This further corroborated the approach of modelling the behaviour of filled elastomers as a function of I_1 only. It also demonstrated that, provided the strain rates were comparable, data that was measured using a given deformation mode could be taken and used to predict the behaviour in alternative deformation modes. This has widespread application not just for the contact analysis shown here but also for predicting the stiffness response of a wide range of other engineering elastomer design applications.

4 CONCLUSIONS

The first observation was that the Hertz solution is surprisingly good at predicting the indentation behaviour of rigid sphere into an elastomer material over a large range of strains. In contrast to what has been reported before, the effect of friction at the rubber against indentor interface becomes significant at large deformations or when applied to the standard indention test for the softest materials. It would be advisable to clarify the surface preparation technique in the standard. The finite element analysis indicates that for softer materials the errors induced were quite large. Equation 4, which had been adopted by the standards, incorporates geometric and material non-linearity terms that are carried over from the work by Scott. It was noted that this also introduces errors in the calculation of modulus from a hardness test. A better approach was found by adopting the Hertz solution. It is proposed that the standard be modified to reflect this. In a real test with the displacement measured between two forces, it is worth noting that the value of modulus calculated from the Hertz solution will tend be a little higher than is predicted using the current standard.

Finite element analysis techniques model this type of contact problem well. The small deviations observed, due to the discontinuous nature of the FEA solution, can be minimised by accurate meshing. This analysis confirmed that small average strains were induced even at apparently high levels of deformation.

The thickness effects were reasonably well represented by the Waters approach if the depth of penetration was less than the radius of the ball. The Waters technique failed to operate at deformations above this value with any degree of usefulness. In fact, the corrections for the finite thickness were in the wrong direction as the inaccuracy of the Hertz solution dominated the behaviour. The effects of friction at the interface became significant when the deflection exceeded the ball radius. A modification of Waters' relation was proposed in Equation 18 that gave a better fit to the finite element results for small indentations. It was suggested that certain suitable minimum thickness values should be incorporated into the standard.

The indentation test of non-linear filled materials can also be modelled. A suitable engineering strain stored energy function was employed to model the typical non-linear elastic behaviour very well. The technique of representing the behaviour of a filled elastomer by a stored energy function of I_1 only was again shown to predict the complex problem of indentation well. If this restriction is to be adopted the functional dependence of W on I_1 can be found from a single stress versus strain measurement. The particular function used here, Equation 11, was particularly suited to the low strains encountered in this indentation measurement. It is also known to be satisfactory (Davies et al. 1994) up to somewhat higher strains found in many engineering applications. However, it should be noted that several other functions could be expected to work as well. Finally, the standard hardness test has a large degree of non-linear geometric effects that the finite element technique seemed quite capable of resolving.

REFERENCES

ASTM: D1415-88. Standard Test procedure for Rubber Property - Indentation Hardness.

BS903: 1995. Physical Testing of Rubber, Part A26. Method for the determination of hardness. British Standards.

Chang, W.V. & Sun, S.C. 1991. Non-linear elastic analysis of the hardness test of rubber like materials. *Rubber Chemistry and Technology* 64: 202-210.

Davies, C.K.L., De D.K. & Thomas, A.G. 1994. Characterisation of the behaviour of rubber for engineering design purposes (I) stress - strain relations. *Rubber Chemistry and Technology* 67: 716.

Gent, A.N. 1958. On the Relation between Indentation Hardness and Young's Modulus. *I.R.I. Transactions* 34: 46-57.

Gere J.M. & Timoshenko, S.P. 1985. *Mechanics of Materials*, Second SI Edition. Boston Mass: PWS Publishers: 28.

Gough, J., Muhr A.H. & Thomas, A.G. 1998. Material characterisation for finite element analysis of rubber components. *Journal of Rubber Research* 1: 222-239.

ISO 48: 1994. Rubber, vulcanised or thermoplastic - Determination of hardness (Hardness between 10 and 100 IRHD). International Organisation for Standardisation.

Lindley, P.B. 1992. *Engineering design with natural rubber*. Revised by Fuller, K.N.G. & Muhr, A.H. Hertford: The Malaysian Rubber Producers' Research Association.

Muhr, A.H. and Thomas, A.G. 1989. Allowing for non-linear

stress-strain relationships of rubber in force-deformation calculations. Part II: relationship of hardness to modulus. *Natural Rubber Technology* 20: 27-32.
Rivlin, R.S. 1956. Large Elastic deformations. *Rheology* Vol. 1, ed Eirich R.S. New York: Academic Press: 351-385.
Scott, J.R. 1948. Improved method of expressing hardness of vulcanised rubber. *Journal of Rubber Research*, 17: 145.
Scott, J.R. 1935. Rationalisation of the hardness testing of rubber. *I.R.I. Transactions*, 11: 224.
Stiehler, R.D. Decker, G.E. & Bullman, G.W. 1979. Determination of Hardness and Modulus of Rubber with Spherical Indentors. *Rubber Chemistry and Technology* 52: 255-262.
Tangorra, G. 1966. Hardness, Modulus, and Thickness. *Rubber Chemistry and Technology* 39: 1520-1525.
Timoshenko, S.P. 1934 *Theory of Elasticity*, McGraw-Hill Book Co., New York: 339-343.
Waters, N.E. 1965. The Indentation of Thin Rubber Sheets by Spherical Indentors. *British Journal of Applied Physics* 16: 557-563.
Yeoh, O.H. 1984. On hardness and Young's modulus of rubber. *Plastics and Rubber Processing and Applications* 4: 141-144.

Constitutive Models for Rubber, Dorfmann & Muhr (eds) © 1999 Taylor & Francis ISBN 90 5809 113 9

Styroflex(R) – The properties and applications of a new styrenic thermoplastic elastomer

J.R.Wünsch & K.Knoll
Business Unit Polystyrene/Styrene, BASF Aktiengesellschaft, Ludwigshafen, Germany

SUMMARY: Styrolux® and Styroflex® are both styrene and butadiene based block copolymers prepared by butyllithium initiated anionic polymerization. Styrolux is a transparent, tough and stiff thermoplastic material for high speed processing. Its specially designed molecular structure allows homogeneous mixing with general purpose polystyrene maintaining the transparency. Styroflex is a new experimental product with the mechanical behavior of a thermoplastic elastomer, e.g. low modulus and yield strength, high elongation and excellent recovery. High transparency and thermal stability give the competitive edge over conventional styrene-butadiene elastomers. Styroflex, Styrolux and general purpose polystyrene form a unit construction system e.g. for transparent film materials and injection molded parts with fine-tunable hardness and toughness.

1. INTRODUCTION

Basic properties of noncrystalline polymers are determined by the value of the glass transition temperature T_g with respect to ambient temperature. Polymers with low T_g are referred to as rubbers and behave as viscoelastic liquids. Typical properties comprise elastic recovery, large elongation at break and softness. To avoid tack and cold flow chemical crosslinking (vulcanization) is required for many applications. On the other hand high T_g resins are stiff and hard, but commonly break when elongated beyond a few %. Well defined block copolymers with at least one block sequence hard-soft-hard offer the chance to combine the advantages of both groups of commonly incompatible materials while maintaining transparency (Fig. 1).

Mechanical Property Range of Block Copolymers

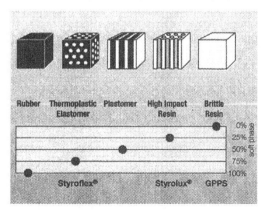

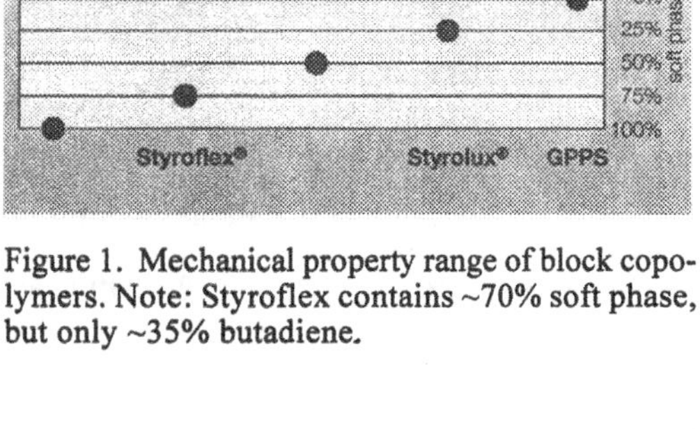

Figure 1. Mechanical property range of block copolymers. Note: Styroflex contains ~70% soft phase, but only ~35% butadiene.

The matrix phase basically determines the bulk properties like mechanical behavior and polarity. Thus thermoplastic elastomers exhibit predominantly elastic deformation. They are characterized by a rubber matrix containing inclusions of the hard spheres or cylinders consisting of high T_g end blocks as molecular anchors. The ratio of plastic to elastic deformation increases with the content of the hard phase. Fairly soft "plastomers" with a 50/50 hard-soft ratio and cocontinuous morphology do not have substantial elastic recovery and no pronounced yield point, but show extended plastic deformation due to shear yielding[1)].

Reduction of the rubber content to 25% in simple symmetrical triblock copolymers leads to high modulus but brittle thermoplastic materials.The challenge was to develop stiff resins with a low rubber content that preserve the ability of plastic deformation[1)] and, in consequence, impact strength.

2. STYROLUX

Styrolux, the well-known and commercial established transparent and tough polystyrene grade of

BASF with styrene butadiene block structure, is used in applications like packaging film, beakers and injection molded parts. This type of polymer was introduced in the late fifties[2)] and has subsequently been improved. It is now a steadily growing specialty in a.m. applications. Roughly 80% of Styrolux is blended with general purpose polystyrene (GPPS)[3)].

In order to understand the role of a transparent and tough polystyrene like Styrolux as blend component, its molecular design is briefly discussed. All Styrolux grades consist of unsymmetrical star block copolymers. When Styrolux without further specification is mentioned later in the text the term always refers to star polymer grades. The Styrolux grades cover a butadiene content from about 20 to 30%, differing in their toughness/stiffness ratio. Styrolux combines a substantial yield point (25-35 MPa) with plastic deformation up to 300%.

2.1 SYNTHESIS, STRUCTURE AND MORPHOLOGY OF STYROLUX.

Transparent and tough polystyrene is prepared by sequential anionic polymerization, where butyllithium may be added in more than one charge. A possible synthesis consists of the formation of a long styrene block in the first step, followed by further addition of butyllithium and styrene, thus yielding short styrene blocks. The molar ratio of short to long chains is significantly larger than 1. In Styrolux, a mixture of styrene and butadiene is finally added, resulting in a butadiene/styrene block with a tapered block transition. This mixture consisting of short, butadiene-rich and long, styrene-rich triblocks is coupled with an oligofunctional coupling agent giving on average an unsymmetrical star polymer with about 4 arms (Fig. 2)[4)]. The coupling reaction is of course a statistical process and

Synthesis of Styrolux

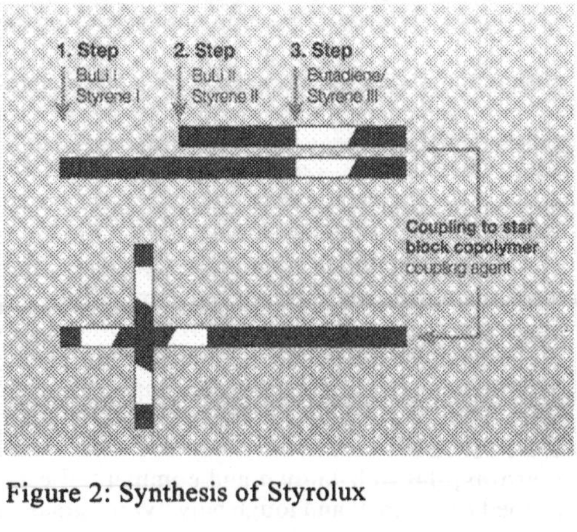

Figure 2: Synthesis of Styrolux

TEM micrographs of Styrolux/GPPS blend (40/60) and neat Styrolux

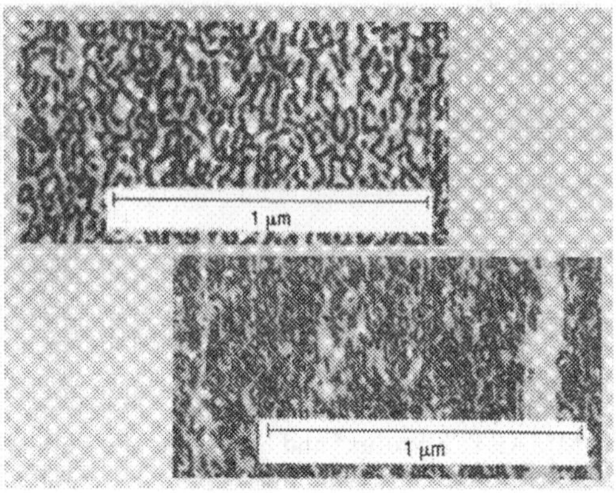

Figure 3.: TEM micrographs of Styrolux/GPPS blend (40/60) (top) and neat Styrolux (bottom).

all possible combinations of long and short arms are observed in an amount predicted by Pascal's triangle, taking the molar ratio into consideration. The morphology of Styrolux does not fit into the common morphology scheme of styrene/butadiene block copolymers with spheres, cylinders and lamellae as basic morphologies. Fig. 3 shows TEM micrographs of a compression-molded Styrolux sample. At first glance, the impression is one of an irregular, wormlike morphology. Styrene and butadiene form an interpenetrating network with styrene as the predominant phase (Fig. 3 top). A closer look reveals that the butadiene "lamellae" are often split and appear to contain small styrene cylinders and spheres. The difference in scale is remarkable (diameter): large styrene "worms": 25 nm, small styrene inclusions in butadiene: 6 nm, butadiene "lamellae": 8 nm. One explanation might be that long and short styrene blocks partly demix. In particular the short styrene block in the core of the star polymer might form a separate phase. [This notion is also supported by DSC measurements. The glass transition temperature of the hard phase is unusually broad and stretches from 105 down to 30°C suggesting at least an inhomogeneous polystyrene phase.]

3. STYROFLEX

Styroflex is a new experimental product mainly based on styrene and butadiene. Our goal was to create a resin for extrusion and injection molding with similar or even improved characteristics compared to plasticized PVC. It should exhibit the mechanics of a thermoplastic elastomer, e.g. low modulus and yield strength, high elongation and excellent recovery and should be suited for high speed processing, especially for thin films. The latter aspect requires an intrinsic high thermal stability in order to avoid gel formation during processing.

3.1 STRUCTURE AND SYNTHESIS

To meet these goals we chose a symmetrical hard-soft-hard structure typical for TPEs with a block length ratio of about 15:70:15[5]. The hard segments consist of polystyrene, but instead of a butadiene soft segment we introduced a statistical SB sequence with a glass transition temperature of approx. -40°C and an S/B ratio of approxi-mately 1. Thus the overall styrene content of Styroflex reaches almost 70% which is in a range known for transparent, impact modified polystyrene like Styrolux. The benefit of this structural variation is a drastically increased molecular weight (140,000 vs. 70,000 g/mol) while retaining the same viscosity or melt flow rate[6-8]. Furthermore, a low 1,2-vinyl content is required for superior thermal stability due to reduced cross-linking.

The polymer is prepared by butyllithium-initiated sequential anionic polymerization in cyclohexane (Fig. 4). In order to generate the statistical SB block the presence of a randomizer is required[9].

There are three basic synthesis routes to Styroflex (Fig. 4): sequential polymerization, coupling of a living S-SB diblock, or bifunctional initiation[10]. Bifunctional coupling agents X such as dichlorodimethylsilane and butanediol diglycidyl ether give almost quantitative coupling yields. Carboxylic esters like ethyl acetate work only well when donor solvents are absent. In the presence of THF the coupling yield drops to 30-40%.

Synthesis routes to Styroflex®

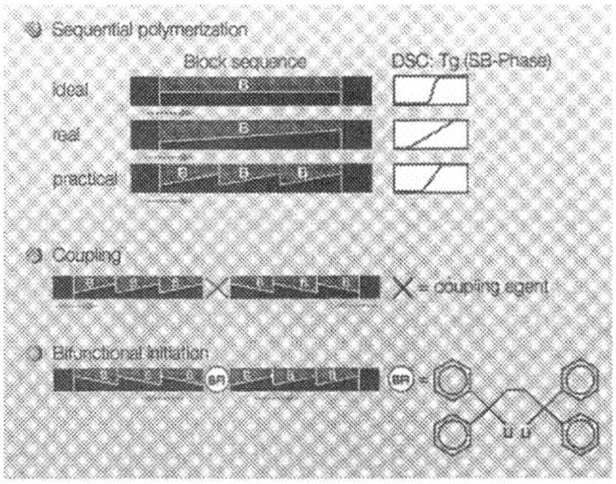

Figure 4.: Synthesis routes to Styroflex. The bars indicate the chemical composition along the polymer chain. The light gray areas within the bars symbolize butadiene, the dark areas styrene.

Essentially ether-free bifunctional initiators for highly bifunctional growing polymers have been developed by BASF[11].

3.2 MORPHOLOGY

Fig. 5 [Fig.6 schematic diagram] shows the TEM micrograph of a compression-molded Styroflex sample. Due to the phase/volume ratio of about 30% polystyrene in the block polymer, the transmission electron micrograph of Styroflex depicts spherical morphology with the SB rubber as matrix and polystyrene as spheres. The fuzzy borders of the spheres indicate an extended interphase typical of a system close to the order-disorder transition (ODT). The repulsive interaction between the polystyrene and the SB phase is greatly diminished compared to a polystyrene/polybutadiene system. However, Styroflex has a higher molecular weight than

Transmission Electron Micrograph of Styroflex®

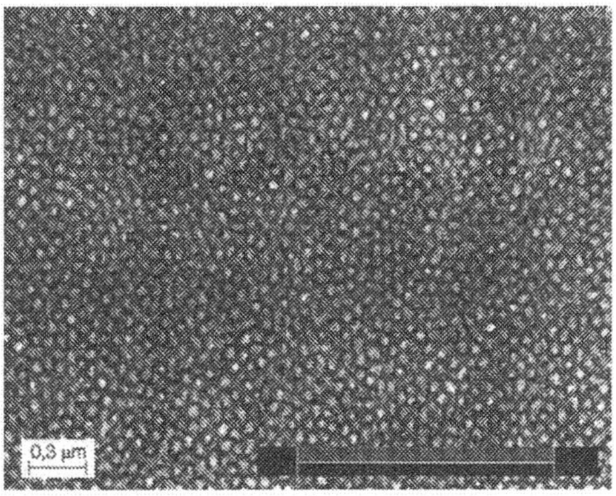

Figure 5.: Morphology of Styroflex

Styroflex – Molecular Architecture

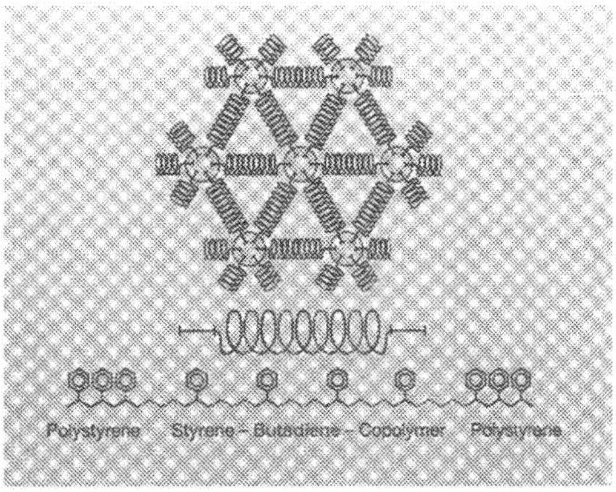

Figure 6.: Molecular Architecture of Styroflex

commercial SBS TPEs which in turn favors phase separation. When cooling from the melt within a few minutes no long range order develops, but the arrangement of the styrene domains is quite regular yet. Styrene domains are surrounded predominantly hexagonally by their neighbors. The domain identity period is about 39 nm.

3.3 THERMAL PROPERTIES

The DSC analysis (Fig. 7) shows the glass transition temperature of the soft phase at around -40°C. The long flat slope up to 90°C indicates an extended interphase ending in a barely separated hard phase. We proved that this is not an effect of the heterogeneity of the SB rubber phase by synthesizing and analyzing the pure rubber block. In this case the glass transition was limited to a temperature range between -40 and -10°C.
DMTA measurements (Fig. 7) on Styroflex show two softening points around -40°C and +90°C, which are in good agreement with the DSC measurement.
According to the thermal stability, Styroflex has properties lying between classical SBS- and hydrogenated SEBS-grades. Because of ist random S/B centre block and low proportion of 1,2-linked butadiene units (vinyl moieties), this thermoplastic elastomer is extremely thermal stable.
To measure thermal stability, a polymer melt at 250°C is forced through a nozzle and the required pressure detected. For conventional SBS grades (linear; star shaped), a rapid pressure uptake is found because of increasing crosslinking of the pure polybutadiene block unit which is caused by shearing and high temperature. The TPE Styroflex does not undergo a pressure rise under the aforementioned conditiones (i.e., there is no crosslinking). Only at higher temperatures (T ≥ 270°C) slow crosslinking is found also with Styroflex. An (arbitrary) measure for "crosslinking" is the time after $\Delta p = 5$ bar.

3.4 RHEOLOGY

Fig. 9 depicts the relationship between the melt viscosities at 190°C and the shear rate for three different triblock copolymers. Styroflex is compared with a Styrolux grade (BASF), which is a symmetrical SBS triblock with a molecular weight (MW) of approx. 70 000 g/mol and a butadiene content of about 26%, and with an SBS-type thermoplastic elastomer with a MW of approx. 70 000 g/mol and a butadiene content of about 70%. It can be seen that the three triblock copolymers behave quite differently. At low shear rates the melt viscosities of both Styroflex and conventional SBS show little rate dependence in contrast to Styrolux, where a decreasing rate is accompanied by a pronounced increase in viscosity, which is characteristic of thixotropic behavior. At high shear rates both Styrolux and Styroflex show significantly more shear thinning compared to the SBS, indicating good processability. Rheological measurements have been performed between 110 and 220°C in order to determine the ODT, which is found at 145°C (Fig. 10). This temperature is well below the favored Styroflex processing temperature (170 - 210°C). Thus, the production of thin films with diminished residual melt history is facilitated. The low ODT is a key advantage over conventional SBS block copolymers.

DSC analysis of Styroflex

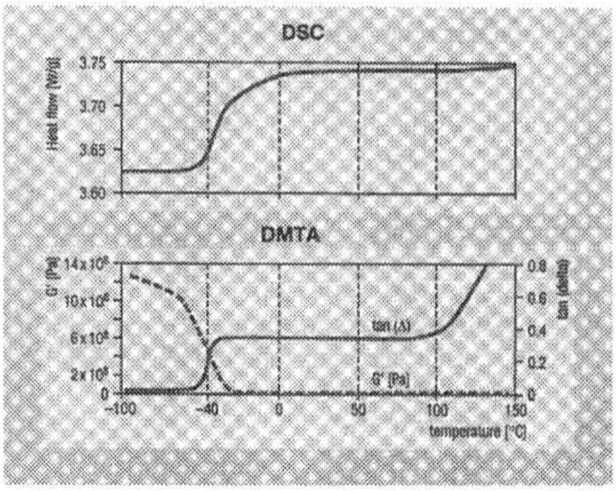

Figure 7.: Top: DSC analysis of Styroflex. Sample quenched from melt; heating rate 20 K/min. Bottom: Dynamic mechanical thermoanalysis (DMTA) of Styroflex.

Thermal Stability of SBS/SEBS Materials and Styroflex

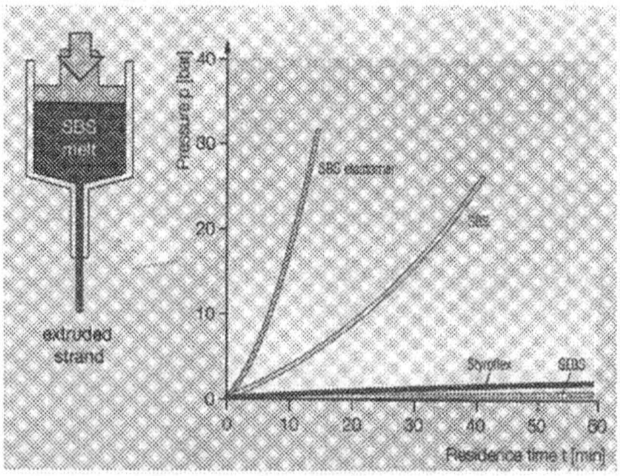

Figure 8.: Thermal stability of SBS/SEBS materials in comparison to Styroflex.

Rheological behavior of different symmetrical triblock copolymers at 190 °C

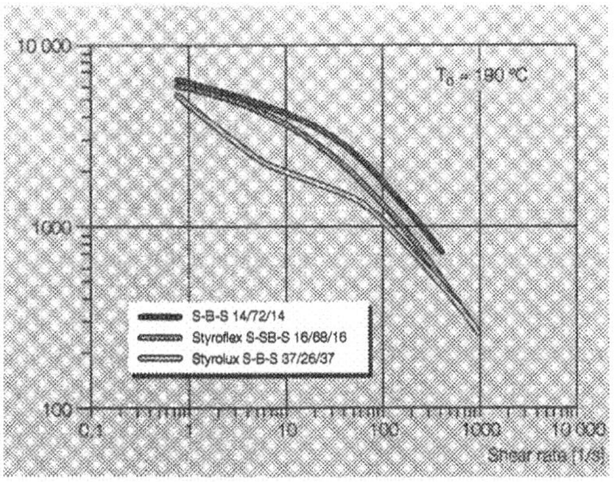

Figure 9.: Rheological behavior of different symmetrical triblock copolymers at 190°C.

Styroflex®: Determination of the Order-Disorder-Transition Temperature by Melt Rheology

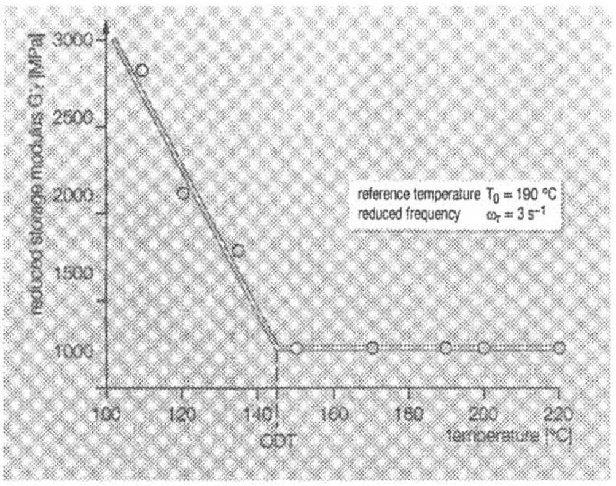

Figure 10.: Rheological determination of Order-Disorder Temperature

3.5 MECHANICAL PROPERTIES

The mechanical behavior of Styroflex is that of a typical thermoplastic elastomer (Fig. 11). It appears to be somewhat harder than a classical SBS TPE. With an annealed specimen, an ultimate elongation at break of approx. 900% can be achieved, whereas typical SBS polymers fail at elongations of about 1000%. Similar behavior is observed with metallocene polyethylene.

At elongations well below failure, Styroflex follows Hooke's law. In contrast Styrolux, a highly transparent and stiff polystyrene with symmetrical SBS triblock structure and a butadiene content comparable to that of Styroflex, has a pronounced yield point followed by plastic deformation. The tensile strength of PVC depends on the annealing time - indicated by the dashed line. Styroflex films achieve a similar hardness to plasticized PVC due to molecular orientation.

Films designed particularly for food wrapping should maintain their smooth, optically attractive surface over an extended period of time even after touching, stapling and other manipulations, which might cause indentations. Therefore a virtually complete recovery of the stretched film is desirable.

Table 1 (Fig. 12) gives an overview about the thermoplastic and elastic properties of Styroflex. Neat Styroflex has a Shore A around 85 and Shore D around 30.The cold temperature according to DIN

Stress-Strain Diagram of Compression-molded Flexible Materials

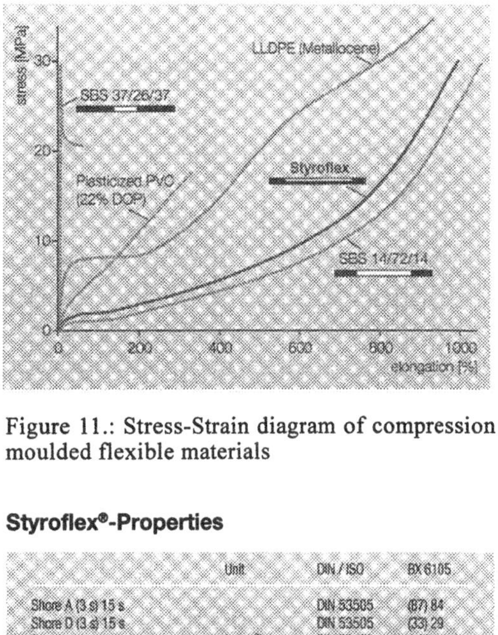

Figure 11.: Stress-Strain diagram of compression moulded flexible materials

Styroflex®-Properties

	Unit	DIN / ISO	BX 6105
Shore A (3 s) 15 s		DIN 53505	(87) 84
Shore D (3 s) 15 s		DIN 53505	(33) 29
Density	[g/cm³]	DIN 53479	0.99
VST/A/B	[°C]		35/n.d.1)
Transparency, 4 mm	[%]	ISO 5033	80
MVR (200/5)	[g/10 min]	DIN 53735	10–15
Elongation at break	[%]	DIN 53455	> 650
Tensile stress	[MPa]	DIN 53455	33
Stress at 100% Elongation	[MPa]	DIN 53455	2.7
Charpy impact, 23°C	[kJ/m²]	ISO 179/1eU	no break
Glas transition temperature	[°C]		–40
Cold temperature	[°C]	DIN 53372	–35
Stiffness in torsion [310 MPa]	[°C]	DIN 53447	–41
Loss in weight 70°C/100 % r. f.	nach 1d [%]		0.01
	nach 3d [%]		0.05
	nach 7d [%]		0.02
Compression set	23 °C [%]	DIN 53517	70
	70 °C [%]	DIN 53517	100

1) not detectable

Figure 12.: Table showing the mechanical properties of Styroflex

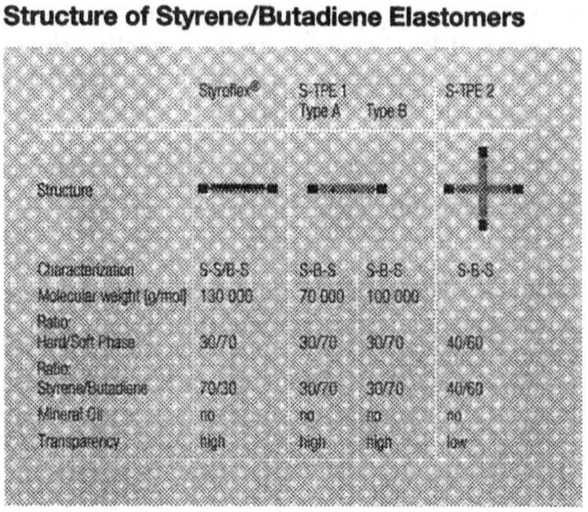

	Styroflex®	S-TPE 1 Type A	S-TPE 1 Type B	S-TPE 2
Structure				
Characterization	S-S/B-S	S-B-S	S-B-S	S-B-S
Molecular weight [g/mol]	130 000	70 000	100 000	
Ratio: Hard/Soft Phase	30/70	30/70	30/70	40/60
Ratio: Styrene/Butadiene	70/30	30/70	30/70	40/60
Mineral Oil	no	no	no	no
Transparency	high	high	high	low

Figure 13.: Structure of Styrene/Butadiene Elastomers

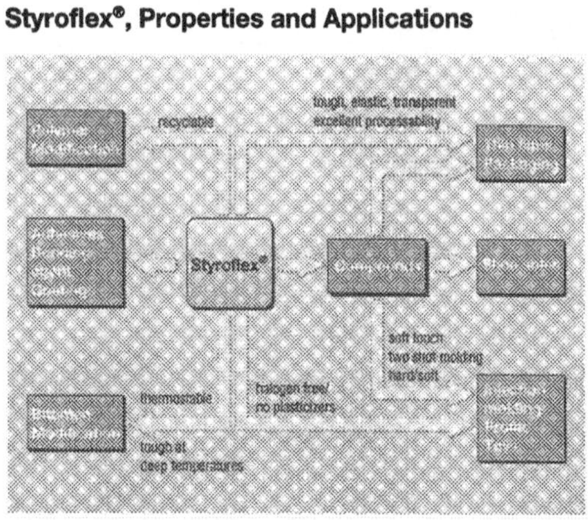

Figure 14.: Properties and applications of Styroflex.

53372 is -35°C and the stiffness in torsion at 310 MPa is found to be -41°C. Because of its structure no loss in weight is detected at 70°C/100% r.H. Showing a melt flow around 14 ml/10min at 200°C/5kg load it can be used either for extrusion or injection molding.

To compare Styroflex with conventional SBS grades (linear; star-shaped) in the molecular structure it was possible to increase the molecular weight in Styroflex over classical SBS which leads to a more optimized structure / property correlationship. [The benefit of this structural variation is a drastically increased molecular weight (140,000 vs. 70,000 g/mol) while retaining the same viscosity or melt flow rate[6-8]]. Again, because of the unique structure of the randomized S/B centre block unit it was possible to obtain the same soft/hard ratio with inverse styrene/butadiene moieties. (Fig. 13)

4. APPLICATIONS

Styroflex can be used in versatile applications (Fig.14). Regrind and neat material is suitable for polymer modification to increase toughness. Because of its toughness also at deep temperatures, combined with the thermostability Styroflex can be used either for asphalt modification or in adhesives or coatings. The neat polymer can be processed to transparent thin films for packaging applications showing excellent elastic behavior. The polymer itself as well as in compounds is further suitable for modification of shoe soles, for injection molding applications like preferentially toys and extrusion of profiles.

Thin films of Styroflex for packaging applications are produced best in coextrusion with EVA containing an antifogging agent. A 12 micron three layer structure shows excellent recovery, high transparency combined with a very high elongation at break. (Fig. 15).

Adapted hysteresis experiments have been performed in order to compare the recovery of Styroflex with other film materials (Fig. 16). The sample is elongated to 200% and released to zero stress at a constant rate of 100% per min. The additional recovery is then monitored for five minutes. The experiment is repeated up to 300 and finally 400%. The measurements prove that the recovery of Styroflex is in the range typical of SBS-TPEs and far better than plasticized PVC. In fact after 30 min. the residual deformation is reduced to only 3%. In comparison, even the least crystalline metallocene linear low-density polyethylene (density 0.903) does not show such a good recovery. The ratio of plastic to elastic deformation increases rapidly with density.

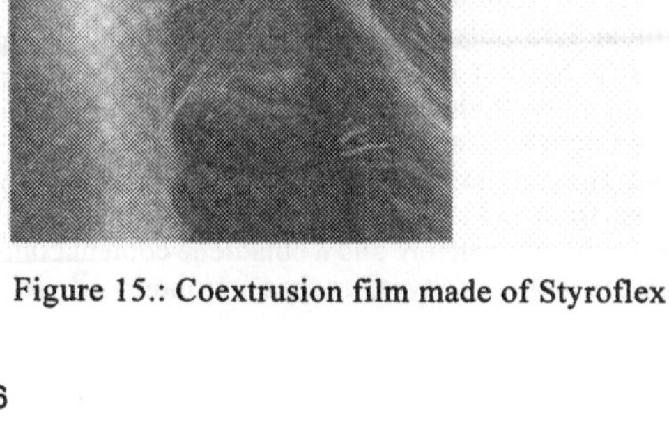

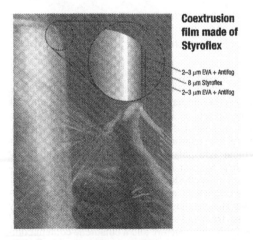

Figure 15.: Coextrusion film made of Styroflex

Hysteresis of film materials

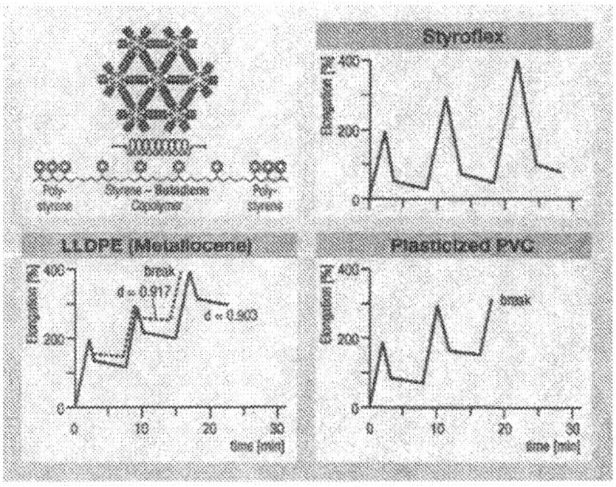

Figure 16: Hysteresis of film materials

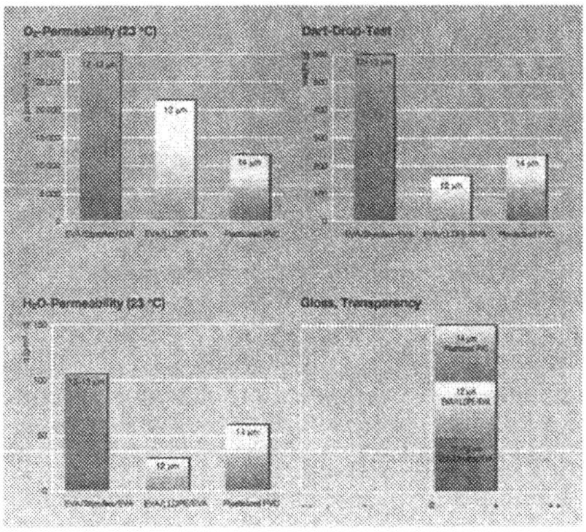

Figure 17.: Properties of thin films made of Styroflex, LLDPE and plasticized PVC

Conventional SBS TPEs behave comparable to Styroflex, however their major drawbacks are poor thermoplastic processability and extensive thermal crosslinking. Thus, they are not applicable for sophisticated extruded and molded parts.
Compared with common cling films, Styroflex gives higher yields per kg due to reduced density and low film thickness. Thus, cling films made of Styroflex meet the strict requirements of today's fresh meat packaging. Because of its high oxygen permeability, the meat keeps its fresh appearance longer (Fig. 17.). In comparison to films made of LLDPE and plasticized PVC, coextrusion films made of Styroflex show further a high water permeability and an excellent puncture resistance. High puncture resistance means, that even sharp bones can be per-

Styroflex®: Polymermodifier

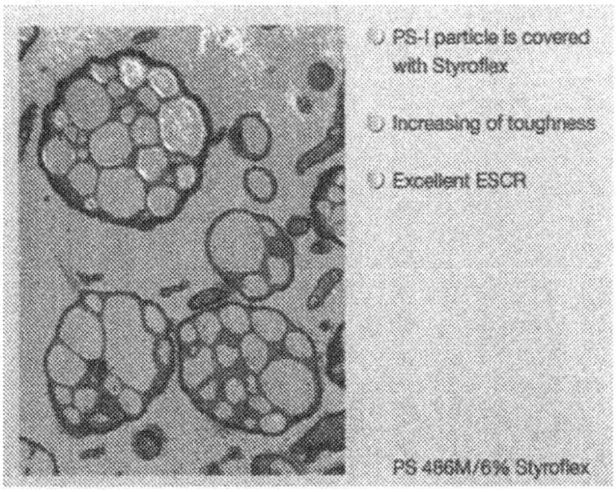

Figure 18.: Styroflex in combination with high impact polystyerne

fectly packed. Transparency and gloss in thin film applications are comparable to LLDPE and plasticized PVC.
Styroflex promotes the toughness and stress cracking resistance in combination with High Impact Polystyrene (PS-I). As shown in the transmission electron microscopy (Fig. 18), the addition of small amounts of Styroflex in high impact PS 486M are sufficient to cover the rubber particle completely which is indicated by the black colour on the top of the particle. Thus, the whole particle behaves like a neat Styroflex particle.This is responsible for the increase of toughness and for the better ESCR data. The properties of such a system which contains 20% Styroflex in PS 486M is shown in comparison to a addition of 20% of a linear thermoplastic SBS and 20% addition of Styrolux 693D (Fig. 19.). The mechanical properties of neat 486M like E-modulus and yield stress are placed at 100%. Adding Styroflex slightly decreases stiffness and yield stress whereas elongation at break is increased tremendously. The influence of conventional SBS in such a system promotes PS-I in the same direction but the performance is not so well balanced like in the case of Styroflex. Styrolux as a classical SBS thermoplast keeps the stiffness in such a blend system with a slight increase in toughness (Elongation at break). Because of its excellent processability and the aforementioned compatibility to polystyrene (general purpose and high impact), Styroflex can also be used in two shot molding applications in combination with styrene containing polymers. Excellent adhesion on PS-I and GPPS is found (Fig. 20). Thus, transparent hard/soft combinations become reality with Styroflex.
Styroflex is the "missing link" between the unpolar TPE-O / TPE-S Elastomers respectively and the polar TPE-U, TPE-V and PVC derivatives. The hig-

Styroflex® Performance as Impact Modifier in High Impact Polystyrene 486 M
injection molded specimens, pre-compounded

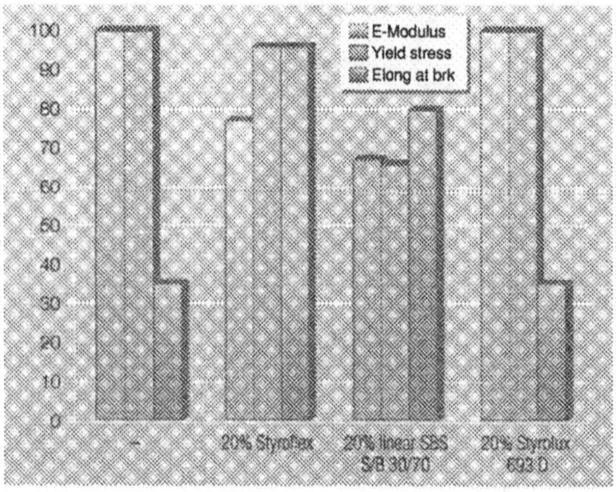

Figure 19. Styroflex: Performance as impact modifier

Styroflex®: Two Shot Molding

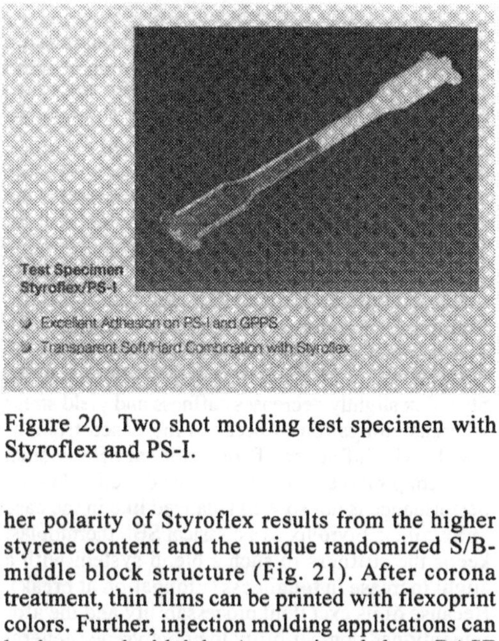

Figure 20. Two shot molding test specimen with Styroflex and PS-I.

Classification of Styroflex versus TPE's

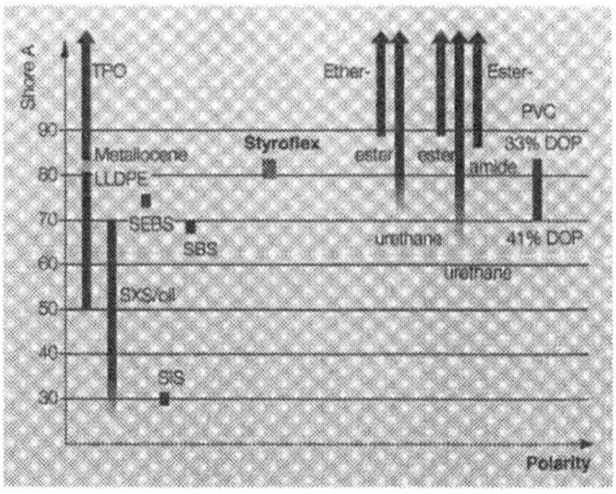

Figure 21.: Classification of Styroflex versus TPE's

Styroflex® – UV-Stability
Indoor: ISO 4892-2 Procedure B
measurements on circular test specimen 60 x 2 mm

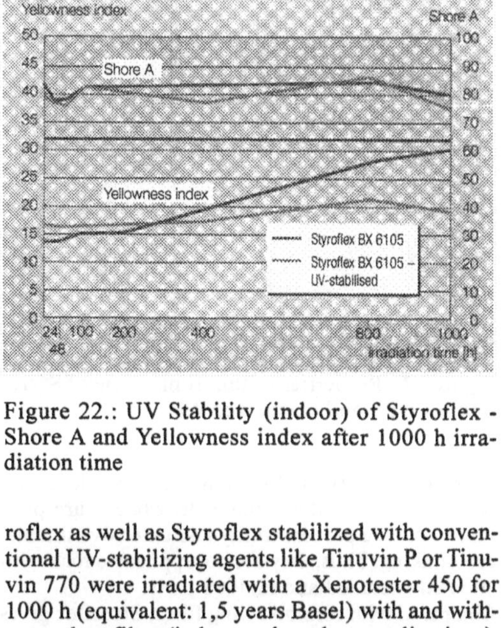

Figure 22.: UV Stability (indoor) of Styroflex - Shore A and Yellowness index after 1000 h irradiation time

her polarity of Styroflex results from the higher styrene content and the unique randomized S/B-middle block structure (Fig. 21). After corona treatment, thin films can be printed with flexoprint colors. Further, injection molding applications can be decorated with inks. As mentioned above BASF offers only neat Styroflex with a Shore A around 85.

The randomized S/B block unit, as already mentioned above gives certain advantages with respect to thermal stability (processability) and polarity (decoration). The figures 22 and 23 confirm that this special structure gives another benefit due to UV stability. Injection molded samples of neat Styroflex as well as Styroflex stabilized with conventional UV-stabilizing agents like Tinuvin P or Tinuvin 770 were irradiated with a Xenotester 450 for 1000 h (equivalent: 1,5 years Basel) with and without a glass filter (indoor and outdoor applications). Surprisingly, no influence on the Shore A was found, independent if a filter was used or not. Thus, UV irradiation did not cause damages in the basic structure. However, without UV stabilization a slight increase in the yellowness index can be detected ($\Delta E \approx 15$). For indoor applications the stabilizing agents are sufficient to keep the yellowing to a low level. The irradiation of Styroflex samples without filter gives ΔE values around 20 within 1000 h. For

Styroflex® – UV-Stability
Outdoor: DIN 53 387
measurements on circular test specimen 60 x 2 mm

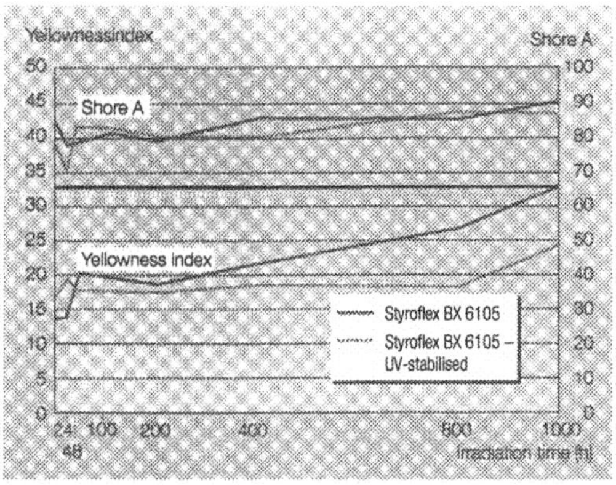

Figure 23.: UV Stability (outdoor) of Styroflex - Shore A and Yellowness index after 1000 h irradiation time

Classification of Styroflex versus TPE's

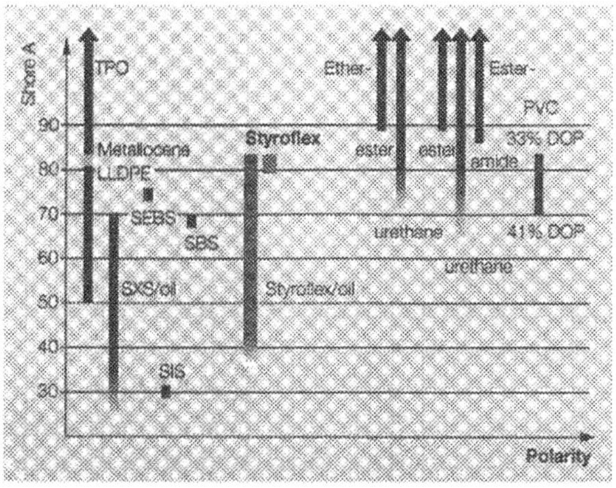

Figure 24.: Classification of Styroflex versus TPE's

a quite long period of time the yellowing can suppressed again with stabilizers. A minor increase of the yellowness index can be detected after 1000 h ($\Delta E \approx 8$).
Due to the know how of compounders it is possible to vary the Shore A hardness of Styroflex in huge range which broadens the versatility of this new thermoplastic elastomer. Of course, it is also possible to vary the polarity by that way. Following figure 24, the picture in the world of TPE's can be expanded. The styrenic TPS-familiy now contains a new member which is located between SBS and SEBS.

Conclusion:
Benefits of Styroflex®, the new S-TPE

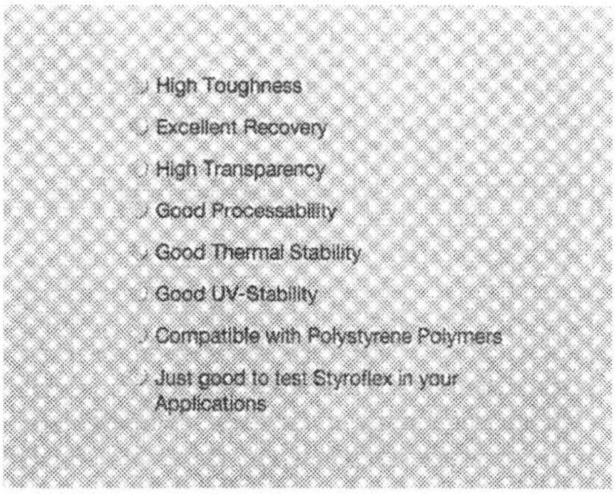

Figure 25.: Benefits of Styroflex - the new S-TPE

5. CONCLUSION

Styrolux and Styroflex are obtained by sequential anionic polymerization of styrene and butadiene. Styrolux is a stiff and tough resin, which retains its transparency when blended with general purpose polystyrene. Although containing only 25% butadiene it shows extended ductility explained by shear yielding processes. The highly unsymmetrical star block structure and the complex morphology result in the favorable mechanical behavior (Fig. 25).
Styroflex, the new BASF experimental styrenic polymer, combines the advantages of SBS elastomers (high toughness, recovery) with the properties of transparent, impact-modified SBS polymers like Styrolux (good processability, thermal stability, high transparency). Together with Styrolux and general purpose polystyrene it forms a unit construction system e.g. for transparent film materials and injection molded parts. This system allows fine-tuning of hardness and toughness and represents the complete range of BASF's styrene/butadiene block copolymers (Fig. 1).

6. REFERENCES

1. Ramsteiner, F., Heckmann, W. Polym. Commun. 1984, 25, 178.
2. Phillips Petroleum US 3078254
3. Fahrbach, G.; Gerberding, K.; Mittnacht, H.; Seiler, E.; Stein, D. DE 2610068.
4. Fahrbach, G.; Gerberding, K.; Seiler, E., Stein, D. DE 2550227.
5. Hashimoto, T. in Holden, G.; Legge, N. R.; Quirk, R. P.; Schroeder, H. E. Thermoplastic Elastomers, 2nd ed.; Carl Hanser: Munich, 1996; Chapter 15A, pp 429.
6. Knoll, K.; Gausepohl, H.; Nießner, N.; Bender, D.; Naegele P. DE-A 4420952.

7. Knoll, K.; Nießner, N. ACS Polym. Prep. 1996, 37(2), 688.
8. Nießner, N.; Knoll, K.; Skupin, G.; Naegele, P.; Beumelburg, C. Kunststoffe 1996, 87, 66.
9. Hsieh, H. L.; Quirk, R. P. Anionic Polymerization, Principles and Practical Applications; Marcel Dekker: New York, 1996, pp 237.
10. Lit.cit.5., pp. 307
11. Bronstert, K.; Knoll, K.; Hädicke, E. EP 0477679

7. BIOGRAPHICAL NOTE:

Dr. Josef R. Wünsch was born 1965 in Augsburg. He studied chemistry at the University of Göttingen where he worked with Prof. Tietze at the Insitut of Organic Chemistry. He obtained his Ph.D. working in the area of Photocycloaddition of Enaminecarbaldehydes with electronpoor π-alkene-systems to build up N-Heterocycles. In 1992 he joined BASF and was involved in the development of nonlinear active polymers for optical computing, the synthesis of SBS-Blockcopolymers as well as syndiotactic Polystyrene. After doing technical service for all polystyrene grades used in packaging apllications, he is now responsible for the product development and marketing of all anionic produced styrene-butadiene blockcopolymers. His special task is the product and marketing development of Styroflex, the first styrene based TPE from BASF.

Constitutive Models for Rubber, Dorfmann & Muhr (eds) © 1999 Taylor & Francis ISBN 90 5809 113 9

Experiences in the numerical computation of elastomers

D. Bartels & U. Freundt
Department of Bridge and Highway Construction, Bauhaus-University, Weimar, Germany

ABSTRACT: In this report a computative determination of stress and deformation behaviour of shear cells under horizontal loading will be introduced. The construction type and the direction of loading will be varied. The load will be increased stepwise or successively until the equivalent stress reaches a required value of 20 N/mm². Through this it can be shown that numerical simulations are efficient tools for the assessment of structural behaviour of elastomer elements.

1 INTRODUCTION

Elastomer structural components portray a high complete initial reversible deformation. This attribute can be used for structural elements, which serve the transfer of high loads in the wake of large deformations. Such an element is for instance a shear cell in which an elastomer element, lying between two steel plates, and displaced with respect to one another, absorbs elastic displacement. The steel plates and the elastomer are joined together by vulcanisation. Two construction types corresponding to Figure 1 will be examined. In variant A the elastomer and steel plates are sealed up firmly together at longitudinal sides. In variant B, the elastomer extends 2 mm away from the steel. In both variants the steel plates overlaps the edges of the elastomer, so that the shear cells can be screwed to the supporting beam. The elastomer layers between the steel plates in both variants have at the initial state an inclination of 45°.

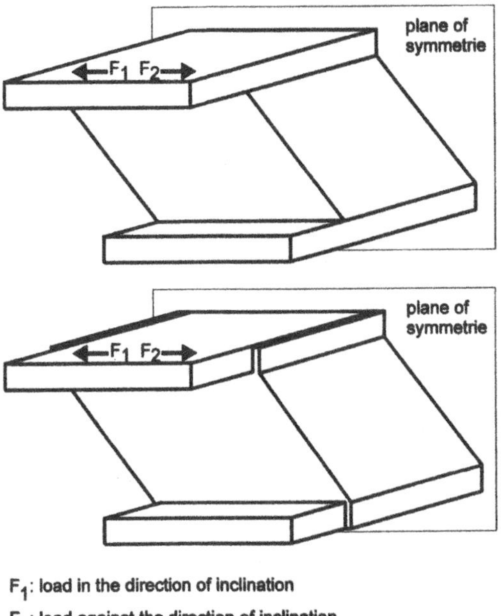

F_1: load in the direction of inclination

F_2: load against the direction of inclination

Figure 1. Construction variants of Shear cell (above: variant A, below: variant B).

In this report a computative determination of stress and deformation behaviour of the described shear cells under horizontal loading and defined boundary values will be introduced. The direction of loading will also in this case be varied according to Figure 1. The load will be increased stepwise or successively until the equivalent stress reaches a targeted value of 20 N/mm².

2 FINITE ELEMENT ANALYSIS

2.1 Modelling

The finite element method has won a recognition as a universal computational tool for the simulation of structures near to reality. The deciding pre-requisite to obtain dependable results in this way, is the exact modelling of material behaviour for a general treedimensional stress and deformation state. The boundary values must be known.

Shear cells are structural elements mainly loaded in the horizontal direction. For the structural simulation constraints will be applied to the overlaps at the sides of the steel plates. The loading will be applied at the same points. The connection between the elastomer and the steel plates is assumed to be rigid.

To solve the problem computatively, mechanical models and physical laws are used. The presented shear cell are identified by their geometry, the main material, elastomer, and their deformation behaviours.

A finite element model with volume elements, which allow large deformation suitable. Physical laws are required in order to formulate the material behaviour (steel and elastomer). The following needed theoretical principles are summarised:

- Balance laws, geometric non-linear, propounded from deformed system;
- Non-linear stress-strain-relationship, in which the fibre extension and displacement are large;
- A physical law, which consists of the relation between forces and deformations, and between stresses and deformations in space, and therefore describes the material behaviour;
- Experimentally obtained parametric values, which describe the material behaviour.

As finite element programmes, which contain the needed mechanical formulation, LS-DYNA3D with explicit solution methods and ANSYS 5.3 with implicit solution methods were tested. Both were suitable, however ANSYS 5.3 was selected.

In ANSYS a possibility for defining hyper-elastic material behaviour with multi-parameter functions, the coupling of degree of freedoms (for instance the definition of rigid parts or the joining of different nets), the solution of system of equations with the Gauß-Elimination methods (Wave-front solver), the automatic or manual load steering and the Newton-Rapson method as non-linear strategy for the solution of the present problem were used. Volume elements like SOLID45, HYPER58 and SOLID92 were successfully used. The former are 8-node elements (hexahedron) and the latter is a 10-node element (tetrahedron). All these elements possess 3 degree of freedom in each node and are explanatory when considering large deformations. Loads, external conditions and material data can be assigned to geometrical parts. The nets can be produced automatically and the net density can be steered with parameters.

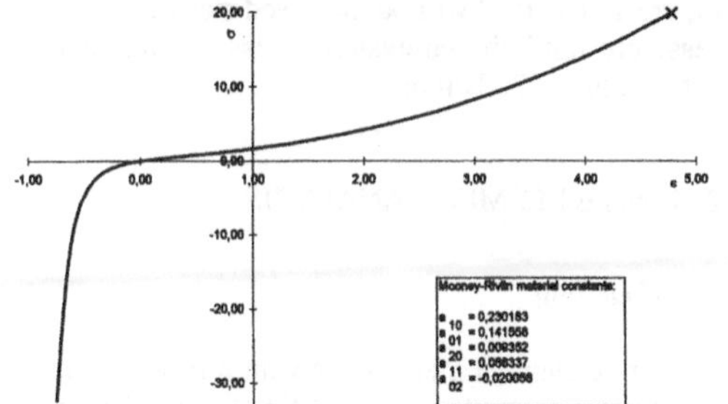

Figure 2. Axial stress-strain relation of elastomer

2.2 *Description of Material Behaviour*

It was explained that, for computations near to the reality, the physical law, which describes the relation between forces and deformations and between stresses and deformations in space must be known.

Here only values from axial tensile and compression test will be used for the elastomer. The formulation for the computation folows from the presentation of Mooney-Rivlin, a multi-parameter approach. The Mooney-Rivlin constitutive law shows a material equation as function of strain-work, which is about a consistent approximation of constitutive equation and for which the parameters from uniaxial test can be determined:

$$\begin{aligned} W = \; & a_{10}(J_1-3) + a_{01}(J_2-3) + a_{20}(J_1-3)^2 \\ & + a_{11}(J_1-3)(J_2-3) + a_{02}(J_2-3) + a_{30}(J_1-3)^3 \\ & + a_{21}(J_1+3)^2(J_2-3) + a_{12}(J_1-3)(J_2-3)^2 \\ & + a_{03}(J_2-3)^3 + \tfrac{1}{2}\kappa(J_3-1)^2 \qquad (1) \end{aligned}$$

with W ... stress function with respect to a unit of undeformed volume;

a_{10} - a_{03} ... Mooney-Rivlin material constants;

J_1 - J_3 ... reduced invariants of the right Cauchy-Green strain tensors;

κ ... Bulk Modulus $= 2\,(a_{10}+a_{01})\,/\,(1-2\nu)$

ν ... Poisson's ratio

If the constants a_{20} - a_{03} are set to zero, the two parameter model is obtained.

The last term in the above equation represents the hydrostatic volume work, performed by the hydrostatic pressure. It is usually assumed that the material is almost incompressible (ν = 0,449), that means there is no hydrostatic volume work performed. The axial stress-strain relation is principally descried in the literature. The applied relation here will be shown for strain between ε = -1 ... 5 in Figure 2, as well as its trend up to the limit tensile strength taken from the tensile test. The adjustment or adaptation computation for the trend of the curve shows that a 5 parameter model is a good approximation.

The elastomer is considered homogeneous and isotropic.

Linear-elastic material behaviour is applied to the steel. The elastic modulus is considered $E = 2{,}1 \cdot 10^5$ N/mm^2 and the Poisson's ratio $\nu = 0{,}3$. Also here homogeneity and isotropy are assumed.

For the determination of loading steps, with which the computations can be aborted, a principle becomes necessary which permits a relation between a three-dimensional stress state and the results of axial tension and compression tests. The stress deviator used for this purpose. In stress state the spherical tensor (i.e. the mean stress) which describes the hydrostatic stress state and produces only change in volume, is separated from the stress tensor. The remaining part, which is the stress deviator, is coupled with the change in shape and therefore a good criterion for the presented considerations. The second invariant of the deviator is interpreted as octahedral shear stress. If this is applied to uni-axial stress state, a uni-axial equivalent stress is obtained which self shows an invariant of general stress state. The use of this knowledge leads to the v. Mises yield conditions, which is shown as a circle on the yield curve.

For the determination of the loading limit in this case the second invariant of the deviatoric stress is related to the uni-axial equivalent stress.

The computation was stopped when the equiva-

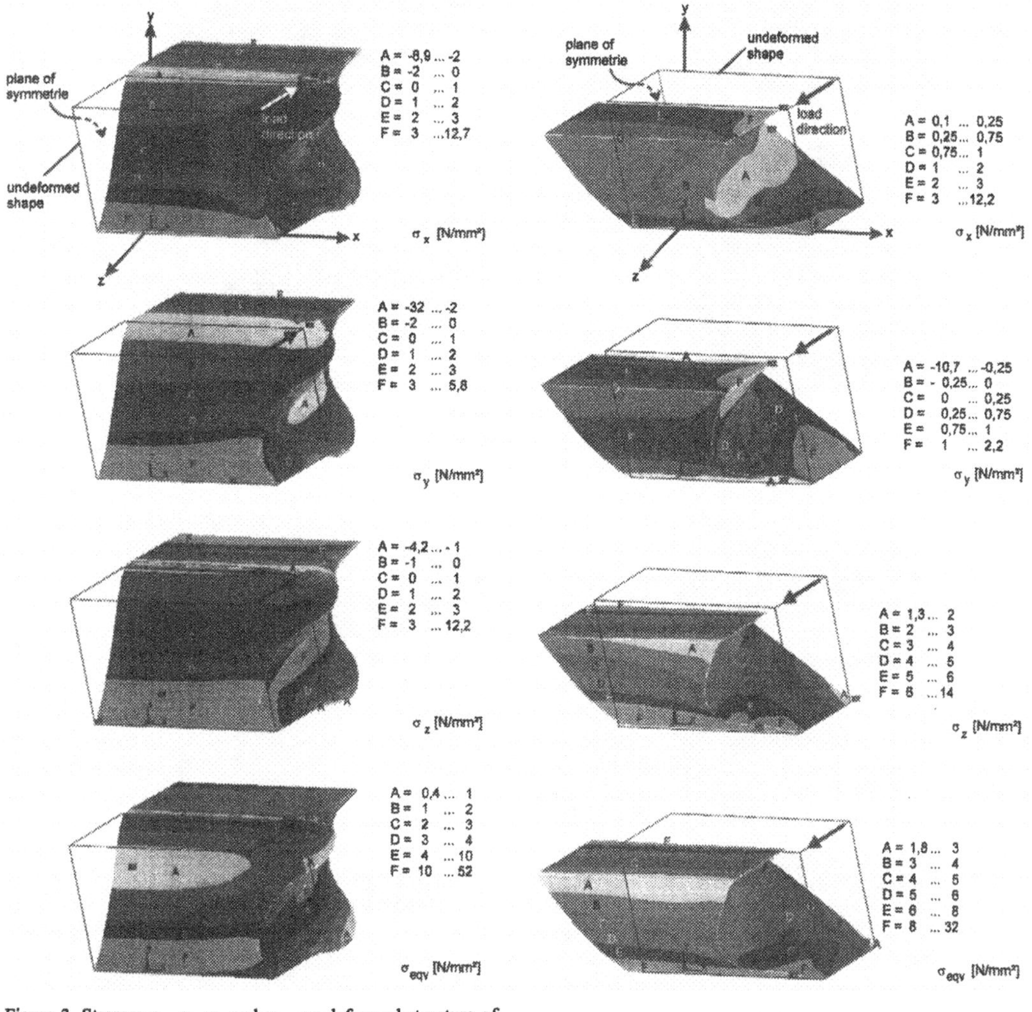

Figure 3. Stresses σ_x, σ_y, σ_z and σ_{eqv} on deformed structure of shear cell variant A.

lent stress reached limiting (or ultimate) value of the tensile strength.

3 RESULTS OF THE COMPUTATION

Four examples are used to discuss the computational results:

- Variant A with horizontal loading ...
  - ... in the direction of inclination
  - ... against the direction of inclination
- Variant B with horizontal loading ...
  - ... in the direction of inclination
  - ... against the direction of inclination

The stresses σ_x, σ_y and σ_z are shown for the four examples each on deformed structure as isosurfaces in grey tone in Figure 3 and 4. The applied force which could be obtained from the figure, was so chosen, as already explained, such that the equivalent stress in the elastomer would reach 20 N/mm^2. The equivalent stress is also shown for documentation.

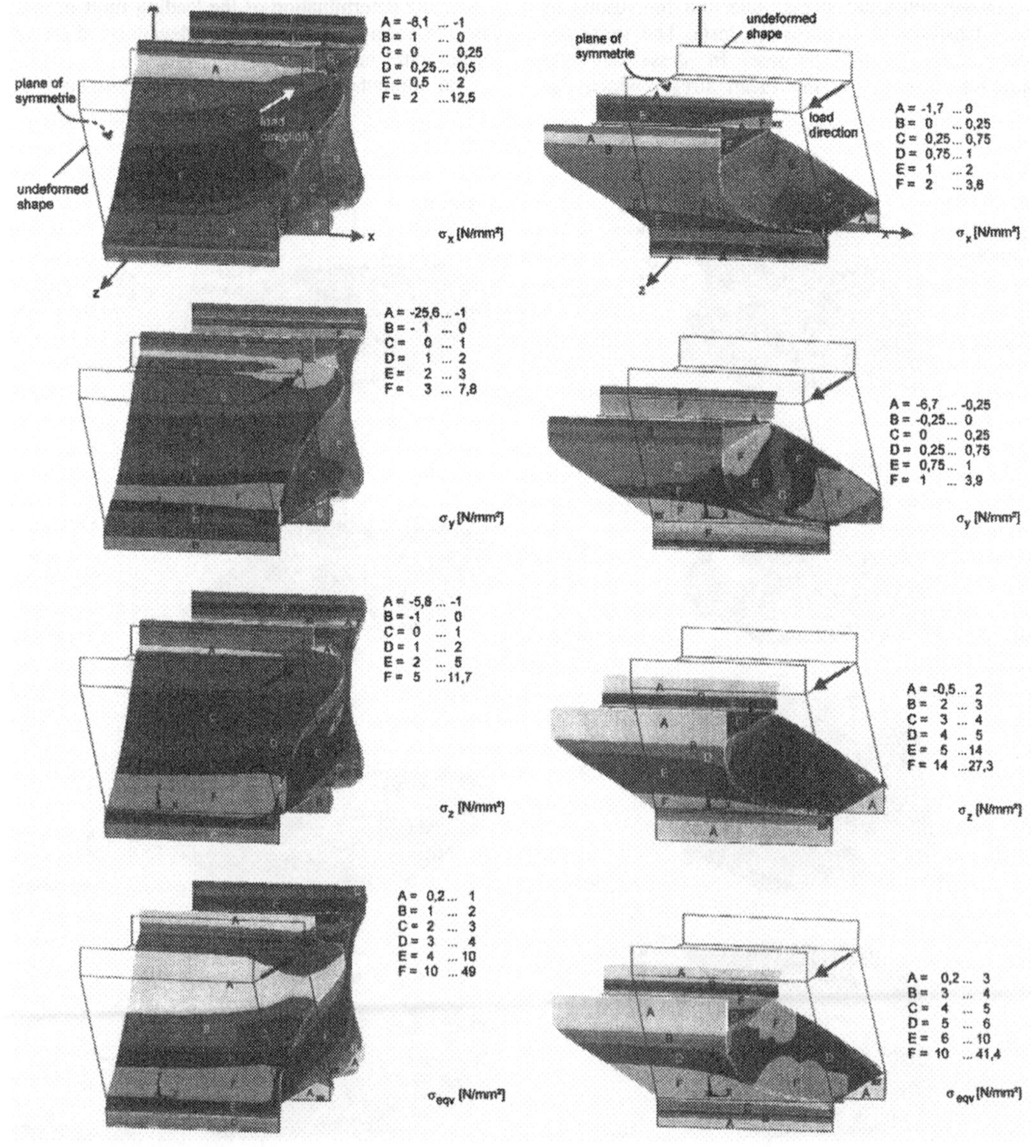

Figure 4. Stresses σ_x, σ_y, σ_z and σ_{eqv} on deformed structure of shear cell variant B.

The following conclusion is drawn from the results:

1. Influence of the load direction
Depending on the load direction, the normal (or axial) stress diagrams in both cell variants are very different. The maximal stress (MX) appears near the strongest constraints on opposite sides. The equivalent stress shows the same tendency. It can be seen from the σ_y diagrams that the different torsion portion show the main influence.

2. The influence of construction type
The type of construction differs only through a 5% extension of the elastomer over the depth. The trends of the resultant axial stress differs very much. This tendency does not depend on the load direction. The stress values show a clear dependency on the load direction. Whiles the targeted equivalent stress value of 20 N/mm^2 is achieved by the application of the same load value when loaded independently from the construction type in the direction of inclination, it requires about 50% more load when loaded against the direction of inclination.

3. Conclussions for load limits
If it is assumed that the steel plates doe not detach themselves from the elastomer, then the failure points in all the four examples, independent from the construction type and load direction, are equal. It lies at the peak edges of the elastomer. It changes only from front to back, depending on the load direction.

This fact is known and confirmed by company's internal experimental reports.

4 SUMMARY

The determined deformations are feasible and the trend is confirmed by practical appearance. It was shown that numerical simulations are capable tools which can be used for the assessment of the structural behaviour of elastomer. The pre-requisite is material modelling very near to the reality. The FEM-Programmes offer almost in all cases material models according to Rivlin and Odgen.
The advantage of the relative smaller number of material parameters was made use of in the present research. The investigation was conducted in the quasi-static parts. Dynamic, time and temperature dependent material behaviour were not investigated. An extension of the material model is therefore required for this purpose.
The literature provide a lot of theories on this. There is however a lack of an experimentally satisfactorily secured material model. An analysis of the present experimentally determined parameters from different experiences is targeted.

REFERENCES

Bathe, K.-J. 1982. *Finite Element Procedures in Engineering Analysis*. New Jersey: Prentice-Hall.

de Borst, R. & P. A. J. van den Bogert & J. Zeilmaker 1988. Modelling and analysis of rubberlike Materials. *Heron*33(1)

Glowinski, R. & P. Le Tallec 1985. Finite elements in nonlinear incompressible elasticity. In J. T. Oden & G. F. Carey (eds), *Special problems in solid mechanics*: 67-93

Imbimbo, M. & J. M. Kelly 1998. Influence of material stiffening on stability of elastomeric bearings at large displacements. *Journal of Engg. Mech.* 124 (9): 1045-1049

Sussman, T. & K.-J. Bathe 1987. A finite element formulation for nonlinear incompressible elastic and inelastic analysis. *Computers and Structures* 26: 357-409

Treloar, L. R. G. 1975. *The physics of rubber elasticity*. Oxford: Clarendon Press

van den Bogert, P. A. J. 1991. *Computational modelling of rubberlike Materials*. Thesis Technical University Delft.

Constitutive Models for Rubber, Dorfmann & Muhr (eds) © *1999 Taylor & Francis ISBN 90 5809 113 9*

Author index

Printed and bound by CPI Group (UK) Ltd, Croydon, CR0 4YY
23/10/2024
01777679-0015